Radio Frequency Radiation Dosimetry

NATO Science Series

A Series presenting the results of activities sponsored by the NATO Science Committee. The Series is published by IOS Press and Kluwer Academic Publishers, in conjunction with the NATO Scientific Affairs Division.

A. Life Sciences	IOS Press
B. Physics	Kluwer Academic Publishers
C. Mathematical and Physical Sciences	Kluwer Academic Publishers
D. Behavioural and Social Sciences	Kluwer Academic Publishers
E. Applied Sciences	Kluwer Academic Publishers
F. Computer and Systems Sciences	IOS Press

1. Disarmament Technologies	Kluwer Academic Publishers
2. Environmental Security	Kluwer Academic Publishers
3. High Technology	Kluwer Academic Publishers
4. Science and Technology Policy	IOS Press
5. Computer Networking	IOS Press

NATO-PCO-DATA BASE

The NATO Science Series continues the series of books published formerly in the NATO ASI Series. An electronic index to the NATO ASI Series provides full bibliographical references (with keywords and/or abstracts) to more than 50000 contributions from international scientists published in all sections of the NATO ASI Series.
Access to the NATO-PCO-DATA BASE is possible via CD-ROM "NATO-PCO-DATA BASE" with user-friendly retrieval software in English, French and German (WTV GmbH and DATAWARE Technologies Inc. 1989).

The CD-ROM of the NATO ASI Series can be ordered from: PCO, Overijse, Belgium

Series 3. High Technology – Vol. 82

Radio Frequency Radiation Dosimetry

and Its Relationship to the Biological Effects of Electromagnetic Fields

edited by

B. Jon Klauenberg
United States Air Force Research Laboratory,
Human Effectiveness Directorate,
Directed Energy Bioeffects Division,
Radio Frequency Radiation Branch,
Brooks Air Force Base,
Texas, U.S.A.

and

Damijan Miklavčič
Faculty of Electrical Engineering,
University of Ljubljana,
Slovenia

Kluwer Academic Publishers

Dordrecht / Boston / London

Published in cooperation with NATO Scientific Affairs Division

Proceedings of the NATO Advanced Research Workshop on
Radio Frequency Radiation Dosimetry and Its Relationship to the Biological
Effects of Electromagnetic Fields
Gozd Martuljek, Slovenia
12-16 October 1998

A C.I.P. Catalogue record for this book is available from the Library of Congress.

ISBN 0-7923-6404-X (HB)
ISBN 0-7923-6405-8 (PB)

Published by Kluwer Academic Publishers,
P.O. Box 17, 3300 AA Dordrecht, The Netherlands.

Sold and distributed in North, Central and South America
by Kluwer Academic Publishers,
101 Philip Drive, Norwell, MA 02061, U.S.A.

In all other countries, sold and distributed
by Kluwer Academic Publishers,
P.O. Box 322, 3300 AH Dordrecht, The Netherlands.

Printed on acid-free paper

CONTENTS

SESSION C: THEORETICAL DOSIMETRY
CHAIR: O. P. GANDHI

SESSION D: EXPERIMENTAL DOSIMETRY
CHAIR: N. KUSTER

SESSION E: CONTACT AND INDUCED CURRENTS
CHAIR: M. ISRAEL

SESSION F: RESPONSES OF MAN AND ANIMALS I
CHAIR: E. R. ADAIR

SESSION G: RESPONSES OF MAN AND ANIMALS II
CHAIR: J. A. D'ANDREA

SESSION H: APPLICATIONS OF DOSIMETRY IN BIOLOGY & MEDICINE
CHAIR: D. MIKLAVČIČ

SESSION I: STANDARDS AND APPLICATIONS
CHAIR: J. M. OSEPCHUK

SESSION J: THE DOSIMETRY HANDBOOK
CHAIR: P. A. MASON

PREFACE

The North Atlantic Treaty Organization (NATO) has sponsored research supporting development of personnel safety standards for exposure to Radio Frequency Radiation (RFR) for over a quarter century. NATO previously recognized that one of the most important tools used in the RFR effects research laboratory is accurate dosimetry when it supported a NATO Advanced Studies Institute (ASI) on *Advances in Biological Effects and Dosimetry of Low Energy Electromagnetic Fields* held in 1981, in Erice, Sicily. That meeting resulted in a NATO ASI publication; *Biological Effects and Dosimetry of Non-ionizing Radiation: Radio frequency and Microwave Energies*[1].

The most recent NATO sponsored program on RFR was an Advanced Research Workshop (ARW) on "Developing a New Standardization Agreement (STANAG) for Radio frequency Radiation" held May 1993, at the Pratica di Mare Italian Air Force Base, Pomezia (Rome) Italy. That ARW produced an ASI proceedings, published in 1995: *Radio frequency Radiation Standards, Biological Effects, Dosimetry, Epidemiology, and Public Health Policy*[2]. The Rome ARW and the Proceedings served as a springboard to the much needed revision of the NATO Standardization Agreement (STANAG) 2345 MED "*Evaluation and Control of Personnel Exposure to Radio Frequency Fields - 3 kHz to 300 GHz*"[3], which was subsequently promulgated in October 1998. One of the published recommendations developed by the Rome ARW was to hold this second ARW focusing on dosimetry and measurements.

The NATO Research and Technology Organization (R&T O) Human Factors and Medicine Panel-026, Task Group 002 "Health Effects of Non-ionizing Radiation in the Military Setting" in 1998 identified Dosimetry and Measurements as a high priority topic needing review and update. Additionally, the two NATO Military Agency for Standardization (MAS) Working Groups dealing with RFR personnel safety (General Medical WG and Radio and Radar Radiation Hazards WG) both endorsed the need for further review of RFR Dosimetry.

The most cited reference on dosimetry is the *Radio Frequency Radiation Dosimetry Handbook*[4-7]. The impetus for the *Handbook* was to bring together dosimetric data to guide researchers in dealing with the complex processes of absorption of RFR in biological tissue. The dosimetric data in the first *Handbook*[4], published in 1976, was limited to the frequency range 10 kHz to 1.5 GHz. The only data provided was for homogeneous spheroidal and ellipsoidal models. Subsequent editions expanded the frequency range and added data on absorption in models irradiated by plane-waves in free space and on or near ground planes. Empirical relations for calculating the rate of energy absorption, some rules of thumb, and data from the literature summarizing metabolic rates, dielectric constants, and conductivities were also included. The 3rd Edition[6] contained a section on dosimetric techniques, including qualitative near-field dosimetry. The 4th Edition[7] was published in October 1986 and updated much of these data to provide convenient access to the information contained in the previous editions. These *Handbooks* have been among the most cited references in scientific reports concerning the biological effects of electromagnetic radiation, indicating their usefulness to the health effects community. There have been many advances in the state-of-knowledge since 1986, as researchers have begun to understand the complexities of tissue absorption of RFR. If biological effects of exposure are to be understood, accurate dosimetry is required in the assessment of exposure to this kind of radiation. New advances in theory, technology, and computation must now be applied to the development of a new edition of the *Handbook*.

The expansion of NATO provides new opportunities for collaborative research and information exchange. Harmonization and alignment of standards for safety are necessary within NATO to ensure interoperability during NATO exercises. Safety standards within NATO and worldwide should be based upon the same scientific data. However, significant differences are seen between the major standards. One of the probable factors for these differences is that the variability in research results upon which standards are based is frequently due to inconsistency in measurement and dosimetry. The output of a transmitter is usually easily established, and energy densities around it can be predicted and, in free field conditions, be measured with some precision. The introduction of animals and support equipment into the field perturbs it, so that measurements or calculations of energy density become very difficult. Power deposition levels within the whole animal and in small areas of the animal are required for meaningful bioeffects research. Until laboratories adopt more standard forms of animal exposure, methods of making density measurements, and reporting experimental results, comparison between different experiments will be impossible. Continuing work is needed on phantom and animal exposure studies aimed at measuring, and then predicting power deposition in individual organs. If dosimetry is not reliable and valid, then the data upon which standards are based should be questioned. A consensus needs to be obtained on a uniform/standardized dosimetry methodology.

The five day Slovenian ARW had four major objectives: (1) Provide an international forum to identify and discuss new technological advances in dosimetry and measurements (2) establish working groups to evaluate these advances; (3) develop a consensus on which advances should be applied to an update of the standard reference *Radiofrequency Radiation Dosimetry Handbook* (Ed 4)[7]; and (4) provide a foundation for a series of lectures to disseminate findings and train scientists. Papers were followed by round-table discussion sessions for comment on presentations and to develop consensus recommendations. The final day consisted of a specialized session of working groups to plan activities to begin revision of the *Radio Frequency Radiation Dosimetry Handbook*.

Our experience with international research and development programs designed to foster consensus on scientific issues has shown that inclusion and involvement are the critical to success. While a significant portion of expertise in the field of RFR research is found in the western world and is concentrated in military establishments, the expansion of NATO requires that experts worldwide be included. The exchange of ideas and information benefited all participants. The directors are especially grateful that scientists from Cooperation Partner and Partner for Peace countries were able to participate as key speakers, working group participants, or as observers. There is a need for international cooperation and a requirement for achieving commonality, compatibility, interchangeability, and interoperability within NATO.

In keeping with NATO objectives for ARWs, this working meeting was designed to assess the state-of-the-art in a given scientific area (RFR dosimetry) and to formulate recommendations for future. Several recommendations were issued by the ARW working groups and are attached as the Appendix. This ARW continues the NATO program goal to enhance security through scientific dialogue and to encourage peaceful exploitation of scientific skills and discoveries. We believe that the NATO Science Program objective of enhancing scientific and technological capabilities of Partner for Peace countries was fulfilled. We hope that this proceedings will stimulate interest and support coordinated research and development in RFR Dosimetry. This ARW proceedings is expected to be a valued resource for developing wider international consensus. No single nation can provide the overarching framework necessary. The consensus developed should be as multi-national as possible including Partner for Peace and other nations as well as NATO member nations. The International EMF Dosimetry Project first organized at this ARW is expected to bring a multinational effort to bear on this topic. We hope that this proceedings will stimulate this effort.

Directors and Editors

B. Jon Klauenberg, Damijan Miklavčič

Organizing Committee

Peter Gajsek, Paolo Vecchia, Stanislaw Szmigielski, Laszlo Szabo

References

1. Grandolfo, M., Michaelson, S.M., and Rindi, A. (eds.) (1983) *Biological Effects and Dosimetry of Nonionizing Radiation: Radiofrequency and Microwave Energies*, Plenum Press, New York and London.
2. Klauenberg, B. J., Grandolfo, M. and Erwin, D. N. (eds.) (1995) *Radiofrequency Radiation Standards: Biological Effects, Dosimetry, Epidemiology, and Public Health Policy*. NATO ASI Series Volume 274, Plenum Press Publishing Corp, New York and London.
3. North Atlantic Treaty Organization (NATO) Military Agency for Standardization. (1995) Standardization Agreement (STANAG) 2345 MED (Edition 2), *Evaluation and Control of Personnel Exposure to Radio Frequency Fields – 3kHz to 300 GHz*, , (ARMY)372-MED/2345.
4. Johnson, C. C., Durney, C. H., Barber, P. W., Massoudi, H., Allen, S. J., and Mitchell, J. C. (1976) *USAF School of Aerospace Medicine, Radiofrequency Radiation Dosimetry Handbook – 1st Edition*, USAFSAM-TR-76-35.
5. Durney, C. H., Johnson, C. C., Barber, P. W., Massoudi, H., Iskander, M. F., Lords, J. L., Ryser, D. K., Allen, S. J., and Mitchell, J. C. (1978)) *USAF School of Aerospace Medicine, Radiofrequency Radiation Dosimetry Handbook – 2nd Edition*, USAFSAM-TR-78-22.
6. Durney, C. H., Iskander, M. F., Massoudi, H., Allen, S. J., and Mitchell, J. C. (1980) *USAF School of Aerospace Medicine Radiofrequency Radiation Dosimetry Handbook – 3rd Edition*, USAFSAM-TR-80-32.
7. Durney, C. H., Massouodi, H., and Iskander, M. F. (1986) *USAF School of Aerospace Medicine Radiofrequency Radiation Dosimetry Handbook - 4th Edition*, USAFSAM-TR-85-73.

ACKNOWLEDGEMENTS

This NATO Advanced Research Workshop would not have been possible without the personal commitment and attention of numerous organizations and individuals. We are indebted to the following sponsoring organizations: NATO Scientific Affairs Division, High Technology Area; NATO Research and Technology Organization, United States Air Force Research Laboratory; United States Air Force Security Assistance Command; United States Air Force European Office of Aerospace Research and Development; the European BioElectromagnetics Association, the Bioelectromagnetics Society, Elletra 2000, Holaday Industries Inc., the University of Ljubljana and the Agricultural Institute of Slovenia. The support and participation of the following Republic of Slovenia Ministries is gratefully acknowledged: the Ministry of Science and Technology; the Ministry Foreign Affairs; the Ministry of Defense; the Institute of Public Health; the, Ministry of Health; and the Ministry of Environment . The beautiful facilities at the Hotel Spik, Gozd Martuljek, Slovenia, made the conference extremely pleasant and the long working days enjoyable. The directors and the participants of this ARW are deeply indebted to numerous individuals for their outstanding contributions that made the ARW such a success. Similarly, the contributions of the local arrangements committee headed by Peter Gajsek of the Institute of Public Health, and technical assistants Marko Puc, David Cukjati, Mojca Pavlin, and Natasa Kitak are greatly acknowledged. The social events and tours are memories of "The Slovene miracle" and "The Land of Green Tourism" that will be treasured. The meeting would not have occurred had not Ms Nancy Schulte, Program Director, High Technology Program given personal attention to our proposal. Her support was invaluable. Two individuals should be singled out for particular thanks for their untiring work before, during, and after the ARW. Ms Debra Jurek, Veridian, for her tireless efforts as executive secretary, ensuring all the many administrative details were accomplished smoothly, especially registration, travel, and accommodations arrangements. All in attendance will fondly remember her personal attention to each individuals needs. Ms Stephanie Miller, Radio Frequency Branch, USAF Research Laboratory, for cheerfully and diplomatically obtaining all the required paper work from the authors, reformatting and editing grammar and syntax of each paper (a gargantuan task) and "coaxing" the senior editor to finish his editorial tasks. Each of the session chairs provided valuable editing suggestions to the papers in their sessions and prepared a summary of the session. Lastly, the careful and scholarly efforts of each of the participants is recognized, appreciated, and clearly evidenced in the Proceedings.

NATO INVOLVEMENT IN RADIO FREQUENCY RADIATION (RFR) RESEARCH AND HEALTH SAFETY

B. J. KLAUENBERG
Air Force Research Laboratory
Human Effectiveness Directorate
Directed Energy Bioeffects Division
Radio Frequency Radiation Branch
Brooks Air Force Base, Texas 78235-5234

1. Introduction

The North Atlantic Treaty Organization (NATO) has encouraged and supported scientific and operational assessment projects concerning effects of RFR on military personnel, ordinance, and fuel for nearly three decades. The need for assessment and harmonization of RFR safety-guidance continues today, as safety standards are updated throughout the world, and as new RFR systems come on line. The majority of RFR systems in use, and in development, are unique to military applications. The adoption of a scientifically acceptable and defensible standard for human exposure to RFR that is applicable across national boundaries is of significant importance to commanders and operators engaged in multinational exercises and operations. Rapidly expanding technologies require that systems hardware and operations are integrated and compatible within NATO.

2. Historical Perspective of NATO Support for Radio Frequency Radiation (RFR) Research

NATO first recognized, in 1970, the need for coordinated evaluation of RFR impact on military operations. The French proposed to the NATO Defense Research Group (DRG) AC/243/Panel III (Physics and Electronics) to form Research Study Group 2 (RSG2) to study the possibility of cooperation in the protection of personnel against RFR. This marked the beginning of cooperative NATO sponsored research and personnel safety standards development for exposure to RFR.

In June 1973, the chairman of RSG2 proposed to the NATO Military Agency for Standardization (MAS) General Medical Working Party (MED) to coordinate actions of the RSG with the MED on the development of a Standardization Agreement (STANAG) concerning personnel safety standards for exposure to RFR. Initially named, "Biological Effects and Protection of Electromagnetic Radiation," RSG2 had its

1

B.J. Klauenberg and D. Miklavcic (eds.), Radio Frequency Radiation Dosimetry, 1-10.
© 2000 *Kluwer Academic Publishers. Printed in the Netherlands.*

first meeting in Aug 1973 in Rijswijk, Netherlands. Member nations included Germany, The Netherlands, Norway, the United States, and the United Kingdom. Canada joined as an active member in Jan 1975.

The group was initially formed as an Exploratory Group. Although, Panel III approved the formation of RSG2 in March 1974, formal approval by NATO was delayed, as NATO was considering formation of a Medical Panel which would include all medical-related activities and research groups. In July 1974, RSG2 was renamed "Biological Effects and Personnel Protection Against Electromagnetic Radiation" and was transferred to a new NATO Panel, AC/243 Panel VIII, "Defence Applications of Human and Bio-Medical Sciences." Panel VIII approved RSG2 and the Terms of Reference (TOR) on 11 Sep 1974.

The Aerospace Medical Panel of the Advisory Group for Aerospace Research and Development (AGARD) cosponsored a lecture series, in 1975, in the Netherlands, Germany, and Norway, on the subject of Radiation Hazards [1] to provide a review and critical analysis of the available information and concepts.

In 1978, several important recommendations where forwarded by RSG2 to the MED including use of frequency-dependent scaling, use of a continuous curve, and elimination of numerous medical procedures and restrictions such as the need for EEG and ophthalmic examinations. These changes were documented in STANAG 2345, "*Control and Evaluation of Personnel Exposure to Radiofrequency Radiation*" (Edition 1) [2], promulgated in 1979.

The TOR for RSG2 was revised in 1980 with the following objectives:

a. To exchange information on national research programs of military significance concerning the protection of personnel against radiofrequency electromagnetic radiations (less than 100 GHz).

b. To develop guidelines for the setting of standards for allowable human exposure and to determine the impact of such standards on military establishments.

c. To formulate standard procedures for field surveys of NIR.

d. To make recommendations concerning the establishment of safety procedures and personnel exposure limits based on its scientific findings.

e. To provide a source of combined medical, engineering/physics expertise in the area of radiofrequency which is available to other NATO organizations with an interest in this area."

At no time did RSG2 consider any non-ionizing electromagnetic radiation other than radio frequency (RF) radiation. It is interesting to note that these objectives are as valid today as they were 20 years ago.

In 1981, Panel VIII sponsored a workshop at the Royal Air Force Institute of Aviation Medicine (IAM), Royal Aircraft Establishment, Farnborough, U.K. to develop and/or compile sufficient knowledge on the long-term effects of pulsed RFR fields to maintain safe procedures and to minimize unnecessary operational constraints. That workshop brought together eighteen scientists from six NATO countries and resulted in USAF/SAM Aeromedical Review 3-81: *A Workshop on the Protection of Personnel Against Radiofrequency Electromagnetic Radiation* [3].

Also in 1981, a NATO Advanced Studies Institute (ASI) on *Advances in Biological Effects and Dosimetry of Low Energy Electromagnetic Fields* was held in Erice, Sicily, Italy. This meeting produced NATO ASI publication: *Biological Effects and Dosimetry of Non-ionizing Radiation: Radiofrequency and Microwave Energies* [4].

In 1984, Panel VIII sponsored another workshop held in Wachtberg-Werthhoven, Federal Republic of Germany, with over 40 scientists from five NATO countries attending. The proceedings of the second workshop were published as USAFSAM-TP-85-14: *Proceedings of a Workshop on Radiofrequency Radiation Bioeffects* [5]. The proceedings begin with an address by Dr. R. Bernotat (GE) Chairman of Panel VIII, wherein he stated "...this panel has been and is responsible for fostering and coordinating research in the RFR area as well as for transferring knowledge to applications." The need for transfer of technological knowledge on safety guidance and the procedures for implementing that guidance to military organizations is greater now then it was a 15 years ago.

Also in 1984, a NATO Advanced Research Workshop (ARW) examined the *Interaction Between Electromagnetic Fields and Cells* [6], in Ettore Majorana Center for Scientific Culture, Erice, Sicily, Italy. The 1984 workshop concluded that "It is important that today's RFR bioeffects research results, emerging from many countries, continue to be disseminated as efficiently as possible for consideration and use by the NATO military organization." The present ARW was organized with this in mind. The theme of the last AGARD lecture series, to focus on RFR, presented in Italy, Portugal, and France in 1985 was *The Impact of Proposed Radiofrequency Standards on Military Operations* [7].

One of the principal benefits derived from the work of RSG2 was fostering of cooperative research efforts between member nations. The first of these cooperative research efforts began in 1974 between France (General (Dr) Servantie) and The Netherlands (Dr Wolthuis). The Canadian Defence Research Establishment (DREO, Ottawa) and the United States Air Force School of Aerospace Medicine (USAF/SAM) established a Cooperative Research and Development Agreement involving biophysics and bioeffects studies. A Memorandum of Understanding Data Exchange Agreement was formalized to study "Measurement of energy deposition in human sized phantoms in HF fields." UK (IAM, Farnborough) and the USA (USAF/SAM) cooperated on "Detection of low level effects in primates in HF fields." Panel VIII closed RSG2 in September 1982. After a 13 year period of dormancy, in 1995 the Technical Representatives to the MED from France, the United Kingdom, and the United States recommended reactivation. This was accomplished this year as a Task Group under the Research and Technology Organization structure is discussed below.

The most recent NATO sponsored program on RFR was a NATO ARW on "Developing a New Standardization Agreement (STANAG) for Radiofrequency Radiation" held 16-21 May 1993, at the Divisione Aerea Studi Richerche e Sperimentazioni (DASRS) at the Aeroporto Pratica di Mare Italian Air Force Base, Pomezia (Rome) Italy. An international group of 47 specialists working in the field of biological effects of electromagnetic fields and standards development attended this Workshop to make presentations and participate in discussions on developing standards for human exposures. The state of knowledge in this field of research was examined in great detail. The program was divided into three major sections: Review of Standards, Scientific Basis for New Standards, and Public Health Policy Concerns. The ARW also produced a NATO ASI Proceedings with over 40 papers: *Radiofrequency Radiation Standards, Biological Effects, Dosimetry, Epidemiology, and Public Health Policy* [8].

International concern for possible biomedical effects from RFR has expanded greatly. The World Health Organization (WHO) has established the EMF Project to evaluate the research and identify knowledge gaps that suggest research needs. Dr Repacholi discusses this program in a paper in this proceedings [9]. The author who serves as Custodial Technical Representative for RFR personnel safety to the MED and the Radio and Radar Radiation Hazards Working Group (RADHAZ) has been appointed Liaison to the International Advisory Committee of the WHO EMF Project. This action will facilitate further coordinated international communication on research and standards setting. Clearly, the ongoing need for communication and activity on issues of RFR health and safety has grown and been recognized by member nations at several levels within NATO and the entire global community.

3. NATO Transition and Expansion

NATO has undergone marked transformation since the dissolution of the Warsaw Treaty Organization in 1989 and the breakup of the Soviet Union in 1991. A significant initiative at NATO has been establishment of close security links with the states of Central and Eastern Europe and those of the former Soviet Union through the North Atlantic Cooperation Council (NACC) and the Partnership for Peace (PfP) program. Recently, an additional program, Cooperation Partner (CP) nations has been implemented. Many of the same nations belong to both the PfP and CP programs.

Twenty-six states have accepted the invitation to participate in the NATO PfP program. Four of these countries – Austria, Finland, Malta, and Sweden – are not members of the NACC, nor are they CP participants, but they do participate in NACC deliberations on PfP issues and take part in other NACC activities as observers.

One of the mechanisms for increasing ties with Partner countries is the ARW program, which requires that meetings be held in the Partner countries. Another program has been inclusion of PfP states in some of the activities of the MED and RADHAZ Working Groups. At these meetings, PfP nations and NATO member nations have exchanged RFR safety standards in an effort to increase communication, to facilitate scientific information exchange, and to become more informed on NATO Standardization issues in preparation for possible admittance into NATO.

4. NATO Standardization Programs

NATO is the world's largest producer of international standardization agreements. The Military Agency for Standardization (MAS) is the principal military agency for standardization within NATO. The MAS was established in London in 1951 and moved to NATO headquarters in 1971 with the goal of meeting the need for "more efficient methods of producing military equipment and of the standardization of parts and end products of military equipment." Its purpose is to facilitate operational, procedural, and materiel standardization among member nations to enable NATO forces to operate together in the most effective manner. Standardization and interoperability between NATO forces make a vital contribution to the combined operational effectiveness of the military forces of the Alliance and enable opportunities to be exploited for making better use of economic resources. NATO STANAGs allow for establishing coordinated minimal standards necessary for achieving commonality, compatibility, interchangeability, and interoperability. These STANAGs must be based on sound scientific research.

The MAS reports directly to the Military Committee (MC), which is the highest military authority in the Alliance under the political authority of the North Atlantic Council and Defence Planning Committee. The MAS includes the Naval Board, Army Board, Air Board, Joint Service Board, and the Terminology Coordinator. Each Board has membership from each NATO nation except Iceland, which has no military, and Luxembourg, whose interests are represented by Belgium.

The formation of the Conference of National Armaments Directors (CNAD) in 1966 had an important impact on the activities of MAS, as CNAD groups began to take on the task of developing standards.

There are currently three main areas of standardization: operational, materiel, and administrative. Operational standardization (doctrine, tactics, and procedures) is the responsibility of the MC and materiel standardization is the responsibility of CNAD. Administrative standardization (terminology) is on a case by case basis.

Standardization is voluntary, and is achieved by agreement, not compulsion. Each NATO member has an equal voice. The MAS is an administrative agency with a consultative alliance. Therefore, MAS cannot enforce any agreement; nor can it require national conformity to any policy. MAS authority with regard to standardization only comes from formal agreement of nations, through their representative, within the rule of consensus.

The Working Groups are the most important element of the MAS. It is at the Working Group level that STANAGs are developed.

4.1. DEVELOPMENT OF NATO STANAGS

NATO STANAGs for procedures and systems are developed and promulgated by the MAS. A STANAG is the official record of agreement among several or all nations to adopt like or similar materiel or procedures. STANAGs are usually implemented by means of national or NATO command documents.

STANAGs first undergo a process of validation, which ensures that the nations and commands agree that a standardization agreement is required. Next, a standing working group or a special panel carries out the development of a draft agreement. The nation leading the drafting process will normally become the custodian responsible for preparation of the draft and circulating it for comment, collating comments received from nations/commands, and producing the final draft in consonance with comments received or agreements reached at meetings. Ratification is the stage where an informally agreed draft is formally accepted by nations. Following ratification, the MAS Chairman's signature formally promulgates the STANAG. The STANAGs are reviewed at least once every two years.

4.2. REVISION OF NATO STANAG 2345

The Standardization Agreement (STANAG) 2345 (Edition 1), "Control and Evaluation of Personnel Exposure to Radio Frequency Radiation" was signed by representatives of NATO member countries in February 1979. The promulgation of this STANAG represented the culmination of deliberations among scientists and health professionals within the NATO community that RSG2 began six years earlier in May 1973.

Edition 1 of STANAG 2345 established criteria for the evaluation and control of personnel exposure to radio-frequency radiation within NATO forces. It defined hazard assessment, allowed for control measures, indicated actions in case of accidental overexposures and established PELs. At the time of adoption, it contained guidance based on the then state-of-knowledge.

In Sep 1991, the Institute of Electronics and Electrical Engineers (IEEE) published C95.1-1991 *Standard for Safety Levels with Respect to Human Exposure to Radio Frequency Electromagnetic Fields, 3 kHz to 300 kHz* [10]. The American National Standards Institute (ANSI) recognized and adopted C95.1-1991 in Nov of 1992.

The 1993 NATO ARW, Developing a New Standardization Agreement (STANAG) for Radio-Frequency Radiation, initiated the process of revising STANAG 2345. The 1993 ARW produced several recommendations for updating the STANAG 2345. Most important was the agreement to use the newly published ANSI/IEEE C-95.1992 standard as a starting point or straw man for updating STANAG 2345. The ANSI/IEEE standard was developed after deliberation by 125 internationally recognized scientist experts, many from NATO countries. The international make-up of the IEEE Standards Coordinating Committee-28 (SCC-28) and the relative recency of the standard were factors in selecting the ANSI/IEEE C95.1-1992 as a foundation for the second edition of STANAG 2345. Thus, the revised STANAG 2345 (Edition 2)[11] is based on the specific absorption rate (SAR), expressed in W/kg, as is the ANSI/IEEE C-95.1992 standard. The STANAG is frequency dependent and covers the range from 3kHz to 300GHz taking into account frequency dependency of SAR and gives advice on limitation of plane-wave field intensity. It is a single-tiered standard, that incorporates the controlled values of the ANSI/IEEE C-95.1992, in recognition of the military operational circumstances under which it will be implemented. It allows use of protective clothing, if demonstrated to be effective under operational conditions. The

STANAG also provides special considerations for induced and contact current guidance. Most importantly, the STANAG serves as the Minimum Acceptable Standard for NATO operations and will not supercede National Standards that have lower (more restrictive) PELs. International consensus among drafting working group members was reached in 1996. Following review by member nations, unanimous recommendation for ratification of STANAG 2345 (Edition 2) by the MED occurred in Apr 1996. Final promulgation by the MAS was 13 October 1997. Since STANAGs are supposed to be reviewed and reaffirmed every two years, the 47th MED, in May 1998, directed the Custodial Technical Representative for STANAG 2345 to form an Ad Hoc working group to determine if the STANAG required amendment in the light of recent changes in international standards. This meeting occurred at Brooks AFB in September 1998 and several recent modifications of the ANSI/IEEE C95.1-1992 standard regarding induced and contact currents were proposed. The proposed changes will be circulated to the nations for comment and a state of consensus will be presented by the Custodial Technical Representative to the 48th MED in May 1999.

5. NATO Scientific Programs

In addition to its well known political and military dimensions, NATO has a "Third Dimension", which seeks to encourage interaction between peoples, to consider some of the challenges facing our modern society, and to foster the development of Science and Technology (S&T). The NATO Science Program was founded in 1957, just eight years after the creation of NATO, in recognition of the crucial role of science and technology in maintaining economic, political, and military strength. The NATO Science Committee held its first meeting in March 1958 making this the 40th anniversary year. The objective of the Science Program is the enhancement of science and technology through a variety of support activities aimed at promoting international scientific cooperation.

Programs supporting research at NATO are undergoing tremendous restructuring. The Scientific Program has been reoriented to point towards addressing scientific and technological problems being encountered by NATO Partner countries. Some of the programs supporting S&T are Collaborative Research Grants, Expert Visits, Linkage Grants, Advanced Study Institutes, and Advanced Research Workshops. The Science for Stability Program (SSP) initiated in 1979 provided 3-5 year R&D project grants to support research in Greece, Portugal, and Turkey to the (then) technological disparities within the Alliance. This successful program ended in 1997. New programs are offered frequently such as the recent Science for Peace (SfP) Program initiative. This program will use many of the mechanisms of the SSP. The SfP Program aims at assisting Partner countries in their transition towards a market-oriented and environmentally sound economy. This program expands the Science Committee's cooperative activities by enabling Partner scientists to engage in collaborative applied R&D projects with NATO scientists. The program objective is to strengthen scientific infrastructure in Partner countries by supporting applied projects that relate to industrial, environmental or security problems. The Partners for Peace are Albania, Armenia, Azerbaijan, Belarus, Bulgaria, Czech Republic, Estonia, Georgia, Hungary, Kazakhstan, Kyrgyzstan, Latvia, Lithuania, Moldova, Poland, Romania, Russian Federation, Slovak Republic, Slovenia, Tajikistan, the former Yugoslav Republic of Macedonia, Turkmenistan, Ukraine, Uzbekistan. In 1998, the Science Program expanded eligibility on a case-by-case basis to Mediterranean Dialogue partner countries, Egypt, Israel, Jordan, Mauritania, Morocco, and Tunisia. This year (1998) was a year of transition, beginning with support available for both NATO and NATO-Partner collaboration and ending with support available exclusively to NATO-Partnership collaborations.

5.1. SCIENTIFIC AFFAIRS DIVISION

The NATO Scientific Affairs Division is currently focusing on five Priority Areas, Environmental Security, High Technology, Disarmament Technologies, Computer Networking, and Science and Technology Policy. Support is available for several activities. The purpose of an ARW is to contribute to the critical assessment of existing knowledge on new important topics, to identify directions for future research, and to promote close working relationships between scientists from different countries and with different professional experience. ARWs are currently supported in the Priority Areas or a General Science Area and must be held in Partner countries.

The High Technology Priority Area is one of the major sponsors of this ARW. This workshop was designed to review the state-of-the-science dealing with Radio Frequency Radiation (RFR) dosimetry, measurements and the relationships between specific absorption rate, power density, and the biological effects of electromagnetic energy. The ARW serves two major purposes: 1) to provide an international forum to determine what theoretical, technological, and scientific events have occurred that should be reflected in a revision of the *The Radiofrequency Radiation Dosimetry Handbook* [12], and 2) to establish working groups to evaluate these advances and determine how to incorporate them into a revision of the *Handbook*. Information about NATO science programs is available at http://www.nato.int/science.

5.2. RESEARCH AND TECHNOLOGY ORGANIZATION (R&T O)

In the first major restructuring of NATO's R&T structure in over three decades the Defense Research Group (DRG) and the Advisory Group for Aerospace Research and Development (AGARD) merged, into the R&T O on 21 November 1996. AGARD, originally founded in 1952 as an agency of the Military Committee concentrated on Aerospace activities. The DRG was established in 1957 by the Committee of Research Directors, which later evolved into the current CNAD. Both AGARD and DRG where established at the initiative of Dr Theodore von Kármán, a leading aerospace scientist. Thus, the common roots of the two organizations have again become one. The RTO is the single focus in NATO for Defense Research and Technology activities, The highest level of national representation at the R&T Board reports directly to both the conference of National Armaments Directors (CNAD) and to the Military Committee. The RTO comprises six Panels, dealing with;

- Studies, Analysis and Simulation; (SAS)
- Systems Concepts and Integration; (SCI)
- Sensors and Electronics Technology; (SET)
- Information Systems Technology; (IST)
- Applied Vehicle Technology; (AVT)
- Human Factors and Medicine; (HFM)

One of the first actions of the Human Factors and Medicine Panel was to recognize DRG legacy status and approve establishment of Task Group-002 (TG-002) "Health Effects of Non-Ionizing Radiation in the Military Setting. TG-002 members include, France, Germany, The Netherlands, Norway, The United Kingdom, and The United States. Projects to be conducted during the three-year program include the following:
1) Research to develop reliable and validated dosimetry and measurement technology. (This ARW on Dosimetry is a program component).
2) Epidemiological studies of military populations exposed to RFR.
3) Research on shock and burn factors that lead to shock and burn hazard from RFR
4) Research into possible effects of RFR at the cellular and organ level using military unique emitters.

Task Group 2 met in London, UK in July 1998 and expanded its Plan of Work to include an evaluation of the possible impact on NATO military operations by the recently published ICNIRP guidelines.[13]

The RTO Administrative arm, the RTA, has provided funds for speakers from several PfP nations to participate in the present ARW on Dosimetry. The RTO maintains a web-site at http://www.nato.int/structur/rto/rto.htm.

Clearly, NATO has a long history of continuing support for scientific investigation into fundamental biological effects of exposure to RFR. This support of research and standardization contributes significantly to the global approach to standardization. Many STANAGs have provided foundation for European Community standardization. There is an increased attention at NATO for standardization to provide increased harmonization. With expansion of NATO and new collaborations with other international organizations concerned with health and safety issues such as the World Health Organization, NATO has become an important mechanism in facilitating international scientific research, data exchange and global harmonization of RFR standards.

6. Acknowledgements

These views and opinions expressed in this paper are those of the author and do not necessarily state or reflect those the United States Air Force, Department of Defense, U.S. Government nor the North Atlantic Treaty Organization.

10

7. References

1. (1975) AGARD Lecture Series No.78 on Radiation hazards, Lecture Series AGARD-LS-78.
2. (1979) NATO STANAG 2345 (MED WP), Control and recording of personnel exposure to radiofrequency radiation, MAS (ARMY) 2345 (79) 060, (Edition 1).
3. Mitchell, J. C. (ed.) (1981) *Proceedings of a Workshop on the Protection of Personnel against Radiofrequency Electromagnetic Radiation*, Research Study Group 2, Panel VIII, Defence Research Group, NATO, at the Royal Air Force Establishment, Farnborough, U.K., 6-8 April, 1981, USAFSAM-TR-81-28.
4. Grandolfo, M., Michaelson, S.M., and Rindi, A. (eds.) (1983) *Biological Effects and Dosimetry of Nonionizing Radiation: Radiofrequency and Microwave Energies*, Plenum Press, New York and London.
5. Mitchell, J.C. (ed.) (1985) *Proceedings of a Workshop on Radiofrequency Radiation Bioeffects*, Defense Research Group, Panel VIII, NATO AC/243, Research Establishment for Applied Science, D-5307 Wachtberg-Werthoven, Federal Republic of Germany, 11-13 Se 1084, USAFSAM-TP-85-14.
6. Chiabrera, A., Nicolini, C., and Schwan, H.P. (eds.) (1985) *Interactions between Electromagnetic Fields and Cells*, Plenum Press, New York and London.
7. (1985) AGARD. The Impact of Proposed Radiofrequency Radiation Standards on Military Operations, Lecture Series AGARD-LS-138.
8. Klauenberg, B. J., Grandolfo, M., and Erwin, D. N. (eds.) (1995) *Radiofrequency Radiation Standards: Biological Effects, Dosimetry, Epidemiology, and Public Health Policy*, NATO ASI Series Volume 274, Plenum Press Publishing Corp, New York and London.
9. Repacholi, M. (1999) International EMF Project: RF Dosimetry and Biological Effects Research, in B. J. Klauenberg and D. Miklavčič (eds.), *Radio Frequency Radiation Dosimetry and Its Relationship to the Biological Effects of Electromagnetic Fields*, Kluwer Academic Publishers, Dordrecht, pp. 21-28.
10. Institute of Electronics and Electrical Engineers (IEEE) C95.1-1991 *Standard for Safety Levels with Respect to Human Exposure to Radio Frequency Electromagnetic Fields, 3 kHz to 300 kHz. 1991.*
11. 11. NATO STANAG 2345 MED (EDITION 2). *Evaluation and Control of Personnel Exposure to Radio Frequency Fields - 3 kHz to 300 GHz*, 13 Oct 97.
12. Durney, C. H., Massoudi, H. and Iskander, M. F. *Radiofrequency Radiation Dosimetry Handbook (Fourth Edition)*, USAFSAM-TR-85-73.
13. Bernhardt, J. H. (1999) The New ICNIRP Guideline: Criteria, Restrictions, and Dosimetric Needs, in B. J. Klauenberg and D. Miklavčič (eds.), *Radio Frequency Radiation Dosimetry and Its Relationship to the Biological Effects of Electromagnetic Fields*, Kluwer Academic Publishers, Dordrecht, pp. 517-526.

UNITED STATES AIR FORCE SUPPORT OF RADIO FREQUENCY RADIATION HEALTH AND SAFETY: BIOEFFECTS, DOSIMETRY, AND STANDARDS

M. R. MURPHY
Air Force Research Laboratory
Human Effectiveness Directorate
Directed Energy Bioeffects Division
8315 Hawks Road
Brooks Air Force Base, TX, 78235, USA

1. The Requirement for Military Attention to the Health and Safety of Radio Frequency Radiation

The United States Department of Defense (DoD) is one of the world's largest developers and users of radio frequency radiation (RFR) emitting systems, with an estimated 8000 different types in the inventory. Technology exploiting RFR for radar, communications, and anti-electronic weapons supports U. S. and allied defense forces globally and is likely to become even more critical in the future [1]. In the use of such systems, humans and the environment invariably incur some exposure to low levels of RFR and military personnel, in particular, run a risk of accidental exposure to higher levels. There are well established bioelectromagnetic interactions from exposures in excess of standardized limits that can pose health and safety concerns for humans, including burns, stimulation of excitable tissue, shock, and increased thermal burden. Since our knowledge of the physical world is never complete, there is always the possibility of yet to be discovered hazards, especially relating to long-term or repeated exposures. For example, some epidemiological studies have suggested greater health risk for military personnel engaged in occupational specialties that provide the possibility of greater RFR exposure [2, 3].

In addition to the established risks of RFR exposure, there are sometimes hypothesized or imagined risks that are exacerbated by media-heightened fear and misunderstanding of electromagnetic fields by the general public. Such concerns can impact military organizations when the siting or use of radars or other RFR systems are opposed, limited, or temporarily enjoined. Active military personnel also are not immune to media hyperbole, and unnecessary worry over the possible hazards of RFR could compromise their performance. Questions and legal claims of health damage due to military exposure to RFR are also sometimes made by retired or separated

11

B.J. Klauenberg and D. Miklavcic (eds.), Radio Frequency Radiation Dosimetry, 11-19.
© 2000 *U.S. Government. Printed in the Netherlands.*

military members, or their surviving family, even relating back to radar exposures during WWII.

The military services have appropriate equipment, trained specialists, and well-established procedures to measure RFR fields and assure that DoD personnel or nearby civilian populations are not overexposed to military RFR emitting systems. By DoD direction, protection measures and operational restrictions must be identified before new RFR emitting systems are fielded [4]. However, day-to-day practical RFR safety procedures rely greatly on the permissible exposure limits found in promulgated exposure standards. The effectiveness of protection against over exposure to existing systems and controls on future systems therefore depends on the validity and appropriateness of the standards being applied. Since the world's database on RFR health effects, as well as the opinions of scientists and policy makers, is ever changing, so too may the standards by which we provide RFR protection. Strong indications of health hazards at RFR exposures lower than the levels currently approved could dictate that standards be made more restrictive, whereas an expanding database, which reduces the uncertainty of RFR bioeffects and hazards, could allow standards to be relaxed.

The military therefore has a strong requirement and obligation to investigate the health and safety of RFR exposure, both because it is the right thing to do to protect personnel and because attention to this area is critical for the efficient approval, fielding, and operation of new RFR emitting systems. Because this requirement exists for all military services, military research activities on RFR health and safety have long been coordinated by a DoD chartered group called the "Tri-Service Electromagnetic Radiation Panel" (TERP). The TERP consists of three members each from the Army, Navy, and Air Force, with one member from each service representing the research, medical, and operational communities, plus one from the U. S. Marine Corps. While the research programs of the U. S. Military services are coordinated, and, since 1994, collocated at Brooks Air Force Base (BAFB), TX, each retains separate identity, funding, and control. Much of this paper focuses on the RFR research program of the U. S. Air Force, because it is that program with which I am most familiar. However, it should be noted that the other U. S. Military services have similarly active programs both historically and currently.

2. Biological Research in Support of RFR Health and Safety

The most important task in evaluating RFR health and safety hazard potential is conducting innovative biological research. The U. S. Air Force began substantial involvement in RFR health and safety research in 1956 upon being assigned responsibility for coordination of the "Tri-Service Program" on the biological effects of microwave energy. This program was managed from Rome Air Development Center, Griffiss AFB, New York, and was directed by Dr George M. Knauf, USAF (MC). During its 5 years, with a budget of over 15 million dollars, the Tri-Service program supported much, now classic, research by Susskind, Michaelson, McAfee, Schwan, and many others [5].

From 1961 to 1968 the Air Force had no RFR bioeffects research program, but, in late 1968, at the request of the Over the Horizon – Backscatter (OTH-B) radar Project Office, the Air Force School of Aerospace Medicine at Brooks AFB, TX, was asked to assess the biological effects of exposure to the operating frequencies of this new radar. Two pioneers of this effort, John Mitchell and William Hurt, have papers in this volume, as does James Merritt, who joined the program a few years later. This small program, which focused on the potential hazards of one military system, was the beginning of the current Air Force Research Laboratory, Radio Frequency Radiation Branch, making this its thirtieth anniversary year.

While greatly larger and broader, the present DoD RFR research programs still focus on contemporary and future military issues and exposure situations. For example, new systems employing novel types of emissions, such as high power microwave (HPM) and ultrawideband (UWB) radiation, are being developed and these emissions must be examined for health and safety impact before the systems are fielded, as well as during development and testing. The U. S. Army, Navy, and Air Force research programs, collocated at Brooks AFB since 1994, now contain the largest, best equipped, and expertly staffed facility for RFR bioelectromagnetics in the world. For the Tri-Services, including military, civilian, contractors, consultants, scientists, technicians, and other support, there are nearly 100 personnel involved in research on the health and safety of RFR at Brooks AFB, Texas.

At the Tri-Service facilities at Brooks AFB, a wide range of RFR exposure parameters are studied, including exposure from microwaves, millimeter waves, HPM, UWB, both pulsed and continuous wave, and acute, chronic, and repeated exposures. (Because of the considerable attention to ELF health and safety issues by other government agencies and the civilian community, the military research programs have chosen to remain alert to the activity in this area, but not to conduct actual biological research on the effects of ELF.) The research is conducted at biological levels of organization from sub-cellular fractions, to cells, to rodents, goats, monkeys, and humans. Biological effects studied include the biochemical, genetic, neural, physiological, behavioral, and cognitive. Mechanistic studies and modeling strive to improve understanding of the generality of the effects and develop predictive power.

The range of research topics addressed by this group is quite broad. Three studies examining the cancer initiating or promoting capability of RFR exposure have been completed [6, 7, 8]. An extensive research program investigating how the human thresholds for perception and pain from exposure to RFR vary with frequency is underway [9]. Several studies on human thermoregulatory response to whole body RFR exposure have been completed recently [10, 11]. Another focus has been the biological effects of pulsed microwaves with very high peak E-fields but low average intensity. A recently completed two year study on the ocular effects of HPM, which exposed the eyes of primates to 1 MW, 1.25 GHz radiation, pulsed at 0.5 to 3 pulses per second, 4 hr/day, 3 days/week, for 3 weeks, found no histopathological changes in the retina [12]. Other activities of the Tri-Service research group focus on the possible hazards of pulses with rapid rise-time and wide frequency-content, so called ultrawideband (UWB) radiation. Our investigations have included tests of cellular biochemistry in human cells, genotoxicity in yeast cells, sensation thresholds in mice, anxiety, central nervous system activity, reproduction, fertility, cardiovascular function [13], teratology [14], behavior in rats, and behavior in monkeys [15]. No acutely hazardous effects of exposures up to 250 kV/m have been found, but some positive bioeffects have been found at lower e-fields and are being evaluated. An investigation of the long-term effects in animals maintained for 18 months after exposure to UWB emissions over a 3-month period has been completed and results are being analyzed.

3. Advances in RFR Dosimetry

For a great many phenomena, knowledge of how the nature of the "effects" varies with the quantity of the "cause" is fundamental to explanation, extrapolation, and prediction. Quantitative knowledge of the dose of RFR energy absorbed in an exposed biological tissue is as essential to understanding the bioeffects of that exposure as knowledge of the dose of a drug or toxin is to the sciences of pharmacology and toxicology. Because direct knowledge of the actual RFR energy received is often unobtainable, dosimetric information must be inferred from measurements made outside the tissue, often when the tissue is not even present. We now know that such knowledge is extremely complex, depending on (1) parameters of the tissue (dielectric values, dimensions and shape, orientation and stability with respect to the field, and the presence of special materials, such as implants); (2) field parameters (frequency, intensity, pulse parameters, time rate of energy deposition, and near vs. far field); and (3) the environment (e.g., reflective or conductive agents in the field).

The science of RFR dosimetry was in its infancy in 1968 when the Air Force bioeffects program began at Brooks AFB. Using a combined stripline and coaxial transmission line approach for exposures, the first published report of RFR research at Brooks AFB [16] contained no mention of dosimetry, only field strength. This situation was corrected the following year by the addition of a Differential Power Measurement System conceived by John Mitchell and build by M. L. Crawford [17], which has since come to be called a TEM or Crawford Cell. The goal of this system, which became the first contribution to RF dosimetry by the current Air Force program, as written by John Mitchell in 1971, was "to measure... the RF power actually absorbed by the animals within the exposure cell". "Biological response can then be related directly to the total power absorbed in the experimental animal as well as the exposure fields in which the animals are placed." (underlined emphasis was in the original report) [18]

Since the development of the TEM cell, many additional advances in RFR dosimetry have been made and sponsored by the Air Force. These include the use of biological phantoms, infrared imaging, and novel RFR interactive chemicals. But probably the most important Air Force contribution to the science of RF dosimetry has been a series of four editions of the Radiofrequency Radiation Handbook, published in 1976, 1978, 1980, and 1986 [19, 20, 21, 22]. A paper in this volume by John Mitchell reviews the history of these highly used and cited handbooks [23].

A recent focus of the Air Force dosimetry efforts has been the use of numerical models to predict localized RFR dosimetry, as well as novel measurement approaches to validate the predictions of the models. A key to this effort was our sponsorship of the work of Camelia Gabiel in assessing the dielectric properties of thirty body tissues for a broad range of frequencies [24]. This modeling effort was further advanced by the coding of tissue types on MRI scans of the rat, goat, and monkey and on the digitized slices of the Visible Man, and assigning them the appropriate dielectric values [25, 26, 27]. Other recent studies on RFR dosimetry at Brooks AFB include work on precision microdosimetry by Andrea Pakhomov [28], an analog dosimetry method for RFR field mapping [29], and studies on ankle SAR measurements by Richard Olsen [30]. A project that has just begun, is the development of an inexpensive RF personal dosimeter. This development effort is based on discoveries by John Kiel and is led by him and his research team at Brooks AFB. We hope that such a dosimeter will improve the estimation of individual RFR exposure in occupational and operational situations and provide a future database for RFR health risk by means epidemiological assessment [31], which will be useful to both the military and the public sector.

Throughout the past 30 years, the U. S. Air Force RFR bioeffects program at Brooks AFB has sought to improve the state-of-the-art of RF dosimetry, not only for its own research efforts, but also to facilitate a high standard of dosimetry in the field at large. While this approach is generally good scientific practice, there is also another reason, namely that facilitating quality dosimetric techniques by others allows us to have confidence in the bioeffects information they obtain. This continuing Air Force goal of both learning from and sharing advances in RFR dosimetry is reflected by its support and participation in this volume on "Radio Frequency Radiation Dosimetry and Its Relationship to the Biological Effects of Electromagnetic Fields" and the conference on which it is based.

4. Contributions to Science-Based RFR Exposure Standards

The research and involvement of the U.S. military services has long been important in the setting of science-based RFR exposure standards. The $10mW/cm^2$ was initially established on the basis of the Tri-Service Program [5, 23], and the work at the Air Force School of Aerospace Medicine in the early 1970's led to the first frequency-dependent standard in the U.S. in 1975 [32].

Today, research findings by the military services are combined with those from other institutions to recommend science-based safety standards for permissible exposure limits to RFR. Such standards extend from system-specific guidance, as is currently in place for UWB systems [33], to service-specific, DoD, national, and international standards. Through an Office of Management and Budget Directive (OMB-119) [34] and the National Technology Transfer Act of 1995 [35], U. S. Government employees are encouraged to participate in national and international non-governmental consensus-based standard-setting bodies rather than draft government-specific standards. The military services, working through the TERP, have fully supported this policy. For most DoD RFR emitting systems, health and safety issues are addressed in a Department of Defense Instruction (DoDI 6055.11), "Protection of DoD Personnel from Exposure to Radio Frequency Radiation and Military Exempt Lasers" [4], issued by the Undersecretary of Defense for Acquisition and Technology, on the recommendation of the Deputy Undersecretary of Defense for Environmental Security. This instruction is written by the TERP based on the open, consensus-based exposure standard developed by the Institute of Electrical and Electronic Engineers (IEEE) [36]. Individual military services promulgate the DoD Instruction in specific-service publications, for example U. S. Air Force Occupational Safety and Health Standard (AFOSH) Std 48-9 [37], which are used by field engineers in evaluating health and safety conditions and ensuring compliance.

During the period 1993-1997, the TERP took the leadership in revising the NATO Standardization Agreement (STANAG) "Control and Evaluation of Personnel Exposure to Radio Frequency Fields - 3 kHz to 300 GHz" [38]. This agreement, STANAG 2345, is based on both the IEEE/ANSI C95.1 standard and DoDI 6055.11. During the process of revising STANAG 2345, Jon Klauenberg and David Erwin of the U. S. Air Force co-organized with Martino Grandolfo a NATO Advanced Workshop [39], very much like the one that is the basis for the current volume, and U.S. military researchers, health physicists, and contractor personnel strongly supported the publication of the most complete work yet on RFR standard setting [40]. As a consequence of the expertise and service provided during the revision of STANAG 2345, Jon Klauenberg of the U.S. Air Force Radio Frequency Branch was designated Technical Representative to the NATO General Medical Working Group and liaison from that group to the Radio and Radar Hazards Working Group, which has included the health and safety STANAG.2345 in their standard, STANAG 1380, by reference.

When new relevant health and safety data are developed or unusual health issues arise, it is the preference of the TERP to bring these issues first to the IEEE Standards Coordinating Committee 28 for consideration, consensus processing, and issuance as a supplement to the IEEE C95.1 standard. John Leonowich discusses in this volume how this process was recently used to address a problem concerning the exposure limits and averaging time for exposure to RFR-induced currents [41].

When the requirements of the military services are military-unique, the TERP develops an "interim guidance" for exposure limitations. Such was the case in 1995 when the TERP issued an "Interim Guidance for Exposure to Ultrawideband Radiation" [33]. This guidance and new research findings have been reviewed and re-endorsed by the TERP each year since. When sufficient data exist, the UWB Guidance will be transferred to the DoD Instruction, and when UWB exposures become more common in the civilian community, the military bioeffects data will be provided to the IEEE SCC28 for consideration.

Because the U. S. Military services operate globally and with many different national partners, uniformity of RFR exposure standards is a desirable goal. Therefore, the RFR research programs and the TERP attempt to facilitate and support worldwide standards harmonization. To this end, Jon Klauenberg, liaison from NATO, and Michael Murphy, liaison from the TERP, participate in the International Advisory

Committee of the World Health Organization Nine Year EMF program, which also pursues a goal of international RFR standards harmonization. Furthermore, the U. S. Air Force Research Laboratory contributes to the WHO EMF program and the Pan American Health Organization as a Collaborating Center on the Biological Effects of Electromagnetic Radiation. Many DoD personnel also provide service on the IEEE Standards Coordinating Committee 28 and its several subcommittees, and Drs Eleanor Adair and John D'Andrea are members of the RFR section of the National Council on Radiation Protection (NCRP). One Army employee, Dr Dave Sliney, serves on the International Commission on Non-Ionizing Radiation Protection (ICNIRP).

5. Summary and Conclusion

The United States DoD has the need and obligation to evaluate the potential health and safety impact of human exposure to the emissions of the RFR systems that it develops and uses. Over the last 42 years, the biological research, dosimetry advances, and standard-setting activities conducted and supported by the U. S. military services have met these requirements, while also contributing to the international scientific RFR database and standards. This tradition is continued today through active biological research and RFR dosimetry programs, extensive standard setting activities focused on international standard harmonization, and the support of international workshops, such as the one represented by the current volume.

6. Disclaimer

The opinions expressed in this paper are those of the author and should not be interpreted as an official position of the United States Air Force, Department of Defense, or Government.

7. References

1. McCall, G. H. and Corder, J. A. (1995) New World Vistas Air and Space Power for the 21st Century – Summary Volume, U. S. Air Force Scientific Advisory Board.
2. Grayson, J. K. and Lyons, T. J. (1996) Brain Cancer, Flying, and Socioeconomic Status: A Nested Case-Control Study of USAF Aircrew, *Aviation, Space, and Environmental Medicine* 67, 1152-1154.
3. Szmigielski, S. (1996) Cancer morbidity in subjects occupationally exposed to high frequency (radiofrequency and microwave) electromagnetic radiation, *The Science of the Total environment* 180, 9-17.
4. Under Secretary of Defense for Acquisition and Technology (1995) Department of Defense Instruction 6055.11, "Protection of DoD Personnel from Exposure to Radiofrequency Radiation and Military Exempt Lasers". Feb. 21, 1995.
5. Michaelson, S. M. (1971) The Tri-Service Program – A Tribute to George M. Knauf, USAF (MC), IEEE *Transactions on Microwave Theory and Techniques* MTT-19, 131-146.

18

6. Toler, J. C., Shelton, W. W., Frei, M. R., Merritt, J. H., and Stedham, M. A. (1997) Long-term, low-level exposure of mice prone to mammary tumors to 435 MHz radiofrequency radiation, Radiation Research 148, 227-234.

7. Frei, M. R., Berger, R. E., Dusch, S. J., Guel, V., Jauchem, J. R., Merritt, J. H., and Stedham, M. A. (1998) Chronic exposure of cancer-prone mice to low-level 2450 MHz radiofrequency radiation", Bioelectromagnetics 19, 20-31.

8. Frei, M. R., Jauchem, J. R., Dusch, S. J., Merritt, J. H., and Stedham, M. A. (1999) Chronic, low-level (1.0 W/kg) exposure of mammary cancer-prone mice to 2450-MHz microwaves, Radiation Research, in press.

9. Blick, D. W., Adair, E. R., Hurt, W. D., Sherry, C. J., Walters, T. J., and Merritt, J. H. (1997) Thresholds for microwave-evoked warmth sensations in human skin", Bioelectromagnetics 18, 403-409.

10. Adair, E. R., Kelleher, S. A., Mack. G. W., and Morocco, T. S. (1998) Thermophysiological responses of human volunteers during controlled whole-body radio frequency exposure at 450 MHz, Bioelectromagnetics 19, 232-245.

11. Adair, E. R. (1999) Thermoregulation: Its Role in Microwave Exposure, in B. J. Klauenberg and D. Miklavčič (eds.), Radio Frequency Radiation Dosimetry and Its Relationship to the Biological Effects of Electromagnetic Fields, Kluwer Academic Publishers, Dordrecht pp. 349-360.

12. D'Andrea, J. A., (1999) Effects of Microwave and Millimeter Wave Radiation on the Visual system, in B. J. Klauenberg and D. Miklavčič (eds.), Radio Frequency Radiation Dosimetry and Its Relationship to the Biological Effects of Electromagnetic Fields, Kluwer Academic Publishers, Dordrecht, pp. 399-406.

13. Jauchem, J. R., Seaman, R. L., Lehnert, H. M., Mathur, S. P., Frei, M. R., and Ryan, K. L. (1998) Ultra-wideband electromagnetic pulses: Lack of effect on heart rate and blood pressure, Bioelectromagnetics, in press.

14. Merritt, J. H. (1999) Teratology, in B. J. Klauenberg and D. Miklavčič (eds.), Radio Frequency Radiation Dosimetry and Its Relationship to the Biological Effects of Electromagnetic Fields, Kluwer Academic Publishers, Dordrecht, pp. 387-396.

15. Sherry, C. J., Blick, D. W., Walters, T. J., Brown, G. C., and Murphy, M. R. (1995) Lack of behavioral effects in non-human primates following exposure to ultrawide band electromagnetic radiation in the microwave frequency range, Radiation Research 143, 93-97.

16. Mitchell, J. C. (1970) A Radiofrequency Radiation Exposure Apparatus, USAF School of Aerospace Medicine Technical Report, SAM-TR-70-43.

17. Crawford, M. L., Hoer, C. A., and Komarek, E. L. (1971) RF Differential Power Measurement System for the Brooks AFB Electromagnetic Radiation Hazards Experiments. National Bureau of Standards Report 9795.

18. Mitchell, J. C. (1971) Modified Exposure System for HF Band RF Radiation Studies, Proceedings of the DoD Electromagnetic Research Workshop, U.S, Navy Bureau of Medicine and Surgery, Washington, D. C. 27-28 January.

19. Johnson, C. C., Durney, C. H., Barber, P. W., Massoudi, H., Allen, S. J., and Mitchell, J. C. (1976) USAF School of Aerospace Medicine, Radiofrequency Radiation Dosimetry Handbook – 1st Edition, USAFSAM-TR-76-35.

20. Durney, C. H., Johnson, C. C., Barber, P. W., Massoudi, H., Iskander, M. F., Lords, J. L., Ryser, D. K., Allen, S. J., and Mitchell, J. C. (1978)) USAF School of Aerospace Medicine, Radiofrequency Radiation Dosimetry Handbook – 2nd Edition, USAFSAM-TR-78-22.

21. Durney, C. H., Iskander, M. F., Massoudi, H., Allen, S. J., and Mitchell, J. C. (1980) USAF School of Aerospace Medicine Radiofrequency Radiation Dosimetry Handbook – 3rd Edition, USAFSAM-TR-80-32.

22. Durney, C, H., Massouodi, H., and Iskander, M. F. (1986) USAF School of Aerospace Medicine Radiofrequency Radiation Dosimetry Handbook - 4th Edition, USAFSAM-TR-85-73. Available at http://www.brooks.af.mil/AFRL/HED/hedr/reports/.

23. Mitchell, J. C. (1999) Historical Perspective of the Dosimetry Handbook: Yesterday, Today, and Tomorrow, in B. J. Klauenberg and D. Miklavčič (eds.), Radio Frequency Radiation Dosimetry and Its Relationship to the Biological Effects of Electromagnetic Fields, Kluwer Academic Publishers, Dordrecht, pp. 551-558.

24. Gabriel, C. (1996) Compilation of the dielectric Proprieties of Body Tissues at RF and Microwave Frequencies. U.S. Air Force Armstrong Laboratory Technical Report, AL/OE-TR-1996-0037. Available at http://www.brooks.af.mil/AFRL/HED/hedr/reports/.

25. Mason, P. (1999) Recent Advances in Dosimetry Measurements and Modeling, in B. J. Klauenberg and D. Miklavčič (eds.), Radio *Frequency Radiation Dosimetry and Its Relationship to the Biological Effects of Electromagnetic Fields*, Kluwer Academic Publishers, Dordrecht, pp. 139-154.

26. Hurt, W. D. (1999) Absorption Characteristics and Measurement Concepts, in B. J. Klauenberg and D. Miklavčič (eds.), Radio *Frequency Radiation Dosimetry and Its Relationship to the Biological Effects of Electromagnetic Fields*, Kluwer Academic Publishers, Dordrecht, pp. 39-52.

27. Ziriax, J. (1999) Dosimetry Measurements and Modeling: Interactive Presentations in the New Dosimetry Handbook, in B. J. Klauenberg and D. Miklavčič (eds.), Radio *Frequency Radiation Dosimetry and Its Relationship to the Biological Effects of Electromagnetic Fields*, Kluwer Academic Publishers, Dordrecht, pp. 559-568.

28. Pakhomov, A. (1999) Precision microdosimetry in microwave-exposed subjects, in B. J. Klauenberg and D. Miklavčič (eds.), *Radio Frequency Radiation Dosimetry and Its Relationship to the Biological Effects of Electromagnetic Fields*, Kluwer Academic Publishers, Dordrecht, pp. 187-198.

29. Hurt, W. D (Ed.) (1996) Radio Frequency Radiation Dosimetry Workshop: Present Status and Recommendations for Future Research. AL/OE-SR-1996-0003.

30. Olsen, R. (1999) SAR Measurements in the Rhesus Monkey Ankle: Implications for Humans, in B. J. Klauenberg and D. Miklavčič (eds.), Radio *Frequency Radiation Dosimetry and Its Relationship to the Biological Effects of Electromagnetic Fields*, Kluwer Academic Publishers, Dordrecht, pp. 371-378.

31. Kiel, J. L. (1999) Molecular Dosimetry, in B. J. Klauenberg and D. Miklavčič (eds.), *Radio Frequency Radiation Dosimetry and Its Relationship to the Biological Effects of Electromagnetic Fields*, Kluwer Academic Publishers, Dordrecht, pp. 227-240.

32. Department of the Air Force (1975) Radiofrequency Radiation Health Hazards Control, Air Force Regulation 161-42.

33. Tri-Service Electromagnetic Radiation Panel (1995) Ultra-Wideband (UWB) Interim Guidelines, Approved May 1995, May 1996; Revised June 1997.

34. Raines, F. D. (1998) U. S. Office of Management and Budget Circular A-119 – Federal Participation in the Development and Use of Voluntary consensus Standards and in conformity Assessment Activities – Revised. (Revised OMB Circular A-119)

35. U. S. Government (1995) National Technology and Transfer and Advancement Act of 1995, Public Law 104-113.

36. Institute of Electrical and Electronics Engineers (1992) IEEE Standard for Safety Levels with Respect to Human Exposure to Radio Frequency Electromagnetic Fields, 3 kHz to 300 GHz, IEEE C95.1, *and IEEE, NEW YORK.*

37. Air Force Occupational Safety and Health Standard (AFOSH) Std 48-9, "Radio Frequency Radiation (RFR) Safety Program," 1 August 1997.

38. North Atlantic Treaty Organization (NATO) Standardization Agreement (STANAG) 2345 MED (Edition 2), Evaluation and Control of Personnel Exposure to Radio Frequency Fields – 3kHz to 300 GHz, *Military Agency for Standardization*, 13 October 1997.

39. Klauenberg, B. J. (1999) NATO Involvement in RFR Research, Health, and Safety, in B. J. Klauenberg and D. Miklavčič (eds.), *Radio Frequency Radiation Dosimetry and Its Relationship to the Biological Effects of Electromagnetic Fields*, Kluwer Academic Publishers, Dordrecht, pp. 1-10.

40. Klauenberg, B. J., Grandolfo, M. and Erwin, D. N. (eds.) (1995) *Radiofrequency Radiation Standard: Biological Effects, Dosimetry, Epidemiology, and Public Health Policy*, Plenum Press, NY, NY.

41. Leonowich, J. (1999) Development of Induced & Contact Current Standards Based on SAR Measurements in the HF and VHF Regions, in B. J. Klauenberg and D. Miklavčič (eds.), *Radio Frequency Radiation Dosimetry and Its Relationship to the Biological Effects of Electromagnetic Fields*, Kluwer Academic Publishers, Dordrecht, pp. 305-312.

INTERNATIONAL EMF PROJECT
Radiofrequency Field Dosimetry and Biological Effects Research

M. H. REPACHOLI
World Health Organization
Geneva, Switzerland

1. Introduction

Electromagnetic fields (EMF) at all frequencies represents one of the most common and fastest growing environmental influences. A complex mixture of exposure to EMF frequencies now occurs to varying degrees to all populations of the world. These exposures will continue to increase with advancing technology. Thus, any small health consequences from EMF exposure could have a major public health impact. It is highly desirable that the mistakes made with commonly available carcinogens, such as cigarette smoke, ionizing radiation and asbestos, should not be repeated.

Radiofrequency (RF) fields are used in many facets of everyday life, such as radio and television broadcast, telecommunications, diagnosis and treatments of disease and in industry for heating and sealing materials. Concerns have been raised that exposure to RF fields may be associated with an increased risk of cancer or other detrimental health effects. With the rapid introduction of mobile telecommunications there has been a focus on problems associated with near field RF dosimetry, especially for the human head exposed to the small radiating antenna from mobile phones.

The International EMF Project was established by the World Health Organization (WHO) to provide a mechanism for resolving the many and complex issues related to possible health effects of EMF exposure. The Project assesses health and environmental effects of exposure to static and time varying electric and magnetic fields in the frequency range 0 - 300 GHz, with a view to facilitating the development of international guidelines on exposure limits. It commenced at WHO in 1996 and is scheduled for completion in 2005.

This paper describes the progress made by the International EMF Project, the research needed for WHO and the International Agency for Research on Cancer (IARC) to properly evaluate health risk from RF field exposure and the important role of dosimetry to this research.

B.J. Klauenberg and D. Miklavcic (eds.), Radio Frequency Radiation Dosimetry, 21-28.
© *2000 Kluwer Academic Publishers. Printed in the Netherlands.*

2. International EMF Project

The International EMF Project has been designed in a logical progression of activities and outputs to allow improved health risk assessments, and identification of any environmental impacts of EMF exposure. The Project objectives are to:

- provide a coordinated international response to the concerns about possible health effects of exposure to EMF;
- assess the scientific literature and make a status report on health effects;
- identify gaps in knowledge needing further research to make better health risk assessments;
- encourage a focused research program in conjunction with research funding agencies;
- incorporate research results into WHO's Environmental Health Criteria (EHC) monographs where formal health risk assessments of exposure to EMF will be made;
- facilitate the development of internationally acceptable standards for EMF exposure;
- provide information on the management of EMF protection programs for national and other authorities, including monographs on EMF risk perception, communication and management; and
- provide advice to national authorities, other institutions, the general public and workers, about any hazards resulting from EMF exposure and any needed for mitigation measures.

3. WHO Research Agenda

The International EMF Project, in collaboration with the International Commission on Non-Ionizing Radiation Protection (ICNIRP), has completed initial international scientific reviews of possible health effects of exposure to electromagnetic fields (EMF). These reviews provided interim conclusions on health hazards from exposure to EMF and gaps in knowledge requiring further research before better health risk assessments could be made. The results of the radiofrequency (RF) review are summarized in the Munich meeting report [1], covering frequencies > 10 MHz to 300 GHz.

The reviews identified research that had raised unresolved questions about whether exposure to low-level RF, particularly over long periods, has any deleterious effects on human health. WHO's EMF Research Agenda [2] has been formulated to try to resolve these questions. The Agenda below resulted from an ad hoc Research Co-ordination Committee meeting held in Geneva 4-5 December 1997. At this meeting, ongoing research was identified that was of sufficient quality and satisfied WHO's requirements for health risk assessment. This was compared with research needed to fill key gaps in knowledge during the scientific reviews. This additional research still needed by WHO then formed the WHO EMF Research Agenda [2] summarized below.

For new studies to be useful to future health risk assessments, the research must be of high scientific quality with clearly-defined hypotheses, estimates of the ability of the study to detect small effects, and use protocols that are consistent with good scientific practice. Accurate dosimetry is essential to allow dose-response relationships to be determined. Quality assurance procedures should be included in the protocol and monitored during the study. WHO's research requirements, publications of the Project, updates on activities and further information, can be found on the home page at: http://www.who.int/emf/.

3.1. DEFINITIONS

The WHO constitution defines *health* as *a state of complete physical, mental and social well-being, and not merely the absence of disease or infirmity*. This definition includes an important subjective component that must be taken into account in health risk assessments. Within the International EMF Project, a working definition of health hazard has been developed: *A health hazard is a biological effect outside the normal range of physiological compensation that is detrimental to health or well being*. In this definition, a *biological effect is a physiological response to exposure*. For a biological effect to lead to an adverse health consequence, it should be *outside the normal range of compensation*, in order to place it beyond normal variation in body responses.

3.2. DETERMINING RESEARCH NEEDS

Criteria used to evaluate health risks by the International EMF Project were adapted from those used by WHO's International Agency for Research on Cancer (IARC) [3]. EMF studies were identified that provided results suggestive of a health risk, but was insufficient for establishing an adverse health consequence. Additional studies were also identified on the basis of unconfirmed effects having implications for health, and replication of key studies. Thus, the overall goal is to promote studies that demonstrate a reproducible effect of EMF exposure that has the likelihood to occur in humans and has a potential health consequence.

While *in vitro* studies can provide important insights into fundamental mechanisms for biological effects from exposure to low-level EMF, *in vivo* studies, whether on animals or human beings, provide more convincing evidence of adverse health consequences.

Epidemiological studies provide the most direct information on risks of adverse effects in human beings. However, these studies have limitations, especially when low relative risks are found. Epidemiological studies are important for monitoring public health impact of exposure, particularly from new technologies.

In these times of scarce budgetary resources it is of importance that the correct mix of priority studies is performed. Obviously, only studies likely to provide useful results should be conducted. In addition to scrutinizing the goals of a proposal, it is important to assess its feasibility and probability that it can detect an effect. Proposed studies should be also evaluated for:

        (i) characterization and/or control of potential confounders,

        (ii)   accuracy of dosimetry

(iii) reproducibility of exposure conditions or measurements and their relevance to human exposures, and

(iv) ongoing quality assurance.

Priority should be given to studies designed to investigate health hazards of concern to the general public, hazards of potential public health importance (based on the size of the populations potentially exposed, the extent of their exposure, and the seriousness of the hypothesized adverse effect), and studies of scientific importance (e.g., testing the relevance of effects observed or mechanisms postulated on the basis of *in vitro* or *in vivo* results).

4. RF Research Priorities

Relatively high-intensity RF fields have been shown to cause adverse health consequences by heating tissues. No adverse health effects have been scientifically confirmed from exposure to low-level RF fields for extended periods, but certain questions have not been thoroughly studied. There is very little information available in the scientific literature to assess any health risks from exposure to pulsed RF fields. Studies are needed that seek to identify any biological effects produced by pulsed RF fields, of both high and low peak pulse intensities. Examples of current and future technologies using pulsed RF fields are telecommunications, civilian and military radar systems, including emerging radar technology such as ultra-wide band radars. Current and future research applicable to mobile telephone systems should focus on the 900-2000 MHz frequency range and appropriate pulsing and modulation patterns. For radars, the frequency and pulsing regimes should be applicable to current and emerging systems.

The following WHO research needs are given in approximate priority order [3]:

(i) Several animal experiments, using various RF exposure regimens, are currently under way, and their results should add to the required database for health risk assessment. However, at least two more, large-scale standard 2-year animal bioassays, such as those typically conducted by the US National Toxicology Program, are needed to test for cancer initiation, promotion, co-promotion and progression. These experiments should expose normal animals and animals initiated with chemical carcinogens to RF fields in the mobile telephone frequency range, using one of the common mobile telephone system pulsing patterns, for 2-6 hours daily. Each study should use a range of intensities (normally 4 different SARs), with the highest being just below the level that may induce temperature changes.

(ii) A large study has suggested that exposure to RF fields increases the incidence of lymphomas in genetically manipulated (transgenic) mice. There is need for at least a further two large studies, using designs similar to (i) above, to clarify the issues raised by this study. Follow-up research is also needed that provides information on the health implications of effects found in transgenic animals.

(iii) Additional studies are needed to test the reproducibility of reported changes in hormone levels, effects on the eye, inner ear and cochlea, memory loss, neurodegenerative diseases and neurophysiological effects. These studies can be performed on animals, but where possible, they should be conducted on human volunteers.

(iv) Analysis of current epidemiological studies of people exposed to low levels of RF has not shown any adverse health effects. However, mobile telephone use is relatively new, and further work is needed. As a general principle, studies on populations exposed to RF at higher levels, though still below the threshold of heating, are more likely to provide information regarding the existence of any health effects, even though such exposure levels may not be representative of general-population exposure. Because of exposure to low levels, causing limitations on exposure assessment, studies of populations exposed to point sources, such as broadcast towers or mobile telephone base stations, are unlikely to be informative about the existence of health effects. Suggestions of an increased incidence of cancer in populations around mobile telephone base stations have not been substantiated.

There needs to be conducted at least two large-scale epidemiological studies with well characterized, higher-level RF exposures to investigate cancers, particularly in the head and neck, and any disorders associated with the eye or inner ear. These studies should preferably be on mobile telephone users or on workers in industries giving high RF exposures provided valid exposure assessments could be developed.

(v) Both epidemiological and laboratory studies are needed to provide basic information that allows better assessments of any health risks from exposure to radar technology, particularly emerging systems such as ultra-wide band radars.

(vi) Well-controlled studies are needed to test people reporting specific symptoms, such as headache, sleep disorders or auditory effects, and who attribute these symptoms to RF exposure. Past human volunteer studies of this type have not successfully linked the symptoms and exposure. Several more controlled investigations should be performed to investigate neurological, neuroendocrine, and immunological effects.

(vii) *In vitro* studies normally have a lower priority than *in vivo* or human studies in health risk assessment. However, such studies can be of great assistance if they are directly relevant to possible *in vivo* effects, and address the issues of RF exposure thresholds and reproducibility for reported positive effects on cell cycle kinetics, proliferation, gene expression, signal transduction pathways and membrane changes. Theoretical modeling investigations can be useful if they support *in vivo* studies by proposing testable basic mechanisms of RF field exposure.

5. RF Dosimetry and Exposure Assessment

5.1. DOSIMETRY

Dosimetry is a critical component of any scientific study assessing effects of RF fields on biological systems. Specific absorption rate (SAR), in watt per kilogram (W/kg), is the fundamental and widely accepted RF dosimetry parameter. Although some consider that the SAR may not be generally applicable to low-level RF field effects, it is still the best dosimetric concept available and should continue to be used. Information about the internal magnetic field should also be provided. For some *in vitro* and *in vivo* models, polarization of the internal electric field might be an important exposure parameter, such as for a monolayer of cells. Signals should be fully described, since the time averaged field strength might not be sufficient.

Determination of SAR and the internal electric and magnetic fields is a complicated function of various exposure parameters, such as the incident field and the physical properties of the biological specimen [4]. Complete analysis of these dosimetric parameters can be performed by numerical simulation techniques, the results of which must be evaluated experimentally. Current computational methods allow assessment of internal electric field distributions with an accuracy of 2-3 dB. Combinations of the most recently developed experimental and computational dosimetric tools enables assessment of the internal field distribution with a precision better than 1 dB [5].

5.2. LABORATORY STUDIES

It is essential for high quality research that accurate assessment of RF field exposure be an integral part of all future studies and that each research team include scientists skilled in RF dosimetry. WHO recommends that future studies have a dosimetric precision of 30% or better. This should be achievable for animal studies and other laboratory investigations for which any dose responses will be determined. If mechanisms of low-level RF interactions are found, accurate dosimetry will allow extrapolation of possible consequences to humans with a greater precision than possible with epidemiological studies.

For both *in vitro* and *in vivo* studies, several conventional exposure systems are commonly used. In many situations however, the field scattered from the exposed object may alter the incident field and source characteristics, thereby creating the possibility of significant errors in dosimetry. Using transverse electromagnetic (TEM) cells to expose specimens in Petri dishes, Burkhardt *et al* [6] found that coupling occurred between the TEM cell walls and the Petri dishes, as well as between the dishes themselves, significantly altering RF absorption in the biological sample.

5.3. EPIDEMIOLOGICAL STUDIES

One of the most difficult areas for assessing RF exposure is in epidemiological studies. There is a great need for good exposure assessment in these studies because the current database of published studies does not provide any useful information to assess health risks.

Key to the improvement of future epidemiological studies is the development of instruments or assessment methods that can conveniently and accurately measure an individual's exposure to RF over an extended period. Great improvements in extremely low frequency field (ELF) exposure assessment have come from the development of personnel dosimeters and data-loggers to record exposures. Unfortunately, accurate RF instrumentation has not kept pace with the needs for RF epidemiology.

So important is the need for good RF exposure assessment, that WHO held, with one of its collaborating centers, the UK National Radiological Protection Board, held a special workshop on this topic from 7-9 September 1998. At this meeting there was extensive discussion about the elements of generic protocols for both ELF and RF exposure assessment for next generation epidemiological studies. This was considered particularly important considering the need to properly assess RF exposure from mobile telephones.

Mobile telephones produce local absorption of RF energy in the head. One of the major problems is the lack of standardized methods for evaluating the local specific absorption rate (SAR). It would be extremely useful if a universally accepted computer modeling technique was available for accurately simulating RF field absorption. Not only would this be useful for researchers, but also for regulators determining compliance with standards.

6. Conclusions

Through its International EMF Project, WHO has identified the research needed to allow better health risk assessments to be made. Precise dosimetry is key to the success of this research. Better instrumentation and improved assessments of exposure are needed if possible subtle health risks are to be identified. The International EMF Project is providing all the assistance it can to ensure that both laboratory and epidemiological studies are conducted with the best available dosimetric methods. When the ultimate goal is to produce health risk assessments where dose-response relationships could lead to the development of exposure limits, accurate dosimetry cannot be overemphasized.

7. References

1. Repacholi, M.H. (1998) Low-level exposure to radiofrequency electromagnetic fields: Health effects and research needs. *Bioelectromagnetics* 9, 1-19.
2. World Health Organization (1998) WHO's Agenda for EMF Research. World Health Organization, International EMF Project. WHO, Geneva, Publication WHO/EHG/98.13.

3. Repacholi, M.H. and Cardis, E. (1997) Criteria for EMF health risk assessment, *Radiat Prot Dosim* **72**, 305-312.

4. Chou, C-K., Bassen, H., Osepchuk, J., Balzano, Q., Petersen, R., Meltz, M., Cleveland, R., Lin, J.C., and Heynick, L. (1996) Radio frequency electromagnetic exposure: Tutorial review on experimental dosimetry, *Bioelectromagnetics* **17**, 195 - 208.

5. Kuster, N. and Balzano, Q. (1996) Experimental and numerical dosimetry, in N. Kuster, Q. Balzano and J.C. Lin (eds.) *Mobile Communications Safety*, London, Chapman Hall, pp. 13-64.

6. Burkhardt, M., Pokovic, K., Gnos, M., and Kuster, N. (1996) Numerical and experimental dosimetry of Petri dish exposure setups, *Bioelectromagnetics* **17**, 483-493.

SUMMARY OF SESSION A: BASICS OF ELECTROMAGNETICS AND DOSIMETRY

Y. GRIGORIEV
Center of Electromagnetic Safety
State Research Center of Russia – Institute of Biophysics
Moscow, Russia

Four reports were presented in session A.

The following basic results were formulated by Yu. Grigoriev in his report *Microwave Effect on Embryo Brain: Dose Dependence and the Effect of Modulation*:
1)The experiments established that a single short-term exposure to CW or pulse-modulated microwaves at the average incident power densities from 0.4 to 10 mW/cm^2 suppressed the imprinting behaviour. 2) Microwaves modulated at 10 Hz were more effective than CW radiation at the same incident power density. 3) The intensity of 40 μW/cm^2 was subthreshold to alter the imprinting behaviour in chicks by a short-term CW irradiation of embryos. However, the threshold for modulated microwaves was below 40 μW/cm^2. 4) It was also demonstrated that a low-intensity modulated microwave signal can be detected and "memorized" by the embryo brain.

Highlights of report *Absorption Characteristics and Measurement Concept* by William Hurt are the following:

Much RF dosimetry information is available because of the recent development of accurate anatomical models of animals and humans, good permittivity data for many different tissue types, and fast computers for running FD-TD programs.

Current needs are:
1)Conformation of SAR distribution by laboratory measurements.
2)Determination of sensitivity of SAR to permittivity.

SAR distribution information should be used in conjunction with thermal diffusion and blood flow characteristics to define resulting temperature distribution. This can be used to better explain thermal induced RF bio-effects.

The distinction between external (outside a human body or considered model) and internal (inside the model) dosimetry has been made in the report by Dina Šimunić *Dosimetry of Pulsed Fields*.

External dosimetry depends on the distance of the object from the source and therefore can be classified into near- and far-field. In the near-field electric and magnetic fields are relevant quantities and in the far-field the power density. The defined quantities of the internal pulsed fields dosimetry are Specific Absorption Rate (SAR) and Specific Absorption (SA). SAR is specific absorbed power per unit mass,

29

B.J. Klauenberg and D. Miklavcic (eds.), Radio Frequency Radiation Dosimetry, 29-30.
© *2000 Kluwer Academic Publishers. Printed in the Netherlands.*

expressed in W/kg and SA is specific absorbed energy per unit mass, expressed in J/kg. Some examples of pulsed fields sources are: radars, digital mobile phones and magnetic resonance imaging equipment. Pulse fields are characterized by duty cycle, pulse repetition frequency and pulse width. Two extreme cases of pulses have been shown: the TEMPO and MRI sequences. The following standards have sections concerning pulsed fields exposure: NATO STANAG 2345 Ed2 1997; IEEE SCC28 1991; ENV 50166-2 1995; ICNIRP 1998. Basic measurement problems which can occur in the far and near field have been considered. Necessary properties and overview of measuring equipment have been given. In conclusion, dosimetry of pulsed fields seems to be a very challenging task and important because of possible non-thermal effects.

Robert Adair in his report *Biophysics Limits on the Biological Effects of Ultrawideband Electromagnetic Radiation* emphasized three points concerning pulsed fields.

1) As a consequence of shielding by induced surface changes, low frequency components of pulsed fields do not enter the body. Hence, even for long external pulses, the internal pulses will be no longer than 20 ns.

2) The significant exposure factor for an electric field pulse is E -τ, where τ is the duration of the internal pulse and E the average internal field. For $E = 100$ kV and $\tau = 1.5$ ns, the momentum transferred to a Biological system holding one charge is equal to that from collision with one water molecule.

3) The attenuation of waves with a millimeter carrier frequency with wavelength $\lambda_{\tilde{n}}$ modulated at a frequency with a wavelength λ_m, is such that the carrier wave is attenuated $I/R^2 \cong 13\%$ in a distance in tissue of about $\lambda_{\tilde{n}}/10$ while the modulated wave is attenuated in a longer distance $\cong \lambda_m / IO$.

MICROWAVE EFFECT ON EMBRYO BRAIN: DOSE DEPENDENCE AND THE EFFECT OF MODULATION

Y. GRIGORIEV and V. STEPANOV
Center of Electromagnetic Safety
The State Research Center of Russia - Institute of Biophysics
Moscow, Russia

1. INTRODUCTION

There are many publications showing that the nervous system is especially sensitive of to EMF. We suppose that EMF effects on the nervous system are critical for the evaluation of possible health hazards from EMF exposure. Earlier we demonstrated that imprinting is a good model for studying effects of low levels of various physical factors on the embryonic nervous system. We are not aware of any studies of EMF effects on imprinting behavior.

The goals of this study were: (1) to develop response spectra for CW and pulse-modulated microwaves with respect to the field intensity, and to determine threshold intensities, and (2) to investigate the principle possibility of detection and "memorizing of a modulated microwave signal by the brain of chick embryo.

2. MATERIAL AND METHODS

Experiments with imprinting were carried out on chicks; a total of 447 chick embryos were used. The embryos were incubated at 37.5-38.0 °C, 55-68% relative humidity for 21-22 days. The embryos were subjected to a short-term exposure to 10-GHz microwaves at various periods of the incubation. Exposures were performed in an anechoic chamber. Parallel controls to each set of experiments were handled in the same way, but received sham irradiation. Tested levels of the incident power density were 0.04, 0.4, 1.0, 8.0, and 10 mW/cm^2.

An imprint-stimulus (either a moving object or light flashes) was presented the first time during the receptive period, 24 hours after hatching. The imprinting behavior was evaluated 24 hours later, i.e., 48 hours after hatching.

Quantitative evaluation of the imprinting behavior was based on the following parameters: latency of the response to the imprint-stimulus, the duration of time interval when the chick remained near the imprint-stimulus, and the number of approaches to and contacts with the stimulus.

31

B.J. Klauenberg and D. Miklavcic (eds.), Radio Frequency Radiation Dosimetry, 31-37.
© 2000 *Kluwer Academic Publishers. Printed in the Netherlands.*

32

A double-blind test procedure was used. Results were statistically evaluated by Student's and Fisher's parametric tests and by a non-parametric Vilkison criterion.

3. RESULTS

EMF effect on memory formation during the early postnatal period. In this first series, we performed three sets of experiments. Embryos were exposed a single time, on the 5th, 16th, or 19th day of incubation (Table 1). The exposure lasted for 30 min at 0.4, 1, 8, or 10 mW/cm^2.

TABLE 1. Conditions of experiments for Series 1.

Set number	Incubation day+EMF exposure	Frequency	EMF Parameters			Number of embryos	
			Conditions of exposure	Power density, mW/cm^2	exposure duration, min	control	experiment
I	5th	10 GHz	CW	1, 8 and 10	30	27	22
II	16th	10 GHz	CW	1	30	9	10
III	19th	10 GHz	CW	0.4	30	18	18

TABLE 2. Imprinting behavior in chicks that were exposed in egg on the 5th day of incubation.

Subsets	Power density, mW/cm^2	Exposure duration, min	Groups	Number of chicks	Number of chick with imprinting
1.1	10	-	control	6	6
		30	experiment	7	0
1.2	8	-	control	11	11
		30	experiment	8	0
1.3	1	-	control	10	9
		30	experiment	7	4

Results of the Set 1 are summarized in Table 2. In subsets 1.1 and 1.2, the imprinting behavior was produced in 100% of control chicks, but in none of those exposed at 10 or 8 mW/cm². Thus, exposure of embryos at these power densities completely suppressed the ability of newborn chicks to develop imprinting. Exposure at a lower power density of 1 mW/cm² had a less pronounced effect: the imprinting response was produced in 9 out of 10 controls, and in 4 out of 7 exposed chicks. In set 2, chick embryos were exposed on the 16th day of incubation. The results were similar to those in subset 1.3. Eight out of 9 chicks developed imprinting in the control group, and 6 out of 10 in the exposed group. Apparently, the day of exposure was not as significant as the incident power density. This allowed us to pool together all the data for exposures at 1 mW/cm² (i.e., from the subset 1.3 and set 2). In this combined group, the imprinting behavior developed in 17 out of 19 controls and in chicks, and in 10 out 17 exposed at 1 mW/cm² (p<0.05).

In set III, embryos were exposed on the 19th day of incubation at the power density of 0.4 mW/cm². The imprinting behavior was established in 17 out of 18 controls, and in 13 out of 18 exposed chicks. This difference was not statistically significant (p<0.1). However, the data combined for all the 3 groups exposed at low intensities (0.4 and 1 mW/cm²) showed a highly significant effect (34 chicks with imprinting out 37 in controls versus 23 out of 35 in the exposed groups, p<0.01).

Therefore, we conclude that the power densities of 8 and 10 mW/cm² produce a strong effect on the development of the embryonic nervous system. A manifestation of this effect is the inability of newborn chicks to develop imprinting behavior. We also infer that lower EMF intensities of 0.4 and 1 mW/cm² can alter the formation of imprinting behavior in some individuals.

"Memorizing" of the microwave signal by embryo brain. This set of experiments was performed on 127 embryos. On the 16th day of incubation, the embryos were exposed to 10 GHz, 0.04 mW/cm² microwaves modulated at 1, 2, 3, 7, 9 and 10 Hz. The exposure duration was 5 min.

We supposed that microwave modulation frequency could be "memorized" by brain, so that after hatching the exposed chicks would prefer light flashes of the same frequency.

The imprinting behavior (actually, in this case it was a "preference" behavior) was tested 48 hours after hatching. Chicks had to choose between two stimuli. One stimulus was light flashes with the same frequency as had been used for modulation during the microwave exposure. The other stimulus was the same flashes, but at a different frequency. The difference in the frequencies of flashes of these two stimuli was always 8 Hz. For example, if one frequency was 10 Hz, the other light flashed at 2 Hz; if one frequency was 5 Hz, the other stimulus was set at 13 Hz, and so forth.

Indeed, chicks showed preference to the imprint stimulus of 9 or 10 Hz if the embryos had been exposed to EMF with 9 or 10 Hz modulation (Figure 1). However, exposure at other modulation frequencies (1, 2, 3, or 7 Hz) was not effective.

34

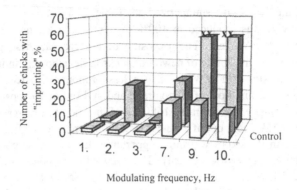

Number of chicks with "imprinting", %

Modulating frequency, Hz

Figure 1. Number of embryos prefering EM modulated signal

Influence of modulated microwaves on imprinting behavior. The above data showed that, for some reasons, the embryonic brain was more susceptive to microwave modulation frequencies of 9 and 10 Hz. Therefore, we decided to compare the efficacy of CW 10-Hz modulated irradiation regimens. The endpoint was formation of the imprinting behavior (same procedures as in the series 1). The light flashes, which were used as an imprint-stimulus, were also delivered at a 10-Hz rate.

A total of 129 embryos were exposed or sham-exposed on the 16th day of incubation, and the imprinting behavior was assessed 48 hours after hatching. In this series, we anticipated stronger effects; therefore, the incident power density was decreased to 40 $\mu W/cm^2$, and the exposure duration was shortened to 5 min. Results of these experiments are presented in Table 3.

TABLE 3. Imprinting behavior in chicks after exposure of embryos at 40 $\mu W/cm^2$ (10 GHz) on the 16th day of incubation.

Set Number	Power density, $\mu W/cm^2$	Exposure duration, min	Exposure regimen	Number of embryos	Number of chicks with imprinting
1.	-	-	Sham	83	81 (97%)
2.	40	5	CW	27	23 (89%)
3.	40	5	10-Hz modulation	19	**9 (47%)**

One can see from the table that CW irradiation at 40 $\mu W/cm^2$ did not produce any changes compared with sham controls. In contrast, the difference between sets 3 and 1 was statistically significant (p<0.01). Exposure to 40 μW/cm², 10-Hz modulated microwaves prevented formation of imprinting behavior in about 50% of cases.

Thus, a 5-min CW microwave exposure at 40 μW/cm² on the 16th day of incubation did not influence formation of the imprinting behavior. The same exposure, but with 10-Hz modulation, caused statistically significant bioeffects, namely it suppressed the ability of chicks to develop imprinting behavior. The data show that, under given experimental conditions, 10-Hz pulse modulation was more effective than the CW exposure regimen.

It was important to find out if using the same frequency for microwave modulation and for the imprint stimulus was a necessary condition for this effect. In the next set, we used microwaves modulated at 10 Hz, but the frequency of light flashes was 2 Hz. The exposure effect was still present and even more profound: The imprinting behavior developed in 6 out of 8 control chicks (75%), but only in 1 out of 8 exposed chicks (15%, p<0.05). Thus, keeping the imprint-stimulus frequency the same as the modulation frequency was not essential; exposure to 10-Hz modulated microwaves suppresses imprinting of both 10- and 2-Hz light flashes.

We also carried out a trial experiment using a 40-Hz modulation of microwaves. Out of 8 chicks that were exposed on the 16th day of egg incubation (5 min at 40 μW/cm²), the imprinting behavior developed in 5 (62%). Because of the small experimental group, this result was not significantly different from the control.

As a next step, we studied how the effect of 10-Hz modulated microwaves depends on the incident power density. The experimental data demonstrated a distinct dose response (Table 4).

TABLE 4. Imprinting in chicks exposed in egg on the 16th day of incubation to 10-Hz modulated microwaves at different incident power densities.

Sets	Exposure regimen	Power density, μW/cm²	Exposure duration, min	Number of chick	Number of chicks with imprinting
I	Sham	-	-	28	22 (78%)
II	CW	40	5	10	8 (80%)
III	Modulated 10 Hz	40	5	12	8 (65%)
IV	Modulated 10 Hz	900	5	7	4 (57%)
V	Modulated 10 Hz	1800	5	6	1 (17%)

We consider it useful to summarize the results of all experiments with exposure at 40 μW/cm^2 in a separate table (Table 5). The experiments presented in this table were performed in different animal groups, during different seasons of the year, and also using somewhat different methods of analysis of imprinting behavior.

TABLE 5. Effect of a 5-min exposure of eggs to 40 μW/cm^2, 10 GHz microwaves on imprinting behavior in newborn chicks.

Modulation	Power density, μW/cm^2	Number of embryos	Number of chicks with imprinting
Sham-exposed			
		36	35 (97%)
		8	8 (100%)
		9	8 (89%)
		28	22 (78%)
		10	8 (80%)
		8	6 (75%)
CW			
	40	18	16 (89%)
		9	7 (78%)
		10	8 (80%)
Modulation			
10 Hz	40	8	3 (37%)
		11	6 (64%)
		12	8 (65%)
10 Hz	40	8	1 (12%) imprint-stimulus - 2Hz
40 Hz	40	8	5 (62%)

The results presented in Table 5 show that intensity of 40 μW/cm^2 may be subthreshold for CW exposure. However, at 10-Hz modulation, even a 5-min exposure is sufficient to produce changes in imprinting behavior.

4. CONCLUSIONS

1. The influence of a short-term exposure to 10-GHz microwaves on imprinting behavior was studied in 447 newborn chicks. CW and pulse-modulated regimens at intensities from 40 μW/cm^2 to 10 mW/cm^2 were tested.

2. The experiments established that a single short-term exposure to CW or pulse-modulated microwaves at the average incident power densities from 0.4 to 10 mW/cm^2 suppressed the imprinting behavior.

3. The effect of exposure on the imprinting behavior alterations showed a clear correlation with the field intensity employed.

4. Microwaves modulated at 10 Hz were more effective than CW radiation at the same incident power density.

5. The intensity of 40 μW/cm^2 was subthreshold to alter the imprinting behavior in chicks by a short-term CW irradiation of embryos. However, the threshold for modulated microwaves was below 40 μW/cm^2.

6. It was also demonstrated that a low-intensity modulated microwave signal can be detected and "memorized" by the embryo brain.

These new findings should be taken into account for assessment of possible health hazards from exposure to modulated microwaves.

ABSORPTION CHARACTERISTICS AND MEASUREMENT CONCEPTS

W. D. HURT
Air Force Research Laboratory
8308 Hawks Road
Brooks AFB, TX 78235-5324

1. Introduction

The radio frequency (RF) region of the electromagnetic (EM) spectrum extends over a wide range of frequencies, from about 3 kHz to 300 GHz. Over the last several years, the use of devices that emit RF radiation has increased dramatically. RF devices include radio and television transmitters, military and civilian radar systems, a variety of communications systems, microwave ovens, industrial RF heat sealers, and various medical devices.

The proliferation of RF devices has been accompanied by increased concern about ensuring the safety of their use. Throughout the world many organizations, both government and nongovernment, have established RF safety standards or guidelines for exposure. Theoretical and experimental methods are used to extrapolate or relate effects observed in animals to similar effects expected to be found in people. Safety standards can be revised if knowledge is obtained about previously unknown adverse RF effects on the human body.

An essential element of the research examining the biological effects of RF exposure is dosimetry: the determination of energy absorbed by an object exposed to the EM fields composing the RF spectrum. Since the energy absorbed is directly related to the internal EM fields (that is, the EM fields inside the object, not the EM fields incident upon the object), dosimetry is also interpreted to mean the determination of internal EM fields. The internal and incident EM fields can be quite different, depending on the size and shape of the object, its electrical properties, its orientation with respect to the incident EM fields, and the frequency of the incident fields. Because any biological effects will be related directly to the internal fields, any cause-and-effect relationship must be formulated in terms of these fields, and not the incident fields. However, direct measurement of the incident fields is easier and more practical than measuring the internal fields, especially in people, so we use dosimetry to relate the internal fields to the incident fields. As used here, the term "internal fields" is to be broadly interpreted as the fields that interact directly with the biological system. In general, the presence of the body causes the internal fields to be different from the incident fields.

B.J. Klauenberg and D. Miklavcic (eds.), Radio Frequency Radiation Dosimetry, 39-52.
© 2000 U.S. Government. Printed in the Netherlands.

The rigorous analysis of a realistically shaped inhomogeneous model for humans or experimental animals is an enormous theoretical task. Because of the difficulty of solving Maxwell's equations, which form the basis of this analysis, a variety of special models and techniques have been used, each valid only in a limited range of frequency or other parameters. Early analyses were based on plane-layered, cylindrical, and spherical models. Although these models are relatively crude representations of the size and shape of the human body, experimental results show that calculations of the average specific absorption rate (SAR), one of the most important dosimetric values, agree reasonably well with measured values. Calculations of the local distribution of the SAR, even though being much more difficult, are now becoming possible. SAR is discussed later in this paper.

2. Interaction of EM Fields with Materials

Electric (E) and magnetic (H) fields interact with materials in two ways. First, the E- and H-fields exert forces on the charged particles in the material, thus altering the charge patterns that originally existed. Second, the altered charge patterns in the materials produce additional E- and H-fields (in addition to the fields that were originally applied). Materials are usually classified as being either magnetic or nonmagnetic. Magnetic materials have magnetic dipoles that are strongly affected by applied magnetic fields, nonmagnetic materials do not. Biological material is almost exclusively nonmagnetic.

In nonmagnetic materials, it is mainly the applied E-field that effects the charges in the material. This occurs in three primary ways:

1. Polarization of bound charges.
2. Orientation of permanent dipoles.
3. Drift of conduction charges (both electronic and ionic).

Materials primarily affected by the first two ways are called dielectrics; materials primarily affected by the third way are called conductors.

The polarization of bound charges is illustrated in Figure 1(a). Bound charges are so tightly controlled by restoring forces in a material that they can move only very slightly. Without an applied E-field, positive and negative bound charges in an atom or molecule are essentially superimposed upon each other and effectively cancel out. When an E-field is applied, the forces on the positive and negative charges are in opposite directions and the charges separate, resulting in an induced electric dipole. A dipole consists of a combination of a positive and a negative charge separated by a small distance. In this case the dipole is said to be induced because it is a result of the applied E-field; when the field is removed, the dipole disappears. When the charges are separated by the applied E-field, the charges no longer cancel. In effect the charge distribution is altered, called polarization charge, which creates new fields that did not exist previously.

The orientation of permanent dipoles is illustrated in Figure 1(b). The arrangement of charges in some molecules produces permanent dipoles that exist regardless of whether an E-field is applied to the material. With no E-field applied, permanent dipoles are randomly oriented because of thermal excitation. With an E-field applied, the resulting forces on the permanent dipoles tend to align the dipole with the applied E-field (Figure 1(b)). This orientation effect is only slight because the thermal excitation is relatively strong, but on the average, there is a net alignment of dipoles over the randomness that existed without an applied E-field. Like induced dipoles, this net alignment of permanent dipoles produces new fields.

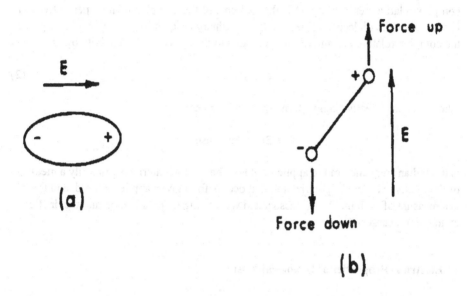

Figure 1. (a) Polarization of bond charges. (b) Orientation of permanent dipoles.

The drift of conduction charges in an applied E-field occurs because these charges are free enough to move significant distances in response to forces of the applied field. Both electrons and ions can be conduction charges. Movement of the conduction charges is called drift because thermal excitation causes random motion of the conduction charges, and the force due to the applied fields superimposes only a slight movement in the direction of the force on this random movement. The drift of conduction charges amounts to a current, and this current produces new fields that did not exist before the E-field was applied.

The two effects--creation of polarization charges by an applied field and creation of new fields by these new charge distributions--for both induced dipoles and orientation of permanent dipoles are taken into account by a quantity called permittivity. Permittivity is a measure of how easily the polarization in a material occurs. If an applied E-field results in many induced dipoles per unit volume or a high net alignment of permanent dipoles per unit volume, the permittivity is high. The drift of conduction charges is accounted for by a quantity called conductivity. Conductivity is a measure of how much drift occurs for a given applied E-field. A large drift means a high conductivity. For sinusoidal steady-state applied fields, complex permittivity is defined to account for both dipole charges and conduction-charge drift. Complex permittivity is usually designated as:

$$\varepsilon^* = \varepsilon_0(\varepsilon' - j\varepsilon'') \quad \text{F/m,} \tag{1}$$

where ε_0 ($8.85 \cdot 10^{-12}$ farad/meter) is the permittivity of free space, $\varepsilon' - j\varepsilon''$ is the complex relative permittivity, ε' is the real part of the complex relative permittivity (ε' is also called the dielectric value), j is the imaginary unit, and ε'' is the imaginary part of the complex relative permittivity. ε'' is related to the effective conductivity by:

$$\varepsilon'' = \sigma/\omega\varepsilon_0, \tag{2}$$

where σ is the effective conductivity in siemens/meter, and

$$\omega = 2\pi f \quad \text{radians/s} \tag{3}$$

is the radian frequency of the applied fields. The ε' of a material is primarily a measure of the relative amount of polarization that occurs for a given applied E-field, and the ε'' is a measure of both the friction associated with changing polarization and the drift of conduction charges.

3. Electrical Properties of Biological Tissue

The permeability of biological tissue is essentially equal to that of free space; in other words, biological tissue is essentially nonmagnetic. The permittivity of biological tissue is a strong function of frequency. Figure 2 shows the average ε' and σ for the human body as a function of frequency. Calculations have shown that the average ε' and σ for the whole human body are equal to approximately two-thirds that of muscle tissue.

The dielectric value generally decreases with frequency. This results from the inability of the charges in the tissue to respond to the higher frequency applied fields, thus resulting in lower dielectric values.

In tissue, the ε'' represents mostly ionic conductivity and absorption due to relaxational processes, including friction associated with the alignment of electric dipoles and with vibrational and rotational motion in molecules.

Energy transferred from applied E-fields to materials is in the form of kinetic energy of the charged particles in the material. The rate of change of the energy transferred to the material is called SAR, the standard quantity for RF dosimetry.

A typical manifestation of average (with respect to time) SAR is heat. The average SAR results from the friction associated with movement of induced dipoles, permanent dipoles, and drifting conduction charges. If there were no friction in the material, the average SAR would be zero.

A material that absorbs a significant rate of energy for a given applied field is said to be a lossy material because of the loss of energy from the applied fields. A measure of the lossiness of a material is ε". The larger the ε", the more lossy the material. In some tables, a quantity called the loss tangent is listed instead of ε". The loss tangent, often designated as tan δ, is defined as:

$$\tan \delta = \varepsilon''/\varepsilon'. \tag{4}$$

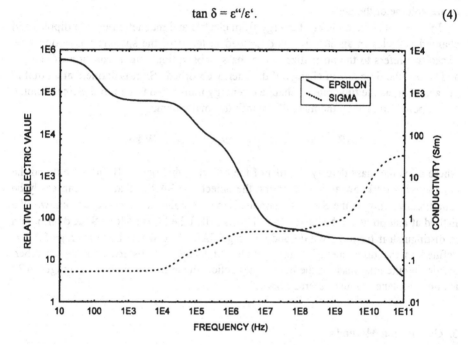

Figure 2. Average permittivity of the human body as a function of frequency.

The loss tangent usually varies with frequency. For example, the loss tangent of distilled water is about 0.04 at 1 MHz and 0.265 at 25 GHz. Sometimes the loss factor is called the dissipation factor. Generally speaking, the wetter a material is, the more lossy it is; the drier it is, the less lossy it is. For example, in a microwave oven a wet piece of paper will get hot as long as it is wet. When the paper dries out, the oven's EM fields will no longer heat it. Much biological tissue is considered to be wet material. Gabriel [1] has measured the permittivity for over 30 tissue types over a frequency range as large as from 10 Hz to 20 GHz.

4. Specific Absorption Rate

For steady-state sinusoidal fields, the time-averaged rate of energy absorbed per unit volume at a point inside an absorber is given by:

$$P = \sigma|E|^2 = \omega\varepsilon_o\varepsilon``|E|^2 \quad W/m^3, \tag{5}$$

where $|E|$ is the root-mean-square (rms) magnitude of the E-field vector at that point inside the material. To find the total rate of energy absorbed by an object, the value of P must be calculated at each point inside the body and summed (integrated) over the entire volume of the body.

In dosimetry, the transfer of energy from electric and magnetic fields to dipoles and charged particles in an absorber is described in terms of the specific absorption rate. "Specific" refers to the normalization to mass; "absorption", the absorption of energy; and "rate," the time rate of change of the energy absorbed. SAR is defined, at a point in the absorber, as the time rate of change of energy transferred to an infinitesimal volume at that point, divided by the mass of the infinitesimal volume.

$$SAR = P/\rho_m = \sigma|E|^2/\rho_m = \omega\varepsilon_o\varepsilon``|E|^2/\rho_m \quad W/kg, \tag{6}$$

where ρ_m is the mass density (kg/m^3) of the object at that point. If the E-field and the conductivity are know at a point inside the object, the SAR at that point can easily be found; conversely, if the SAR and conductivity at a point in the object are known, the E-field at that point can be calculated. This is called the local SAR or SAR distribution to distinguish it from the whole-body average SAR. The whole-body average SAR is defined as the time rate of change of the total energy transferred to the absorber, divided by the total mass of the body. In practice, the term "whole-body average SAR" is often shortened to just "average SAR."

5. Calculation Methods

In principle, the internal fields in any object irradiated by EM fields can be calculated by solving Maxwell's equations. In the past, this has been very difficult and could be done only for a few very special cases (idealized models, such as planar slabs, spheres, infinitely long cylinders, spheroids, or ellipsoids). Because of the mathematical complexities involved in calculating SAR, a combination of techniques was used to obtain SAR for various models as functions of frequency. Each of these techniques provided information over a limited range of parameters. Combining the information thus obtained gave a reasonably good description of SAR as a function of frequency over a wide range of frequencies for a number of useful models.

The theoretical techniques used to calculate the SAR in models of humans and animals can be divided into three basic approaches, based on the degree of complexity of the model's shape. One-dimensional models are the simplest and are particularly useful at higher frequencies where body curvature can be neglected. Such models, however, cannot predict body resonance, which occurs in models of finite sizes. Two-dimensional models are basically single- or multi-layered infinite cylindrical geometries suitable to simulate limbs, thighs, or arms. Three-dimensional models include both idealized shapes (such as spheres, spheroids, and ellipsoids) and block models. Although the latter models are the most complicated and can be solved only numerically, they give the best estimates of the average SAR, the resonance frequency, and the SAR distribution.

Although planar models do not represent humans well, analyses of these models have provided important qualitative understanding of energy-absorption characteristics. When a planewave is incident on a planar dielectric object, the wave transmitted into the dielectric attenuates as it travels and transfers energy to the dielectric. For very lossy dielectrics, the wave attenuates rapidly. This characteristic is described by skin depth: the depth at which the E- and H-fields have decayed to e^{-1} (0.368) of their value at the surface of the dielectric. Skin depth is also the depth at which the Poynting vector has decayed to e^{-2} (0.135) of its value at the surface. At higher frequencies, the skin depth is very small; thus, most of the energy from the fields is absorbed near the surface. As an example, for humans and animals at 2450 MHz the skin depth is about 2 cm; at 10 GHz, it is about 0.4 cm.

Other models--spheres, cylinders, spheroids, ellipsoids, and block models (cubical mathematical cells arranged in a shape like a human or animal body)--have been used to represent the human or animal body for calculating and measuring energy absorbed during plane wave irradiation [2-5]. Especially important for nonplanar objects are the effects of the polarization of the incident fields. The orientations of incident E- and H-fields with respect to the irradiated object have a very strong effect on the strength of fields inside the object. This orientation is defined in terms of the polarization of the incident fields. The incident-field vector--E, H, or k--that is parallel to the long axis of the body defines polarization for objects of revolution (circular symmetry about the long axis). The polarization is called E-polarization if E is parallel to the long axis, H if H is parallel to the long axis, and K if k is parallel to the long axis. This definition is illustrated in terms of prolate spheroids in Figure 3.

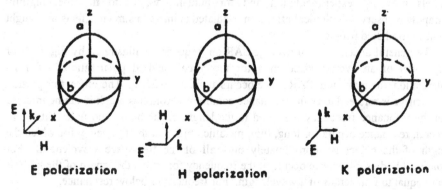

E polarization   H polarization   K polarization

Figure 3. Polarization of the incident field with respect to an irradiated object.

With the availability of improved computer technology, numerical techniques such as the admittance method [6], the impedance method [7], and the finite-difference time-domain (FD-TD) method [8, 9] have been developed for the solution of EM field interactions with objects having irregular geometries and inhomogeneous dielectric composition. The admittance and impedance methods are applicable to objects that are small compared to the wavelength of the exposure fields since they are based on dividing the medium into electrical current nodes or meshes. The FD-TD numerical approach is unique because it involves discrete, time-domain computations of differential equations applicable for all size objects within the limits of the speed and memory of the computer used for making the calculation. Imperfect boundary condition information does not degrade the accuracy of these techniques significantly. These and other advantages have made these numerical analysis techniques popular for solving related, non-biological, EM field problems. Recent adaptations of the FD-TD method to complex biological problems have utilized models divided into many thousands of rectangular cells. One problem associated with the implementation of the FD-TD method is the need for a very powerful computer to solve problems involving many thousands of cells. However, currently available workstations now have sufficient power to run many of the models and are approaching that required to run even the largest models such as the full-sized human.

6. SAR versus Frequency

SAR is an important quantity in dosimetry both because it gives a measure of the energy absorption that can be manifest as heat, and because it gives a measure of the internal fields which could affect the biological system in ways other than through ordinary heating. The internal fields, and hence the SAR, are a strong function of the incident fields, the frequency, and the properties of the absorber. Since any biological effects would be caused by internal fields, not incident fields, the ability to determine internal fields or SARs in people and experimental animals for a given radiation exposure situation is very important. Without such a determination in both animal models used for experimentation and for humans, we could not meaningfully extrapolate observed biological effects in irradiated animals to similar effects that might occur in irradiated people.

The general dependence of average SAR on frequency is illustrated by Figure 4 for the model of an average-sized man for the three standard polarizations. For E-polarization, the maximum SAR value occurs at about 80 MHz. Therefore, this portion of the curve is refereed to as the resonance condition. From this graph it can be inferred that the resonance frequency is related to the length of the body, and indeed it is. In general, resonance occurs for long, thin, metallic objects at the frequency for which the length of the object is approximately one-half of the free-space wavelength. For biological bodies, resonance occurs at the frequency for which the length of the body is about equal to four-tenths of a wavelength. For frequencies below resonance, the SAR varies approximately as f^2; just beyond the resonance frequency, SAR varies as f^{-1}.

Figure 4 also indicates that below the resonance frequency, the SAR is generally higher for E-polarization, intermediate for K, and lowest for H-polarization. Again, this is generally true. These characteristics can be explained by two qualitative principles:

1. The SAR is higher when the incident E-field is more parallel to the body than perpendicular to it.
2. The SAR is higher when the cross section of the body perpendicular to the incident H-field is larger than when it is smaller.

The average SAR is higher for E-polarization because the incident E-field is more parallel to the body than perpendicular to it, and the cross section of the body perpendicular to the incident H-field is relatively larger (see Figure 3). For H-polarization, however, the incident E-field is more perpendicular to the body than parallel to it, and the cross section of the body perpendicular to the incident H-field is relatively smaller. Both conditions contribute to a lower average SAR. The average SAR for K-polarization is intermediate between the other two because the incident E-field is more perpendicular to the body, contributing to a lower SAR. However, the cross section perpendicular to the incident H-field is larger, contributing to a larger SAR.

48

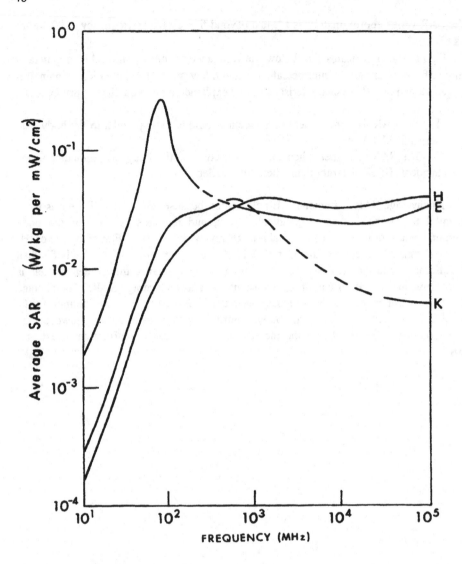

Figure 4. Whole-body average SAR versus frequency for an average man.

When a man is standing on a perfectly conducting ground plane, with E-polarization the ground plane has the effect of making the man appear to be electrically about twice as tall. This lowers the resonance frequency to approximately half of that for free space. For a man on such a ground plane, the graph of SAR versus frequency for E-polarization would therefore be similar to the one in Figure 4, but shifted to the left by approximately 40 MHz. Another important qualitative characteristic is that when the incident E-field is mostly parallel to the body, the average SAR goes up if the body is made longer and thinner. The data shown in Figure 4 are used to help set the RF safety standard for permissible exposure levels.

7. Measurement Methods

Generally, accepted methods of measurement of SAR include the measurement of the rate of temperature rise within the exposed object or the measurement of the internal E-field strength. The temperature rise may be characterized by a whole-body-averaged (calorimetric) measurement, a point measurement (via a thermometer implanted in the body being exposed), or thermographic camera analysis of bisected phantom models that have been exposed to large RF fields. The internal E-field strength may be measured with an implantable E-field probe.

Average SAR may be measured using calorimetric methods. In the past, such methods have been used predominantly with small animals or animal models [10-13], recently, however, calorimetric twin-well methods have been successfully used to measure SAR in a full-size human model [14]. The heart of the measurement system is the calorimeter device itself, and gradient-layer devices are commonly used. Gradient-layer calorimeters have a convenient voltage output signal that is proportional to the rate of heat energy flowing out of the device. However, the Dewar-flask method of calorimetry is a relatively simple, straight-forward way of determining the whole-body average SAR of small-bodied animals [15]. The calorimetric technique of determining a whole-body average temperature requires that the cadaver be immersed in a Dewar-flask containing a medium, such as water, at a known temperature; then the temperature of the cadaver, following irradiation, can be determined by noting the final temperature of the cadaver/medium mixture.

Certain temperature probes can be used successfully to make SAR measurements. The minimum requirements are that the temperature sensor and associated leads should be nonperturbing to the EM fields, and the SAR should be large enough to produce a measurable temperature rise during a period of less than about 20 seconds. The first requirement is usually satisfied by using highly resistive material or fiber optics, instead of metal components, for the temperature-sensing element leads [16]. The second requirements entails the measurement of SARs no lower than a few mW/g. This lower limit exists because the resolution of most temperature probes is typically 0.01 to 0.1°C, and the longest practical duration of irradiation that allows reasonably accurate SAR measurement is typically 5 to 30 seconds. Irradiation of an inert, lossy dielectric object for longer durations causes local "hot spots" to lose their thermal energy to the surrounding area via conduction, diffusion, and convection. In living biological systems, active thermo-regulation also degrades accuracy. It is acceptable to use an ordinary metal-wire temperature probe for RF dosimetry when the probe is not in place during irradiation, but is in place immediately before and after. The use of metallic probes during irradiation is not acceptable. Even when the probe's metallic leads appear to be oriented orthogonal to the incident E-field, the depolarization of fields within finite sized dielectric objects induces errors. The objective is to measure the time rate of irradiation-induced temperature rise ($\Delta T/\Delta t$) at a specific location in tissue or phantom material. The SAR, which is proportional to ($\Delta T/\Delta t$), can then be determined.

A method for rapid evaluation of the distribution of SAR throughout an entire planar surface of a biological object or a model that is composed of heterogeneous dielectric materials is described by Cetas [17], D'Andrea [10], and Johnson [18]. The method involves the use of a thermographic camera for recording the rate of rise of temperature in a plane that bisects an entire model (phantom) or a cadaver of a biological subject under study. The temperature distribution before, and immediately after brief, high-power irradiation, is observed on the precut surface on each half of the bisected model. This is done to prevent cooling by evaporation or flow of the wet synthetic tissue out of its shell (usually composed of synthetic fat or rigid plastic foam). This procedure is also used to measure heating patterns on the surface of subjects exposed to RF fields [19].

Miniature isotropic, implantable E-field probes with high impedance feed lines, which have been commercially available for a number of years [20], have been used to measure SAR distributions in phantom models and in living, anesthetized animals [21, 22]. These probes have much higher sensitivity than thermal probes and are especially suitable for measuring E-fields within simulated or actual biological tissues of moderate to high water content, i.e., brain and muscle. While it is possible to measure SARs of the order of 1 mW/g using sensitive and precise thermal measurements ($\Delta T/\Delta t \approx$ 0.1°C/30 s), it is well within the domain of E-field probes to measure SARs as low as 10 μW/g [23]. Recent improvements in decoupling the diodes from the high resistance line have led to the realization of true rms sensors [24]. To reduce the magnitude of the effort of taking data throughout a volume of tissue, data can be taken while the probe is scanned through the volume. Since the E-field probe has a response time of the order of a few milliseconds, continuous line scans of the internal E-field may be dynamically recorded by means of a robot that generates position data as the probe is moved along a path. Extensive data in an object can, thus, be plotted in a relatively short period of time, and the possibility of missing a local peak is reduced.

8. Conclusion

RF dosimetry is the process of determining the SAR that results from exposure to EM fields. Approximations of whole-body average SAR are calculated by using homogeneous cylindrical and prolate spheroidal models. The reliability of these values has been confirmed with measurements made in the laboratory. Taken together with biological data, experts have been able to establish RF safety standards. Iterative numerical techniques such as FD-TD are being used to solve Maxwell's equations for anatomically accurate block models. These results not only give the average SAR, but the SAR distribution, so that better mechanisms for explaining RF bioeffects can be generated. This improved understanding of dosimetry and RF bioeffects can then be used to evaluate current RF safety standards and, if appropriate, update existing RF safety standards.

9. References

1. Gabriel, C. (June 1996) *Compilation of the Dielectric Properties of Body Tissues at RF and Microwave Frequencies, AL/OE-TR-1996-0037,* Armstrong Laboratory, Brooks Air Force Base, TX 78235.

2. Durney, C.H., Massoudi, H. and Iskander, M.F. (October 1986) *Radiofrequency Radiation Dosimetry Handbook (Fourth Edition), USAFSAM-TR-85-73,* USAF School of Aerospace Medicine, Brooks Air Force Base, TX 78235.

3. *IEEE Recommended Practice for the Measurement of Potentially Hazardous Electromagnetic Fields-- RF and Microwave, IEEE Std C95.3-1991,* Institute of Electrical and Electronics Engineers, Inc., New York, NY (August 21, 1992).

4. Hurt, W.D. (December 4-8, 1988) Specific absorption rate measurement techniques, in *Proceeding of the Twenty-second Midyear Topical Meeting on Instrumentation,* Health Physics Society, San Antonio, Texas, pp. 139-151.

5. Hurt, W.D. (1997) Dosimetry of radiofrequency (RF) fields, in K. Hardy, M. Meltz, and R. Glickman (eds.), *Non-Ionizing Radiation: An Overview of the Physics and Biology,* Health Physics Society 1997 Summer School, Medical Physics Publishing, Madison, Wisconsin.

6. Armitage, D.W., LeVeen, H.H., and Pethig, R. (1983) Radiofrequency-induced hyperthermia: computer simulation of specific absorption rate distributions using realistic anatomical models, *Phys. Med. Bio.* **28,** 31-42.

7. Orcutt N. and Gandhi, O.P. (1988) A 3-D impedance method to calculate power deposition in biological bodies subjected to time varying magnetic fields, *IEEE Trans. Biomed. Eng.* **BME-35,** 577-583.

8. Kunz K.S. and Luebbers R.J. (1993) *The Finite Difference Time Domain Method for Electromagnetics,* CRC Press, Inc., Boca Raton, Florida.

9. Taflove, A. (1995) *Computational Electrodynamics:The Finite-Difference Time-Domain Method,* Artech House, Inc., Norwood, Massachusetts.

10. D'Andrea, J.A., Emmerson, R.Y., Bailey, C.M., Olsen, R.G., and Gandhi, O.P. (1985) Microwave radiation absorption in the rat: Frequency-dependent SAR distribution in body and tail, *Bioelectromagnetics* **6,** 199-206.

11. Olsen, R.G. (June 1-5, 1986) Localized specific absorption rate (SAR) in a full-sized man model near a shipboard monopole antenna: Effects on near-field, reradiating structures and of whole-body resonance, *Eighth Annual Meeting—Abstracts of the Bioelectromagnetics Society,* 34.

12. Allen, S.J. and Hurt, W.D. (1979) Calorimetric measurement of microwave energy absorption in mice after simultaneous exposure of 18 animals, *Radio Sci.* **14,** 1-4.

13. Blackman, C.F. and Black, J.A. (1977) Measurement of microwave radiation absorbed in biological systems, 2, analysis of Dewar-flask calorimetry, *Radio Sci.* **12,** 9-14.

14. Olsen, R.G. and Griner, T.A. (1989) Outdoor measurements of SAR in a full-size human model exposed to 29.2 MHz near-field irradiation, *Bioelectromagnetics* **10,** 162-171.

15. Padilla, J.M. and Bixby, R.R. (1986) *Using Dewar-flask Calorimetry and Rectal Temperatures to Determine the Specific Absorption Rates of Small Rodents, USAFSAM-TP-86-3,* School of Aerospace Medicine, Brooks Air Force Base, TX 78235.

16. Hochuli, C. (1981) *Procedures for Evaluating Nonperturbing Temperature Probes in Microwave Fields, FDA 81-8143,* Food and Drug Administration, Rockville, MD 20857.

17. Cetas, T. and Conner, W.G. (1978) Practical thermometry with a thermographic camera—calibration, transmission, and emittance measurements, *Rev.-Sci. Instn.* **49,** 245-254.

18. Johnson, C.C. and Guy, A.W. (1972) Nonionizing electromagnetic wave effects in biological material and systems. *Proc. IEEE* **60,** 692-718.

19. Walters, T.J., Blick, D.W., Johnson, L.R., Adair, E.R., and Foster, K.R. (submitted) Heating and pain sensation produced in human skin by millimeter waves. *Health Phys.*

20. Cheung, A. (Dec. 1976) Experimental calibration of a miniature electric field probe within simulated muscular tissues, *Selected Papers of the USNC/URSI Annual Meeting, Boulder, CO, October 20-23, 1975, vol. II, DHEW Publication (FDA) 77-8011,* 324-327.

21. Bassen, H.I., Herchenroeder, P., Cheung, A., and Neuder, S.M. (1977) Evaluation of implantable

electric field probes within finite simulated tissues, *Radio Sci.* **12**, 15-23.

22. Stuchly, S. (1987) *Specific Absorption Rate Distribution in a Heterogeneous Model of the Human Body at Radiofrequencies, Report PB87-201356,* Ottawa University, Ontario.

23. Balzano, Q., Garay, O., and Manning, Jr., T.J. (1995) Electromagnetic energy exposure of simulated users of portable cellular telephones, *IEEE Trans. Veh. Tech.* **VT-44**, 390-403.

24. Pokovic, K., Schmid, T., and Koves, N. (June 23-25 1996) *E*-field probes with improved isotropy in brain simulating liquids, *ELMAR Proceedings,* Zadar, Croatia.

DOSIMETRY AND DENSITOMETRY OF PULSED FIELDS

D. ŠIMUNIĆ
University of Zagreb, Faculty of Electrical Engineering and Computing
Unska 3, HR-10000 Zagreb, Croatia

1. Introduction

The dosimetry of electromagnetic fields in the radiofrequency part of the spectrum (3 kHz to 300 GHz) is a discipline oriented to determine the amount of absorbed energy in an object exposed to electromagnetic fields. Special attention of RF dosimetry has been focused to the pulsed fields dosimetry, because of a variety of applied sources using pulsed fields, and especially magnetic resonance imaging [1], digital mobile phones [2], hand-held traffic radars [3], marine-used radars [4] and radars using very short, high-peak ultra-wideband pulses [5].

The word dosimetry refers to the absorbed energy in an object; densitometry encompasses external field measurements. Due to the fact that it is still not possible to measure the induced fields inside the living object (i.e. human being, which is the aim) directly under the exposure conditions, the dosimetry and densitometry are closely related. Therefore, in order to perceive the real absorbed energy, it was necessary to include accurate numerical computations. Hence, the other distinction is of numerical and experimental dosimetry.

2. Pulsed Environment

In the pulsed environment, the duty cycle (DC) or duty factor is a relevant quantity. Duty cycle is defined as a ratio of a pulse duration (PD) to the pulse repetition time (PRT) of a periodic pulse train (Figure 1). Pulse repetition time is a reciprocal of pulse repetition frequency, giving the number of pulses per unit of time.

$$DC = \frac{PD}{PRT} \qquad (1)$$

B.J. Klauenberg and D. Miklavcic (eds.), Radio Frequency Radiation Dosimetry, 53-62.
© 2000 *Kluwer Academic Publishers. Printed in the Netherlands.*

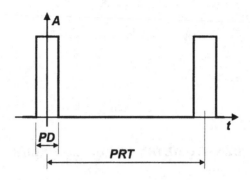

Figure 1. The basic quantities describing pulsed environment

Some specific examples of pulsed fields sources show that duty factors, rise time, pulse duration and power level of the applied sequences can be quite different. For instance, irregular microwave (3 GHz) TEMPO pulse [5] of peak power of 700 MW may be represented as a rectangular pulse of 200 MW for 80 ns. Pulse repetition rate is 0.125 pulses per second, which gives duty factor of 10^{-8} and the average power of 2 W. Therefore, even though the incredible peak power of 700 MW is being transmitted, the whole-body average SAR in exposed rats is 0.036 W/kg per watt of transmitted power, which gives 0.072 W/kg at full power.

As another extreme of pulsed field sources is taken Magnetic Resonance Imaging (MRI). As an illustration, the sequence which is considered to be the "worst" concerning high energy deposition is the multislice turbo spin-echo time sequence (Figure 2). In the complex 1.5 T MR imaging the RF time sequence consists of 1 pulse of 90 degrees and 16 pulses of 180 degrees, which is repeated 16 times, in order to get the resolution of 256 pixels. The pulse repetition rate is 71, which with the pulse duration of 0.79 ms gives a duty cycle of 0.056. Of course, this duty factor is much greater than the TEMPO pulses.

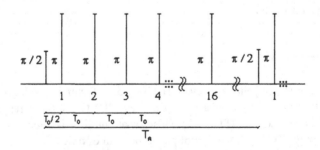

Figure 2. The time sequence in multislice turbo spin-echo imaging

3. Standards and Recommendations

3.1. IEEE C95.1, 1991

Dosimetry of pulsed fields in the IEEE Standard for Safety Levels with Respect to Human Exposure to Radio Frequency Electromagnetic Fields, 3 kHz to 300 GHz, IEEE C95.1-1991, [6] is based on the two basic rules: the peak temporal electric field value and the limitations when to use peak Maximum Permissible Exposure (MPE). If there are less than five pulses with a pulse-repetition period of less than 100 ms, the peak MPE should be used, and it is defined as MPE multiplied by averaging time in seconds divided by a factor of five and pulsewidth, also in seconds. Then is energy density limited as:

$$Peak \quad MPE = \frac{MPE \cdot Avg.Time}{5 \cdot Pulsewidth} \tag{2}$$

The limits on peak power are the values obtained by consideration of a well-established scientific base of data that includes the auditory effects in humans [7] and radio-frequency energy-induced unconsciousness in rats [8]. In the microwave range this means that the specific absorption over any six-minute period for exposure to a single pulse is limited to the spatial averaged value of 28.8 J/kg and corresponding spatial peak value of 576 J/kg.

3.2. ENV 50166-2, 1995

European prestandard ENV 50166-2 "Human exposure to electromagnetic fields / High frequency (10 kHz to 300 GHz)" [9] given by European Committee for Electrotechnical Standardization (CENELEC) gives the exposure limit value for pulsed fields in terms of peak electric (V/m) and magnetic field (A/m) strengths and peak power density (W/m²). The SAR averaged over any 6-minutes time interval and over the whole body should be less than 0.4 W/kg for workers and 0.08 W/kg for general population. For pulses of duration less than 30 µs at a frequency above 30 MHz the peak SA averaged over any 10 grams of tissue should be less than 10 mJ/kg for workers and 2 mJ/kg for general population. The duration of 30 µs has been chosen because of the auditory effect perceived in humans for pulses of duration less than 30 µs.

3.3. NATO STANAG 2345, 1997

In NATO STANAG 2345 [10] entitled "Evaluation and Control of Personnel Exposure to RF Fields (3 kHz to 300 GHz)" the relevant dosimetric measure is a whole body SAR of 0.4 W/kg. Permissible Exposure Level (PEL) is an external quantity and refers to time-averaged exposure values obtained by spatial averaging. PEL is given as electric and magnetic field and power density. Two basic rules to be applied are introduced if exposed to pulsed fields:

a. temporal peak electric field for one pulse should be less than 100 kV/m for each pulse;

b. energy density W (J/m²) should be multiplied with a factor of 5 during any 100 milliseconds period of the RF pulse time exposure if:

- the number of pulses with maximum energy is less than 5 during any time period equal to the averaging time
  OR
- the pulse durations are less than 100 milliseconds.

Energy density W contained in a pulse is given by:

$$W = S_{peak} \cdot PD \qquad (3)$$

where S_{peak} is the peak power density and PD is, as before, pulse duration.

For n_p pulses in the considered time period, it can be written:

$$W = S_{peak} \cdot PD \cdot n_p \qquad (4)$$

The rules are introduced in order to preclude absorbed energy higher than the limits for decreasingly short pulse durations. Therefore, if the single pulse is emitted, then the PEL is reduced by a factor of five times below the value that normal time averaging would permit.

3.4. ICNIRP Guidelines, 1998

Guidelines for Limiting Exposure to Time-Varying Electric, Magnetic and Electromagnetic Fields (up to 300 GHz) given by International Commission on Non-Ionizing Radiation Protection (ICNIRP) from 1998 [11] define the reference levels for exposure to time varying electromagnetic fields in terms of electric and magnetic fields in the high-frequency range 10 MHz - 10 GHz. The factor between the occupational and general public exposure is 2.2 (square root of 5, because of the relation of field strength and power density). In the microwave range 10 - 300 GHz, the general public reference levels are defined by the power density and the safety factor between the occupational and general public exposure is 5. The reference levels of pulsed fields are defined in terms of peak field strength and peak power. The peak field strength as averaged over the pulse width should not exceed 32 times the field strength reference levels of continuous waves. The peak power as averaged over the pulse width should not exceed 1000 times the reference levels of continuous waves. Similarly to ENV 50166-2, the SA should be less than 4-16 mJ/kg in the frequency range for frequencies above 30 MHz and for localized exposure of the head for pulses of duration less than 30 μs.

4. Densitometry

The quantities of densitometry depend on the distance of the object from the source (near and far field dosimetry). In the near field exposure relevant quantities are electric and magnetic fields; in the far field the relevant quantity is power density.

4.1. MEASUREMENT CONSIDERATIONS

Selection of the method and instruments depends on the frequency, output source power, modulation type, duty factor, pulse width, pulse-repetition frequency, intermittency, e.g. scanning beams, spurious frequencies including radiated harmonics and number of radiating sources. Before starting the measurements, the items that have to be addressed include the check of the previously mentioned source characteristics and the check of propagation characteristics, such as distance of source to test site, type of antenna and properties including gain, beamwidth, orientation, scanning program, physical size with respect to the distance to the area being surveyed, polarization of the E- and H-fields, existence of absorbing or scattering objects likely to influence the field distribution at the test site, as suggested in [12]. Finally, it is unavoidable to be aware of the characteristics of measuring device, given in the next chapter, as well as of quantities and values given in guidelines and standards that will be followed.

In an especial case, when performing near field measurements, care should be taken of two important points, i.e. spacing between measurements, which should be relatively small, if all minima and maxima are to be taken into account and presence of operator or probe close to the radiation source, which could cause serious disturbances of the reactive fields.

4.2. PROPERTIES OF MEASURING DEVICE

Measuring devices for field strength or power density measurements consist of three main parts: probe, connecting leads and instrumentation. The probe includes field sensing elements: for an electric field it is a dipole and for a magnetic field it is a loop. Isotropic probe comprises three field sensing elements.

The questions to be asked before a decision whether a measuring device is appropriate are:

- has the instrument an isotropic probe, which means that the response is not directional or polarized?
- what kind of response to other radiation, such as ionizing radiation, artificial light, sunlight or corona discharge has the instrument?
- is the probe response only to a specific parameter (only electric or only magnetic field)?
- is the response time, i.e. the time required for the instrument to reach 90% of its value when exposed to a step function of continuous wave energy, known?

58

- is it frequency selective or a broadband instrument?
  - the very important comment is that the peak values should be added linearly and the RMS values geometrically
- what is the stability of the instrument?
- what is the out-of-band response?
- what is a dynamic range of the instrument?
- are the probe dimensions less than $\lambda/10$ at highest operating frequency (the perturbation of the original field)?
- do the leads from the sensor to the meter significantly perturb the field at the sensor?
- do the leads from the sensor to the meter extract energy from the field?
- is the whole instrument producing significant scattering of the electromagnetic field?
- is the instrument with the particular probe separately calibrated for the electric and for the magnetic field?
- is the instrument supplied with a comprehensive handbook which includes a clear statement of the performance, with an especial attention to any restrictions in its application (e.g. pulsed fields, multiple frequency sources, near-field measurements)?

4.3. MEASUREMENT PROBLEMS

The problems that can be present during measurements are environmental influences, which encompass mainly temperature and humidity; measuring arrangement; field interference (presence of the operator or a metallic surface of the equipment) and inadequate interference immunity of the measuring instrument.

4.4. MEASURING EQUIPMENT

The measuring equipment that can be used for measuring pulsed fields is the following: wide-band antenna and power meter or spectrum analyzer, electric field-strength meter and magnetic field-strength meter. It is necessary that the peak power density or the peak field strength is measured by a peak monitoring instrumentation. On the basis of peaks, one can calculate the average power density or the average field strength. The special case of rapidly changing field, e.g. that of a scanning radar antenna, introduces the dependence of the duty cycle of field on the radiation pattern and scanning frequency of the antenna. Therefore, in addition to other requirements to the instrument (the most important is a probe isotropy), the response time of the instrument should be shorter than the pulse width (usually 1 μs for typical air traffic control radar) in order to allow instrument to show the peak power density without stopping the scanning. Due to the very high peak values, the dynamic range of approximately 60 dB is needed, as well as a protection from the overload.

4.4.1. *Wide-Band Antenna and Power Meter*

At frequencies above 300 MHz, a receiving wide-band antenna and power meter can be used to measure the power-flux density. The power-flux density S, defined in (W/m²), depends on the measured power (P_{meas}) and the effective antenna area (A_{eff}), which can be written as:

$$S = \frac{P_{meas}}{A_{eff}} \qquad (5)$$

Such a measurement gives time-averaged quantity. The peak values for pulsed signals can be recalculated by knowing the characteristic properties, and especially the inertia of the response, of the power meter.

4.4.2. *Electric Field-Strength Meter*

The field-sensing element can be dipole or monopole, and the output voltage or current is a measure of the electric field intensity. If the meter is to be used for far-field measurements only, then the electrically small loop can be used, as well. In the latter case, the magnetic field intensity is measured in the direction perpendicular to the plane of the loop and the electric field components can be mathematically converted from the magnetic field components from the known relation for the far-field:

$$E_i = H_i \cdot Z_0 \qquad (6)$$

where Z_0 is the free-space impedance.

4.4.3. *Magnetic Field-Strength Meter*

The field sensing element is an electrically small induction coil or loop antenna. If the instrument is to be applied in far-field measurements only, then the same principle as above can be applied, i.e. the dipoles as electric-field sensing elements and then convert them to magnetic field from (6).

4.5. PRINCIPLES OF THE INSTRUMENTATION

Currently used electromagnetic fields measuring equipment employs two kinds of detectors, thermocouplers and diodes. For the special case of measuring short high-peak pulses, such as generated by a scanning radar antenna, both have advantages and disadvantages as discussed in [13]. Thermocouple based instrumentation has a slow-response-time square law detectors, which integrate the power over time, giving a true RMS value. Unfortunately, it is not always possible to stop radar rotation because of operational requirements (e.g. air traffic control). The disadvantages of using thermocouplers as detectors are:

(1) their sensitivity to overload from short term peak signals
(2) their slow response time which is typically 1.5 s [14], while radar pulses are usually much faster

(3) their poor sensitivity (typically 10 V/m), which is too low for measuring pulses at distant locations [15]

Diode detectors have not been used in electric field meters for measuring high-peak short pulses, because the diode response changes with the amplitude of applied electric field strength. But the diode as a detector has the advantages:

(1) a faster response time

(2) larger dynamic range - it can respond to peak electric fields over a 60 dB dynamic range without being overloaded

(3) the protection from overload is possible

Knowledge of the dynamic response of the electric field-strength meter utilizing a diode detector can be used to generate correction curves necessary for a compensation of the non-ideal physical response of the diode. Correction values can be incorporated in the electronics. Thus, the disadvantages of the diode detector can be compensated for resulting in a practical assessment instrument for scanning radar type devices and possibly open the door to many new measuring applications for the diode-based detectors [16].

5. Dosimetry

The relevant quantities of dosimetry of pulsed fields are Specific Absorption (specific absorbed energy per unit mass, defined in [J/kg]) and Specific Absorption Rate (specific absorbed power per unit mass, defined in [W/kg]). SAR is the time rate at which radio-frequency electromagnetic energy is imparted to an element of mass of a biological body.

The dosimetry concerning a human body dosimetry can be pursued only by numerical computations. Several numerical methods have already been used for calculating absorbed energy in the complex model of the human body: Method of Moments (MoM), e.g. NEC2 [17], most-widely spread Finite-Difference Time-Domain (FDTD) method [18] and Finite Element Method (FEM) [19].

As an illustrative example of applying FEM for a calculation of absorbed energy and power in the simulated human body is taken that of [20]. The first step was to calculate the maximum SA for one 180 degree pulse, because the sequence in MRI consists of a number of the so-called 90 and 180 degrees pulses, as already mentioned in the section 2.1 of this text.

For the whole sequence the total maximum SA is calculated in the following manner:

$$SA|_{tot\,max} = n_s \cdot n_p \cdot SA|_{180°} \qquad (7)$$

where n_s is number of imaged slices and n_p number of pulses. The total maximum Specific Absorption Rate (SAR) is calculated taking the averaging time in account:

$$SAR|_{tot\,max} = \frac{SA|_{tot\,max}}{averaging\quad time} \qquad (8)$$

According to the mentioned standards and recommendations in the section 3. of this paper, at frequency of 64 MHz the averaging time is 6 minutes. On the other hand, the RF pulse sequence in MRI lasts much less (order of seconds) and is not repeated. Therefore, it is obvious that for this special application of MRI, the SAR could be considered as an artificial quantity and only the SA shows the real absorption. Concerning the repetitive pulsed fields for which the exclusions (low number of pulses or short pulse duration) are not valid, the dosimetry comprises equally both SA and SAR.

Even though several numerical techniques have already been mentioned for treating problems of pulsed fields, for the especial case of the short ultra-wide-band pulses [21], the time-domain numerical techniques represent the only solution.

The experimental dosimetry is very developed as a support for "in vivo" and "in vitro" studies, as well as for compliance testing especially at frequencies of mobile telecommunications equipment [22, 23].

6. Conclusions

Dosimetry of pulsed fields is very important, because of possible non-thermal interaction of various equipment of emerging technology, encompassing variety of sources, with human beings.

The paper gives the relation of several global standards and recommendations to the pulsed field dosimetry. Furthermore, the philosophy of pulsed field measurements is pointed out, as well as concomitant questions to be addressed.

In conclusion, the dosimetry is a very challenging task for both: numerical computations and measurements.

7. References

1. Bottomley P.A. and Andrew E.R. (1978) RF magnetic field penetration, phase shift and power dissipation in biological tissue: Implications for NMR imaging, *Phys. Med. Biol.* **23**, 630-643.
2. Hansson Mild K., Oftedal G., Sandstrom M., Wilen J., Tynes T., Haugsdal B., and Hauger E. (1998) *Comparison of Symptoms Experienced by Users of Analogue and Digital Mobile Phones: A Swedish-Norwegian Epidemiological Study*, Arbetslivsrapport, Sweden, ISSN 1401-2928.
3. Balzano Q., Bergeron J.A., Cohen J., Osepchuk J.M., Petersen R.C., and Roszyk L.M. (1995) Measurement of equivalent power density and RF energy deposition in the immediate vicinity of a 24 GHz traffic radar antenna, *IEEE Transactions on Electromagnetic Compatibility* **37**, 183-191.
4. Royal Norwegian Navy Materiel Command (1998) *Investigation into a Possible Causal Link Between High Frequency Electromagnetic Fields and Congenital Malformations*, Report no: 633-71331-100-002E
5. Raslear T.G., Akyel, Y., Bates, F., Belt, M., and Lu S.T. (1993) Temporal bisection in rats: The effects of high-peak-power pulsed microwave irradiation, *Bioelectromagnetics* **14**, 459-478.
6. IEEE C95.1-1991 (1991) *Standard for Safety Levels with Respect to Human Exposure to Radio Frequency Electromagnetic Fields, 3 kHz to 300 GHz*, Standards Coordinating Committees
7. Lin J.C. (1977) On microwave-induced hearing sensation, *IEEE Transactions on Microwave Theory and Techniques* **25**, 605-613.

62

8. Guy, A.W. and Chou, C. K. (1982) Effects of high intensity microwave pulse exposure of rat brain, *Radio Science*, **17**, 169-178.

9. ENV 50166-2 (1995) Human exposure to electromagnetic fields: High frequency (10 kHz to 300 GHz), CENELEC.

10. NATO (1997) Evaluation and control of personnel exposure to radio frequency fields - 3 kHz to 300 GHz, STANAG 2345 MED (EDITION 2).

11. ICNIRP Guidelines (1998) Guidelines for limiting exposure to time-varying electric, magnetic and electromagnetic fields (up to 300 GHz), *Health Physics* **74**, 494-522.

12. IEEE C95.3-1991 (1992) *IEEE Recommended Practice for the Measurement of Potentially Hazardous Electromagnetic Fields - RF and Microwave*, Standards Coordinating Committees.

13. Bassen H.I. and Smith G.S. (1983) Electric field probes - A review, *IEEE Transactions on Antennas and Propagation* **31**, 710-718.

14. General Microwave, RAHAM Radiation Hazard Meters (1992) Technical data of radiation hazard meters 81, 83, 91 & 93.

15. LORAL, Microwave Narda (1994) Technical specifications of electric field probes, Models 8721 & 8723.

16. Šimunić D. and Koren Z.T. (1997) An electric field measurement of a scanning radar antenna, *Microwave Journal*, 124-136.

17. NEC2 (1995) NEC2, available by anonymous ftp from ftp.netcom.com

18. Gandhi O.P., Gao B.Q. and Chen J.-Y. (1993) A frequency dependent finite-difference time-domain formulation for general dispersive media, *IEEE Transactions on Microwave Theory and Techniques* **41**,

19. Šimunić D., Wach P., Renhart W. and Stollberger R. (1996) Spatial distribution of high-frequency electromagnetic energy in human head during MRI: Numerical results and measurements, *IEEE Transactions on Biomedical Engineering* **43(1)**, 88-94.

20. Šimunić D. (1998) Calculation of energy absorption in a human body model in a homogeneous pulsed high-frequency field, *Bioelectrochemistry and Bioenergetics* **47**, 221-230.

21. Gandhi O.P. (1994) Some recent applications of FDTD for EM dosimetry: ELF to microwave frequencies, in B. Jon Klauenberg, Martino Grandolfo and David N. Erwin (eds), *Radiofrequency Radiation Standards: Biological Effects, Dosimetry, Epidemiology, and Public Health Policy*, Plenum Press, New York and London, Published in cooperation with NATO Scientific Affairs Division, pp. 55-79

22. Burkhardt M., Pokovic K., Gnos M., Schmid T. and Kuster N. (1996) Numerical and experimental dosimetry of Petri dish exposure setups, *Bioelectromagnetics* **17(6)**, 483-493.

23. Balzano Q., Garay O. and Steel F.R. (1978) Energy deposition in simulated human operators of 800 MHz portable transmitters, *IEEE Transactions on VehicularTechnology* **27(4)**, 174-188.

BIOPHYSICS LIMITS ON THE BIOLOGICAL EFFECTS OF ULTRAWIDEBAND ELECTROMAGNETIC RADIATION

ROBERT K ADAIR

Department of Physics, Yale University and Veridian Inc.
PO Box 208121, New Haven, CT 06517-8121

1. Introduction

Very large amplitude pulsed fields are important in many technical applications. Accordingly, the biological effects of such fields is of considerable practical interest. Even as many parameters are required to describe such pulsed fields, the set of such fields that are different in important ways is large and it is impractical to examine the biological effects of the whole set experimentally. Hence, we examine a representive set of such pulses in some detail with the view that the analyses of these pulses will provide a useful overview to the consideration of most specific pulses.

Electromagnetic pulses can be described in the *time-domain* in terms of the amplitude, $E(t)$, as a function of time or in the *frequency domain* where the amplitude, $E(\omega)$ is described as a function of frequency. Typically, different mechanisms are best considered in terms of one or the other domains and it is important to understand the relationships between the two descriptions. The two different descriptions – each logically complete – are related through Fourier transforms. We consider the character of those transforms and the power and limitations of their use.

We begin by considering radiative electromagnetic field pulses individually. For circumstances where the pulses with a width $\Delta t < 1$ μs, are repeated at a rate of $\nu_p \leq 1000$ pulses per second, the repetition period, $\tau = 1/\nu \gg \Delta t$ and any effects of a series of pulses will be usually be simply equal to the sum of the effects of the individual pulses and we need not consider special affects of the ELF pulse frequency.

For such fields that might have biological effects, the maximum electric field strengths, $E_{max} \approx 100$ kV/m and the magnetic fields accompanying the electric fields will be such that $B_{max} = E_{max}/v > 0.33$ mT in free space where $v = c = 3 \cdot 10^8$ m/s.

We will discuss two characteristic pulses of this kind. One, described by A. W. Guy in a report[1] from the Bioelectromagnetics Research Laboratory at the University of Washington, is a pulse with a rise time of about 10 ns and a width of about 1 μs. The other is that produced by the *Sandia Pulse Generator* and has a rise time of about 100 ps and a width of about 1 ns. Both of these sources produce polarized fields where the E-field lies in a specific direction, normal to the direction of emission. For simplicity, we will always assume that we are considering the electric fields in that direction and the magnetic fields in a direction perpendicular to the E-field and the direction of propagation. The amplitudes of the electric fields of the two pulses as a function of time, $\nu(t)$, are shown in Fig. 1.

Each pulse can also be described as in terms of the variation of the electric field

B.J. Klauenberg and D. Miklavcic (eds.), Radio Frequency Radiation Dosimetry, 63-72.
© 2000 *Kluwer Academic Publishers. Printed in the Netherlands.*

64

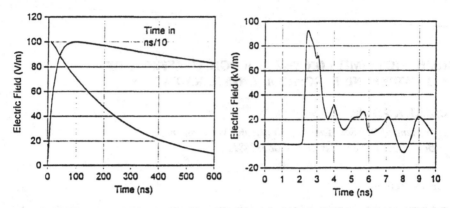

Figure 1: Figure 1. At the left, the amplitude, $E(t)$ of the electric field as a function of time for the "BRL" pulse, at the right for the "Sandia" pulse.

with as a function of frequency through Fourier transforms;

$$E(t) \xleftrightarrow{transform} E(\omega) \tag{1}$$

For the pulsed fields that concern us, the description in terms of time, $E(t)$, is that of a set of real values of E for each value of time, t, – a set of infinite numbers. The description in terms of frequency, $E(\omega)$, where $\omega = 2\pi\nu$, is a set of complex numbers expressing the magnitude and phase of E for each value of the frequency, ω – two sets of infinite numbers.

The transformations take the Fourier Integral forms,

$$E(\omega) = \frac{1}{\sqrt{2\pi}} \int_{-\infty}^{\infty} E(t)\, e^{-i\omega t}\, dt \ \text{ and } \ E(t) = \frac{1}{\sqrt{2\pi}} \int_{-\infty}^{\infty} E(\omega)\, e^{i\omega t}\, d\omega \tag{2}$$

The absolute values of the amplitudes, $E(\omega)$ are shown in Fig. 2.

A simple heuristic discussion of the Fourier transformation of a pulsed field: It is useful to gain some insight into the character of the Fourier transform by considering an especially simple mechanical example of a force which varies with time. Here we can consider that the transform describing that force as a function of frequency will be manifest in the effects of the force on systems that respond only to a specific frequency – resonant systems. We take the resonant system as an iron weight of mass M suspended from a non-magnetic spring with a resonant frequency ν and apply the force vertically as from a magnetic field acting on the iron. Only if the force oscillates at a frequency ν will the system oscillate with a significant amplitude hence the spring-weight system

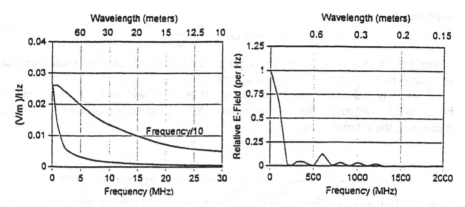

Figure 2: Figure 2 The frequency distributions for the pulsed field shown in Fig. 1; at the left for the BRL pulse, at the right for the Sandia pulse. The absolute values are shown with the phases suppressed.

acts as a frequency analyzer. But if the field is turned on for a very short time exerting a force F on the weight over a time Δt much shorter than the natural vibration time, τ, of the spring and weight – the weight will be set into vibration with an energy equal to $(F \cdot \Delta t)^2/2M$. If the force were half as great exerted for twice as long, the energy would be the same – as long as the time were still much shorter than the natural vibration period. And the force will activate the system no matter the resonant frequency, ν, if the period of vibration $\tau = 1/\nu \gg \Delta t$. However, if the force is weak and the time long, $\Delta t \gg \tau$, very little energy will be transmitted to the spring-weight combination even as the force will oppose the motion of the weight, thus subtracting energy, about one-half of the time. Hence, a pulsed field over a time Δt contains significant, and equal frequency components up to frequencies of $\nu < 2\pi\Delta t$, but only very small amplitudes for $\nu > 2\pi/\Delta t$.

2. The Modification of the Pulse by the Admittance of the Body

As a consequence of the conductivity of tissue, the interior of the body is shielded from external static electric fields by the effects of surface charges induced by the external field. When the external field is reversed, the charges reverse and in that reversal constitutes a current through the body inducing a field $E_{in} = i\rho$ where E is the local field in the body, i the current density and ρ the local resistivity of the body tissue. For a human being subjected to an external field E_{ext}, the maximum induced field can be

66

expressed as,

$$E_{in} \approx E_{ext} \frac{a\epsilon_0\omega\rho}{1 + a\epsilon_0\omega\rho} \qquad (3)$$

where $\rho \approx 1 \ \Omega m$ and $a \approx 1000$. Consequently, even long external pulses will translate to internal pulses that are not much longer than 10 ns.

Conversely, pulses that are much narrower than 10 ns, will not be much attenuated by the body conductivity and we will assume that such internal pulses will not differ much from the external pulses.

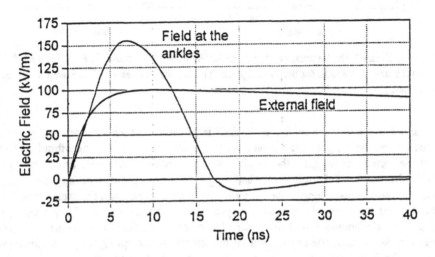

Figure 3: The maximum electric field, $E_{int}(t)$, in the ankle tissue of a man exposed to the 100 keV BRL pulse.

Fig. 3 shows the field in the ankle induced by the BRL pulse as calculated by Guy. Though the external pulse has a width of about 1 μs, the internal pulse width is only about 10 ns. For an external pulse with a rise time of 10 ns or less, the internal pulse will never be wider than about 10 ns no matter the length of the external pulse.

3. Momentum Transfer from a Pulse and from Thermal noise

The only effect of electromagnetic fields on matter is through forces exerted on charges. Hence, we can reduce the effects of the pulses to momentum transfers to charged elements (impulses), where that momentum transfer from one pulse with a width of Δt, and an amplitude of E, acting on a charge, Q, will be[2],

$$\Delta p_E = EQ\,\Delta t = 1.6 \cdot 10^{-23} \ \mathrm{N\,m/s} \qquad (4)$$

where the numerical values follow from the description of the Sandia pulse; $E = 100$ kV/m, $\Delta t = 1$ ns, and the charge Q is taken as one as for a singly charged ion.

To place this in context we consider the thermal momentum of a a water molecule that will collide with the charged system. That molecule will have a mean kinetic energy of 3/2 kT and a corresponding momentum of $\sqrt{3m\,kT} = 2 \cdot 10^{-23}$ Nm/s, where $m = 3 \cdot 10^{-26}$ kg is the mass of the molecule. That is the momentum transfer from the Sandia electric pulse will be about the same as from a typical collision with a water molecule.

The BRL pulse, in the ankle, will represent an impulse about ten times larger and it, thus, about equal to the fluctuation of 100 mean collisions with water molecules.

But there are larger molecular segments that might hold larger charge. We can gain some idea of effects of such pulses on larger systems by considering the hemoglobin molecule which is known to have a large dipole moment, $\mu_h = 1.6 \cdot 10^{-27}$ C m. The length of the molecule is $L_h = 6.8 \cdot 10^{-9}$ m and the mass is about 10^{-22} kg ($\approx 65,000$ amu). Then, if the effective charge centers are separated by $L_h/2 = 3.4 \cdot 10^{-9}$, the effective charge at each center would be $2e = 3.2 \cdot 10^{-19}$ C. If, as illustrative example, each half of the molecule were more tightly bound internally than the halves to each other, the canonical electric field pulse would give each part an opposite momentum of $3.2 \cdot 10^{-23}$ N m/s while the mean momentum of each segment from thermal agitation would be about $7 \cdot 10^{-22}$ kg·m/s – much larger than the impulse from the electric field pulse.

Even the internal field from the BRL pulse, as measured in the ankles under conditions where the admittance is especially high, will not affect the large molecule more than the regular effects of normal thermal agitation.

Angular impulses: In this discussion we have been considering linear impulses. But, for many situations such as for the water molecule and the hemoglobin molecule, the molecule (or larger system) is electrically neutral overall but holds a substantial permanent dipole moment. For such systems, the electric field effect will generate a torque on the system. Labeling the dipole moment as μ, the maximum angular impulse will be $\Delta E \cdot \mu$ to be compared with the mean angular momentum of the system from thermal agitation, $\overline{(I\omega)}_{kT} = \sqrt{2I\,kT}$. Taking the dipole moment of the water molecule as $6 \cdot 10^{-30}$ mC and the maximum moment of inertia as, $I_w \approx 3.3 \cdot 10^{-47}$ kg·m^2, the angular impulse from the field will be, $E\,\Delta t\,\mu_w = 6 \cdot 10^{-34}$ kg·m^2/s while the mean angular momentum from thermal agitation will be nearly the same at $\sqrt{2I_w\,kT} = 5 \cdot 10^{-34}$ kg·m^2/s.

Similarly for hemoglobin, where we take $I_h \approx 8 \cdot 10^{-40}$ kg m^2, we find the impulse from the field as $E\,\Delta t\,\mu_w = 1.6 \cdot 10^{-31}$ kg·m^2/s while the mean angular momentum from thermal agitation will be about ten times greater at $\sqrt{2I_h\,kT} = 2.6 \cdot 10^{-30}$ kg·m^2/s.

Pressure (Brownian) fluctuations: We can look at the mean thermal momentum from another viewpoint by considering the fluctuations in the momentum transfers generated by collisions with free molecules in the local environment – collisions of the kind that generate the Brownian movements of particles in liquids.

In particular, let us consider momentum transfers to a sphere of radius 10^{-9}m that would hold 125 water molecules. At one atmospheric pressure, the compressive force on the sphere will be about $1.25 \cdot 10^{-12}$ N, which, over 1 ns, will be provided by about 55 collisions with water molecules. Statistical fluctuations will lead to a mean momentum transfer in any particular direction of $1.2 \cdot 10^{-22}$ kg·m/s, about equal to the momentum transfer of the BRL pulse on a system holding 1 charge and ten times the momentum transfer of the Sandia pulse.

Conversely, the momentum transfer from the electrical field pulse to a system holding 1 charge will be about the same as the pressure fluctuation during the one nanosecond period of the pulse if the system has a volume similar to that of a sphere of radius 10 A.

Hence, for plausible situations, the change in momentum of a biological system generated by the impulse from the electric field pulse will be no greater than the mean momentum of the system from thermal agitation. But the electrical impulse will be generated 600 times in a second, while the molecular collision will take place about 10^{10} times a second. The impulse from the field during the nanosecond pulse should be compared with the largest stochastic impulse to be expected in 1,666,666 nanoseconds which will be of the order of $\log_e(1, 666, 666) \approx 15$ times the mean stochastic impulse expected in one nanosecond. Thus we do not expect that any single biological system can "observe" the electric field signal over thermal noise.

4. Relaxation Times

Since the pulse rate is considered of possible importance, we ask if there is some kind of reinforcement of the effects of previous pulses on the effect of a pulse. Any extra energy transferred to a system by the pulse will be dissipated after a certain "relaxation time." We can estimate this relaxation time dimensionally as,

$$\tau = \left[\frac{1}{4}\right] \frac{C\rho}{\kappa} a^2 \approx [1.5 \cdot 10^6] a^2 \tag{5}$$

where a is a characteristic length. Here, $C\rho$ ($4.18 \cdot 10^6$ J/m^3) is a heat capacity, κ (0.6 W·m^{-1}deg$^{-1}s^{-1}$), is a thermal conductivity and the numerical values in parentheses are those of water. The time-constant, τ is naturally shape dependent; the numerical factor of 1/4 holds for a sphere with a radius a measured in meters.

Although the calculation of τ is made from macroscopic considerations, the results are probably good within an order of magnitude on the molecular level. For a water molecules with an effective radius of $a = 2A$, Eq. 6 gives $\tau = 6 \cdot 10^{-14}$ s, while the inverse of the collision frequency can be estimated as $\tau = a/\sqrt{3kT/m} = 4.4 \cdot 10^{-13}$ seconds. For a hemoglobin molecule with an effective radius of $a \approx 3 \cdot 10^{-9}$ m, we can expect that $\tau \approx 10^{-11}$ seconds. Hence, after a pulse, the system will be back in equilibrium long before the next pulse takes place 1.6 ms later. Hence, the primary effect of a pulse cannot be expected to affect the reception of a pulse that takes place a millisecond later – indeed, for any plausible pulse rate, the pulses must act independently in their primary interaction.

5. Macroscopic Coherent Effects

The energy per unit area of the characteristic $\Delta t = 10^{-9}$ second pulse, with an amplitude of $E = 100$ kV/m, is $\epsilon_0 E^2 c \Delta t = 27$ mJ/m^2. The energy absorption length is about 1 cm, so we can consider that in one nanosecond 2.7 mJ/kg is deposited in a 1 cm thick layer of material. Taking the thermal capacity as that of water, this increases the temperature by about $6.5 \cdot 10^{-7}$ $^{\circ}$C which, taking the coefficient of thermal expansion as that of water, increases the volume by about $1.3 \cdot 10^{-10}$ thus generating an acoustic pulse. Of course, for electromagnetic field pulse rates of 600 pps, there will be 600 such acoustic pulses a second. The beam intensity will then be 16 W/m^2 (1.6 mW/cm^2) and the SAR, in the cm of penetration, will be about 1.6 W/kg. And, neglecting thermal conduction, the rate of temperature rise in that absorption layer would be about $4 \cdot 10^{-4}$ $^{\circ}$C/s.

That energy deposition per pulse is rather smaller than the theshold of about 400 mJ/m^2 noted for animal (and human) detection of microwave pulses, and the temperature rise is smaller than the corresponding detection threshold level of $5 \cdot 10^{-5}$ $^{\circ}$C temperature rise. Moreover, it seems that the auditory system only records acoustic pulses that resonate in the brain cavity which for humans is in the range between 7 and 10 kHz (and is, of course, higher for smaller animals), hence we cannot expect that such 600 pps signals will be detected by man – or by other animals. And without a sensory system able to integrate small signals from a very large number of detection elements, we must conclude that such coherent effects can not initiate biological responses.

6. Pulses Made up of Centimeter and Millimeter Waves

In the analysis of the biological effects of electromagnetic waves where the wave lengths are larger than the dimensions of the body, it is usually desirable to consider the body as a whole though certain gross aspects of the body, such as the constrictions at the ankles, must be considered. In the previous section, we have discussed pulses that, in the frequency domain, are made up of largely of wavelengths of a meter or more and have relied on information from such whole-body analyses to consider the effects of such pulses.

However, pulses containing substantial intensities of much higher frequencies – and much smaller wavelengths – are also of interest. Here, we discuss a representative pulse dominated by wavelengths of about 2.5 cm. Although the general character of the discussion extends to mm waves, since the wavelengths are then much smaller than characteristic human dimensions, we consider effects over small regions where the body can be considered as a plane surface struck at some angle by the electromagnetic waves.

Roughly speaking, the attenuation length of short wavelength electromagnetic waves in tissue is of the order of 1/10 of a wave length. As a consequence of the spectrum of frequencies – and wavelenths – that make up the original wave, and the different absorptions of those components, the pulse changes "shape" as it penetrates the body and we address the character of that penetration.

In this discussion we consider, again, a specific pulse to illustrate general effects[3,4]. This pulse can be described as a *carrier* wave modulated by a *modulation* amplitude. For the pulse considered here, the carrier wave is a 10 GHz wave, $E_0 \sin \omega t$ where $\omega = 2\pi \cdot 10^9$ s^{-1} and the modulation amplitude $A(t)$ is a step function with a value

70

$A(t) = 1$ for the pulse duration of 1 ns, and zero elsewhere. Formally, the pulse in the time domain takes the form,

$$E(t) = A(t) \cdot E_0 \sin \omega t \quad \text{where } A(t) = 0 \text{ except } A(t) = 1 \text{ for } t_0 < t < t_f \quad (6)$$

For the canonical pulse, we take the turn-on time as $t_0 = 0$ and the turn-off time as $t_f = t_0 + 10^{-9}$ s. Hence, this particular pulse is exactly 10 carrier wavelengths long and is defined so that it begins when the carrier wave amplitude is zero. Although the specific results reflect the special character of the canonical pulse where the repetition rate is taken as $\nu_p = 10^8$ s^{-1} and the modulation rise-time and fall-times are very short and taken as effectively zero, the important conclusions are general.

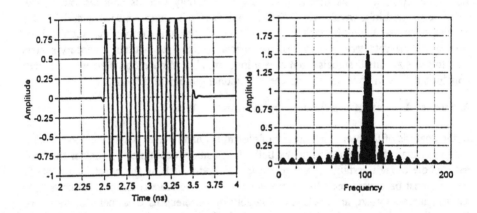

Figure 4: The left-hand diagram (a) shows the input pulse, $E(t)$, as it might be measured just after it enters a plane surface of tissue. The right-hand diagram (b) shows the frequency domain pulse, $E(\omega)$.

The diagram of Fig. 4a shows an initial pulse at the surface of the skin (0 depth) expressed as function of time, $E(t, 0)$. The variation with frequency, $E(\omega, 0)$, is determined by a Fourier transform which takes a discrete character when representing an infinite series of pulses in the time-domain;

$$E(\omega) = \sum_k a_k \sin(k\omega_p t) + b_k \cos(k\omega_p t) \quad \text{where } \omega_p = 2\pi\nu_p \text{ and}$$

$$a_k = 2\nu_p \int_{t_0}^{t_0 + \tau_p} E(t) \cos(k\omega_p t) \, dt, \quad b_k = 2\nu_p \int_{t_0}^{t_0 + \tau_p} E(t) \sin(k\omega_p t) \, dt \quad (7)$$

The values of the $\sqrt{a^2 + b^2}$ as a function of k are shown in Fig. 4b where the absolute amplitudes are shown for expositional simplicity.

In the course of passage of the pulse through matter, the pulse is absorbed and refracted. At a depth x,

$$E(\omega, x) = E(\omega, 0)\, e^{i\omega n(\omega)x/c} = E(\omega, 0)\, e^{-\omega n_i(\omega)x/c}\, e^{i\omega n_r(\omega)x/c} \qquad (8)$$

where $E(\omega, 0)$ is the amplitude at an initial depth, just below the surface of the material, and $n(\omega) = n_r(\omega) + i\, n_i(\omega)$ is the complex index of refraction of the tissue which is taken as having the properties of water. The absorbing material is taken as having the electromagnetic properties of water, also as per reference 1. Fig. 1a shows the pulse at an initial depth, $x = 0$, just below the surface of the medium. We consider the pulse below the surface to avoid considering reflection.

Then $E(t, x)$, the time dependence of the pulse at a depth x, can be determined by applying the inverse Fourier transformation to $E(\omega, x)$,

$$E(t, x) = \frac{1}{\sqrt{2\pi}} \int_{-\infty}^{\infty} E(\omega, x)\, e^{i\omega t}\, d\omega \qquad (9)$$

Fig. 5 shows both the time-domain and frequency domain pulse after passing through 1 cm of tissue. Note the severe attenuation of the higher frequency amplitudes and the lesser attenuation of the lower frequency components.

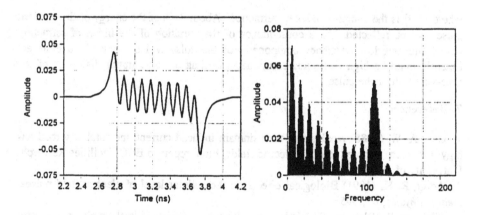

Figure 5: (a) At the left, the time-domain amplitude, $E(t)$ is shown for the pulse after it penetrates 1 cm of tissue. (b) At the right the frquency domain spectra is shown.

The relatively large initial and final segments of the pulse are called, "Brillouin precursors" inasmuch as their existence was first postulated by Brillouin[5] in 1914. It is important to note that these Brillouin precursors are not so much *generated* by the passage of the pulse through the water as they are a residual that remains after the rest of the pulse was more strongly absorbed.

The energy of the pulse at a depth of 1 cm is about 0.8% of the initial energy; the maximum electric field strength, E is reduced by a factor of about 12, and the maximum rate-of-change of the field, dE/dt, is reduced by about a factor of 20. Every characteristic of the pulse at depth is contained, albeit attenuated, in the initial pulse. Consequently, we do not find it credible that the strongly attenuated pulse – albeit with a different shape – can induce biological effects beyond that of the initial pulse.

As can be expected, the portion of the energy density carried by the higher frequencies, is small initially and falls off very quickly. For the canonical initial pulse shown in Fig. 1a, about half of the pulse energy is carried by frequencies greater than twice the carrier frequency. At $x = 0.25$ cm the energy carried by the high frequency amplitudes is only about $1.3 \cdot 10^{-6}$ of the original pulse energy, a depth of 0.5 cm, the ratio is $4.3 \cdot 10^{-9}$. The validity of these ratios is not affected by the neglect of dispersion.

Reflection: In this discussion, for simplicity of exposition, we have neglected the reflection of the pulse at the surface. Because of the change in index of refraction at the surface, the transmitted field, $E(\omega, 0)$, will be reduced from the incident field in air, $E_i(\omega)$, by the frequency dependent factor,

$$E(\omega, 0) = E_i(\omega) \frac{2}{1 + n(\omega)} \qquad (10)$$

where $n(\omega)$ is the complex index of refraction. About 87% of the energy of the incident pulse will be reflected. As a consequence of the variation of the index of refraction with frequency, low frequency components of the pulse will be transmitted more efficiently than the high frequency components leading to some initial distortion of the time-variation of the pulse.

7. References

1. Guy, A. W. (1988) Analysis of time domain induced current and total absorbed energy in humans exposed to EMP electric fields, Final report to ERC Facilities Research, University of Washington, Seattle.
2. Adair, R. K. (1991) Biological effects on the cellular level of electric field pulses, *Health Physics*, **61**, 395.
3. Albanese, R., Blaschak, J., Medina, R., & Penn, J. (1994) Ultrashort electromagnetic signals: biophysical questions, Safety Issues and Medical Opportunities. *Aviation Space & Environmental Medicine* **65**, A116-A120
4. Adair, R. K. (1995) Ultrashort microwave signals: a didactic discussion. *Aviation Space & Environmental Medicine* **66**, 792-794.
6. Brillouin, L. (1914), Uber der Fortpflanzung des Lichtes in disperdierenden Medien. *Ann. Phys.* **44**, 203.

SUMMARY OF SESSION B: THE DIELECTRIC PROPERTIES OF TISSUES

C. GABRIEL
Microwave Consultants Ltd.
17B Woodford Road
London, E18 2EL
United Kingdom

The study of the frequency dependence of the dielectric properties of biological materials has a long history. Systematic studies were carried out in the 1920's and 1930's on various aqueous molecular solutions and cell suspensions and on some biological tissues. The interpretation of these data in terms of mechanisms of interaction and molecular and cellular structures counts among the early successes achieved.

The field of dielectric spectroscopy made big strides in the 1950's owing to the availability of stable, high frequency sources developed in the previous decade. Further major development took place in the 1960's and 1970's because of the then availability of automated measurement systems and data analysis procedures.

Since the 1970's, progress in this area of research, particularly with respect to measurement on tissues, has been impressive for two main reasons. The first is because of developments in theoretical dosimetry, which have produced numerical procedures capable of estimating the internal electromagnetic field in a full-scale human model with high-resolution anatomical details. Detailed and accurate knowledge of the dielectric properties of body tissues has been an essential prerequisite for this development. The second reason why progress has been significant is because of the emergence of swept frequency network analysers and their adaptation for dielectric investigation. This has greatly simplified the measurement procedure and also allowed determinations of permittivity and conductivity to be made rapidly and over a wide frequency range.

The impetus to these developments has come from several points of origin. Principal of these is probably the escalation in interest in the biological effects and health hazards arising from exposure to electromagnetic fields. Another area in which dielectric studies are increasingly playing an important role is reaction kinetics, and how they may be influenced by microwave heating. And finally the century-old discipline of using dielectric spectroscopy to study interaction mechanisms has been further advanced by the latest experimental and theoretical developments.

To follow are four papers on the dielectric properties of tissues. The first paper, by C. Gabriel, describes developments in the measurement and interpretation of the

B.J. Klauenberg and D. Miklavcic (eds.), Radio Frequency Radiation Dosimetry, 73-74.
© 2000 *Kluwer Academic Publishers. Printed in the Netherlands.*

dielectric properties of tissues.   This is preceded by the necessary dielectric theory. The second paper, authored by Land, Gorton and Hamilton, describes microwave and thermal modelling of body regions and provides some new dielectric data.   This is followed by a contribution on the dielectric properties of skin by Lahtinen, Nuutinen and Alanen, where the importance of considering skin as a layered structure is emphasised.   Finally, there is a paper by Szabo and Bakos who discuss the dielectric properties of lens material and their relevance to cataractogenesis consequent upon exposure to radiowaves and microwaves.

THE DIELECTRIC PROPERTIES OF TISSUES

C. GABRIEL
Microwave Consultants Ltd.
17B Woodford Road
London, E18 2EL
United Kingdom

1. Introduction

It is generally true to say that the dielectric properties are intrinsic parameters that determine the effects of electric fields on matter. This leads to the statement that dielectric properties (relative permittivity ε' and effective conductivity σ) play a dominant role in the overall consideration of interaction between electromagnetic fields and matter and in related applications in numerous disciplines including electromagnetic dosimetry.

This paper will look at the dielectric properties of biological materials and the extent to which our current knowledge contributes to understanding the mechanisms of interaction and satisfies the need to quantify the interaction of electromagnetic fields and people as in numerical dosimetry. A few aspects of the subject will be discussed including dielectric spectroscopy, the variability of the dielectric properties of biological tissues reported in the literature and the accuracy and precision with which measurements are made. In view of its relevance to dosimetry, a comment is included on the conductivity of the body at low frequencies. The paper will start with a brief overview of the basic dielectric theory the purpose of which is to introduce the terminology and to provide a basis for explaining experimental data. The discussion will be illustrated with data on biological tissues from recent studies and data on a solid, dry material developed to simulate the dielectric properties of a high water content tissue.

2. Molecular Origin of the Dielectric Properties

The most important effect arising from the interaction of an electric field with a dielectric material is polarization. Polarization occurs when internal charge in the material moves in response to an external electric field. At the most fundamental level, electronic and atomic displacement take place in all dielectric materials. Molecules with permanent, induced or transient induced dipoles exhibit their own specific

B.J. Klauenberg and D. Miklavcic (eds.), Radio Frequency Radiation Dosimetry, 75-84.
© 2000 *Kluwer Academic Publishers. Printed in the Netherlands.*

polarization, for example, dipole orientation. Biological materials contain free and bound charges including ions, polar molecules and an internal cellular structure. The effect of an electric field is twofold: to trigger several polarization mechanisms, each governed by its own time constants, and to cause ionic drift. The net result is the establishment of both displacement and conduction currents. For this reason, biological materials are classified as lossy dielectric materials.

At any time, the total polarization is the vector sum of all contributions such that

$$P = \sum_{i}^{n} P_i \tag{1}$$

Irrespective of the molecular process, the polarization P is related to the dielectric displacement D, the internal electric field strength E and the dielectric properties as follows: $D = \varepsilon_0 E$ in vacuum, and $D = \varepsilon_0 \varepsilon E$ in a medium of permittivity ε relative to ε_0 which is that of free space. The latter expression may also be written as

$$D = \varepsilon_0 E + P \tag{2}$$

which describes the polarisation P as a material-specific displacement vector. The dependence of P on E can take several forms, the simplest being a scalar proportionality such that $P = \varepsilon_0(\varepsilon-1)E$ which defines the relative permittivity in terms of the polarisation per unit field as

$$\varepsilon = \frac{P}{\varepsilon_0 E} + 1 \tag{3}$$

The relative permittivity ε can be a tensor, a complex parameter or simply a real number depending on the directionality of the response and the phase difference between the displacement vector and the electric field. Assuming a linear, isotropic behavior, the relative permittivity of biological materials ε is a complex parameter $\hat{\varepsilon}$ expressed as

$$\hat{\varepsilon} = \varepsilon' - j\varepsilon'' \tag{4}$$

where $j^2 = -1$. The real part ε' determines the component of the displacement current which is out-of-phase with the driving field, while the imaginary part ε'' relates to the in-phase or power loss component and is referred to as loss factor. Considering displacement and ionic currents, a biological material is characterized by an effective conductivity σ and an effective loss factor ε'' such that $\varepsilon'' = \sigma/\varepsilon_0\omega$ and $\sigma = \sigma_d + \sigma_i$ where the subscripts refer to displacement and ionic parameters. In this paper loss factor and conductivity always refer to the effective or total values since, in practice, these are the parameters that determine the effect of the field on the material and these are the parameters measured. The conductivity is expressed in siemens per meter (S/m) when ε_0 is expressed in (F/m) and ω in radians per second. The dielectric properties are usually presented as ε' and ε'' values, or ε' and σ values, as a function of frequency and, to a lesser extent, temperature.

3. Frequency Dependence of the Dielectric Properties of Tissues

The time dependence of the polarization is due to the various physical interactions and the time dependent response of the material to them. This is reflected in the frequency dependence of the permittivity. Three main interaction mechanisms, each governed by its own kinetics, determine the main features of the dielectric spectrum of a tissue. Three main spectral regions known as the α, β and γ dispersions are predicted from known interaction mechanisms. These dispersions have been identified experimentally in the frequency range from hertz to gigahertz (Figure 1). The γ dispersion, in the gigahertz region, is due to the polarization of water molecules. The β dispersion, typically in the hundreds of kilohertz region, is due mainly to the polarization of cellular membranes that act as barriers to the flow of ions between the intra and extra cellular media. Other contributions to the β dispersion come from the polarization of protein and other organic macromolecules. The low frequency α dispersion is associated with ionic diffusion processes at the site of the cellular membrane. Figure 1 also shows that tissues have finite ionic conductivities. In its simplest form, each dispersion is characterized by a single time constant τ and exhibits the following frequency dependence

$$\hat{\varepsilon} = \varepsilon_\infty + \frac{\varepsilon_s - \varepsilon_\infty}{1 + j\omega\tau} \tag{5}$$

This is the well-known Debye expression in which ε_∞ is the permittivity at field frequencies where $\omega\tau \gg 1$, and ε_s the permittivity at $\omega\tau \ll 1$. The magnitude of the dispersion is described as $\Delta\varepsilon = \varepsilon_s - \varepsilon_\infty$. The presence in a material of several mechanisms with relaxation times distributed around τ can be described in terms of a deviation from Debye behavior and may be expressed as

$$\hat{\varepsilon}(\omega) = \varepsilon_\infty + \frac{\Delta\varepsilon}{1 + (j\omega\tau)^{(1-\alpha)}} \tag{6}$$

This is the semi-empirical Cole-Cole expression in which the parameter α is introduced to describe the broadening of the dispersion.

The spectrum of a tissue may be described mathematically in terms of multiple Cole-Cole dispersion and an ionic conductivity term such that

$$\hat{\varepsilon}(\omega) = \varepsilon_\infty + \sum_n \frac{\Delta\varepsilon_n}{1 + (j\omega\tau_n)^{(1-\alpha_n)}} + \sigma_i / j\omega\varepsilon_0 \tag{7}$$

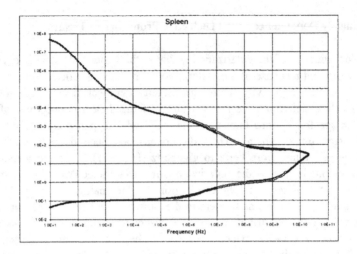

Figure 1. Permittivity and conductivity of ovine spleen tissue measured at 37°C, presented here as an example of the spectrum of a high water content tissue.

The dielectric spectrum of a high water content tissue, such as the data in Figure 1, is adequately described by Equation 7 with 4 Cole-Cole terms whereby 14 parameters are needed to determine the permittivity and conductivity of the tissue over 10 frequency decades.

4. Parametric models for the dielectric spectrum of tissues

Fitting Equation 7 to experimental data provides a combination of tissue-specific parameters which can be used to generate dielectric data that are in line with the literature relating to the tissue. In Figure 2, the experimental data from various sources are compared to the prediction of the model for a high water content tissue. The parameters for several tissue types have been reported in [1] and more recently published on the World Wide Web [2].

It is important to realize that the predictions of the model are only as good as the data used to develop it in the first place. With respect to the data reported in [1-2] the following reservations have been noted:

—The predictions of the model can be used with confidence for frequencies above 1 MHz.

— At lower frequencies, where the literature values are scarce and have larger than average uncertainties, the model should be used with caution in the knowledge that it provides a 'best estimate' based on present knowledge.

— It is important to be aware of the limitations of the models particularly where there are no data at all to support their predictions.

The comment on 'best estimate' triggers the question of what it means for biological materials. Is there a true value for the dielectric properties of a tissue?

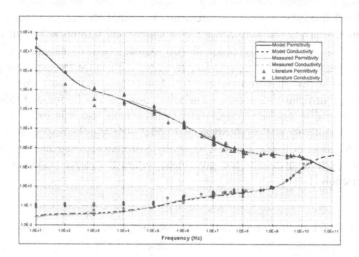

Figure 2. Permittivity and conductivity of a high water content tissue (liver): prediction of the model (black filled and dotted lines), experimental data at 37 °C (grey filled and dotted lines) and data from the literature

5. Experimental Measurements: Precision, Accuracy and Truth

Statistical analysis can be applied to experimental procedures to determine the precision and accuracy of measurement provided that the measured parameter has a well defined, true value and the errors in instrumentation and methodology are known from an independent source. This is not the case with the measurement of the dielectric properties of biological tissues where the literature shows significant differences in the values reported from different studies. This is mostly due to natural, unpredictable variability in structure and composition between samples from different animals and from different species. Moreover, the samples are affected by differences in the handling and storage prior to the measurement. However, much greater consistency is achieved within a set of measurements on a given tissue sample. From past experience, in the frequency range from a few hundred to a few thousand megahertz, multiple measurements on a tissue sample would fall within about ±5% of the average. By comparison, the stated accuracy of the widely used coaxial contact probe procedure is between 1 and 2% when assessed by repeat measurement on liquids of well-established dielectric properties [3]. The procedure involves the measurement of the reflection coefficient of the probe and the numerical deduction of the dielectric properties of the sample based on a theoretical model of the impedance of the probe and the prior calibration of the equipment using a reference material of known dielectric properties. These aspects of the procedure, together with residual instrumental errors, have been identified as the main sources of uncertainty. Several authors have attempted formal

80

estimation of errors from these sources [4-5]. These are interesting studies, best used as means to optimize the measurement procedure with respect to, among other things, the frequency range of interest, the nature of the sample to be measured and the dimensions of the probe. Ultimately, repeat measurement of appropriate standard samples can always be used to assess the performance of a system and inform on systematic and random errors. As previously reported [6], liquids used to simulate tissue in experimental dosimetry can be characterized to within 1-2% in the frequency range of a few hundred to a few thousand megahertz. Solid tissue equivalent materials are inherently more difficult to measure. Dielectric data, at 20 °C, for a filled-plastic material, developed to simulate a high water content tissue, are reported in Figure 3. The average permittivity and conductivity at 1 and 2 GHz and the respective standard deviations are given in Table 1. At 1 GHz the two systems used to carry out the measurement perform optimally and give mean values that agree within the stated standard deviation (Table1).

TABLE 1. Dielectric data from Figure 3 : average and standard deviation of five measurements.

	1 GHz				2 GHz			
	ε'	±	σ	±	ε'	±	σ	±
System 1	47.8	1.39	1.13	0.06	42.9	1.17	1.51	0.075
System 2	46.5	2.1	1.08	0.10				

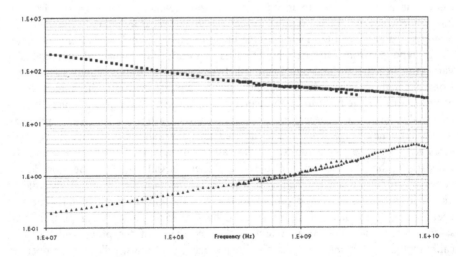

Figure 3. Permittivity and conductivity, at 20 °C, of a filled-plastic material intended to simulate a high water content tissue.

6. Systematic Change in the Dielectric Properties of Tissues

Changes in the dielectric properties of biological material occur in response to changes in physiological conditions due to metabolic activity or pathological conditions. Depending on the nature of the change part of or sometimes the whole of the dielectric spectrum may be affected. For example, metabolic changes that occur following death affect the ionic environment of cells and affect ionic phenomena suggesting that there should be differences between the *in vivo* and *in vitro* dielectric spectrum below a few kHz because of the effect on the α dispersion. Such differences have been observed experimentally within hours of death [7-8]. The dependence of the β dispersion on the physical and physiological state of the cell membrane is well established. Shape, size and capacitive impedance of the cellular membrane are the dominant factors. The destruction of the cellular membrane is manifested by significant changes in the β dispersion. An example of such an effect is given in Figure 4 which shows the permittivity and conductivity spectrum of muscle before and after the destruction of the cellular membrane through freeze/thaw injury. Systematic differences in the β dispersion were also reported between healthy and cancerous tissue [9], and between blood containing cells of different configuration [10]. Of direct relevance to dosimetry is the anisotropic dielectric response of muscle fibres that depend on the direction of the fibre with respect to the electric field [11]. Skin, bone and brain are also anisotropic with respect to their dielectric response [12-15]. Apart from the anisotropy, differences in structure and water content of skin from sole, palm and forearm gives them different dielectric properties as shown in Figure 5, larger differences were observed between the three dry skin-types [16]. Differences in the dielectric properties of skin as a function of age have not been investigated but the observed differences between the different types of skin suggest that similar differences are possible as the skin ages.

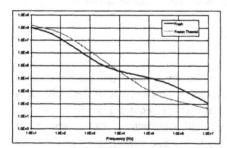 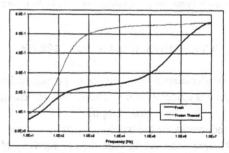

Figure 4. Effect of freeze-thaw injury on the permittivity (left) and conductivity (right) of muscle tissue. The conductivity on the right side graph is in S/m . The data have not been corrected for electrode polarisation and are presented to show the differential due to damage to the cell membrane.

82

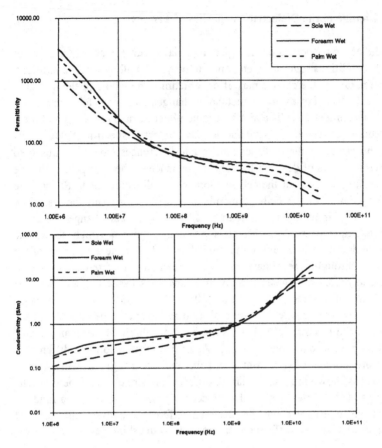

Figure 5. Permittivity and conductivity of moistened (wet) skin from the palm, sole and forearm.

7. Conductivity of tissue below 100 Hz

The assessment of exposure of people to low frequency electric and magnetic fields requires calculation of induced current density levels in various parts of the body. It is often the case that simplified models are used which require knowledge of the conductivity of the whole or parts of the body [17-19]. Conductivity values reported in [2] were used to calculate the conductivity of the whole and various parts of the body (Table 2) by volume averaging the data for individual tissues using a voxel anatomical human model, NORMAN [20], developed at the National Radiological Protection Board (NRPB).

TABLE 2: Conductivity, in S/m, of the whole and parts of the body obtained by volume averaging the conductivity values in reference [2]

	Whole body	Head	Torso	Arm	Leg	Neck
50 Hz	0.216	0.254	0.223	0.195	0.196	
10 kHz	0.276	0.285	0.256		0.238	0.222
100 kHz	0.288	0.30	0.332		0.239	0.243

At 50 Hz the calculated conductivity values are comparable to the commonly used estimate of 0.2 S/m. At 100 kHz the conductivities of the whole body and of the torso are respectively over 40 and 60% higher.

8. Non-linear Phenomena

The dielectric phenomena described in this paper occur from interaction with weak fields eliciting linear responses. Theoretical considerations and experimental observations predict non-linear molecular and cellular polarisation phenomena at high field strength. Field strengths of the order of 10^6 V/m may be capable of initiating polarisation mechanisms that affect the cellular function. Higher fields may cause the dielectric breakdown within the membrane ultimately leading to cell destruction. Under controlled conditions, such high-field effects are the focus of numerous applications in biotechnology [21]. The hypothesis that weak fields may trigger non-linear responses is being investigated [22-24]. The generation of harmonics provides a means of monitoring this effect and is conceptually equivalent to a dielectric permittivity with components of harmonics of the fundamental. The approach was easy to overlook by experimentalists using instrumentation designed to filter out currents and voltages at frequencies other than the fundamental and thus forcing linearity on the system. While the study of harmonics is no doubt a subtle and clever tool, it has yet to prove its effectiveness in monitoring physiological responses to weak fields.

9. References

1. Gabriel, S., Lau, R. W. and Gabriel, C. (1996) The dielectric properties of biological tissues: 3. Parametric models for the dielectric spectrum of tissues, *Phys. Med. Biol.* **41**, 2271-2293.
2. Gabriel, C. (1996) *Compilation of the Dielectric Properties of Body Tissues at RF and Microwave Frequencies*, Final Technical Report, US Air Force, TR-1996-0037, www.brooks.af.mil/HSC/AL/OE/OER/Title/Title.html
3. Gabriel, C., Chan, T. Y. A., and Grant, E. H. (1994) Admittance models for open ended coaxial probes and their place in dielectric spectroscopy, *Phys. Med. Biol.* **39**, 2183-2200.
4. Jenkins, S., Hodgetts, T. E., Clarke, R. N., and Preece, A. W. (1990) Dielectric measurements on reference liquids using automatic network analysers and calculable geometries, *Meas. Sci. Technol.* **1**, 691-702.
5. Wei, Y-Z and Sridar, S. (1991) Radiation-corrected open-ended coaxial line technique for dielectric measurements of liquids up to 20 GHz, *IEEE Trans. MTT* **39**, 526-531.

84

6. Gabriel, C. (1997) Interaction of the body with the radio emissions from hand-held transceivers – IBREHT, Section 4, NRPB, Chilton, Oxon, UK.
7. Schawn, H.P. (1957) Electrical properties of tissues and cell suspensions. *Adv. Biol. Med. Phys.* **5**, 147-209.
8. Surowiec, A., Stuchly, S. S., and Swarup, A. (1985) Radiofrequency dielectric properties of animal tissues as a function of time following death, *Phys. Med. Biol.* **30**, 1131-1141.
9. Fujimoto, E. and Kinouchi, Y. (1996) Tissue diagnosis through nonuniformity estimated by bioimpedance P-182B, BEMS, Eighteenth Annual Meeting, B.C. Canada.
10. Irimajiri, A., Ando, M., Matsuoka, T., Ichinowatari, T., and Takeuchi, S. (1996) Dielectric monitoring of rouleaux formation in human blood: A feasibility study, *Biochem. Biophys. Acta* **1290**, 207-209.
11. Dawson, T. W. and Stuchly, M. A. (1998) Effects of skeletal muscle anisotropy on human organ dosimetry under 60 Hz uniform magnetic field exposure, *Phys. Med. Biol.* **43**, 1059-1074.
12. Ivanchenko, I.A., Andreyev, E.A., Lizogub, V.G., and Aveshnikova, L.V. (1994) Space-time distribution of normal and pathological human skin dielectric properties in the millimeter wave range, *Electro- and Magnetobiol.* **13**, 15-25.
13. Saha, S. and Williams, P.A. (1992) Electrical and dielectric properties of wet human coetical bone as a function of frequency, *IEEE Trans. Biomed. Eng.* **39**, 1298-1304.
14. Nicholson, P.W. (1965) Specific impedance of cerebral white matter, *Experimental Neurology* **13**, 386-401.
15. Ranck, J.B. and BeMent, S.L. (1965) The specific impedance of the dorsal columns of cat: An anisotropic medium, *Experimental Neurology* **11**, 451-463.
16. Gabriel, C. (1997) Comments on 'Dielectric properties of skin' , *Phys. Med. Biol.* **42**, 1671-1674.
17. NRPB (1993) Board Statement on Restrictions on Human Exposure to Static and Time Varying Electromagnetic Fields and Radiation., NRPB, Chilton, Oxon, UK.
18. CENELEC (1995) *Human exposure to electromagnetic fields*, ENV50166-1 and ENV 50166-2, CENELEC, Brussels.
19. ICNIRP Guidelines (1998) Guidelines for limiting exposure to time-varying electric, magnetic, and electromagnetic fields, *Health Physics* **74**, 494-522.
20. Dimbylow, J.P. (1996) The development of realistic voxel phantoms for electromagnetic field dosimetry. *Proc. of an International Workshop on Voxel Phantom Development*, NRPB Report, NRPB, Chilton, Oxon, UK.
21. Pethig, R. (1996) Dielectrophoresis: Using inhomogeneous AC electric fields to separate and manipulate cells, *Critical Reviews in Biotechnology* **16**, 331-348.
22. Woodward, A.M. and Kell, D.B. (1990) On the nonlinear dielectric properties of biological systems, *Bioelectrochem. and Bioenergetics* **24**, 83-100.
23. Weaver, J.C. and Astumian, R.D. (1990) The response of living cells to very weak electric fields: The thermal noise, *Science* **247**, 459-462.
24. Moussavi, M., Schwan, H.P., and Sun, H.H. (1994) *Med. &Biol. Eng. & Comput.* **32**, 121-125.

INVESTIGATIONS OF TISSUE MICROWAVE AND THERMAL PROPERTIES FOR COMBINED MICROWAVE AND THERMAL MODELLING OF BODY TISSUE REGIONS

D. V. LAND, A. J. GORTON and G. HAMILTON
Department of Physics and Astronomy
University of Glasgow
Glasgow, G12 8QQ, U.K.

1. Introduction

The modeling of electromagnetic field heating in human and animal bodies assumes its most complex form over the ultra-high frequency through the microwave frequency range, from about 300 MHz to about 10 GHz. Over this range, body anatomical and metabolic structures and tissue thermal pattern dimensions are comparable to radiation wavelengths, both outside and inside the body. At lower frequencies, fields are tending to quasi-static forms, ohmic heating is the dominant loss mechanism, and the field patterns of the exposure environment become the major concern. At higher frequencies, rapidly rising dielectric losses limit exposure to the superficial tissues, becoming anatomically and electromagnetically a skin-depth problem.

Quantifiable modeling of radiation exposure of body regions requires tissue geometry, tissue electromagnetic and thermal properties, and tissue heat transfer mechanisms within and surrounding the region to be known. Both dynamic and steady state conditions must be recognized for the radiation-body system exposure. The aggressive but controlled electromagnetic heating of tissues for hyperthermia induction is an example of the need to understand both of these conditions. The aim is to achieve and hold a particular elevated tissue temperature for an extended period, while the natural response of the body is to rapidly and massively increase blood perfusion to maintain normal temperature. The complete radiation, tissue, heat removal, temperature measuring system must be modeled realistically and dynamically if successful, stable and safe heating is to be achieved. Accidental radiation exposure is, by contrast, usually highly time dependent, so the dynamic condition will often apply. However, since there is no control of the exposure, the worst case, the steady state, must usually be assumed.

Through both the antenna reciprocity principle and the thermodynamic principle of detailed balancing, the natural high-frequency and microwave thermal radiation emission from body tissues depends on the same tissue dielectric, thermal and geometrical properties as does tissue heating through radiation exposure. Further, at any particular frequency, the relative generation of thermal radiation throughout a tissue region and the relative heating of the tissue, have exactly the same form for a given field impressing or detecting antenna. This detailed relating of the active and passive

B.J. Klauenberg and D. Miklavcic (eds.), Radio Frequency Radiation Dosimetry, 85-96.
© 2000 *Kluwer Academic Publishers. Printed in the Netherlands.*

radiation-tissue interactions uniquely allows microwave radiometry to be used for inherently safe, non-invasive, *in-vivo* testing of basic tissue electromagnetic and thermal model.

2. Modeling for Microwave Radiometry

In tissues the deposition of high-frequency electromagnetic power as heat can conveniently be represented by two mechanisms, ohmic heating by conduction currents, and dielectric absorption loss. At realistic power levels these effects are linear, allowing heating power density to be expressed in terms of the conductivity, angular frequency, dielectric loss-factor and the radiation electric field as

$$\tfrac{1}{2}\left(\sigma + \omega\varepsilon''\right)\mathbf{E}^2 \tag{1}$$

The power loss can also be expressed through the power attenuation constant for radiation propagation through the tissue:

$$2\alpha = \omega\sqrt{2\mu_0\varepsilon'\left[\sqrt{1+\left(\left(\sigma/\omega+\varepsilon''\right)/\varepsilon'\right)^2}-1\right]} \tag{2}$$

At a particular frequency, a source antenna will produce a distribution of $\mathbf{E}^2$ dependent on the antenna properties and the coupled tissue volume properties, and proportional to the radiated power. If this antenna is used to view the thermal radiation from the same tissue region, antenna reciprocity requires the spatial response to the radiation to be the same as this deposited power density distribution. In microwave radiometry this power normalized $\mathbf{E}^2$ distribution is termed the weighting function of the antenna-tissue system.

At microwave frequencies, the thermal power radiated from a matched source is kBT, where k is Boltzmann's constant, B is the power bandwidth of the radiation measuring system, and T is the Absolute temperature of the source. Applied to an antenna-tissue system for which there is a spatially dependent weighting function $w(\mathbf{r})$, attenuation factor $2\alpha(\mathbf{r})$, and temperature distribution $T(\mathbf{r})$, the received matched noise power giving an effective source temperature T_e can be expressed as

$$kBT_e = \int 2\alpha(\mathbf{r})w(\mathbf{r})T(\mathbf{r})dv \tag{3}$$

The integral is taken over the whole volume of material coupled to the antenna [1].

Applying this to the simplest possible case of plane wave propagation normal to the surface of a semi-infinite region of uniform material, the effective radiation temperature seen by an impedance matched radiometer system is

$$T_e = 2\alpha \int \exp(-2\alpha z)T(z)dz \tag{4}$$

Practical radiometry antennas can only approach this ideal weighting function behavior, and real body regions are a complex mixture of tissues, but this does express the major features of radiometric temperature measurement [2, 3].

3. Tissue Thermal Model

For the purposes being considered here, the major features of body tissue temperature patterns can be modeled by considering heat to be supplied by arterial blood perfusion and metabolic activity, transported to the skin surface by conduction through the subcutaneous tissue, and lost from the surface by natural cooling mechanisms. This assumes the Pennes model for heat transfer, in which arterial blood fully reaches temperature equilibrium with the bulk tissue at the capillary level [4]. Experience has shown that, for most parts of the resting body, arterial blood perfusion is the major heat source, and metabolic heat supply appears to give a rather smaller contribution [5].

Since it is likely that there is a general interdependence of perfusion and metabolic activity, the whole heat supply can reasonably be expressed in terms of an effective perfusion, which will be only slightly higher than the actual perfusion. In terms of the blood perfusion and specific heat capacity, and the arterial blood and tissue temperatures, the thermal power density deposited is $w_b C_b (T_a - T_t)$. With the tissue thermal conductivity K, the equilibrium Fourier heat diffusion equation for tissues can then be expressed for a one-dimensional heat flow as

$$\frac{d^2 T(z)}{dz^2} + \frac{w_b C_b}{K}\left(T_a - T(z)\right) = 0 \qquad (5)$$

At rest, heat loss from the skin surface is primarily by infrared radiation and by convection, with a very small contribution from insensible perspiration. For normal values of skin to ambient temperature difference, the heat loss per unit area dependence on temperature can be linearized to Newtonian cooling as $H(T_s - T_{amb})$ [6]. The heat loss coefficient or surface conductivity H is well determined for smooth body surfaces.

If this thermal model is applied to the previous uniform tissue region, the temperature distribution is simply

$$T(z) = T_a - (T_a - T_s)\exp\left(-\sqrt{w_b C_b / K}\, z\right) \qquad (6)$$

with the surface temperature given by

$$T_s = (T_a - T_{amb})/(1 + H/\sqrt{w_b C_b K}) + T_{amb} \qquad (7)$$

Though again a considerable simplification for many body regions, these simple relations do show the principal features of tissue temperature variation.

88

4. Combined Microwave and Thermal Model

If the two simple tissue models used above are combined, the arterial blood temperature, the microwave radiometric temperature, the surface temperature and the ambient temperature can be related in the form

$$T_a - T_{mw} = \frac{H}{\left[\sqrt{w_b C_b K} + w_b C_b / 2\alpha\right]}\left(T_s - T_{amb}\right) \tag{8}$$

These four temperatures are accessible for measurement. The arterial temperature is close to 0.3°C above oral temperature. A properly calibrated and operated radiometer system will provide the microwave temperature. Careful infrared thermometry can provide the surface temperature, and the ambient temperature can be controlled and measured. The arterial-microwave temperature difference is usually small, in the range 1°C - 7°C, so careful inter-modal calibration is needed [5]. Plots of the arterial-microwave versus surface-ambient temperature differences will lie on loci determined by the tissue properties of microwave attenuation, thermal conductivity and blood perfusion [7, 8].

The tissue perfusion is obviously independent of the attenuation and conductivity parameters. The attenuation and conductivity are, however, likely to be related through common dependence on major tissue constituents. If this commonality of these aspects of tissue behavior is utilized, the temperature variation then has only a two-parameter dependency. With the common geometry of anatomy and the common tissue temperature pattern, the related behavior of attenuation and conductivity forms the third link between the microwave and the thermal models of a body region.

The characteristic lengths of solutions to the heat transport equation and the microwave attenuation lengths,

$$\sqrt{K/w_b C_b} \quad \text{and} \quad 1/2\alpha , \tag{9}$$

are also very similar for realistic tissue properties. This is a further indication that the microwave and thermal systems should be strongly coupled for analysis.

More sophisticated models than the simple analytical cases above have been developed for specific body regions, using finite element computational modeling to provide flexibility [5]. Two and three different tissue types have been introduced, using layered regions of skin, fat, muscle and bone as appropriate. Finite thickness and cylindrical regions have been investigated for modeling fingers, hands, wrists and knees. For soft tissues, a linear relationship based on water content has been used for the attenuation and conductivity dependency [7]. Bone has been taken to have single, distinct attenuation and conductivity values. Antenna field patterns have been modeled and measured in tissue simulating media, and effective attenuation factors introduced to allow for real antenna weighting functions in different tissues [9, 3]. These more complex microwave and thermal models show changes in some features, such as the lower core temperature of the extremities of the limbs, but the general forms of the microwave to surface temperature relationships, however, remain very similar to those of the simple model.

5. Tissue Dielectric Properties at Microwave Frequencies

Good knowledge of the microwave dielectric properties of the full range of human tissues is essential if any form of interaction of microwave radiation with the body is to be understood, applied, or interpreted. The investigations of tissue dielectric behavior summarized here have been carried out to provide information for microwave radiometric modeling and to improve the interpretation of microwave temperature data for medical applications. The measurements have mostly been made at frequencies close to 3.2 GHz, the center of the radiometer measurement band, where there is a good combination of tissue penetration distance and spatial resolution of thermal features [2]. The investigations have concentrated on soft tissues, both human and animal, in order to determine and understand the dielectric behavior as the tissue water content varies [10, 11].

The dielectric behavior of soft tissues at microwave frequencies is usually considered to be determined by the relative contents of a low permittivity, low loss component of fat or protein, and a high permittivity, higher loss, water component. Pure fats and proteins have relative permittivities of a little over 2 and loss factors of about 0.1 at these frequencies. Water is present in tissues as electrolytes, mostly very similar to physiological saline, a 0.9% sodium chloride solution. Two factors determine the saline dielectric behavior: the Debye relaxation of the polar water molecules at about 30 GHz *in-vivo*, and the sodium chloride ionic conductivity [12, 13]. At the low frequency end of the microwave region, the loss due to the dielectric relaxation increases with frequency, while the loss due to the ionic conductivity σ/ω decreases. At about 3 GHz the two loss components are equal and at a minimum, giving a minimum in the loss-tangent and indicating an optimum frequency for microwave imaging [2]. Figure 1(a) shows this behavior, with both conduction and dielectric losses shown as a loss factor.

Since the water dielectric behavior must dominate over that of fat or protein at the real tissue water contents of 10% to 85%, tissue permittivities and loss factors will tend to rise with increasing water content, and their loss tangent to show a minimum near 3 GHz. In general, this is true. At the extremes of water content, fat tissue at about 10%-20% and blood at about 85%, the permittivity and loss factor values are close to those given by the more realistic mixture modeling relations [11, 14, 15, 16, 17]. The loss-tangent minimum is also well shown [2]. The middle to high water content tissues, muscle, the organs and tumors, show good correlation between loss factor and permittivity, with small differences between tissues [11,16]. Figure 1(b) is an example of this for female breast tissues, which can show a relatively wide range of water contents. There appear to be small differences between corresponding tissues from different species, though this may in part be due to different post-mortem handling [11].

Foster and Schwan [18] and Campbell [16] have provided comprehensive reviews of published data. Gorton [11] has recently added considerably to the body of data on human and animal soft tissues, and has also related his findings to existing data and to mixture modeling methods (Table 1).

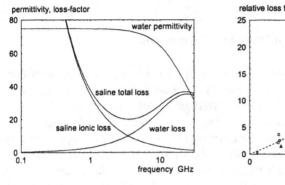

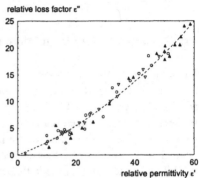

(a)   (b)

Figure 1. (a) Dielectric behaviour of water and physiological saline with frequency. (b) Permittivity and loss factor variation for female breast tissue at 3.2 GHz: O normal tissue; □ benign tumour tissue; ▲ malignant tumour tissue. ((b) Adapted from Campbell and Land, 1992)

TABLE 1. Measured mean values of relative permittivity, loss factor and water content for human tissues at 3.2 GHz and 20°C. (Adapted from Gorton, 1996)

Human Tissue Type	Relative Permittivity	Relative Loss Factor	Water Content %
Skeletal muscle	49.3±1.4	16.7±0.7	77±3
Liver	41.7±1.2	13.5±0.4	75±1
Kidney cortex	49.7±1.2	17.3±0.6	79±1
Kidney medulla	53.9±1.9	18.8±1.0	
Heart	49.8±1.0	14.8±0.4	81
White connective	40.8±2.3	14.8±0.4	72±3
Uterus	49.1±1.5	15.8±1.5	80±2
Spleen	55.4±1.5	17.2±0.4	83±1
Pancreas	49.2±1.1	16.6±0.6	
Brain grey matter	54.9±1.7	12.5±0.5	86±2
Brain white matter	31.8±3.0	7.3±0.6	75±1
Breast fat	6.5±3.5	1.0±1.1	16±5
Breast tissue	13 to 45	3 to 15	
Breast tumors	50 to 54	15 to 17	80 to 84

However, when the dependence of permittivity and loss factor on water content is examined, a wide spread of individual measurement values is found in many cases. This spread seems to be similar for different investigators and different preparation and measurement techniques, and is much larger than can be explained by any deficiency of technique. The spread of values appears to increase with increasing tissue water content, and is seen most clearly in the loss factor data (Figure 2). In some high water content samples the measured loss factor considerably exceeds the maximum level predicted by any of the mixture relations [11, 16,17].

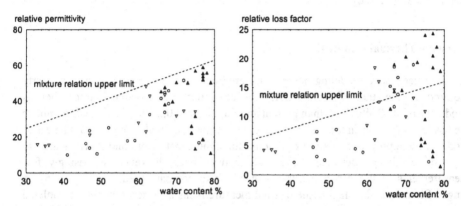

Figure 2. Variation of relative permittivity and loss factor of female breast tissue with water content. O normal tissue; □ benign tumor tissue; ▲ malignant tumor tissue. (Adapted from Campbell and Land, 1992)

This behavior of the higher water content tissues has recently been investigated in some detail, using forced variation of the water content, variation of temperature, variation of frequency, and tissue simulating gels [11]. It seems most likely that the cause of the variation in dielectric behavior is the variation of water molecular binding to surfaces of the tissue structures over distances comparable to the measuring field region, which is often only about 1-2 mm across. The mixture relation then breaks down because of the variation in properties of one of the components. The effect of the water binding is to cause a considerable reduction in the relaxation frequency of the water molecules for a significant fraction of the tissue water [19]. Extended mixture equation analysis suggests that in the higher water content tissues about 0.4 grams of water is bound per gram of protein solids. This bound water can be expected to have a wide range of relaxation frequencies, with an effective value that may be below 1 GHz [11]. Other possible causes of dielectric behavior variation, such as blood cell resonance or electrolyte variations, are only important at very much lower frequencies, and their effects at microwave frequencies are simply too small [15].

The effect of this variation of free to bound water content for large volumes of tissue is not yet clear. It is likely that over wavelength distances in bulk muscle and organ tissues there will be a strong averaging effect, and such volumes of these tissues will show quite well defined permittivity and loss factor values close to the average of values of the small volume measurements [20]. The situation for tumor tissue is less certain, and small tumors may well show marked variations in dielectric properties [16].

For the microwave radiometric modeling, where it is properties over wavelength and attenuation length distances that are important, the average dielectric properties for the average water contents of the soft tissues have been used. For bone the dielectric behavior, though less widely studied, appears to be well defined and close to that expected from mixture modeling for its constituents and low water content. For 3 GHz modeling, bone has been taken as having a relative permittivity of 4.8±0.6, and a loss factor of 0.85±0.2 [20].

6. Tissue Thermal Conductivity

For the steady state modeling being considered here, the only tissue thermal property required is the thermal conductivity. Since water has a rather larger thermal conductivity, $0.59 \text{ Wm}^{-1}\text{K}^{-1}$, than protein and fat, at about $0.2 \text{ Wm}^{-1}\text{K}^{-1}$, it will dominate the tissue behavior as in the dielectric case. Tissue thermal conductivity should be easily modeled by appropriate mixture equations, since the constituent conductivities are not expected to change between tissues. There have been, however, surprisingly few measurements of tissue conductivity with which to compare predicted values [6]. A range of human and animal tissue thermal measurements have recently been completed to provide data for combined microwave radiometric and thermal model [21]. The human tissue thermal conductivity results are summarized in Table 2. The measurements on animal tissue show inter-species differences in thermal properties to be very small.

These results are generally similar to existing published data where they exist. They show satisfactory agreement with Maxwell and Hashin-Shtrikman type water-protein mixture equations, particularly if tissue protein is taken to have a thermal conductivity of $0.25 \text{ Wm}^{-1}\text{K}^{-1}$.

From some limited measurements, and mixture modeling of its constituents, bone is taken to have a thermal conductivity of about $0.7 \text{ Wm}^{-1}\text{K}^{-1}$.

TABLE 2. Thermal conductivities and water contents for human tissues. (Adapted from Hamilton, 1998)

Human Tissue Type	Thermal conductivity $Wm^{-1}K^{-1}$	Water Content %
Breast fat (post operative)	0.20	13
Breast fat (post mortem)	0.21	15
Muscle: abdominal, soleus	0.48	76
Muscle: quadriceps, thigh	0.54	79
Liver	0.51	75
Kidney	0.54	79
Brain: white matter	0.49	79
Brain: grey matter	0.55	86
Spleen	0.52	83
Pancreas	0.45	67

7. Application of Combined Microwave and Thermal Model

The microwave radiometric and thermal models of a body region are linked by:
1) a common anatomy,
2) a common temperature pattern,
3) a relation of the radiation propagation constant to the thermal conductivity through a common dependence on tissue water content.

It has been found convenient to include the effect of the weighting function of the measuring antenna in the attenuation-thermal conductivity relation, through the use of an effective attenuation constant for propagation in any particular tissue type. The working relation used for this modeling of the soft tissue behavior at 3.2 GHz was 2α =380 K. For bone, $2\alpha = 25$ m^{-1} and K = 0.7 Wm^{-1}K^{-1} have been used.

The combined modeling has been applied to measurements made over several areas of the human body to validate the modeling and investigate its potential for clinical use [5, 7, 8]. One and two dimensional, single, multi-layer and finite thickness models have all been used. Two cases have been chosen to illustrate the use of the technique, both using microwave temperature measurements taken with 3.2 GHz radiometers and surface temperatures taken with infra-red thermometers.

Figure 3(a) illustrates the application for the analysis of temperature data taken on the anterior surface of the upper leg over the quadriceps muscles of normal subjects. In most subjects, and particularly in females, the muscle tissue will be viewed through a layer of subcutaneous fat of various thicknesses. The temperature plot shows that the effective tissue behavior of the region does tend to lie between that expected for fat and for muscle. The effective perfusion varies between about 0.1 and 0.35 kg.m^{-1}s^{-1}. This perfusion range is in good agreement with published data on muscle perfusion found by invasive techniques [22, 23].

Figure 3(b) illustrates the application of the technique to the analysis of temperature scans taken across both breasts of a 22 year old, normal, female subject. The two scans shown have been taken transversely at nipple level from lateral to medial edge of each breast. There are indications of fat tissue in the lateral quadrants, with a change to denser tissue in the medial quadrants. There is a large drop in effective perfusion from the edge of the breast tissues at the chest wall to just lateral of the nipple in both breasts. Experience has shown that this is one of the most common forms of temperature and perfusion variation for the female breast, but that there are also considerable variations between subjects and many subjects show significant cyclical pattern changes [5]. These estimated perfusion values are, again, in good agreement with findings of the only invasive measurements that have been published [24].

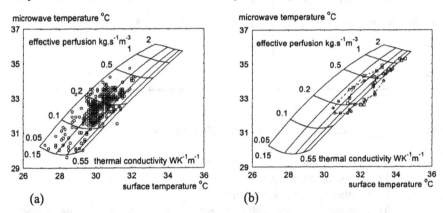

(a)   (b)

Figure 3. Illustrations of combined microwave radiometric and thermal model: (a) temperatures measured over the quadriceps muscles of normal subjects; (b) temperature loci measured over left ○ and right □ breasts of a normal female subject. ((a) adapted from Kelso and Land 1994)

8. Conclusion

Analysis of passive, radiometric measurements of natural high frequency and microwave thermal radiation from the human body is suggested as a technique for understanding the in-vivo behavior of tissue regions for exposure to electromagnetic radiation. The interpretation of these measurements can be greatly simplified by recognizing that electromagnetic and thermal models of a tissue region are not independent, but share a common dependence of properties on tissue type, particularly through the tissue water content, as well as sharing a common anatomy and a common temperature distribution.

Measurement data on tissue microwave dielectric properties and thermal properties have been considerably extended to provide the information needed for combined microwave and thermal model. This modeling technique has been successfully applied to interpret microwave and surface temperature measurements taken over several body regions.

9. References

1. Leroy, Y., Mamouni, M., Van de Velde, J. C., Bocquet, B., and Dujardin, B. (1987) Microwave radiometry for noninvasive thermometry, *Automedica* **8**, 181-202.
2. Land, D. V. (1987) A clinical microwave thermography system, *Proc. IEE* **134**, 193-200.
3. Land, D. V. (1992) Simplified nonresonant perturbation method of measuring aspects of performance of UHF and microwave antennas for biomedical applications, *Electron. Lett.* **28**, 1190-1192.
4. Pennes, H. H. (1948) Analysis of tissue and arterial blood temperatures in the resting human forearm, *J. Appl. Physiology* **1**, 93-122.
5. Kelso, M. B. (1995) *A Study of the Use of Combined Thermal and Microwave Modelling of Body Regions for Microwave Thermography*, PhD Thesis, University of Glasgow.
6. Draper, J. W. and Boag, J. W. (1971) The calculation of skin temperature distributions in thermography, *Phys. Med. Biol.* **16**, 201-211.
7. Land, D. V., Brown, V. J., and Fraser, S. M. (1991) Clinical testing of combined thermal and microwave radiometric tissue modelling, *J. Photo. Sci.* **39**, 166-169.
8. Kelso, M. B., Land, D. V., and Sturrock, R. D. (1995) Recent investigations of combined microwave and thermal modelling of body tissue regions for the interpretation of microwave thermographic images, *IEE Digest 1995/041* **8**, 1-6.
9. Land, D. V. (1988) Application of the nonresonant perturbation technique to the measurement of high frequency fields in biological phantom materials, *Electron. Lett.* **24**, 70-72.
10. Gorton, A. J. and Land, D. V. (1995) Dielectric tissue measurements using a coaxial probe with a quarter wave choke, *IEE Digest 1995/041*, **10**, 1-6.
11. Gorton, A. G. (1996) *Measurements and Analysis of the Microwave Dielectric Properties of Human and Animal Tissues*, PhD Thesis, University of Glasgow.
12. Hasted, J. B. (1972) Liquid water: Dielectric properties, in F. Franks (ed.) *Water: A Comprehensive Treatise*, Plenum, New York.
13. Hasted, J. B. (1973) *Aqueous Dielectrics*, Chapman and Hall, London.
14. Schepps, J. L. and Foster, K. R. (1980) The UHF and microwave dielectric properties of normal and tumour tissues: Variation in dielectric properties with tissue water contents, *Phys. Med. Biol.* **25**, 1149-1159.
15. Schwan, H. P. and Foster, K. R. (1980) RF-field interactions with biological systems: Electrical properties and biophysical mechanisms, *Proc. IEEE* **68**, 104-113.
16. Campbell, A. M. (1990) *Measurements and Analysis of the Microwave Dielectric Properties of Tissue*, PhD Thesis, University of Glasgow.
17. Campbell, A. M. and Land, D. V. (1992) Dielectric properties of female human breast tissue measured *in-vitro* at 3.2 GHz, *Phys. Med. Biol.* **37**, 193-210.
18. Foster, K. R. and Schwan, H. P. (1989) Dielectric properties of tissues and biological materials: a critical review, *Biomed. Eng.* **17**, 25-104.
19. Schwan, H. P. (1965) Electrical properties of bound water, *N.Y. Acad. Sci.* **125**, 344-354.
20. Cook, H. F. (1951) The dielectric behaviour of some types of human tissues at microwave frequencies, *Br. J. Appl. Phys.* **2**, 295-300.
21. Hamilton, G. (1998) *Investigations of the Thermal Properties of Human and Animal Tissues*, PhD Thesis, University of Glasgow, in preparation.

22. Lassen, N. A., Lindbjerg, J., and Munck, O. (1964) Measurement of blood flow through skeletal muscle by intramuscular injection of xenon 133, *Lancet* **1**, 686.
23. Shepherd, R. C. and Warren, R. (1960) Studies of the blood flow through the lower limb of man by the nitrous oxide technique, *Clin. Sci.* **20**, 99-105.
24. Beaney, R. P. et al. (1988) Positron emission tomography for *in-vivo* measurement of regional blood flow, oxygen utilisation, and blood volume in patients with breast carcinoma, *Lancet* **1**, 131-134.

DIELECTRIC PROPERTIES OF SKIN

T. LAHTINEN[1,2], J. NUUTINEN[1] and E. ALANEN[1]
[1]Department of Oncology
Kuopio University Hospital
[2]Department of Applied Physics
University of Kuopio
FIN - 70210 Kuopio, Finland

1. Introduction

Knowledge of the dielectric properties of skin is important when electromagnetic (EM) fields are interacting with the human body or when EM fields are applied therapeutically or diagnostically. However, in published works the definition of skin is often not given. Neither is the detailed description of the measurement system and the applied probes always well presented. This results in problems in interpreting the published dielectric data for the skin. We shall first summarize the structure of the skin from a dielectric point of view. Then, we describe methods, mainly related to radio frequencies, to measure the permittivity of skin.

2. Layered Structure of the Skin

The outermost structure of the skin is stratum corneum with a typical thickness of 20 μm. The water content is about 20 % w/w increasing approximately linearly with depth and reaching a maximum of 70 % at about 0.1-0.2 mm [1]. Stratum corneum is a "brick and mortar" mosaic structure composed of dead epidermal cells, called corneocytes. Together with a well-organized lipid matrix they form the very resistant protective layer of human body. Due to the corneocytes the protein content of the stratum corneum is high. The proteins are moistured by the water transfer through the skin. Microscopically the skin surface is rough which has important consequences for dielectric measurements.

Epidermis is composed of closely packed epidermal cells. Although several anatomical layers, based on the definition of cell shape, can be found, they cannot be separated dielectrically. Basal cells of the epidermis at the depth of 0.1-0.2 mm are responsible for the high cellularity of the epidermis. The mature cells gradually move upwards the skin surface and form the corneum. Dielectric properties are determined by the membranes of epidermal cells as well as intra- and extracellular water.

97

B.J. Klauenberg and D. Miklavcic (eds.), Radio Frequency Radiation Dosimetry, 97-101.
© 2000 Kluwer Academic Publishers. Printed in the Netherlands.

Dermis forms the largest anatomical structure of the skin. The thickness varies from site to site but is typically 1-2 mm. The main components are collagens with about 70 % of the dry mass of the skin solids [2] and extracellular water with about 70 % of the wet mass of the dermis [2]. Two major vascular regions can be found. The upper plexus consists of fine vascular network under the epidermis while the deep plexus with larger arteries and veins lies at the dermis-subcutaneous fat interface. The cellularity of the dermis is low. Dielectrically the most important component is extracellular water either free or bound onto the surface of proteins. The amount of bound water depends on the amount of proteins but typically the proteins may bind 0.2-0.4 g of motionally restricted or bound water per 1 g of protein [3].

3. Measurement of the Dielectric Properties of Different Skin Layers

3.1. DIELECTRIC PROPERTIES OF TISSUES AS A FUNCTION OF FREQUENCY

Generally the dielectric properties of biological tissues with increasing frequency are characterized by three major dispersions α, β and γ as the frequency increases from a few Hz to 100 GHz [4]. Since the α-dispersion at low frequencies is mainly considered to be associated with the relaxation of ions related to charged membranes, the α-dispersion in the skin could be found with stratum corneum and underlying epidermis [8]. The existence of this dispersion with dermis with a low number of cells is unknown. At radio frequencies the β-dispersion is related to the Maxwell-Wagner effect due to the cell membranes. Due to a low number of cells in the skin the magnitude of the β-dispersion is considered slight. Above 1 GHz where the γ-dispersion related to tissue water content is dominating, the tissue dielectric properties are rather independent on tissue structure. According to Schwan [4] the majority of the cellular water at high radio and microwave frequencies is identical with normal water from a dielectric point of view. Between the β- and γ-dispersions a slight δ-dispersion related to protein bound water and/or partial rotation of polar sidechains can be found. With the skin the δ-dispersion is important since about 70 % of the dry weight of the skin is proteins (collagens). Thus the amount of bound water is also high.

3.2. OPEN-ENDED COAXIAL PROBE AND REFLECTION COEFFICIENT

If low frequencies are not considered, an open-ended coaxial probe and a reflection method have been used most often with skin dielectric measurements. Figure 1 illustrates a typical block diagram of the reflection method for the frequency range from 300 kHz to 3 GHz. The measurement arrangement is based on a standard network analyzer and a S-parameter test set. From the measured reflection coefficient the permittivity and conductivity of the skin can be calculated.

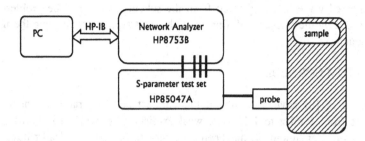

Figure 1. A typical block diagram based on the reflection principle [5].

The open-ended coaxial probe measures the frequency-dependent dielectric properties of different skin layers. At low frequencies the poorly conducting stratum corneum between the well-conducting dermis and the electrode can be considered as a capacitance. The capacitance is then mainly dependent on the properties of stratum corneum [6]. In the low MHz range, up to 10 MHz, the measured value is related to this capacitance and therefore, the probe measures mainly the dielectric properties of stratum corneum. In the high MHz range, above 50 MHz, the skin can be considered as a multilayer structure where the complex permittivity of each layers consists of the dielectric constant and the conductivity terms [11]. The field strength at different depths in tissue is dependent on the probe dimensions, especially on the distance between the inner and outer conductor (Figure 2).

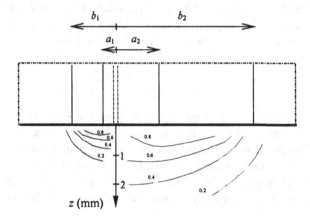

Figure 2. Isopotential lines of the field strength at different depths for two probes with different spacing between the inner and the outer conductor (From Lahtinen *et al* [7]). The symbols a and b are the radii of the inner and outer conductors, respectively.

Figure 2 illustrates that if the typical thickness of the skin is 1-2 mm, the effect of subcutaneous fat has to be considered with open-ended coaxial probes since some part of the reflected waves are originated from the subcutaneous fat. The problem can be partly avoided by the selection of a probe with small separation between the inner and outer conductors.

3.3. STRATUM CORNEUM

Martinsen *et al* [8] have investigated the conductance of stratum corneum at EM frequencies from 1 mHz to 1 kHz. A weak α-dispersion below 10 kHz detected by Yamamoto and Yamamoto [9] has been explained to be due to a dead nature and low conductivity of stratum corneum. Dielectric properties of stratum corneum from the kHz range to GHz range are not well known. Recently, Alanen *et al* [6] has used an open-ended coaxial probe and reflection technique in the low MHz range and found that the moisture content of the most superficial structures of the skin i.e. principally of stratum corneum is the main determinant for the measured dielectric properties.

Gabriel [10] has compared measurements with open-ended coaxial probe for normal skin and moistured skin and points out that with normal skin the uppermost layer (stratum corneum) is a dominating one, but with moistured skin the effect of the layers is reduced such that the measured value is a combination of the properties of epidermis/dermis and stratum corneum.

3.4. EPIDERMIS AND DERMIS

In order to determine the dielectric properties of the main part of the skin, epidermis and dermis, a model to eliminate the effect of subcutaneous fat is needed. This has been done for frequencies in the range from several tens of MHz to 1 GHz. The real dielectric constant of the skin ε_l can be determined from the formula:

$$\varepsilon_1 = \frac{\varepsilon_p - \varepsilon_2 e^{-gd}}{1 - e^{-gd}} \tag{1}$$

where ε_P is the dielectric constant of the skin seen by the probe and ε_2 is the dielectric constant of subcutaneous fat [11]. d is the thickness of the skin and g a probe-dependent constant. Since the model is unable to eliminate the effect of stratum corneum, the dielectric constant of stratum corneum is included in ε_l. However, the influence is slight when probes with deep penetration of the EM fields are used. With the reflection principle the permittivity of the healthy human skin is 52 ± 1 at 300 MHz [11].

3.5. SUBCUTANEOUS FAT

Although subcutaneous fat is not a part of the skin the interface between skin and fat is not sharp and the skin thickness and thus the measured dielectric properties of the skin vary from site to site. The dielectric constant of subcutaneous fat can be calculated from measurements with three probes of different sizes at several hundreds of MHz. The method is based on a three-layer model of stratum corneum-epidermis/dermis-subcutaneous fat [12]. Typical values for the permittivity of subcutaneous fat are of the order of 15-20 for female breast tissue fat.

4. Conclusions

Due to a multilayer structure it is always important to specify what one considers with "the skin" when the dielectric properties of the skin are discussed. It would be also helpful if the field strengths of the applied probes in tissue or at least the dimensions of the probes are presented. Due to different measurement arrangements one system can be sensitive to stratum corneum while the other to dermis and/or subcutaneous fat.

Finally, in the practical measurements one needs careful attention to additional factors like the moisture content of the skin, posture of the investigated subject or standardizing the measurement technique, i.e. pressure of the probe between the electrode and skin.

5. Acknowledgments

Support by Technology Development Centre of Finland is gratefully acknowledged.

6. References

1. Warner, R.R., Myers, M.C., and Taylor, D.A. (1988) Electron probe analysis of human skin: Determination of the water concentration profile, *J. Invest. Dermatol.* **90**, 218-224.
2. Uitto, J. and Perejda, A.J. (1987) *Connective Tissue Disease: Molecular Pathology of the Extracellular Matrix*, Marcel Decker, New York.
3. Pethig, R. and Kell, D.B. (1987) The passive electrical properties of biological systems: Their significance in physiology, biophysics and biotechnology, *Phys. Med. Biol.* **32**, 933-970.
4. Schwan, H.P. (1957) Electrical properties of tissue and cell suspensions, *Adv. Biol. Med. Phys.* **5**, 147-209.
5. Tamura, T., Tenhunen, M., Lahtinen, T., Repo, T., and Schwan, H.P. (1994) Modeling of the dielectric properties of normal and irradiated skin, *Phys. Med. Biol.* **39**, 927-936.
6. Alanen, E., Lahtinen, T., and Nuutinen, J. (1998) Penetration of electromagnetic fields of open-ended coaxial probe at different frequencies in skin measurements, *Proc. X Int. Conf. Electrical Bio-impedance, April 5-9, Barcelona, Spain,* pp. 251-254.
7. Lahtinen, T., Nuutinen, J., and Alanen, E. (1997) Dielectric properties of the skin, *Phys. Med. Biol.* **42**, 1471-1472.
8. Martinsen, O.G., Grimnes, S., and Sveen, O. (1997) Dielectric properties of some keratinised tissues. Part 1: Stratum corneum and nail in situ, *Med. Biol. Eng. Comput.* **35**, 172-176.
9. Yamamoto, T. and Yamamoto, Y. (1976) Electrical properties of the epidermal stratum corneum, *Med. Biol. Eng.* **14**, 151-158.
10. Gabriel, C. (1997) Comments on "Dielectric properties of the skin", *Phys. Med. Biol.* **42**, 1671-1673.
11. Alanen, E., Lahtinen, T. and Nuutinen, J. (1998) Variational formulation of open-ended coaxial line in contact with layered biological medium, *IEEE Trans. Biomed. Eng.* **45**, 1241-1248.
12. Alanen, E., Lahtinen, T., and Nuutinen, J. (1998) Measurement of dielectric properties of subcutaneous fat with open-ended coaxial sensor, *Phys. Med. Biol.* **43**, 475-485.

DIELECTRIC PROPERTIES OF HUMAN CRYSTALLINE LENS: CATARACTOGENIC EFFECTS OF RFR

L. D. SZABÓ and J. BAKOS
National Research Institute for Radiobiology and Radiohygiene
Budapest, P.O.Box 101., H-1775, Hungary

1. Introduction

The scientific database of radiofrequency radiation (RFR) compiled from researches over 50 years indicates that the predominant effects of electromagnetic fields in the region from 100 kHz to 300 GHz are the thermal effects. Exposure of biological systems to RFR energy of sufficient intensity may lead to temperature elevation when the rate of energy absorption exceeds the rate of energy dissipation.

2. The Eye and the Crystalline Lens

The eyes are complex sense organs. Each eye has a layer of receptors in the retina, and a lens system for focusing light on theses receptors. The crystalline lens has a transparent structure. In the eye light is refracted at the anterior and posterior surfaces of the lens.

The eye has been of the greatest interest in the discussion on the hazardous effects of RFR, as this organ has a unique feature in human body. The crystalline lens in eye is a continuously growing tissue at normally excellent transparency. The cells of the lens epithelium produce the lens fibres throughout life, and are situated underneath the elastic lens capsule. The lens capsule is a true basement membrane produced by the epithelial cells, which elongate so that they reach from pole to pole to form the lens fibres, and they lose their cell nuclei (Figure 1). In the adult human lens the epithelium can only be seen in the anterior and equatorial positions of the subcapsular region. The lens has no nerves or blood vessels. Its biochemical metabolism is arranged through active and passive transportation of different chemical substances. The complex enzymatic systems and transportation routes are of the greatest importance to withhold transparency, and one can easily understand how temperature increase will interfere with these good equilibrated systems [1, 2, 3, 4].

103

B.J. Klauenberg and D. Miklavcic (eds.), Radio Frequency Radiation Dosimetry, 103-107.

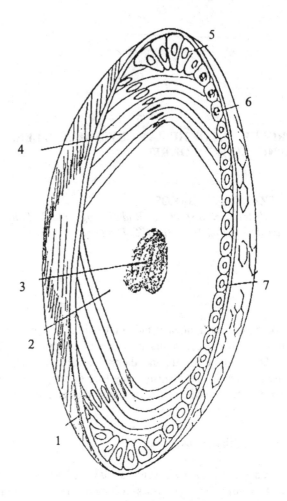

Figure 1. Structure of crystalline lens.
1. Capsule of lens, 2. Old fibres, 3. Nucleus of lens, 4. Young fibres, 5. Elongated equatorial cells, 6. Germinative region, 7. Epithelial cells

3. Dielectric Properties of Human Crystalline Lens

The crystalline lens is dielectric biomaterial in which the interactions with RFR leads to the polarization of bound charges and orientation of charged dipoles. Schwan experimentally identified the three different frequency dispersions in biomaterial, which may be attributed to different mechanisms. The regime of α-dispersion $0 - 0.1$ MHz is very important to the understanding of low frequency electric field effects. The β-dispersion encompasses a region of frequencies between $0.1 - 10$ MHz. The γ-dispersion shows up around 1 GHz. The conductivity shows a huge and sharp increase around 10 GHz. This effect is mainly the result of polarization of tissue water. Also contributions from polarization of proteins and bound water which exhibit a broad spectrum of relaxation frequencies from 10 MHz to 3 GHz, are present [5].

The absorption of RFR and subsequent distribution in the body are strongly dependent on the size and orientation of body and the frequency and polarization of the incident RFR. The penetration depth of high-frequency radiation can be derived from the frequency dependent dielectric tissue data. For muscle tissue, the penetration depth (distance over which about 2/3 of RFR energy is adsorbed) measures about 10 cm at 30 MHz and 3 cm at 3 GHz. Concerning penetration into tissue, microwaves of frequency 20 GHz practically behave like infrared radiation. Since most interactions in the RFR range is due to the rate of energy deposition per unit mass, the parameter of specific absorption rate (SAR) is used for descriptions. The initial rise in temperature increase, when heat losses are neglected (in crystalline lens of eye), is directly proportional to the SAR. The absorbed energy can be calculated from the field strength in tissue, taking dielectric tissue data and conductivity into account [6].

4. Cataractogenesis Induced by RFR

First of all cataract formation is not a part of aging but something abnormal. The mechanisms behind cataract formation are complex. Exposure to ionizing and ultraviolet radiation primarily will affect the germine epithelium in the lens equator, changing the mitotic rate and affecting fibre formation The RFR results in a temperature increase by locally absorbed and deposited energy, which will result in a coagulation of proteins and an alteration in epithelial cell metabolism and function, which will give secondary changes in the lens. High power RFR is mainly emitted by radar equipment, microwave ovens and industrial heaters. RFR at thermal levels causes a number of changes in the human eye, and mainly in the lens. At a very high level of the irradiation is so strong that it is not bearable for a human being. Most commonly experienced levels of irradiation will be when the heat sensation is negligible but if there is a risk for a temperature elevation in the lens then this will slowly result in tissue damage [7].

5. Experimental Studies

Since 1948 a number of experimental studies on rabbits, monkeys, dogs and rats have been performed. One of the purposes of these studies performed has been to estimate the power levels below which no cataracts or other lens changes can be induced by RFR. Guy and coworkers published the exact geometry of the near and far field irradiation [8].

The cataractogenic agents can also combine in the induction of cataracts. Experimental studies where subthreshold irradiation at microwaves has been used in diabetic rabbit, have shown that the cataract development accelerates and that some

106

cataractogenic agents can combine in the induction of cataract, even if each of them is given in a subcataractogenic dosage [9].

6. Epidemiological Studies

The induction of opacities and cataracts have been investigated in several case-control studies on workers exposed to high intensity of RFR, with contradictory results. The first documented incident report of cataracts was made in 1952 [10]. Zaret has reported a number of occurrences of cataracts in workers involved in radar [11]. Recently Bergqvist published a review of epidemiological studies relevant to ocular effects [12].

In this review eight studies have included endpoints related to ocular effects such as lens opacities and cataracts. After an examination of methodological aspects of these studies only six of theirs were included in the review.

Clearly and Pasternack performed a cross-sectional study of 1295 individuals working at 16 microwave installations. Non-exposed individuals were selected from the same location and environment [13]. The above mentioned authors related by a linear regression analysis the occurrence and/or intensity of some lens changes with an exposure score. The duration of microwave work and exposure score contributed significantly to the lens opacity.

Overall the another five studies are clearly not demonstrating an effect of RFR exposure on cataracts [12]. Further epidemiological studies of RFR-related ocular effects are needed [14, 15].

7. References

1. Tengroth, B.M. (1983) Cataractogenesis induced by RF and MW energy, in M. Grandolfo et al. (eds.), *Biological Effects and Dosimetry of Non-ionizing Radiation*, Plenum Press, New York and London, pp. 485-500.
2. Carpenter, R.L. and Van Ummerson, C.A. (1968) The action of microwave radiation on the eye, *J. Microwave Power* **3**, 3.
3. Radnót M. and Köteles G.J. (1969) The effect of irradiation on the eye lens, *Atomic Energy Review*.
4. Carpenter, R.L., Biddle, D.K. and Van Ummerson, C.A. (1960) Opacities in the lens of the eye experimentally induced by exposure to microwave irradiation, IRE Trans on Medical Electronics, **ME-7**, 152.
5. Schwan, H.P. (1957) Electrical properties of tissue and cell suspensions, in J.H. Lawrence and C.A. Tobias(eds.), *Advances in Biological and Medical Physics*, Academic Press, New York, pp. 147-209.
6. Bernhardt, J.H. and Vogel, E. (1984) Electromagnetic fields: Biophysical interaction mechanisms, in R.Matthes(ed.), *Non-ionizing Radiation*, Markl-Druck, München, pp. 230-244.
7. Michaelson, S.M. (1983) Biological effects and health hazards of RF and MW energy: fundamentals and overall phenomenology, in M. Grandolfo et al.(eds.), *Biological Effects and Dosimetry of Non-ionizing Radiation*, Plenum Press, New York and London, pp. 337-352.
8. Guy, W., Lin, J.C., Kramar, P. and Emery, A.F. (1975) Effects of 2450 MHz radiation on the rabbit's eye, *IEEE Trans. Microwave Theory* MTT-23.
9. Hackwin, O. and Kock, H.R. (1975) Combined noxious influences, in Bellows, J.G. (ed.) *Cataract and Abnormylities of the Lens*, Gune and Stratton, Inc. pp. 243-254.

10. Hirsch, F.G. and Parker, J.T. (1952) Bilateral lenticular opacities occurring in a technician operating a microwave generator, *A.M.A. Arch. Ind. Occup. Med.* **6**, 512-517.

11. Zaret, M.M. (1975) Blindness, deafness, and vesticular disfunction in a microwave worker, *Eye, Ear, Nose Throat Monthly* **54**, 291-294.

12. Bergqvist, U. (1997) Review of epidemiological studies, in Kuster, N., Balzano, Q. and Lin, J.C. (eds.) *Mobile Communications Safety*, Chapman and Hall, London, pp. 147-170.

13. Cleary, S.F. and Pasternack, B.S. (1966) Lenticular changes inn microwave workers. *Archives of Environmental Health* **12**, 23-29.

14. World Health Organization (1993) Electromagnetic fields (300 Hz – 300 GHz). *Environmental Health Criteria* **137**, Geneva, WHO.

15. Duchene, A., Lakey, J.R.A., Repacholi, M. (1991) *IRPA Guidelines on Protection against Non-ionizing Radiation.* Pergamon Press, New York.

SUMMARY OF SESSION C: THEORETICAL DOSIMETRY

O. P. GANDHI
Department of Electrical Engineering
University of Utah
Salt Lake City, UT 84112, U.S.A.

This session consisted of four papers with emphasis on numerical methods and models that can be used to obtain SAR distributions for electromagnetic safety assessment and for providing dosimetric information for biobehavioral animal experiments. In an overview of the developments in the field, it was noted that theoretical dosimetry has progressed tremendously from the earlier relatively crude block models with resolutions of 2-5 cm to present day models based on the MRI scans with resolutions on the order of 1-5 mm. This may be seen for models of the human body that have been developed at the University of Utah (paper by Gandhi), at NRPB (paper by Dimbylow) and the model based on scans of the "Visible Man" developed by Mason et al. described in their paper in this section. A much needed development (see the paper by Mason et al.) has been the creation of the heterogeneous, anatomically-based models of experimental animals such as a Sprague-Dawley rat, pigmy goat and rhesus monkey. Obtained from the MRI scans of these animals, these models are beginning to be used for dosimetric information for a variety of exposure conditions that are used for biological experiments.

The development of these extremely realistic anatomically-based models has, of course, been prompted by the availability of new, highly efficient numerical techniques such as the finite-difference time-domain (FDTD) method and faster computers with huge memories in the last 120 years. Today the FDTD is the method of choice for bioelectromagnetic computations as may be seen from all of the papers given in this session. Further improvement in this approach just beginning to appear in the bioelectromagnetics literature will involve use of a non-uniform rather than uniform grid so that the coupled region of interest may be modeled with higher submillimeter resolution while an expanding grid to larger grid dimensions or a uniformly coarser grid is used for the weakly coupled regions. This helps to reduce the overall number of voxels for which the EM fields are to be calculated thereby reducing the memory requirement for the computing work stations by a factor of 10 or more while retaining or even improving the resolution of modeling for the coupled region of interest.

It is impressive to see the applicability of these theoretical dosimetry tools to a wide range of exposure conditions from plane wave RF exposures (see e.g. papers by Gandhi and Dimbylow), to calculations of induced currents and electric fields at low

B.J. Klauenberg and D. Miklavcic (eds.), Radio Frequency Radiation Dosimetry, 109-110.
© 2000 *Kluwer Academic Publishers. Printed in the Netherlands.*

frequencies including power frequency EMFs, to RF and switched gradient magnetic fields of Magnetic Resonance Imagers (MRI) and a myriad of other exposure conditions. Were it not for these spectacular advances in numerical dosimetry, it would not be possible to use these techniques for SAR compliance testing and better designs of personal wireless devices that are being introduced into the society at a rapid pace.

Lest we become complacent about the developments in theoretical dosimetry, there are some unsolved problems that still remain. First and foremost is the problem of compliance testing for realistic, spatially non-uniform EM fields encountered in real life. Since these fields in real-life work environments are highly variable from one location to another, it becomes a major computer-intensive effort to evaluate if the EM fields thus encountered are compliant with the safety guidelines that are invariably set in terms of whole-body-averaged and peak local SARs. New tools sometimes partially approximate (see e.g. Rudakov) are, therefore, needed for a more rapid determination of compliance for such real-life exposure conditions.

NUMERICAL AND EXPERIMENTAL METHODS FOR DOSIMETRY OF RF RADIATION: SOME RECENT RESULTS

O. P. GANDHI
Department of Electrical Engineering
University of Utah
Salt Lake City, UT 84112, U.S.A.

1. Introduction

Considerable progress has been made in the development of numerical methods for dosimetry of exposure to electromagnetic fields from extra low frequencies (ELF) to microwave frequencies. Even though the finite-difference time-domain (FDTD) method has been used extensively, several other methods most notably the method of moments (MOM), the conjugate gradient-fast Fourier transform (CG-FFT) and the quasi-static impedance or admittance methods have also been used for a number of exposure conditions [1-3]. Concomitant with the development of these methods, increasing resolution anatomically-based models of the human body have also been developed, initially from anatomical sectional diagrams and, more recently, from MRI scans of living humans [1-5] or cadavers such as "visible man" and "visible woman" developed by National Library of Medicine. With the advent of larger memory PCs, computing work stations, and parallel processors, it is possible to run anatomically-based whole-body models with resolutions on the order of a few millimeters e.g. 2 × 2 × 3 mm resolution model of the human body [6] and submillimeter resolution models for partial body exposures e.g. head and neck for the case of wireless telephones [7]. Anatomically-based models of the commonly used animals such as the rat and rhesus monkey are also being developed [8] and used for dosimetric calculations because of the importance of such subjects for laboratory experiments [9].

Over the years, experimental phantoms have also been developed to understand coupling of EM fields to models of the biological bodies [10]. While most of these models do an excellent job of modeling the external shape of the exposed bodies, detailed modeling of the internal heterogeneities of the body is very difficult and has only been attempted on a very limited scale [11, 12] and in a relatively crude manner. Simple homogeneously-filled models have, therefore, been used more often [13-17]. Though these simplistic models are incapable of giving accurate SAR distributions, they do allow use of actual near-field sources such as wireless telephones including their internal structures. And, they have been shown to give peak 1-g SARs needed for SAR compliance testing within an uncertainty of ± 20 percent [15-17].

B.J. Klauenberg and D. Miklavcic (eds.), Radio Frequency Radiation Dosimetry, 111-121.
© 2000 *Kluwer Academic Publishers. Printed in the Netherlands.*

2. Some Recent Applications of Numerical Techniques for Dosimetry

2.1. CONDITIONS OF MAXIMUM ABSORPTION IN THE HEAD FOR PLANE-WAVE EXPOSURES [6]

It is important to know the frequencies and the polarization of incident waves for which maximum electromagnetic energy is absorbed in the head in general and brain in particular because of the biological effects that this may cause. We have used the FDTD method with the previously-described 1.974 × 1.974 × 3.0 mm resolution anatomically-based model of the human body [1-3] to pinpoint the conditions of maximum absorption in the head or the so-called head-resonance conditions. For a vertically-polarized plane wave, the resonant frequencies for the adult human head and neck are 207 MHz and 193 MHz for the isolated and grounded conditions of the model, with absorption cross sections that are 3.38 and 2.57 times the shadow cross section of this region, respectively [6]. Figure 1 gives the calculated variations of the whole-body, head and neck and brain averaged SARs with frequency for the isolated condition of the model of the human adult [9]. Table 1, on the other hand, gives the highlights of the calculated results both for the high resolution model as well as for the coarser model where 3 × 3 × 2 cells of the original model are combined to obtain an easier-to-run model with voxel dimensions of 5.922 × 5.922 × 6.0 mm along the three orthogonal axes. Also given in this Table 1 are the results calculated for scaled models of 10- and 5-year old children. As expected, the resonant frequencies both for whole-body resonance and for head resonance are higher for models of 10- and 5-year old children as compared to that for the adult human.

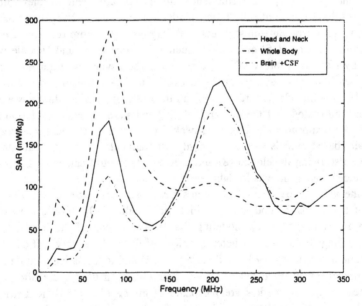

Figure 1. **E-Polarization,** isolated exposure - The whole-body, head and neck, and brain-averaged SARs for the adult human model

TABLE 1. Summary of the SARs and absorption cross section enhancement factors (S) calculated for models of the human adult and scaled models of 10- and 5-year old children [6]. Shown in parentheses are the resonant frequencies for which the various parameters are calculated

	1.974 × 1.974 × 3.0 mm adult model	5.922 × 5.922 × 6.0 mm adult model	Scaled 10-year old child model	Scaled 5-year old child model
Isolated Model				
Whole body averaged SAR (mW/kg)	282.4 (78 MHz)	283.0 (78 MHz)	354.7 (104 MHz)	377.4 (126 MHz)
Head and neck averaged SAR (mW/kg)	220.4 (207 MHz)	227.6 (207 MHz)	246.8 (263 MHz)	291.0 (322 MHz)
Whole body enhancement factor S	3.88 (78 MHz)	3.87 (78 MHz)	3.89 (104 MHz)	3.58 (126 MHz)
Head nd neck enhancement factor S	3.38 (207 MHz)	3.27 (207 MHz)	3.26 (263 MHz)	3.13 (322 MHz)
Grounded Model				
Whole body averaged SAR (mW/kg)	357.5 (47 MHz)	359.0 (47 MHz)	370.3 (65 MHz)	383.7 (73 MHz)
Head and neck averaged SAR (mW/kg)	169.0 (193 MHz)	182.4 (193 MHz)	219.7 (244 MHz)	235.5 (325 MHz)
Whole body enhancement factor S	4.91 (47 MHz)	4.92 (47 MHz)	4.10 (65 MHz)	3.63 (73 MHz)
Head and neck enhancement factor S	2.57 (193 MHz)	2.62 (193 MHz)	2.90 (244 MHz)	2.53 (325 MHz)

2.2. COMPARISON OF INDUCED 60 Hz EXOGENOUS FIELDS WITH ENDOGENOUS FIELDS IN THE HUMAN BODY

Power frequency electromagnetic fields (EMFs) are alleged to have deleterious effects on the human health. To understand potential mechanisms of interaction, it may be informative to compare the induced exogenous fields in the human body with endogenous fields in the various organs and regions of the body due to the electrophysiological functions occurring in a living human. For example, time-varying electrical potentials and current dipoles are associated with a beating heart, and response of the brain to various stimuli, and for neuromuscular activity. We have used [18] the measured time-domain potentials at 64 locations on the surface of a large canine heart, considered comparable to those of a human heart, to calculate the induced electric fields and current densities within the various regions and organs of the human body. For these calculations, we have used the previously described 1.974 × 1.974 × 3.0 mm resolution, 30 tissue-segmented model of the human body [1-3] which has been modified into a somewhat coarser model with voxel dimensions of 5.922 × 5.922 × 6.0 mm (nominal 6.0 mm resolution) by combining cells as described above. The admittance method described in the literature [19] with successive over-relaxation

(SOR) is used to calculate the voltage variation through torso and head as a function of time. From this time-domain voltage variation, values of electric field E(t) and current density J(t) are obtained allowing for the maximum values of these quantities to be calculated within the tissues of interest. Since the voltage variations, as well as the calculated E(t) and J(t), have an associated spectra with most of the energy in the frequency band 0-100 Hz, the frequency analysis is used to obtain E(f) and J(f) in the various organs so that average, minimum, and maximum values within specified bands 0-40, 40-70, and 70-100 Hz are obtained. To check the accuracy of the calculations, the computed variations of the surface potentials for the chest are compared with the EKG waveforms and isopotential surface maps. A good agreement is obtained for both the computed and measured results. The endogenous fields and current densities thus obtained for the various regions and organs of the body are compared with the corresponding induced quantities for power frequency EMFs proposed in the various safety guidelines [20, 21]. An electric field of 25 KV/m and a magnetic field of 1 mT have been proposed as the limit values in [20] and exposures with field strengths of 2.5-3.0 times smaller have been proposed as the safety guideline for occupational exposures at 60 Hz. We compare in Tables 2 and 3 the calculated organ-averaged endogenous electric fields and current densities with the induced (exogenous) E and J for the same organs for EMFs that may be encountered under high voltage transmission lines (E = 10 KV/m, B = 33 μT) and for uniform magnetic fields of 1 mT suggested as the limit value in [20]. Since the endogenous fields have a broader spectrum, comparison is made with |E(f)| and |J(f)| contained in the frequency band 40-70 Hz which brackets the 50/60 Hz power frequencies used in the various countries of the world. From Tables 2 and 3, one can see that the organ averaged exogenous fields for all the major organs, except for tissues in close proximity to and including the heart, are considerably larger than the endogenous fields in the same tissues due to the beating of the heart. This is a very interesting result which certainly needs further investigation.

2.3. SARS AND INDUCED CURRENTS FOR EXPOSURE TO TIME-VARYING MAGNETIC FIELDS OF MRI

Magnetic resonance imaging is becoming an increasingly important tool for medical diagnostic applications. Newer techniques are leading to the use of higher static magnetic fields, more rapidly switched gradient fields, and higher radiofrequency magnetic fields. Use of the increasingly stronger electromagnetic fields is causing concern about patient safety [22]. We have used the above-described 5.974 × 5.974 × 6.0 mm resolution, 30-tissue anatomically-based model of the human body to calculate the SARs and the induced current density distributions for RF and switched-gradient magnetic fields used for MRI, respectively [23]. For SAR distributions, the finite-difference time-domain (FDTD) method is used including modeling of 16-conductor birdcage coils and outer shields of dimensions that are typical of body and head coils and a new high-frequency head coil proposed for the 300-400 MHz band. The dimensions of the 16-rung birdcage coils considered for the SAR calculations are

TABLE 2. Comparison of organ-averaged endogenous and organ-averaged exogenous electric fields, in mV/m

Tissue	Endogenous 40–70 Hz	HV power line, $|E| = 10$ kV/m $|B| = 33\mu$ T	Uniform B(1 mT), polarized from front to back
Brain	0.04	11.1	11.5
CSF	0.03	3.2	2.8
Pineal gland	0.02	8.5	4.8
Pituitary gland	0.06	20.6	3.4
Eye Humor (2)	0.02	3.7	2.1
Lungs	5.73	8.1	27.7
Heart	25.4	4.4	12.4
Liver	2.15	11.4	25.1
Stomach	3.75	11.4	11.2
Pancreas	1.64	10.2	17.1
Intestine	0.52	12.2	16.6
Kidneys	0.54	10.2	13.8
Bladder	0.13	11.3	18.7
Prostate	0.10	9.6	21.6
Testicles	0.04	0.3	34.6

TABLE 3. Comparison of organ-averaged endogenous and organ-averaged exogenous current densities, in mA/m^2

Tissue	Endogenous 40–70 Hz	HV power line, $|E| = 10$ kV/m $|B| = 33\mu$ T	Uniform B(1 mT), polarized from front to back
Brain	0.006	1.9	2.0
CSF	0.018	4.8	4.6
Pineal gland	0.004	1.5	0.8
Pituitary gland	0.002	3.5	0.6
Eye Humor (2)	0.016	5.6	3.5
Lungs	0.46	0.6	2.5
Heart	11.1	2.2	6.2
Liver	0.26	1.4	3.3
Stomach	0.41	1.3	5.6
Pancreas	0.22	1.5	2.0
Intestine	0.051	1.3	8.3
Kidneys	0.12	2.8	5.9
Bladder	0.020	1.9	3.7
Prostate	0.010	1.1	2.4
Testicles	0.004	0.7	3.8

116

given in Table 4 and the salient features of the calculated SARs for these coils are given in Table 5. Since a variety of pulse sequences are used for various imaging modalities, a duty cycle of 1/25 typical of some of these procedures is used for the SARs given in Table 5. SARs at 64, 128, and 170 MHz have been found to increase with frequency (f) as f^k where k is on the order of 1.1-1.2. Since the FDA (U.S.) and NRPB (U.K.) safety guidelines are set up in terms of the peak SARs for any 1 kg of tissue [24, 25], the data in Table 5 may be used to calculate the maximum

RF coil currents and/or magnetic fields or duty cycles that must not be exceeded to stay within the safety guidelines.

TABLE 4. The dimensions of the 16-rung birdcage coils considered for the SAR calculations

	Body coil A	Head coil B	High frequency head coil C
Diameter of the cage with rungs (cm)	56.0	28.5	30.6
Diameter of the shield (cm)	65.5	65.5	36.4
Length of the rungs (cm)	56.0	39.3	19.0
Length of the shield (cm)	56.0	56.0	21.2
Frequencies (MHz)	64, 128, 170	64, 128, 170	300, 350, 400

This anatomically-based model of resolution 5.974 × 5.974 × 6.0 mm has also been used to calculate the currents induced in the human body due to switched-gradient magnetic fields. Because of the low frequencies associated with the switched gradient magnetic fields, the quasi-static impedance method developed by our laboratory [26] has been used for calculation of induced current densities that are compared with the following induced current density limits (J) suggested to prevent muscular reactions in the NRPB safety guidelines [25]

$$J \leq 400 \frac{mA}{m^2} \, for \tau \geq 120 \mu s \tag{1}$$

$$J\tau \leq 48 \times 10^{-3} mA \frac{\sec}{m^2} \, for \tau < 120 \mu s \tag{2}$$

where τ is the duration of the induced current pulse in seconds. For an applied gradient magnetic field for maximum dB/dt = 22 T/s produced by a Maxwell pair of single turn loops of time duration τ = 100 μs, we calculate maximum induced current densities (J) as high as 386 mA/m² [23]. From Equation 2, somewhat higher dB/dt on the order of 27.4 T/s could thus be used for τ = 100 μs switched gradient fields. This also compares

TABLE 5. Salient features of the FDTD-calculated data for the 16-rung birdcage coils A, B, C of Table 4. Assumed for the calculations is a current of 1.0 A (RMS) for each of the rungs which are fed with a progressive phase shift of 22.5° to obtain circular polarization. A duty cycle of 1/25 is assumed for the SAR calculations.

Frequency (MHz)	Body Coil A			Head Coil B			Head Coil C		
	64	128	170	64	128	170	300	350	400
	Empty Coils								
H⊥ center with no shield (A/m)	6.03	6.52	6.87	12.44	12.42	12.62	8.10	8.36	8.50
H⊥ center from Biot-Savart's Law (A/m)	6.43	6.43	6.43	12.62	12.62	12.62	11.77	11.77	11.77
H⊥ center with outer shield (A/m)	2.57	2.60	2.64	10.49	10.75	11.13	3.58	3.67	3.85
	Calculated SARs (W/kg) for the 5.922 × 5.922 × 6.0 mm resolution model of the human								
Whole body averaged SAR	0.14	0.25	0.44	0.19	0.35	0.55	0.15	0.15	0.13
Maximum SAR for 1 kg* tissue	0.37 (1.0 kg)	0.83 (0.98 kg)	2.17 (0.98 kg)	3.47 (1.0 kg)	4.72 (0.96 kg)	6.78 (1.03 kg)	2.67 (1.11 kg)	2.57 (1.11 kg)	2.20 (1.13 kg)
Maximum SAR for 100 g* tissue	1.52 (101 g)	2.13 (90 g)	4.37 (102 g)	7.21 (100 g)	7.92 (96 g)	9.90 (10 g)	4.17 (116 g)	3.88 (112 g)	3.29 (115 g)

* Actual weights given in parentheses

favorably with $dB/dt \leq 2400/\tau(\mu s)$ for $12 \leq \tau \leq 120$ μs suggested in the FDA safety guidelines [24].

3.   SARs and Radiation Patterns of Handheld Wireless Telephones

The FDTD method has been used for calculations of SAR distributions and far-field radiation patterns of personal wireless telephones [1, 2, 7, 17, 27-32]. Some of the recent developments [17] pertain to the use of Pro-Engineer CAD files of cellular telephones for realistic description of the device, and expanding grid formulation of the FDTD method [33] for finer resolution and a more accurate representation of the coupled region including the antenna. Since the expanding grid method allows an increasingly coarser representation of the more-distant less-coupled region, together with the truncation of the model of the head, this procedure leads to a factor of 20 saving of the computer memory needed for SAR calculations. Automated SAR and radiation pattern measurement systems are used to validate both the calculated 1-g SARs and radiation patterns for several telephones including some research test samples. Given in Table 6 is the comparison of the numerical and measured peak 1-g SARs for 10 telephones, five each at 835 and 1900 MHz, respectively. The 1-g SARs for these telephones which use different operational modes (AMPS, TDMA, CDMA, etc.) and antenna structures (helical, monopole, or helix-monopole) are found to vary widely; for our test samples, from 0.13 to 5.41 W/kg. Nevertheless, there is an excellent agreement (within ±20% or 1 dB) between calculated and measured values. Similarly, there is also a good agreement between calculated and measured values of the radiation patterns of the telephones held against the model of the human head [32].

TABLE 6. Comparison of the measured and FDTD-calculated peak 1-g SARs for 10 wireless devices, five each at 835 and 1900 MHz, respectively.

Cellular Telephones at 835 MHz

	Time-averaged radiated power mW	Experimental method W/kg	Numerical method W/kg
Telephone A	600	4.02	3.90
Telephone B	600	5.41	4.55
Telephone C	600	4.48	3.52
Telephone D	600	3.21	2.80
Telephone E	600	0.54	0.53

PCS Telephones at 1900 MHz

	Time-averaged radiated power mW	Experimental method W/kg	Numerical method W/kg
Telephone A'	125	1.48	1.47
Telephone B'	125	0.13	0.15
Telephone C'	125	0.65	0.81
Telephone D'	125	1.32	1.56
Telephone E'	99.3	1.41	1.25

4. Concluding Remarks

From the foregoing, it can be seen that the numerical methods have reached a level of maturity and are thus being applied for a myriad of bioelectromagnetic applications. Numerical methods may, for example, be used to design applicators both for RF and for switched-gradient magnetic fields for MRI. For personal wireless devices, the numerical methods offer an advantage in that several alternative antenna designs may be evaluated during the design stage to define the designs with the lowest SARs and hence better radiation efficiencies since 44-50% of the power is otherwise wasted by absorption in the head, neck, and hand [27-30].

5. References

1. Gandhi, O.P. (1998) FDTD in Bioelectromagnetics: Safety assessment and medical applications, in A. Taflove (ed.) *Advances in Computational Electrodynamics: The Finite-Difference Time-Domain Method*, Artech House, Inc., Norwood, MA, pp. 613-651.
2. Gandhi, O.P. (1995) Some numerical methods for dosimetry: Extremely low frequencies to microwave frequencies, *Radio Science* **30**, 161-177.
3. Lin, J.C. and Gandhi, O.P. (1996) Computational methods for predicting field, in C. Polk, an E Postow (eds) *Biological Effects of Electromagnetic Fields*, 2nd Edition, CRC Press Inc., Boca Raton, FL, pp.337-402.
4. Dimbylow, P.J. (1996) Development of realistic voxel phantoms for electromagnetic field dosimetry, 1-7, in P. J. Dimbylow (ed.) *Voxel Phantom Development, Proceedings of an International Workshop*, National Radiological Protection Board, Chilton, Didcot, U.K.

5. Zubal, I.G., Harrell, C.R., Smith, E.O., Rattner, Z., Gindi, G.R. and Hoffer, P.H. (1994) Computerized three-dimensional segmented human anatomy, *Med. Phys. Biol.*, **21**, 299-302.

6. Tinniswood, A.D., Furse, C.M., and Gandhi, O.P. (1998) Power deposition in the head and neck of an anatomically-based human body model for plane wave exposures, *Physics in Medicine and Biology* **43**, 2361-2378.

7. Tinniswood, A.D., Furse, C.M., and Gandhi, O.P. (1998) Computations of SAR distributions for two anatomically-based models of the human head using CAD files of commercial telephones and the parallelized FDTD code, *IEEE Transaction on Antennas and Propagation* **46**, 829-833.

8. Mason, P.A., Ziriax, J.M., Hurt, W.D., Walters, T.J., Ryan, K.L., Nelson, D.A., and D'Andrea, J.A. Recent advances in dosimetry measurements and modeling, this book.

9. Tinniswood, A. and Gandhi, O.P., (1999) Head and neck resonance in a rhesus monkey -- A comparison with results from a human model, *Physics in Medicine and Biology* **44**, 695-704.

10. Stuchly, M.A. and Stuchly, S.S. (1996) Experimental radiowave and microwave dosimetry, in C. Polk and E. Postow (eds.) *Handbook of Biological Effects of Electromagnetic Fields*, Second Edition, CRC Press Inc., Boca Raton, FL, pp.301-342.

11. Stuchly, M.A., Kraszewski, A., Stuchly, S.S., Hartsgrove, G.W., and Spiegel, R.J. (1987) RF energy deposition in a heterogeneous model of man: Near-field exposures, *IEEE Trans. Biomedical Eng*, **34**, 12.

12. Cleveland, R.F. and Athey, T.W. (1989) Specific absorption rate (SAR) in models of the human head exposed to UHF portable radios, *Bioelectromagnetics* **10**, 173-186.

13. Chatterjee, I., Gu, Y.G., and Gandhi, O.P. (1985) Quantification of electromagnetic absorption in humans from body-mounted communication transceivers, *IEEE Transactions on Vehicular Technology*, **34**, 55-62.

14. Stuchly, M.A., Spiegel, R.J., Stuchly, S. S., and Kraszewski, A. (1986) Exposure of man in the near field of a resonant dipole: Comparison between theory and measurements, *IEEE Transactions on Microwave Theory and Technique* **10**, 173-186.

15. Balzano, Q., Garay, O., and Manning, T. (1995) Electromagnetic energy exposure of simulated users of portable cellular telephones, *IEEE Transactions on Vehicular Technology* **44**, 390-403.

16. Schmid, T., Egger, O., and Kuster, N. (1996) Automated E-field scanning systems for dosimetric assessments, *IEEE Transactions on Microwave Theory and Techniques* **44**, 105-113.

17. Gandhi, O.P., Lazzi, G., Tinniswood, A., and Yu, Q.S. (1999) Comparison of numerical and experimental methods for determination of SAR and radiation patterns of handheld wireless telephones, *Bioelectromagnetics, Supplement* **4**, 93-101.

18. Hart, R.A. and Gandhi, O.P. (1998) Comparison of cardiac-induced endogenous fields and power frequency induced exogenous fields in an anatomical model of the human body, *Physics in Medicine and Biology* **43**, 3083-3099.

19. Armitage, D.W., LaVeen, H.H., and Pethig, R. (1983) Radiofrequency induced hyperthermia: Computer simulation of specific absorption rate distribution using realistic anatomical models, *Physics in Medicine and Biology* **28**, 31-42.

20. ACGIH, *1996-97 Threshold Limit Values and Biological Exposure Indices*, available from the American Conference of Governmental Industrial Hygienists (ACGIH), Technical Affairs Office, Kemper Woods Center, Cincinnati, Ohio 45240.

21. ICNIRP (1998) Guidelines for limiting exposure to time-varying electric, magnetic, and electromagnetic fields (up to 300 GHz), *Health Physics* **74**, 494-522.

22. Magin, R.L., Liburdy, R.P., and Persson, B. (eds.) (1992) *Biological Effects and Safety Aspects of Nuclear Magnetic Resonance Imaging and Spectroscopy*, **649**.

23. Gandhi, O.P. and Chen, X.B. (1999) SARs and induced currents for an anatomy-based model of the human for exposure to time-varying magnetic fields of MRI, *Magnetic Resonance in Medicine* **41**, 816-823.

24. Athey, T.W. (1992) Current FDA guidance for MR patient exposure and considerations for the future, *Annals of the New York Academy of Sciences* **649**, 242-257.

25. NRPB, *Board Statement on Clinical Magnetic Resonance Diagnostic Procedures*, **2**, National Radiological Protection Board, Chilton, Didcot, Oxon OX110RQ, U.K.

26. Gandhi, O.P., DeFord, J.F., and Kanai, H. (1984) Impedance method for calculation of power deposition patterns in magnetically-induced hyperthermia, *IEEE Transactions on Biomedical Engineering* **BME-31**, 644-651.

27. Dimbylow, P.J. and Mann, S.M. (1994) SAR calculations in an anatomically-based realistic model of the head for mobile communication transceivers at 900 MHz and 1.8 GHz, *Physics in Medicine and Biology* **39**, 1537-1553.

28. Jensen, M.A. and Rahmat-Samii. (1995) EM interaction in handset antennas and a human in personal communications, *Proc. IEEE* **83**, 7-17.

29. Okoniewski, M. and Stuchly, M.A. (1996) A study of handset antennas and human body interaction, *IEEE Transactions on Microwave Theory and Techniques* **44**, 1855-1864.

30. Gandhi, O.P., Lazzi, G., and Furse, C.M. (1996) Electromagnetic absorption in the human head and neck for mobile telephones at 835 and 1900 MHz, *IEEE Transactions on Microwave Theory and Techniques* **44**, 1884-1897.

31. Luebbers, R., Chen, L., Uno, T., and Adachi, S. (1992) FDTD calculations of radiation patterns, impedance and gain for a monopole antenna on a conducting box, *IEEE Transactions on Antennas and Propagation* **40**, 1577-1582.

32. Lazzi, G., Pattnaik, S.S., Furse, C.M., and Gandhi, O.P. (1998) Comparison of FDTD-computed and measured radiation patterns of commercial mobile telephones in presence of the human head, *IEEE Transactions on Antennas and Propagation* **46**, 943-944.

33. Gao, B.Q. Gandhi, O.P. (1992) An expanding grid algorithm for the finite-difference time-domain method, *IEEE Transactions on Electromagnetic Compatibility* **34**, 277-283.

ELECTROMAGNETIC FIELD CALCULATIONS IN AN ANATOMICALLY REALISTIC VOXEL MODEL OF THE HUMAN BODY

P. J. DIMBYLOW
National Radiological Protection Board
Chilton
Didcot
OX11 0RQ
England

1. Introduction

This paper presents a profile of recent work at the National Radiological Protection Board on electromagnetic field dosimetry using an anatomically realistic voxel model of the human body. This phantom consists of ~ 9 million, 2 mm voxels segmented into 37 tissue types. The exact dimensions of the voxels were scaled so that the height (1.76 m) and the mass (73 kg) would agree with the new values of reference man in ICRP 66. Hence the phantom is known as NORMAN (**normalized man**).

The Finite-Difference Time-Domain (FDTD) method has been used to calculate the whole-body averaged SAR from 1 MHz to 1 GHz for plane wave irradiation of the adult phantom and for scaled versions representing 10, 5 and 1 year old children. It is not computationally tractable to perform FDTD calculations directly at a cell size of 2 mm. Therefore, the phantom was rescaled to produce 6 mm, 1 and 2 cm models depending on the frequency. Coupled calculations, for localized SAR, have also been performed when the source and body have to be modeled together such as for mobile 'phones and radio man-packs.

At lower frequencies, where the quasistatic approximation is valid, the electromagnetic field problem can be reduced from a vector to a scalar representation and this enables calculations to be performed at the 2 mm resolution of the fully detailed phantom. Cells of discrete tissue type rather than a weighted mixture of properties required for larger cell sizes can be used. This is important for distinguishing the current densities in various tissue types, in particular the brain, spinal cord and retina. Induced current densities for both applied low frequency magnetic and electric fields have been calculated.

This scalar potential method has also been applied to calculate localized SAR in the ankle as a function of current through the ankle.

123

B.J. Klauenberg and D. Miklavcic (eds.), Radio Frequency Radiation Dosimetry, 123-131.

2. Development of the Voxel Phantom, 'NORMAN'

The raw MRI data were taken from a series of continuous partial body scans of a single subject who was scanned in 6 blocks, i.e., head, thorax, abdomen, thighs, knees and feet. The blocks of data were conjoined by rescaling, translation and rotation for form an entire body. The 8-bit grey scale images were segmented unambiguously as belonging to one of 37 different tissue types [1]. Wherever possible this segmentation was carried out semi-automatically using grey scale thresholding. In most cases this was a reasonably straightforward process but in a few situations, e.g. the thyroid, the organ had to be inserted manually using anatomical textbooks. The tissue types are skin, fat, muscle, tendon, bone, trabecular bone, blood, brain, spinal cord, cerebrospinal fluid, eye - lens, oesophagus, stomach - contents, duodenum, small intestine, lower large intestine, upper large intestine, pancreas, gall bladder, bile, liver, spleen, kidney, bladder, urine, prostate, testis, male breast, thymus, thyroid, adrenals, heart, lung, air and background domain.

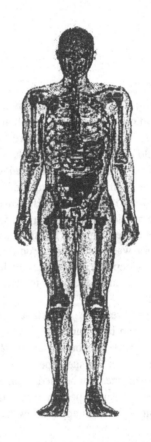

Figure 1. An illuminated 3D image of NORMAN

An evaluated review of the dielectric properties of all the tissue types in NORMAN was performed by Gabriel [2, 3, 4, 5]. A 4-Cole-Cole dispersion model was fitted to the data for each tissue type to parameterize the conductivity and permittivity as a function of frequency.

The adult phantom was normalized to be 1.76 m tall and to have a mass of 73 kg, the new values for 'reference man' in ICRP 66 [6]. The height fixes the vertical voxel dimension and the horizontal dimensions are then fixed by the mass. The total domain of the phantom and adjacent air is a 3D array of 148 voxels from front to back, 277 from side to side and 871 voxels high.

Figure 1 shows an illuminated, rendered 3D image of NORMAN. Each voxel greyscale has an associated opacity which can be varied from 0 to 1 to make specific tissues more transparent so that you can 'look through' the body. In this case only the skin, skeleton and digestive system are shown; all other organs have been made transparent. The opacity of the skin and bones has been set to 0.5.

3. Whole-Body Averaged SAR

The FDTD method [7] has been applied to an anatomically realistic model of the body to provide a comprehensive set of whole-body averaged SAR values for adult, 10-y, 5-y and 1-y old phantoms; grounded and isolated in air from 1 MHz to 1 GHz for plane wave exposure [8]. External electric field values can be derived from the whole-body averaged SAR values corresponding to particular restrictions on that quantity. These investigation levels can then be used to assess compliance with these restrictions on exposure for plane wave irradiation under far field conditions.

The perfectly matched layer (pml) based boundary conditions of Berenger [9] were implemented with an optimized geometric grading condition [10] of the layer absorption. A Huygens surface was implemented in the FDTD code to allow the description of arbitrary incident fields, to separate the scattered field that is required for the boundary conditions from the total field required for the FDTD formulation and also to connect the pml layers to the inner region of the domain. Electric and "magnetic" currents are defined on the Huygens surface which produce the correct total fields inside the surface but just the scattered fields outside the surface.

It is not computationally tractable to perform FDTD calculations directly at a cell size of 2 mm. Therefore, the phantom was rescaled to produce 6 mm, 1 cm and 2 cm models with the properties of the rescaled cells being taken as the volume average of the basic component voxels. As the frequency increases smaller cell sizes are required so that an adequate sampling of the waveform is performed. The period of the wave is proportional to the inverse of the frequency and so at the lower frequencies more time steps are required. Therefore, 2 cm resolution was used at the lowest frequencies whilst 6 mm was used at the highest and 1 cm at intermediate frequencies.

Figure 2 shows the whole-body averaged SAR as a function of frequency when the phantom is grounded through the feet. The whole body resonance can clearly be seen. When the phantom is isolated in air, this occurs when the height of the phantom is ~

$\lambda/2$, where λ is the wavelength in air. When the phantom is grounded, the reflection in the ground plane halves the resonant frequency, i.e. the height $\sim \lambda/4$. However, the body is quite a 'fat', irregularly shaped antenna and the distribution of SAR in the body depends also on the anatomy and frequency dependent dielectric properties and so the above conditions are approximate guidelines. The calculated resonant frequencies for the adult, 10-, 5- and 1-year old phantoms are 35, 55, 70 and 100 MHz when grounded.

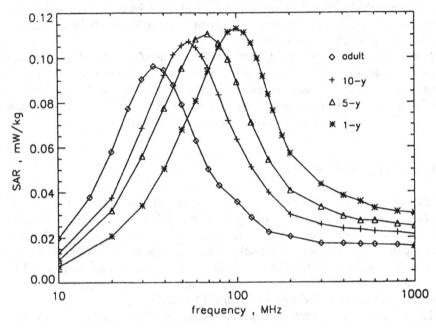

Figure 2. **The whole-body averaged SAR when the phantom is grounded for an incident electric field of 1 V/m (rms).**

4. Localised SAR from Coupled Sources

An important application of the computation of the interaction of electromagnetic fields with people is the evaluation of the power deposition in the head from hand-held telephones. FDTD calculations of SAR have been performed [11] in the head with a 2 mm resolution for a generic mobile communication transceiver represented by a quarter-wavelength monopole on a metal box. The antenna was mounted either at the centre or corner of the top face of the box. The frequencies examined were 900 MHz and 1.8 GHz. Three irradiation geometries were considered, viz - a vertical handset in front of the eye and vertical and horizontal orientations at the side of the ear. The maximum SAR values produced by the generic transceiver for the horizontal orientation at the side of the head which is the most typical position, averaged over 10 g of tissue at 900 MHz and 1.8 GHz are 2.1 and 3.0 Wkg^{-1} per W of radiated power.

However, if one were to consider all possible operational conditions, the placement of the transceiver in front of the eye will give 3.1 and 4.6 Wkg^{-1} per W averaged over 10 g of tissue at 900 MHz and 1.8 GHz, respectively.

A somewhat larger coupled problem has been to calculate the SAR from a radio man-pack from 30 MHz to 75 MHz. In this case the whole of the body and the back mounted radio plus whip antenna had to be modeled together at a 1 cm resolution.

5. SAR in the Ankles

The ankle region has a narrow cross-section and contains little high conductivity muscle, comprising mainly of low conductivity bone, tendon and fat. Consequently there is a channeling of the current through the high conductivity muscle, which produces high localized values of SAR. The cells in the FDTD calculations were rescaled to 6 mm, 1 cm and 2 cm depending on the frequency considered. This procedure will underestimate the local maximum SAR in the ankle because of the blending of high and low conductivity tissues. To overcome this, quasistatic calculations of the SAR in the ankle at the basic 2 mm resolution of NORMAN for a unit injected current at the top of the tibia were performed [12].

The finite-difference method for the solution of the quasistatic potential equation is described in detail in [13]. A finite-difference method was used to define the potential, at nodes where cells meet in terms of potentials, at the 6 neighboring nodes and averaged conductivities, along the edges between the central and neighboring nodes. An initial solution is defined on the 3D mesh of cells describing the leg. The solution is then iteratively refined using the computational molecule and the associated boundary conditions.

The bottom 220 slices of the whole body model, NORMAN were extracted to make a voxel model of the lower right leg. The voxels are 2.0207 mm high and have horizontal dimensions of 2.077 mm. In total there are $\sim 2.8 \times 10^6$ voxels in the leg model which comprises of skin, fat, blood, muscle, trabecular bone, cortical bone and tendon. The narrowest section of the leg contains 80 voxels to produce a cross-sectional area of 34.56 cm^2.

The original segmentation of NORMAN grouped together fat, tendons, ligaments as a general background tissue. However, it is important to differentiate between fat and tendons/ligaments because the dielectric property review has revealed that fat has very low conductivity whilst tendon and ligaments are about half-way between fat and muscle. The generalized connective tissue was replaced by tendon around the ankles and wrists except for a 1 cell-thick layer of subcutaneous fat.

128

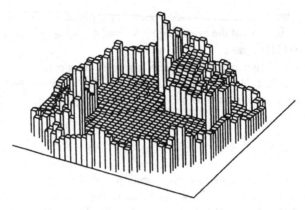

Figure 3. **Histogram of the SAR across a slice near the base of the tibia and fibula. The front of the leg faces the bottom left -hand corner.**

Figure 3 shows a histogram of the SAR across a slice near the base of the tibia and fibula. The front of the leg faces the bottom left-hand corner. The SAR is proportional to conductivity and the low values in the bone can clearly be seen with the maximum values in the two muscle groups, fore and aft, surrounding the bones.

6. Induced Current Densities from Low Frequency Fields

The current density in a fine resolution (2 mm) anatomically realistic voxel model of the human body for uniform magnetic fields has been investigated [14]. Both the impedance method and the scalar potential finite difference method were used to provide mutual corroboration.

The outer layer of the eye in NORMAN was not differentiated from the humour at low frequencies. The retina is important because of the induction of phosphenes. Therefore, because the humour and sclera have quite different conductivities the outer layer of the eye was reclassified as sclera. The rear part of this shell was then considered to be the retina for the analysis of induced current density.

In the impedance method [15], the target body is split into cuboid cells, which are then further differentiated into a 3-D network of impedances. The Scalar Potential Finite Difference (SPFD) method [16, 17] incorporates the applied magnetic field source as a vector potential term in the electric field. This equation for the electric field is then transformed into a scalar potential form which is then solved using finite differences.

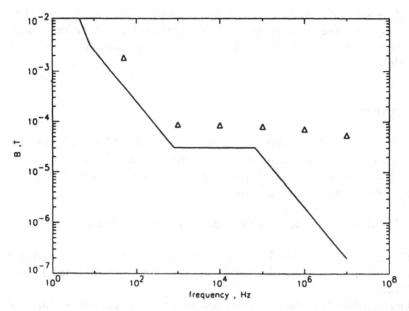

Figure 4. **Magnetic field values (symbols) required to produce the ICNIRP current density restrictions for occupational exposure. The line is the ICNIRP reference level.**

The highest values of current density will tend to be at the boundary of the section normal to the incident magnetic field. However, this pattern will be modified by the conductivity of the tissues in the section and current flow from adjacent regions. High values of current density occur around the edge of the thorax in the intercostal muscles, neck, urine and intestine. The skin, subcutaneous fat, ribs and lung have low conductivity and so the current density is relatively low despite their position on the periphery. There is an enhancement of current density at the top of the legs due to the shape effect whereby an indentation such as at the armpits and at the top of the legs or in a particularly high conductivity tissue will produce a tightening of the 'current lines' and an increase in the current density. Therefore there is a need to average over a finite area and indeed ICNIRP [18] has chosen 1 cm^2. The ICNIRP guidelines intend to avoid the effects of induced electric currents on functions of the central nervous system such as the control of movement and posture, memory, reasoning and visual processing. Therefore, when investigating maximum current densities in the body in relation to standards, it is appropriate to consider only the brain + spinal cord and retina. Figure 4 shows the derived external magnetic flux density levels that would produce the ICNIRP recommended current density restrictions for occupational exposure averaged over 1 cm^2 in brain, spinal cord or retina. In all cases they are larger than the ICNIRP reference levels, so the latter provide conservative estimates of the induced current density.

Work is in progress to calculate induced current densities in the 2 mm resolution voxel phantom, NORMAN for low frequency electric fields. The method employs the finite difference solution of the quasistatic potential equation in a series of nested grids. As one progresses away from the body the steplength of each grid doubles so that the perturbation in the applied field at the outer boundary produced by the body is negligible.

7. Conclusions

The anatomically realistic, fine resolution voxel model, NORMAN linked with appropriate numerical methods to solve the electromagnetic field equations provide a comprehensive toolcase to solve problems in non-ionizing radiation dosimetry. Examples of applications undertaken are -

The FDTD method has been used at higher frequencies to calculate-

- the whole-body averaged SAR from 1 MHz to 1 GHz for plane wave irradiation of the adult phantom and for scaled versions representing 10, 5 and 1 year old children.
- localized SAR from closely coupled sources such as mobile 'phones from 900 to 1800 MHz and radio man-packs from 2 to 75 MHz.

At lower frequencies, where the quasistatic approximation is valid, scalar potential formulations have been performed at the 2 mm resolution of the fully detailed phantom to calculate-

- induced current densities for uniform low frequency magnetic fields from 50 Hz to 10 MHz.
- induced current densities for uniform low frequency electric fields from 50 Hz to 10 MHz.
- localized SAR in the ankle as a function of current through the ankle from 0.1 to 80 MHz.

8. Acknowledgement

The author wishes to acknowledge the partial financial support of the UK Ministry of Defence (MOD).

9. References

1. Dimbylow P. J. (1996) The development of realistic voxel phantoms for electromagnetic field dosimetry, *Proc Int Workshop on Voxel Phantom Development: National Radiological Protection Board Report*, pp. 1-7.
2. Gabriel C. (1995) Compilation of the dielectric properties of body tissues at RF and microwave

frequencies, Report prepared for the NRPB by Microwave Consultants Ltd.

3. Gabriel C., Gabriel S., and Corthout E. (1996a) The dielectric properties of biological tissues: 1. Literature Survey, *Phys. Med. Biol.* **41**, 2231-2249.

4. Gabriel S., Lau R. W., and Gabriel C. (1996b) The dielectric properties of biological tissues: 2. Measurements in the frequency range 10 Hz to 20GHz, *Phys. Med. Biol.* **41**, 2251-2269.

5. Gabriel, S., Lau, R. W., and Gabriel C. (1996c) The dielectric properties of biological tissues: 3. Parametric models for the dielectric spectrum of tissues, *Phys. Med. Biol.* **41**, 2271-2293.

6. ICRP (1994) *Human Respiratory Tract Model for Radiological Protection*, ICRP Publication 66, table B.6, p. 189.

7. Taflove, A. (1995) *Computational Electromagnetics - The Finite-Difference Time-Domain Method*, Artech House, London.

8. Dimbylow P. J. (1997) FDTD calculations of the whole-body averaged SAR in an anatomically realistic voxel model of the human body from 1 MHz to 1 GHz, *Phys. Med. Biol.* **42**, 479-490.

9. Berenger, J. P. (1994) A perfectly matched layer for the absorption of electromagnetic waves, *J. Comp. Phys.* **114**, 185-200.

10. Berenger, J. P. (1996) Perfectly matched layer for the FDTD solution of wave-structure interaction problems, *IEEE Trans. Antennas and Propagat.* **44**, 110-117.

11. Dimbylow, P. J. and Mann, S. M. (1994) SAR calculations in an anatomically realistic model of the head for mobile communication transceivers at 900 MHz and 1.8 GHz, *Phys. Med. Biol.* **39**, 1537-1553.

12. Dimbylow, P. J. (1997b). The calculation of localized SAR in a 2 mm resolution anatomically realistic model of the lower leg, *Radiat. Prot. Dosim.* **72**, 321-326.

13. Dimbylow, P. J. (1988) The calculation of induced currents and absorbed power in a realistic, heterogeneous model of the lower leg for applied electric fields from 60 Hz to 30 MHz, *Phys. Med. Biol.* **33**, 1453-1468.

14. Dimbylow, P. J. (1998) Induced current densities from low frequency magnetic fields in a 2 mm resolution anatomically realistic model of the body, *Phys. Med. Biol.* **43**, 221-230.

15. Gandhi O. P., DeFord J. F., and Kanai, H. (1984) Impedance method for calculation of power deposition patterns in magnetically induced hyperthermia, *IEEE Trans. Biomed. Eng.* **31**, 644-651.

16. Dawson, T. W., De Moerloose, J., and Stuchly, M. A. (1996) Comparison of magnetically induced ELF fields in humans computed by FDTD and scalar potential FD codes, *ACES Journal* **11**, 63-71.

17. Dawson T. W. and Stuchly, M. A. (1997) A comparison of analytical and numerical solutions for induction in a sphere with equatorially varying conductivity by low-frequency uniform magnetic fields of arbitrary orientation, *Proc of 1997 Symposium of the Applied Computational Electromagnetics Society*, 533-540.

18. ICNIRP (1998) Guidelines for limiting exposure to time-varying electric, magnetic, and electromagnetic fields (up to 300 GHz), *Health Physics* **74**, 494-522.

THEORETICAL METHODS OF EVALUATION OF ABSORBED DOSE OF NON-IONIZING RADIATION IN NONUNIFORM MEDIA

M. L. RUDAKOV
Baltic State Technical University
198005, St-Petersburg
1-St Krasnoarmeyskaya St., Bld.1, Dept. A9
Russian Federation

1. Introduction

The significant scientific and public interest focused on the problem of electromagnetic safety, as well as technical development and social and political changes in the world require further improvement of corresponding standards and methods of exposure analysis.

Nowadays a lot of national and international standards exist. As for RF range, the most evident differences are between European and American standards on one side and the standards of ex-soviet countries, with Russia as a typical representative, on the other side [1, 2]. The most complete western standards such as ANSI, VDE, CENELEC prestandard and others [3-6] take into account well established and strongly pronounced biological effects (overheating or stimulation of excitable tissues). Practically, the western ideology of standardization is based on short-term hazardous action of EM field. For RF range reference is made to thermal effect due to induced currents and dielectric losses (for frequencies less than about 0.1 MHz the effect of tissue stimulation becomes more critical). As soon as basic parameters (induced current density or SAR) are chosen and basic restrictions are established, the permissible levels of incident EM field (E-, H-field intensities, power density) may be derived with the help of physical and mathematical modeling in wide frequency range. Principally this approach allows to "transform" biological problem of interaction with EM field into problem of current and SAR distribution identification within biological body.

The RF hygienic standards of Russia are set on the basis of complex studies. While setting of permissible exposures they take into consideration also the long-time influence of effects of RF exposure, including minor ones, which may occur at non-thermal levels. Also, the long-term exposure to weak EM fields supposed to lead the cumulating of biological effect that may cause specific diseases. The "eastern" approach may be regarded as more profound from point of view of safety, but in practice it is much more complicated – it demands a large set of medical and biological researches, it does not permit us to interpolate/extrapolate data obtained to other exposure conditions

133

B.J. Klauenberg and D. Miklavcic (eds.), Radio Frequency Radiation Dosimetry, 133-140.
© 2000 *Kluwer Academic Publishers. Printed in the Netherlands.*

(next frequency range, for instance). As a result of such approach the permissible levels of field intensities in Russian standards are locally constant in frequency ranges of their validity with "steps" up to ten times at the boundaries of adjanced ranges.

In hygienic practice sometimes it is necessary to evaluate SAR distribution in body organs and biological tissues without any reference to corresponding standards (hyperthermia in medical applications, industrial safety in the workplaces etc.). Also, sometimes the specialist may not need highly detailed or accurate SAR data. Thus, for such situations we need approximate methods of SAR calculation that allow us to estimate very rapidly body heating and/or labor conditions.

This report contains the following topics:
1. Description of a proposed approximate method.
2. Empirical expressions for dielectric parameters of biological tissues.
3. Transformation of heterogeneous body into an "equivalent homogenous" one.
4. Modification of Fredholm surface integral equation for SAR calculations.
5. A numerical example.

2. Method Proposed

The comparison of different numerical SAR calculation methods seems to be a very difficult task because of the wide range of procedures that have been used. Strictly speaking, for an engineer or a hygienist the accuracy and stability of the results obtained with computer are the only criteria of comparison. But, of course, software realizations of the same method may differ significantly from each other. That is why, even authors of fundamental publications are reticent in method comparisons. In this report we shall not stop on this problem, but some notes are to be done.

From the point of view of electrodynamics mathematical model of interaction RF field with biological body may be designed in following forms:
1. differential equation (finite element (FE) method, finite difference (FD) method, finite difference time domain (FDTD) method e t.c.);
2. volume integral equation, solved with conjugate gradient method, method of moments, fast Fourier transformation e t.c.;
3. combination of differential and integral equation (boundary element method, for instance);
4. multiple multipole (MMP) method;
5. complex impedance method.

Today the FDTD procedure is often regarded as the most successful and the one with the greatest scope for the future.

Depending on the frequency and body area, an approximately rigid separation into subparts (blocks) is carried out with different geometrical shapes and dielectric characteristics in three dimensional space. As a rule, the biological body is divided into cubic blocks.

In near-field exposure conditions when E- and H-field may be analyzed separately, the surface integral equation method (SIEM) seems to be very effective. Step-by-step procedure is the following:

1. heterogeneous body is transformed into homogenous;
2. electric and/or mathematical magnetic charges are introduced at the body-air boundary;
3. charge densities are determined from Fredholm integral equation of the 2nd kind.
4. internal electric field and SAR are calculated.

3. Empirical Expressions for Dielectric Parameters of Biological Tissues

Determining of dielectric constant and specific conductivity of various biological tissues is the first step when calculating the SAR. Nowadays these characteristics are described in many publications [7-13], but there are some problems in practical use of published data:

1. tabulated data and frequency dependencies are difficult to incorporate into programming,
2. there are some differences in characteristics of the same tissues in various publications.
3. dielectric properties have been obtained in various temperature conditions.
4. dielectric properties of tissues with low water content may vary over a wide range.

In order to overcome aforesaid problems the empirical expressions have been obtained with the help of methods of applied statistics. As an example below are relatively simple and easy programmable expressions for dielectric constant ε_r (1) and for specific conductivity σ (2) (expressions valid in frequency range 10 - 100 MHz – dielectric heating in industry). Peak error when using (1) and (2) does not exceed 5 %.

$$\varepsilon_r(f) = c_1 \cdot f^{a_1} \ , \tag{1}$$

$$\sigma(f) = c_2 \cdot f^{a_2} \ , \quad \text{S/m}, \tag{2}$$

where f is in MHz.

Factors c_1, c_2, a_1, a_2 are given in Table 1 [14].

TABLE 1. Factors c_1, c_2, a_1, a_2 for expressions (1), (2).

Tissue or organ	c_1	a_1	c_2	a_2
muscle and skin	350.687	− 0.34080	0.48689	0.11053
fat and bone	140.513	− 0.52360	0.00484	0.57081
marrow	58.824	− 0.46852	0.00484	0.57081
brain	791.134	− 0.50031	0.18534	0.24922
blood	243.000	− 0.25527	0.03556	0.65321
eye	230.781	− 0.31012	0.31688	0.09018
stomach and intestine	172.329	− 0.28214	0.19654	0.65321
liver	595.950	− 0.45299	0.36946	0.10638
kidney	1137.785	− 0.55091	0.61141	0.08873
spleen and pancreas	900.000	− 0.47712	0.48828	0.10721
lung	156.372	− 0.28035	0.53601	0.10721
heart muscle	873.636	− 0.44997	0.56529	0.04012

4. "Equivalent Homogenous" Biological Object

The term "equivalent" means that there are no changes in external electric fields when we replace heterogeneous media with homogenous one. Mean value of absorbed power also remains the same. The procedure of transformation looks very attractively, but relatively simple analytical expressions have been obtained only for objects of canonical geometrical forms (sphere, cylinder, sheroid), irradiated with plane wave [15]. In case of arbitrary shaped object in near-field exposure conditions Lichtenecker Law may be recommended to estimate equivalent complex permittivity of media (3):

$$\dot{\varepsilon}_r = \prod_{i=1}^{n} \dot{\varepsilon}_{ri}^{V_i} \quad , \tag{3}$$

where V_i – specific volumetric partition of i-dielectric component.

5. Fredholm Surface Integral Equations

As mentioned above, electric and magnetic charge densities on body-air boundary are to be found from Fredholm surface integral equations of the 2nd kind. In case of dominating electric field the well-known integral equation looks like (4):

$$\delta(Q)-\frac{\lambda}{2\pi}\oint_S \frac{\delta(M)\cos(\vec{r}_{QM},\vec{n}_Q)}{r_{QM}^2}dS_M=2\varepsilon_o E_n^0(Q),\qquad(4)$$

where $\delta(Q)$ – electric charge density in point Q; M – variable integrating point; r_{QM} – distance between points Q and M; S – surface of biological object with charge density in question; n_Q – outer unit vector to surface S in point Q; $En^0(Q)$ – normal component of outer electric field in point Q; $\lambda = (\varepsilon_r - \varepsilon_o)/(\varepsilon_r + \varepsilon_o)$, ε_r – dielectric constant of homogenous biological media; $\varepsilon_o = 8.85\cdot10^{-12}$ F/m.

While solving equation (4) numerically one has to overcome two main problems: instability of results calculated when λ is close to 1 ($\varepsilon_r \gg \varepsilon_o$) and nucleus singularity when point M is in coincidence with point Q. These problems have been overcome with transformation of initial equation (4) with the function $F(Q)$ that satisfies the following conditions:

$$F(Q)=\frac{\int_S K(Q,M)\left[\int_S K(Q,M)dS_Q\right]dS_M}{\int_S\left[\int_S K(Q,M)dS_Q\right]^2 dS_M},\quad \oint_S F(Q)dS_Q=1,$$

where $K(Q, M)$ is the nucleus of equation (4).

Omitting the intermediate expressions, the final formula to calculate δ has the form (5):

$$\delta(Q)-\frac{\lambda}{2\pi}\oint_S\delta(M)\left[K(Q,M)-F(Q)\oint_S K(Q,M)dS_Q\right]dS_M=f(Q),\qquad(5)$$

where f(Q) is the right part of equation (4).

Results of calculations of δ with formula (5) are quite stable even when order of discretization is rather high.

Then, having charge density one can calculate internal electric field, induced currents, current density and specific absorption within a biological body.

6. Numerical Example

Leakage EM field of HF plastic welder has been investigated with rated frequency 27.12 MHz. Owing to particularity of heating with bar-type electrodes only electric field has been taken into account. The geometry of the sealer structure and other exposure conditions have been modeled to correspond to situation, described by Chen, Gandhi and Conover [16].

Parameters of tissue types used in creating the human model at 27.12 MHz (see (1) and (2)) are given in Table 2.

TABLE 2. Parameters of biological tissues at 27.12 MHz

No	Tissue type	Mass density ρ, kg/m^3	Mass, m, kg	ε_r	σ, S/m
1	Muscle	1050	30	114	0.7012
2	Bone	1200	10.5	21	0.0318
3	Fat	1200	8	21	0.0318
4	Blood	1000	4.5	105	0.3071
5	Stomach and intestine	1000	6	68	0.3033
6	Liver	1030	1.5	134	0.3033
7	Kidney	1020	$2 \cdot 0.18$	185	0.8194
8	Pancreas	1030	0.1	186	0.6956
9	Spleen	1030	0.18	186	0.6956
10	Lung	330	$2 \cdot 0.38$	30 *	0.1733*
11	Heart	1030	0.332	198	0.6453
12	Brain	1050	1.38	152	0.4218
13	Skin	1000	3.5	114	0.7012
14	Marrow	1050	2.6	13	0.0211

* with 67 % air

Total weight of biological object has been equal to 69.7 kg. Other parameters have been used in calculations: equivalent dielectric constant and specific conductivity $\varepsilon_r = 79$, $\sigma = 0.264$ S/m. Height of biological body is chosen to be 1.75 m. Total body surface $S = 1.84$ m^2.

When calculating the total body surface S has been divided into 315 surface elements.

In Table 3 there are some results, compared with data obtained with FDTD method for similar input data.

TABLE 3. Layer-averaged SAR for barefoot operator

Body part	Height above floor, cm	SAR in mW/kg, calculated with SIEM	SAR in mW/kg, calculated with FDTD method[16]
ankle	0.09	2000	2500
calf	0.3	594	500
knee	0.47	800	954
hip	0.63	466	409
belly	1,0	299	250
chest	1.32	180	91
neck	1.55	369	659
head	1.68	160	227

SIEM yields to results that are less than results obtained with FDTD method. It seems that fact is caused with two reasons. Firstly, in FDTD method the internal E component induced with external magnetic field is taken into consideration (this component is ignored in SIEM method). Secondly, FDTD method demands much more detailed description of dielectric characteristics of body tissues when subdividing body into volume blocks. Nevertheless SIEM gives *a qualitative* behavior of the SAR distribution. (Other calculations with isolated floors, wet floors have been done too).

7. Conclusions

1. Reducing of realistic biological body into homogenous one may be recommended for express estimation of absorbed power. Certainly, when SAR distribution in various organs is of special interest (hyperthermia, for instance) we can not do this transformation.

2. Surface integral equation in form of (5) gives stable results. SIEM may be recommended also for fast estimation of SAR distribution influenced with the exposure conditions (various floor covers in workshops, clothing and shore-wearing of irradiated person). The possible field of application of SIEM is quite wide – irradiation from high-voltage power lines, residential electric appliances, industrial induction furnaces e t.c. The only critical factor is a frequency of EM field. The upper limit may be estimated as

40 - 50 MHz. One can use SIEM as a first step of SAR calculations. If more information about SAR is desired, more accurate method must be used.

8. References

1. GOST SSBT 12.1.006-84*. *RF Electromagnetic Fields. Permissible Levels at Workplaces and Requirements for Control* (in Russian).
2. Sanitary norms and rules 2.2.4/2.1.8.055-96. *Electromagnetic RF Radiation* (in Russian).
3. ANSI C95.1-1991 (1991) *ANSI Safety Levels with Respect to Human Exposure to RF Electromagnetic Fields, 3 kHz to 300 GHz".*
4. DIN VDE 0848-2 (1991*). Shutz von personen im Frequenzbereich von 30 kHz bis 3000 GHz.*
5. European Prestandard ENV 50166-2. *Human exposure to electromagnetic fields. High frequency (10 kHz to 300 GHz).*
6. IRPA/INIRC-1990. (1990) *Guidelines on Limits of Exposure to RF Electromagnetic Fields in the Frequency Range from 100 kHz to 300 GHz.*
7. Schwan, H.P. (1957) Electrical properties of tissues and sells, *Advan. Biol. Med. Phys.* **5**, 147-209.
8. Grant E. H., Keefe S. E., and Takashima S. (1968) The dielectric behavior of aqueous solutions of bovine serum albumin from radiowave to microwave frequencies, *J. Phys. Chem.* **72**, 4373 - 4380.
9. Schwan, H. P. et al. (1970) Electrical properties of phospholipid vesicles, *Biophys. J.*, **10**, 1102 - 1119.
10. Yamamoto, T., and Yamamoto, Y. (1976) Dielectric constant and resistivity of epidermal stratum correum, *Med. Biolog. Engin.* **14**, 494 - 499.
11. Johnson, C. C., and Guy, A. W. (1972) Nonionizing electromagnetic wave effects in biological materials and systems, *Proc. IEEE.* **60**, 692 - 718.
12. Schwan, H. P., and Li, K. (1953) Capacity and conductivity of body tissues at ultrahigh frequencies, *Proc. IRE.* **41**, 1735 - 1740.
13. Stuchly, M. A. and Stuchly, S.S. (1980) Dielectric properties of biological substances – Tabulated, *J. Microwave Power* **15**, 19 - 26.
14. Rudakov, M.L. (1997) Empirical expressions of dielectric characteristics of biological tissues for high-frequency range, *Electrichestvo.* **9**, 75-77 (in Russian).
15. Rudakov, M.L. (1997) Calculations of mean absorbed power density in heterogeneous biological objects, *Gigiena i sanitaria.* **5**, 61- 63 (in Russian).
16. Chen, J.Y., Gandhi, O.P., and Conover, D.L. (1991) SAR and induced current distributions for operator exposure to RF dielectric sealers, *IEEE Trans. on EMC.* **33**, 252-261.

RECENT ADVANCEMENTS IN DOSIMETRY MEASUREMENTS AND MODELING

P. A. MASON[1], J. M. ZIRIAX[2], W. D. HURT, T. J. WALTERS[1], K. L. RYAN[3], D. A. NELSON[4], K. I. SMITH[5], AND J. A. D'ANDREA[2].
Air Force Research Laboratory, Human Effectiveness Directorate, Directed Energy Bioeffects Division, Brooks AFB, TX, 78235, [1]Veridian Engineering, Inc., San Antonio, TX, 78216, [2]Naval Health Research Center Detachment at Brooks, AFB, TX, 78235, [3]Trinity University, San Antonio, TX, 78212, [4]Michigan Technological University, Houghton, MI, 49931, [5]The University of Texas Health Science Center at San Antonio, San Antonio, TX, 78284.

1. Introduction

Whole-body specific absorption rate (SAR) values provide useful information about energy deposition resulting from exposure to radio frequency radiation (RFR). However, whole-body SAR values do not reveal possible localized "hot spots". Although differences in regional temperatures have been measured in animals during RFR exposure [1-3], the use of temperature probes to make empirical measurements of these "hot spots" can be extremely time consuming and are invasive in nature [4].

Advances in tissue permittivity measuring capabilities permit the application of the mathematical models, such as the Finite-Difference Time-Domain (FD-TD) model [5-7], to assess the amount of energy absorbed by geometrically-complex biological models during RFR exposure. Nonuniform energy absorption after RFR exposure has previously been predicted using spherical and block models [8-14]. The goal of this ongoing project is to produce computer models that could dynamically integrate the: 1) electrical characteristics of tissue, 2) distributions of tissue types in the Sprague-Dawley rat, rhesus monkey, phantom "green" monkey [15], pigmy goat, and man, 3) and the specific characteristics of an RFR exposure. Once these computer models are produced, they will be used to predict localized and whole-body SAR values. The Mie theory (developed by Professor Gustav Mie in 1908) and FD-TD code predict SAR values. However, temperature fields are not predicted directly by these SAR results. Rather, SAR values are used as input for a thermal model program to describe heat transfer over time. Thermoregulatory parameters that are being incorporated into our dosimetry models are some of those (e.g., modes of heat transfer including conduction, convection) that were incorporated into the spherical model of the human head [16] and block model of the squirrel monkey [17]. In developing models, it is essential that the

141

B.J. Klauenberg and D. Miklavcic (eds.), Radio Frequency Radiation Dosimetry, 141-155.

142

predicted results be verified by empirical measurements. Our predicted results are compared to temperature measurements made with implanted probes, temperature-sensitive paint, or infrared thermography. These convergent technologies validate the use of computational codes to predict SAR values and demonstrate that a small change in the animal's posture and orientation with respect to the field may greatly vary the localized SAR values. These variations in localized SAR values may influence certain physiological responses.

2. Model Development

2.1. SPHERES

Computer-generated spheres were 20, 66, and 105 mm in diameter and each 1 mm cubed voxel within the sphere was assigned a dielectric value corresponding to 2/3 of that for muscle [18-20]. This value was used in the Dosimetry Handbook (Figure 3.35 in [21]) as the average permittivity of the human body.

2.2. ANIMALS

The computer modeling efforts in our laboratory began in 1993. At that time, accurate anatomical models of the Sprague-Dawley rat, pigmy goat, and rhesus monkey were required in a digital format. After reviewing the literature, we concluded that these models were not available. In collaboration with Mobile Technologies, Inc. (San Antonio, TX) and the Research Imaging Center at the University of Texas Health Science Center at San Antonio, we obtained T1-weighted magnetic resonance imaging (MRI) scans of these animals. Detailed description of how the MRI scans of the Sprague-Dawley rat (370 gm), pigmy goat (20 kg), and rhesus monkey (7.1 kg) were acquired has been described in an earlier publication [22]. A 3-dimensional model of the phantom "green" monkey in a digital format was also needed. The phantom monkey was filled with a mixture having the dielectric properties of 2/3 muscle [23] and imaged in both a sitting and squatting position. MRI scans of the phantom monkey were acquired in a similar manner to that described above. It was necessary to convert the MRI scans, which did not adequately code tissue types, into a format that would allow tissue types to be manually encoded with a color. All MRI scans were converted to 24-bit tagged image file (TIF) images for use within Adobe Photoshop™ (Mountain View, CA) software on the PC. Within Photoshop™ all images were aligned so as to eliminate some of the scanning artifacts (e.g., echo due to respiration) previously reported [22]. Each of the selected tissue types was assigned permittivity and conductivity values obtained from Gabriel [18-20]. These dielectric values are also available on the World Wide Web (WWW) at: www.brooks.af.mil/AFRL/HED/hedr/reports/dielectric/home.html. A digital version of the fourth edition of the Radiofrequency Radiation Dosimetry handbook [21] is also available at: www.brooks.af.mil/AFRL/HED/hedr/reports/handbook/home.html.

Each tissue was assigned a specific Red-Green-Blue (RGB) color value. RGB 24-bit color rather than 8-bit color was used since some software programs modify colors in the 8-bit images. All pixels on each image were then "painted" manually with the RGB colors representing the appropriate tissue types. There were no internal structures in the phantom monkey; however the MRI scans did reveal air bubbles throughout the phantom. These bubbles were eliminated during the "painting" process. MRI atlases [24, 25] and atlases of the rat [26, 27], goat [28], and monkey [29], and skeletons of the rat and monkey were used to identify the location of each tissue type.

Due to the size constraints of the General Electric Signa MRI scanner, the goat's front and back legs were stretched outwards. In order to intubate the monkey during scanning, its neck was tilted backwards. We have developed algorithms that aid in moving and rotating body parts and have successfully rotated the monkey's head from its original tilted back position to that consistent with those of our RFR-exposed monkeys. However, these rotations still require extensive manual interaction to ensure correct tissue alignment.

Since the completion of our rat dosimetry model, other rat dosimetry models have been developed [30]. Our rat was positioned in a manner that was consistent with an anesthetized rat lying on a table. This positioning has been used in numerous RFR studies in our laboratory. In order to duplicate the posture of an awake animal, Lapin and Allen [30] placed either an anesthetized or *post mortem* rat on a plastic torso support and then acquired 0.8 x 0.8 x 2.0 mm voxels using CT scanning. A *post mortem* model is useful in that carcasses have been exposed and then placed in a calorimeter to obtain whole-body SAR values [31-33]. Unsupported anesthetized or *post mortem* rats were also imaged. There were no differences between abdomen shape for an unsupported anesthetized rat and an unsupported *post mortem* rat. This relationship in abdomen shape also existed for the supported anesthetized and supported *post mortem* rat. However, differences between anesthetized and *post mortem* rats were observed in terms of internal organ structure. Distention of the tissue in the digestive tract was observed within one hour after death. In the future, additional animal dosimetry models may be developed that will allow the researcher to choose the model which correlates with the particular states (e.g., awake, anesthetized, euthanized, supine, sitting) used to obtain empirical measurements.

2.3. MAN

Efforts to produce a man dosimetry model were initiated in 1996. At that time, the following man models were being developed or used. The model described by Sullivan *et al.* [34] was comprised of 1.3 cm cubical cells. The model was developed using cross-sectional diagrams of the human body at spacings 1 inch apart [35]. Furse and Gandhi [36] used a 6 mm resolution man model developed from MRI scans to calculate electrical fields and induced currents. The human was actually scanned at a 3 mm resolution. Due to the immense size of the dataset and the limited computer processing power available to run such a dataset, voxels were combined by taking the dominant tissue in each group to produce the 6 mm cubical cells. Dawson et al. [14] developed a

3.6 mm cubed model of a man to compute electrical fields. This model was developed using a combination of head and torso MRI images obtained from Yale Medical School [37] and MRI and CT images of the arms and legs from the Visible Human project (National Library of Medicine, www.nlm.nih.gov/research/visible/visible_human. html). A 7.2 mm cubed man was produced by applying a 3 x 3 x 3 median filtering algorithm to the above dataset. The National Radiological Protection Board (Chilton, UK) developed a man dosimetry model using MRI scans. The axial scans were approximately 10 mm apart and the data volume was rescaled and interpolated to produce 2-mm cubical voxels [38]. The desire to have a 1 mm cubed voxel model of the entire man led us to use the photographic images of the Visible Human and computer-segmented images (CieMed, a collaboration between National University of Singapore and John Hopkins University) of this dataset to develop a man dosimetry model. Images from the Visible Man project are now used as a standard anatomy reference. This dataset consists of 1878 axial images, each voxel being 1 mm cubed. This resolution permits the inclusion of fine detail, such as small blood vessels and nerves, in the dosimetry model. Structural detail and voxel size are both important factors to consider when selecting a dosimetry model. MRI and anatomical atlases of the human [24, 25, 39, 40] numerous WWW sites pertaining to the Visible Human project, and human cadavers were used to aid in identifying the location of each tissue type. The tissues were color coded in the same manner as described above. The Visible Man was missing a tooth and one testicle, so these were added during the color-coding process. As described above for the monkey's head, the feet of the man are currently being rotated upwards and inwards. This will change each foot from a drop foot position to that required to properly ground the man for plane-wave exposure simulations. Similar adjustments to feet were reported by Gandhi and Furse [41].

2.4. ERROR CHECKING AND 3-DIMENSIONAL RECONSTRUCTION

Images were checked for unknown pixel values and for pixel values located in incorrect locations (i.e., heart pixel value located in the leg) using WaveAdvantage™ (Visual Numerics, Inc., Houston, TX) software. Throughout the "painting" process, images were reconstructed 3-dimensionally using Slicer™ (Fortner Research LLC, Sterling, VA) software on the PC and VoxBlast™ (Fairfield, IA) software on a Silicon Graphics Indigo workstation to ensure correct anatomical placement of each tissue type. To predict electrical fields and SAR values, each TIF image file was converted to an 8-bit binary file and imported into a mathematical model. Processing the relatively small models (e.g., rat) may be accomplished on a stand-alone workstation; however, parallel computer systems will be utilized to process the larger models (e.g., monkey, goat, phantoms, man).

As discussed in a previous publication [22], our images should be useful to a variety of researchers and instructors. We foresee even more uses for the datasets than those previously reported, since the tissue and organ types are now color-coded for identification. As with the MRI grayscale datasets, the color-coded datasets are available via anonymous file transfer protocol (ftp) to merlin.brooks.af.mil. Images are located in the /pub/dosimetry_models directory.

3. SAR Predictions

Our goal is to predict localized and whole-body SAR values in heterogenous models having complex geometries. Currently, FD-TD is one of the best solutions available and has been widely accepted by the RFR community [34, 42-44]. Our FD-TD code was modified from that published by Luebbers et al. [6] and Kunz and Luebbers [7]. The modified code reads our binary model files and tissue permittivity file and outputs 3-dimensional data. The RFR source was changed from pulsed to continuous wave. The FD-TD results for the spheres were verified by comparing them with the results from the Mie procedure. Shown in Figure 1 are the E-field values (Volts/meter) predicted by the FD-TD [6, 7] and Mie [45] computational models for a 20 mm diameter sphere exposed in the far field to 1800 MHz. There is good agreement in the results predicted by the two models.

SAR values for the 66 and 105 mm diameter spheres exposed in the far field to 2.06-GHz were predicted by the FD-TD computational model [6, 7], as shown in Figure 1. Arrows represent the direction of exposure. The higher SAR values were located in the center of the 66 mm sphere and along the leading edge of the 105 mm sphere.

SAR values were predicted by the FD-TD code for a rat exposed in the far field to 2.06-GHz in the KHE and KEH polarizations. For KHE polarization, the long axis of the rat (head to tail) was in the direction of propagation. The second largest dimension of the rat was width and that was parallel to the H field. In the KHE polarization, there were higher SAR values surrounding the nasal cavity, in the brain, and along the dorsal surface posterior to the cervical region than in the KEH polarization (see images in Section 5.2.2).

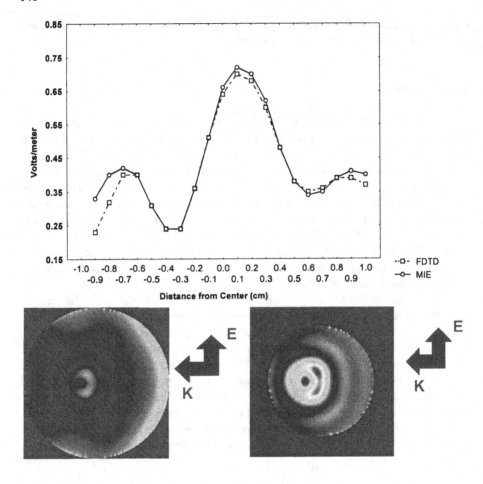

Figure 1. E-field values (Volts/meter) predicted by the FD-TD and Mie computational models for a 20 mm diameter sphere exposed in the far field to 1800 MHz. The two lower images represent the distribution of SAR values predicted by the FD-TD model for 66 and 105 mm diameter spheres exposed in the far field to 2.06 GHz. The lighter areas represent the higher SAR values.

In contrast, the KEH polarization produced higher SAR values in the areas surrounding the carotid arteries and masseter muscle regions and in the brachial muscle regions than those produced for the KHE polarization. Both polarizations revealed high SAR values in the testicles, but with slightly different patterns.

One of the limitations in predicting SAR values is the number of tissues for which dielectric values are available. Currently, there are approximately 40 tissues available [18-20], but there are at least 369 separate body regions isolated on the dataset of the VisibleMan (CieMed, www.ciemed.iss.nus.sg). Although many of these were redundant in terms of dielectric values (e.g., c2 vs. c3 vertebral body, proximal and middle phalanges), additional dielectric values for unique tissue types would be useful. This would be especially relevant for SAR predictions at higher frequencies, which require smaller voxel sizes in the block models. These smaller voxel sizes would permit finer resolution among body parts, such as separating the dura from the pia matter.

4. Thermal Transfer

A critical element of RFR dosimetry is the modeling or prediction of tissue temperatures from localized SAR values. Tissue heating models have two separate and distinct applications: (1) experimental validation of SAR calculations, commonly done by measuring tissue temperature rate-of-rise in laboratory animals under controlled exposure conditions, and (2) the extrapolation of thermal effects to humans, on the basis of observed effects (temperatures or SAR) in laboratory animals.

Predictive models which utilize numerical schemes such as the FD-TD algorithm are capable of estimating local tissue absorption rates with very high spatial resolution and numerical precision. However, the accuracy of the estimates can be determined only by comparison with empirical measurements. While there are several methods for measuring SAR (either directly or indirectly), obtaining local tissue SAR in a live animal requires measuring the rate of tissue temperature increase under the prescribed exposure conditions. A thermal model is needed to allow comparison of measured temperatures with calculated SAR data. Additionally, a reliable thermal model is necessary if the ultimate goal of an investigation is the determination of the thermal effects (either local or whole-body) resulting from RFR exposure. Such models must account for inter-species differences in thermoregulatory mechanisms if the objective is the extrapolation of data from laboratory animals to humans.

The simplest thermal model neglects all heat transfer mechanisms except RFR energy deposition. A simple energy balance yields a linear relationship between SAR and the rate of temperature increase:

$$SAR = C \, dT/dt \qquad (1)$$

where C is the specific heat of the tissue, dT is temperature rise, and dt is exposure time. For modest temperature increases, C can generally be assumed to be constant (i.e., independent of temperature) for a given tissue type. An average value of 3.5 kJ/kg/K is appropriate for the determination of whole-body SAR (Equation 7.7, in [21]). However C varies greatly between tissue types, ranging from 1.6 kJ/kg/K for bone (cortical and cancellous) to 3.8 kJ/kg/K for muscle [46]. When determining local SAR values in a heterogeneous animal, it can be misleading to use a single coefficient to correlate SAR with rate of temperature increase across tissue types.

The linear extrapolation model has limited applicability, as it neglects many potentially significant heat transfer mechanisms, including thermal conduction, surface cooling or heating (convection and evaporation, radiation), metabolic heat generation, and advection (heat transfer by blood flow). In many situations of interest, the blood flow effects are the most significant of these mechanisms, and also the most difficult to model. Blood flow effects can also confound inter-species comparisons, due to differences in perfusion rates and the fact that some species have anatomical structures (e.g., the rat tail) which serve a thermoregulatory function by cooling the arterial supply under conditions of thermal stress.

Estimation of local tissue temperatures from SAR calculations in heterogeneous animals is being accomplished using a modified bio-heat transfer model similar to that employed by Iskander and Khoshdel-Milani [47]. Blood flow effects are described by a "conduction-like" term in which the advective cooling is proportional to the difference between the tissue temperature and the arterial temperature. This term is separate from thermal conduction, which is also retained in the model. Thermal properties (specific heat, thermal conductivity) of the respective tissues are obtained from the literature (e.g., 48, 49] Blood perfusion rates for each organ and species are also determined from the literature.

Thermal analysis is performed using the anatomical information in the 24-bit RGB TIF files. These raster image files must be vectorized so that a finite element mesh can be generated from the geometry. The file conversion is performed using a commercial program, TracTrix (Trix Systems AB, Vegby, Sweden) which produces an IGES (International Graphics Exchange Specification) file. The vector file can then be imported directly into a finite element program for meshing and analysis. Local SAR values, metabolic heating rates, and boundary conditions (which account for surface convection, evaporation and radiation) are specified within the finite element code. Analysis is currently being performed using commercial finite-element software (ANSYS, Ansys Inc., Canonsburg, PA, USA) on a Sparc10 workstation (Sun Microsystems). Current efforts are directed at implementing the model on a personal computer using a commercial finite-difference code (Thermal Analysis System, Harvard Thermal Inc., Harvard, MA, USA). Although the model is being implemented on two-dimensional heterogeneous tissue geometries, future work will include developing a three-dimensional anaylsis capability [50].

5. Empirical Measurements

Validation of predicted SAR values is an ongoing essential component of this research and requires the use of a variety of techniques. Mathematically, we compared the SAR values predicted by the FD-TD code to those predicted by the Mie theory. The predicted SAR values are compared to data obtained from empirical measurements obtained using infrared thermography, heat-sensitive paint, and implanted temperature probes.

5.1. SPHERES

Spheres were 66 and 105 mm in diameter and composed of material having the dielectric properties of 2/3 muscle [23]. This material was encased in two halves of a Styrofoam™ shell. The Styrofoam™ regions in contact with the dielectric material had been coated with epoxy (Bob Smith's Industries, Atascadero, CA). This coating attenuated the amount of water that would be extracted from the dielectric material into the Styrofoam™ shell during the pre-exposure time period in which the temperature of the spheres was equilibrating to that of the exposure chamber. Each half was covered with silkscreen and the two halves were held together with tape. During pilot experiments, the two halves were separated by gauze pads. However, the gauze pads would stick to one another when the two halves were separated to show the internal temperatures after exposure. This would result in the removal of the surface layer of the dielectric material since it had adhered to the gauze. Replacing the gauze pads with silkscreen eliminated this problem. A Radiance1 infrared camera system and ImageDesk software (256 x 256 indium antimonide sensor array sensitive over the 3-5 μm waveband, Amber Engineering, Inc., Goleta, CA) were used to record temperature gradients in each half immediately after exposure. Spheres were exposed in the far field to 2.06-GHz at a power density of 1.7 W/cm^2. The temperature and humidity of the anechoic chamber were 22°C and 58%, respectively. Exposure durations for the 66 and 105 mm diameter spheres were 30 and 60 sec, respectively. Data were not corrected for the influence of heat flow after termination of exposure. Data analysis was accomplished using Transform and Plot (Fortner Research, Sterling, VA) and Excel (Microsoft, Redmond, WA) software. Results are shown in Figure 2.

150

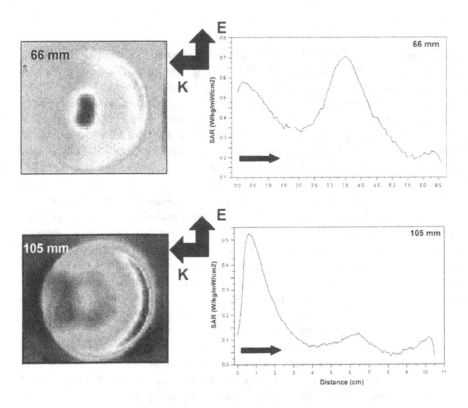

Figure 2. Temperature gradients recorded by infrared thermography across 66 and 105 mm diameter spheres after exposure in the far field to 2.06-GHz radiation. The greater temperature increases (darker regions) were recorded in the center of the 66 mm sphere and along the leading edges of the 105 mm sphere. The arrows indicate the direction of exposure with respect to each plot. Normalized SAR values ranging from 0.02 to 0.7 W/kg/mW/cm^2 were calculated from these temperature changes and were consistent with those predicted by the Mie theory and FT-TD code.

5.2 RAT

5.2.1 *Surface Painting*

Temperature-sensitive paint was used to measure the distribution of energy in an RFR-exposed rat. The paint (Type 27, Matsui International Co, Inc, Gardena, CA) changes color at 33°C. This rat was euthanized by CO_2 inhalation, the medial dorsal surface of the rat was shaved, and the paint was then applied to the skin. During pilot experiments, it was discovered that only application of the pigment was required, rather than the combination of the pigment, binder, and fixer as stated in the instructions. Comparison of normal and infrared video images showed that the thermal-sensitive properties of the pigment alone was the same as that of the pigment, binder, and fixer combination. However, the pigment has a tendency to flake when applied by itself. For this reason, the pigment was only applied to the dorsal surface of the rat. Temperature probes were implanted, as described above, in the following regions: nose, sinus, forehead, left ear, right ear, neck, spine, rectum, tail (base, middle, and tip). The rat was then exposed in the far field in the KHE polarization (i.e., nose pointed towards the horn) to 2.06-GHz with an incident field of 1.7 W/cm^2. Anechoic chamber temperature was 22-24°C.

The initial changes in the color of the temperature-sensitive paint occurred in the nose, ears, dorsal surface in the cervical and thoracic regions, and at the base of the tail (see Figure 3). These changes in paint color were consistent with the temperatures measured by the implanted probes. Our present methodology in using the temperature-sensitive paint provided qualitative and quantitative results. Although these latter results were for a single temperature threshold (33°C), more refined quantitative results might be obtained by layering paints having different temperature sensitivities. The infrared imaging system described below is much more expensive, but is much easier to use than the temperature-sensitive paint.

5.2.2 Infrared Imaging

A rat was anesthetized with urethane (1.6 gm/kg, i.p.) and the entire body were shaved. Rats weighing between 360 and 380 gm are used to be consistent with that of the dosimetry rat model (370 gm). Inside the anechoic chamber, a lubricated rectal temperature probe (Model 101, Vitek) was inserted 5 cm beyond the anal sphincter. A Radiance1 infrared camera system (Amber Engineering, Goleta, CA) was hung from the ceiling in the chamber in order to view the dorsal surface of the rat. The rat was exposed in the far field in the KHE polarization to 2.06-GHz using the same equipment and parameters as described above. The infrared images showed that the nose, ears, dorsal surface in the cervical and thoracic regions, and the base of the tail were the first to become warm (see Figure 3). These changes corresponded to increases in surface temperature as measured by ImageDesk software during the first 15 sec of exposure.

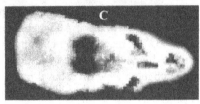

Figure 3. Images shown above are from: A) infrared thermography, B) FD-TD SAR predictions for a rat exposed in the same KHE orientation, and C) a rat painted with temperature-sensitive paint. The greater temperature increases occurred in the nose, ears, dorsal surface in the cervical and thoracic regions, and at the base of the tail.

5.2.3 *Temperature Probes*

The section in this book by Dr. Thomas J. Walters and co-authors [51] discusses in depth the procedures involved in measuring temperatures in animals during RFR exposures. Additional results are reported by Walters *et al.* [52].

6. Future Applications

The models discussed above could be used in research settings as a guide to designing clinical devices (e.g., hyperthermia applicators) [17], communication systems, and RFR exposure paradigms. The results of experimental exposures would be used to validate and refine the model. In an occupational setting, these models could serve as a forensic tool to aid the reconstruction of accidental exposures. Using the computer models to predict localized SAR values may lead to establishing new safety standards or modification of existing safety standards. As part of the determination of safety guidelines for equipment design and maintenance, these models could be used as a bridge between empirical data and actual exposure conditions. Verified, mathematical models will provide a comprehensive picture of the RFR exposure process, for less time and labor than empirically-based techniques. As such, computational dosimetry models could meet the time and labor constraints required by real-world RFR applications.

7. Notes

Experimental procedures were approved by the Animal Care and Use Committee at Armstrong Laboratory, Brooks AFB. The animals used in this study were procured, maintained and used in accordance with the Animal Welfare Act and the "Guide for the Care and Use of Laboratory Animals" prepared by the Committee on Care and Use of Laboratory Animals of the Institute of Laboratory Animal Resources - National Research Council. Views presented are those of the authors and do not reflect the official policy or position of the Department of the Air Force, Department of the Navy, Department of Defense, or U.S. Government. Trade names of materials and/or products of commercial or nongovernment organizations are cited as needed for precision. These citations do not constitute official endorsement or approval of the use of such commercial materials and/or products.

8. Acknowledgments

Research was funded by U.S. Air Force Contracts F33615-90-C-0604, F41624-96-C-9009, HQ Human Systems Center, Air Force Materiel Command, Brooks AFB, TX, 78235-5000, Air Force Office of Scientific Research awards (1995, 1996) to Mr. William Hurt, and Naval Medical Research and Development Command under Work Unit 61153N MR04101.001-1603. Technical assistance from SSgt. James Belcher, Ms. Beth Brewer, Mr. Ethan Foster, Ms. Meredith Gilbert, Mr. Don Hatcher, Ms. Pam Henry, Mr. David Jentsch, Ms. Deanna Lane, Mr. George Lantrip, Mr. Kevin Mylacraine, Ms. Kim Obenshain, SRA Lea Ann Paulus, Ensign Robert Post, Ms. Janet Roe, Mr. John Villacis, and Ms. Kara Watt was greatly appreciated.

9. References

1. Lin, J.C., Guy, A.W., and Kraft, G.H. (1973) Microwave selective brain heating, *J. Microwave Power* **8**, 275-286.
2. Chou, C.K., Guy, A.W., McDougall, J.A., and Lai, H. (1985) Specific absorption rate in rats exposed to 2,450 MHz microwaves under seven exposure conditions, *Bioelectromagnetics* **6**, 73-88.
3. Ward, T.R., Svensgaard, D.J., Spiegel, R.J., Puckett, E.T., Long, M.D., and Kinn, J.B. (1986) Brain temperature measurements in rats: A comparison of microwave and ambient temperature exposures, *Bioelectromagnetics* **7**, 243-258.
4. Chou, C.K., Bassen, H., Osepchuk, J., Balzano, Q., Peterson, R., Meltz, M., Cleveland, R., Lin, J.C., and Heynick, L. (1996) Radiofrequency electromagnetic exposure: Tutorial review on experimental dosimetry, *Bioelectromagnetics* **17**, 195-208.
5. Yee, K.S. (1966) Numerical solution of initial boundary value problems involving Maxwell's equations in isotropic media, *IEEE Trans. Ant. Prop.* **AP-14**, 302-307.
6. Luebbers, R., Hunsberger, F.P., Kunz, K.S., Standler, R.B., and Schneider, M. (1990) A frequency dependent finite-difference time-domain formulation for dispersive materials, *IEEE Trans. Electromagn. Compat.*, **EMC-32**, 222-227.
7. Kunz K.S. and Luebbers R.J. (1993) *The Finite Difference Time Domain Method for Electromagnetics*. CRC Press, Inc., Boca Raton, FL.
8. Shapiro, W.R., Lutomirski, R.F., and Yura, H.T. (1971) Induced fields and heating within a cranial structure irradiated by an electromagnetic plane wave, *IEEE Trans. Microwave Theory Tech.* **MTT 19**, 187-196.

9. Kritikos, H.N. and Schwan, H.P. (1972) Hot spots generated in conducting spheres by electromagnetic waves and biological implications, *IEEE Trans. Biomed Eng.* **BME-19**, 53-58.

10. Kritikos H.N. and Schwan, H.P. (1975) The distribution of heating potential inside lossy spheres, *IEEE Trans. Biomed Eng.* **BME-22**, 457-463.

11. Ohlsson, T. and Risman, P.O. (1978) Temperature distribution of microwave heating - Spheres and cylinders, *J. Microwave Power* **13**, 303-310.

12. Rukspollmuang, S. and Chen, K-M. (1979) Heating of spherical versus realistic models of human and infrahuman heads by electromagnetic waves, *Rad. Sci.* **14**, 51-62.

13. Hagmann, M.J., Gandhi, O.P., D'Andrea, J.A., and Chatterjee, I. (1979) Head resonance: Numerical solutions and experimental results, *IEEE Trans. Microwave Theory Tech.* **MTT-27**, 809-813.

14. Dawson, T.W., Cupata, K., and Stuchly, M.A. (1997) Influence of human model resolution on computed currents induced in organs by 60-Hz magnetic fields, *Bioelectromagnetics* **18**, 478-490, 1997.

15. D'Andrea, J.A., Thomas, A., and Hatcher, D.J. (1994) Rhesus monkey behavior during exposure to high-peak-power 5.62-GHz microwave pulses, *Bioelectromagnetics* **15**, 163-176.

16. Kritikos, H.N. and Schwan, H.P. (1979) Potential temperature rise induced by electromagnetic field in brain tissues, *IEEE Trans. Biomed Eng.* **BME-26**, 29-34.

17. Spiegel, R.J., Fatmi, M.B.E., and Ward, T.R. (1987) A finite-difference electromagnetic deposition/thermoregulatory model: Comparison between theory and measurements, *Bioelectromagnetics* **8**, 259-273.

18. Gabriel, C. (1996*) Compilation of the Dielectric Properties of Body Tissues at RF and Microwave Frequencies, Air Force Material Command*, **AL/OE-TR-1996-0037**, Brooks Air Force Base, TX.

19. Gabriel, C., Gabriel, S., and Courthout, E. (1996) The dielectric properties of biological tissues: 1. Literature survey, *Phys. Med. Biol.* **41**, 2231-2250.

20. Gabriel, S., Lau, R.W., and Gabriel, C. (1996) The dielectric properties of biological tissues: 2. Measurement in the frequency range 10 Hz to 20 GHz, *Phys. Med. Biol.* **41**, 2251-2269.

21. Durney, C.H., Massoudi, H., and Iskander, M.F. (1986) *Radiofrequency Radiation Dosimetry Handbook.* **USAFSAM-TR-85-73**, Brooks Air Force Base, TX.

22. Mason, P.A., Walters, T.J., Fanton, J.W., Erwin, D.N., Gao, J.H., Roby, J.W., Kane, J.L., Lott, K.A., Lott, L.E. and Blystone, R.V. (1995) Database created from magnetic resonance images of a Sprague-Dawley rat, rhesus monkey, and pigmy goat, *FASEB J.* **9**, 434-440.

23. Chou, C.K., Chen, G.W., Guy, A.W., and Luk, K.H. (1984) Formulas for preparing phantom muscle tissue at various radiofrequencies, *Bioelectromagnetics* **5**, 435-441.

24. Barrett, C.P., Anderson, L.D., Holder, L.E., and Poliakoff, S.J. (1994) *Primer of Sectional Anatomy with MRI and CT Correlation*, Williams & Wilkins, Baltimore, MD.

25. Mai, J.K., Assheuer, J., and Paxinos, G. (1997) *Atlas of the Human Brain*. Academic Press, London, England.

26. Olds, R.J. and Olds, J.R. (1979) *A Colour Atlas of the Rat - Dissection Guide*. Wolfe Medical Publications, Ltd., Ipswich, England.

27. Popesko, P., Rajtova, V., and Horak, J. (1992) *A Color Atlas of Anatomy of Small Laboratory Animals. Vol. II: Rat, Mouse, and Hamster*. Wolfe Publishing Ltd., London, England.

28. Hopkins, Sr., C.E., Hamm, Jr., T.E., and Leppart, G.L. (1972) *Atlas of Goat Anatomy. Part II. Serial Cross Sections. Edgewood Arsenal Report*, **1972-EA-TR-4626**, Edgewood Arsenal, MD.

29. Dalrymple, G.V. (1965) *An Atlas of Cross-Sectional Anatomy of the Macaca Mulatta for use in Radiobiological Experiments*. **USAF-TR-65-32**, Brooks Air Force Base, TX.

30. Lapin, G.D. and Allen, C. (1997) Requirements for accurate anatomical imaging of the rat for electromagnetic modeling. *Proceedings of the 19th International Conference*, IEEE/EMBS, Chicago, IL, 2480-2483.

31. Blackman, C.F. and Black, J.A. (1977) Measurement of microwave radiation absorbed by biological systems. 2. Analysis by Dewar-flask calorimetry, *Radio Sci.* **12**, 9-14.

32. Allen, S.J. and Hurt, W.D. (1979) Calorimetric measurements in microwave energy absorption by mice after simultaneous exposure in 18 animals, *Radio Sci.* **14**, 1-4.

155

33. Chou, C.K., Guy, A.W., and Johnson, R.B. (1984) SAR in rats exposed in 2450-MHz circularly polarized waveguide, *Bioelectromagnetics* **5**, 389-398.
34. Sullivan, D., Gandhi, O., and Taflove, A. (1988) Use of the finite difference time-domain method in calculating EM absorption in man models, *IEEE Trans. Biomed Eng.* **BME-35**, 179-185.
35. Eycleshymer, A.C. and Schoemaker, D.M. (1911) *A Cross-Section Anatomy*, Appleton, New York.
36. Furse, C.M. and Gandhi, O.P. (1998) Calculation of electric fields and currents induced in a millimeter-resolution human model at 60 Hz using the FDTD method, *Bioelectromagnetics* **19**, 293-299.
37. Zubal, I.G., Harrell, C.R., Smith, E.O., Rattner, Z., Gindi, G.R., and Hoffer, P.H. (1994) Computerized three dimensional segmented human anatomy, *Med. Phys. Biol.* **21**, 299-302.
38. Dimbylow, P.J. (1995) The development of realistic voxel phantoms for electromagnetic field dosimetry, in P.J. Dimbylow (ed.), *Voxel Phantom Development, Proceedings of an International Workshop held at the National Radiological Protection Board*, Chilton, UK, 1-7.
39. Netter, F.H. (1989) *Atlas of Human Anatomy*, CIBA-GEIGY Corp., Summit, New Jersey.
40. Spitzer, V.M. and Whitlock, D.G. (1998) *Atlas of the Visible Human Male*, Jones and Bartlett Publishers, Sudbury, MA.
41. Gandhi, O.P. and Furse, C.M. (1995) Millimeter-resolution MRI-based models of the human body for electromagnetic dosimetry from ELF to microwave frequencies. in P.J. Dimbylow (ed.), *Voxel Phantom Development, Proceedings of an International Workshop held at the National Radiological Protection Board*, Chilton, UK, 24-31.
42. Lau, R.W.M. and Sheppard, R.J. (1986) The modeling of biological systems in three dimensions using the time domain finite-difference method: I. The implementation of the model, *Phys. Med. Biol.* **31**, 1247-1256.
43. Sullivan, D.M. (1992) A frequency-dependent FDTD method for biological applications, *IEEE Trans. Microwave Theory Tech.* **MTT-40**, 532-539.
44. Pontalti, R., Cristoforetti, L., and Cescatti, L. (1983) The frequency dependent FDTD method for multi-frequency results in microwave hyperthermia treatment simulation, *Phys. Med. Biol.* **38**, 1283-1298.
45. Penn, J. and Cohoon, D. (1978) *Analysis of a Fortran Program for Computing Electric Field Distribution in Heterogeneous Penetrable Nonmagnetic Bodies of Arbitrary Shape Through Application of Tensor Green's Functions*, **USAFSAM-TR-78-40**, Brooks Air Force Base, TX.
46. Gordon, R.G., Roemer, R.B., and Horvath, S.M. (1976) A mathematical model of the human temperature regulatory system - Transient cold exposure response, *IEEE Trans. Biomed. Eng.*, **BME-23**, 434-444.
47. Iskander, M.F. and Khoshdel-Milani, O. (1984) Numerical calculations of the temperature distribution in realistic cross sections of the human body, *Int. J. Radiation Oncology Biol. Phys.* **10**, 1907-1912.
48. Bowman, H.F., Cravalho, E.G., and Woods, M. (1975) Theory, measurement, and application of thermal properties of biomaterials, *Ann. Rev. Biophys. Bioeng.* **4**, 43-80.
49. Chato, J.C. (1985) Selected thermophysical properties of biological materials, in A. Shitzer and R.C. Eberhart (eds.), *Heat Transfer in Medicine and Biology* **2**, 413-418.
50. Müller, M., Sachse, F., and Meyer-Waarden, K. (1996) Creation of finite element models of human body based upon tissue classified voxel representations, *Proc. First Users Conf. Nat. Lib. Med. Visible Human Project*, 1-5.
51. Walters, T.J., Mason, P.A., Ryan, K.L., Nelson, D.A., and Hurt, W.D. A comparison of SAR values determined empirically and by FD-TD modeling, in B.J. Klauenberg and D. Miklavčič, Eds., *Radio Frequency Radiation Dosimetry and Its Relationship to the Biological Effects of Electromagnetic Fields*, Kluwer Academic PublishersB.V., Dordrecht, The Netherlands, pp. 207-216.
52. Walters, T.J., Ryan, K.L., Belcher, J.C., Doyle, J.M., Tehrany, M.R., and Mason, P.A. (1998) Regional brain heating during microwave exposure (2.06-GHz), warm-water immersion, environmental heating, and exercise, *Bioelectromagnetics*, **19**, 341-353.

SUMMARY OF SESSION D: EXPERIMENTAL DOSIMETRY

N. KUSTER and T. J. WALTERS*
ETH Zurich, Lab of EMF & Microwave Elect., IFH-ETZ, ETH Gloriastr. 35, Zurich CH-8092, Switzerland
Veridian, Inc., San Antonio, TX, 78216

Pokovic, et al. represents an articulate state-of-the-art review of the methods and considerations that must be accounted for during near-field evaluations of RF transmitters, with an emphasis on cellular telephone handsets. The paper provides a review of safety standards and compliance requirements, as well as a discussion on the sources of uncertainty in compliance testing. A section on the experimental methods for performing near-field dosimetry in phantoms and free-space provides a balanced review of the pros and cons of each method. This is followed with a review of numerical methods for modeling near-field conditions including: Finite-Difference Time-Domain (FD-TD); the Finite-Element (FE) method; as well as several methods for numerical evaluation of complex antenna structures. The final section reviews special considerations for precise assessment of SAR for compliance testing, e.g., anatomical variability of head shape, effect of hand position, etc..

Pakhomov, et al. detail a technique for measuring local SAR within *in-vitro* exposure systems such as those used for nerve or brain slice preparations. The technique involves the use of a microthermocouple (MTC) to measure local temperature change in response to microwave irradiation during very brief irradiation periods, and then calculating SAR from the temperature change. While the use of thermography to indirectly determine SAR is well established, the available non-perturbing thermal probes lack the precision or response rate of the MTC, and thus are unable to measure the temperature change during the very brief linear portion of the heating curve, requiring that a line be fitted to the curve. This technique offers an attractive alternative to current techniques used to perform dosimetry in similar exposure systems (e.g. in vitro tissue baths, cell culture flasks).

Khizhnyak and Kiskin detail the use of infrared (IR) thermography for mapping SAR in biological tissue phantoms and biological systems exposed to millimeter-wave (mm-wave) irradiation. Because of the shallow penetration depth at mm-wave wavelengths, this method is ideally suited for performing dosimetry. In addition, the authors point out that the non-contact nature of the method eliminates measurement artifact. The IR systems described employ a focal plane array to collect the IR emissions, which is capable of adaquate sampling rates and thermal sensitivity to capture the initial, linear portion of the heating curve. Experiments employing this method are described. These experiments demonstrate the non-uniformity of mm-wave

157

B.J. Klauenberg and D. Miklavcic (eds.), Radio Frequency Radiation Dosimetry, 157-158.

fields produced under near-field exposure using various horn antennas. The possible mechanisms for the complexity of the fields and the difficulty of predicting these fields numerically are discussed.

Walters, et al. use both experimental (thermographic) and computer modeling (FD-TD) to demonstrate the profound influence of orientation of the biological system relative to the RFR source on local SAR in the rat. Additionally, the differences in local SAR in a number of anatomical sites, primarily in the brain, are examined. The authors compare the local SAR's from specific anatomical sites generated by the two methods. For the most part, the two methods produce good agreement, except when large "hot" or "cold" spots are predicted by the FD-TD method. In these cases, thermal conduction artifactually results in under- or overestimation of local SAR, respectively. However, the authors point out that while this may be a shortcoming of using thermographic methods to estimate local SAR, it emphasizes that SAR alone can not predict the local thermal environment produced by RFR exposure. Future computer models will also need to account for thermal transfer.

Vecchia and Polichetti give a basic overview about dosimetry at low frequency RF fields. The rationale of various safety standards around the world at lower frequencies and sources of exposure are described. The paper focuses on the exposure characteristics of proximity readers, which might be relevant due to their rather large dissemination. Based on experimental data, it is found that for, some devices and configurations compliance with the basic limits must be shown since the reference levels are exceeded in some cases. It is concluded that since all standards are based on the same literature using different conservative safety factors, it is sufficient to comply with one of the safety standards to exclude major health risks. The author recommends work toward harmonization of the differing standards.

Kiel discusses the quest for a personal dosimeter, specifically a molecular dosimeter. The requirements of a molecular dosimeter, as well as the possible candidates, are discussed. The artificial polymer diazoluminomelanin (DALM) is claimed to fit the requisite profile. The remainder of the paper describes the detailed experimental and computational approaches that clearly support the potential for the use of DALM as a personal dosimeter. Finally, DALM is proposed to provide a model for the basis for designing sensors that will respond to different chemical and physical microenvironments.

Markov focuses on dosimetry of magnetic fields in the radiofrequency range used for therapeutic applications. In particular, the determination of the magnetic field, impedance, and standing wave ratio for specific body targets using a commercially available device are reported.

Bernardi, et al. give an overview of dosimetric assessments for wireless technologies. It compares studies performed numerically and experimentally. However, since the variation in phone modeling, phantom modeling, and distance between source and phantom is considerable, a comparison between the different studies remains difficult. Studies on well defined benchmark setups are discussed. Dosimetry related to wireless local area networks is discussed with exposure of the eye of particular relevance.

EXPERIMENTAL AND NUMERICAL NEAR-FIELD EVALUATION OF RF TRANSMITTERS

Katja Poković, Michael Burkhardt, Thomas Schmid and Niels Kuster

Swiss Federal Institute of Technology Zurich

8092 Zurich, Switzerland

1 Introduction

Precise evaluation of the near-field generated by transmitters operating in the radio frequency (RF) range (30 MHz to 6 GHz) is still an engineering challenge not only requiring sophisticated tools but also involving different procedures with a multitude of parameters to be considered. This paper reviews the state-of-the-art of near-field evaluation with emphasis on compliance testing of handheld RF transmitters with safety limits. Similar tools and techniques can be utilized for the analysis and optimization of antennas operating in complex environments as well as for evaluation of special electromagnetic compatibility and interference problems occurring in the near-field of transmitters. Another area of applications is the analysis, evaluation and optimization of exposure setups used in RF safety research to investigate possible basic, therapeutic or adverse health effects from non-ionizing radiation.

2 Safety Standards and Compliance Requirements

In the frequency range of mobile communications, the physical unit used to define the safety limits of the major international and national safety standards around the world (e.g., [1], [2]) is the Specific Absorption Rate (SAR). SAR can be expressed in terms of the induced electric field strength or the temperature rise in tissue by:

$$SAR = \frac{dP}{dm} = \frac{\sigma}{\rho}E^2 = c\frac{dT}{dt} \tag{1}$$

where E is the root-mean-square value of the induced electric field strength, dT/dt the temperature rise, ρ the tissue density, σ the dielectric conductivity and c the specific heat capacity.

SAR limits are defined for the whole-body averaged as well as for local absorption. The latter, the spatial peak SAR limit, is also a spatially averaged volume in order to distinguish between non-hazardous strong local heating of the skin (e.g., RF burns) and possibly serious hazards caused by

159

B.J. Klauenberg and D. Miklavcic (eds.), Radio Frequency Radiation Dosimetry, 159-186.

exposure of larger tissue volumes. For example, [2] defines a spatial peak SAR value of 1.6 mW/g and a volume of 1 g tissue mass in the shape of a cube whereas [1] defines a limit of 2 mW/g and a volume of 10 g of continuous tissue mass of any shape. [3] and [4] require spatial averaging over 10 g with the shape of a cube. Because of the strong attenuation due to the skin effect at higher frequencies, the size and shape of the averaging volume is of great importance. For example, an averaging cube of 10 g instead of 1 g tissue mass reduces the requirement by more than a factor of two at higher frequencies. The defined SAR limits are not only volume averaged but also time averaged values, the latter of which introduces a relation between the absorbed power and the induced heat. The differences of time and volume averaging requirements between the guidelines/standards, which have significant implications with respect to compliance requirements, indicate that the various commissions follow different rationales and interpretations of the scientific basis, although the standards/guidelines are based on the same scientific bases.

The assessment of the basic limits requires considerable resources and in practice can only be conducted in a laboratory environment. To enable on-spot compliance checks, exposure limits in terms of incident field strengths have been defined which can be measured with standard field probes. The rationale of these derived or secondary limits is that if they are met then the basic limits are also not exceeded (worst case scenario). In quasi far-field conditions (i.e., field impedance ≈ 377 Ohms), it is sufficient to evaluate either the electric, magnetic or poynting field strength for demonstration of compliance. In the near-field of transmitters, however, compliance with the basic limits can only be demonstrated on the basis of incident field strength measurements if both the electric and magnetic field strength limits are met. Poynting field measurements do not have much significance with respect to characterizing the absorption in the near-field. The reason is that the field impedance can significantly deviate from 377 Ohms and whether the electric or magnetic coupling mechanism is dominant depends on the particular situation.

In the very close proximity of transmitters, field strengths are often considerably greater than the derived limits, such that compliance can only be demonstrated by assessing the basic limits directly. In many cases, the limits of whole-body average SAR are intrinsically met, since the average output power is less than 1 W and the users are heavier than 12 kg. For the spatial peak SAR, such worst-case considerations are not applicable for cellular phones, which typically have a maximum antenna input power of greater than a few mW, so that compliance can only be demonstrated by experimental or numerical evaluations.

During recent years various organizations have begun to develop standard procedures for compliance testing of handheld RF transmitters. The current status of the most important documents is summarized in Table 1.

Table 1: Basic compliance test requirements proposed in the USA [5], Europe [3] and Japan [4].

	FCC [5]	CENELEC [3]	ARIB STD-T56 [4]
based on	NCRP/ANSI [6, 2]	ENV[7]	RCR STD-38[8]
group	uncontrolled env.	general public	condition G
whole-body av. SAR	0.08 mW/g	0.08 mW/g	0.08 mW/g
spatial peak SAR	1.6 W/kg	2 W/kg	2 W/kg
averaging time	30 min.	6 min.	6 min.
averaging mass	1 g	10 g	10 g
volume shape	cube	cube	cube
phantom requir.	not defined	reasonable cross-section of users	several proposed phantoms
device position	standard	4 positions	normal

Progressive health agencies require that if a device passes the compliance test that it is unlikely that any user will be exposed to levels exceeding the safety limits when the device is properly used, i.e., in the fashion it was designed and marketed for. On the other hand, the test procedure should not be too conservative, i.e., unnecessarily inhibit the advancement of the technology. Although these are obvious and straight forward criteria, the implementation of such a procedure is everything but a trivial task. The reason is due to the complexity of the task and the small margins of current devices with respect to the safety limits (see Figure 11). Near-field evaluations involve many different procedures with a multitude of parameters to be considered and therefore require considerable engineering in order to keep the total uncertainty reasonably low. The different uncertainty components can be grouped into three main classes of uncertainty:

- The assessment uncertainty (measurement uncertainty or simulation uncertainty): This is the uncertainty for assessment of the spatial peak SAR value in a given SAR distribution within a given setup (e.g., head phantom). The uncertainty must be determined in such a manner that it is valid for all evaluations.

- The phantom uncertainty: This is the deviation of the technical setup (head phantom) with respect to definitions in the standard. Such could be a description of a standard phantom or the statement of a minimum coverage in percentage of the total user group. The assessment of this uncertainty requires knowledge about the dependence of the absorption on anatomical variations as well as about the affects of accessories such as optical glasses or jewelry. The

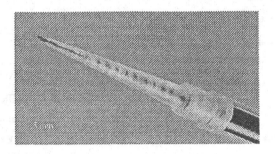

Figure 1: Thermistor temperature probe with a temperature range of 0°-60°. Sensitivity of temperature rise measurements is <0.15 mK/s (10 s sampling time) and RF susceptibility is <0.6 mK/s with 1 kV/m over 50 mm.

uncertainty of the phantom can be assessed once, such that it is valid for all RF transmitters operating in the near-field.

• The source uncertainty: This is the uncertainty of the spatial peak SAR assessed with a particular phone or a numerical representation of the phone compared to the phones produced in mass production for the market (output power, frequency response, modulation, amplifier, matching network, antenna, manufacturer tolerances, etc.). The uncertainty of the position with respect to the phantom can also be considered as part of the source uncertainty. The source uncertainty is especially crucial, since it involves the object under test and is therefore dependent upon it. The most simple way to handle this problem is by assessing the spatial peak SAR of different samples of the mass production and using appropriate statistics to assess the source uncertainty.

The ability to perform dosimetric evaluations with low uncertainties will be the main criteria for the selection of an evaluation technique for standardized compliance testing of RF transmitters with safety standards, in addition to investment and labor costs.

3 Experimental Techniques

3.1 Temperature Probes and Radiation Thermometry

A very straight forward but nevertheless elegant design of a *temperature probe* for measurements in RF electromagnetic environment was presented by Bowman [9] in the mid seventies. It is based on a high-resistance thermistor as sensor which is connected to four resistive leads (approx. 160

Table 2: Comparison of commercially available temperature measurement systems.

Physical principle	Fluorescent decay	Cavity resonator	Thermistor
Sensitivity (T) (sampling time)	±0.1°C (1 s)	±0.1°C	±0.005°C (0.1 s)
Sensitivity (dT/dt) (10 s of exposure)	±15 mK/s	±10 mK/s	±0.15 mK/s
Sensitivity to RF exposure (835 MHz)			
- resistive lines parallel to E-field	-	-	<0.5 mK/s[a]
- resistive lines normal to E-field	-	-	<0.1 mK/s[a]

[a]Measured by exposing the first 6 cm of the line from the tip to an incident field of 950 V/m.

kOhms/cm) whereby the current is induced in two of these and the voltage monitored across the other two. By that, the probe becomes largely RF transparent but not completely insensitive to RF exposures and temperature changes of the leads (Table 2). Nevertheless, precise temperature rise measurements can be conducted when the necessary precautions for a well designed setup are made.

In [10] a standard voltmeter module was suggested to drive the sensor and to evaluate the signal. Lately, it has been shown that the sensitivity can be increased up to ±0.15 mK/s (10 s evaluation time) by using specialized electrometer grade amplifiers and software for filtering [11].

In the eighties temperature sensors utilizing optical effects were developed and commercialized for applications such as temperature control in high voltage transformers, industrial microwave heating, hyperthermia, etc. The effects employed were temperature dependent fluorescent decay of phosphorescent layers or interferometric microshift of cavity resonators. Since the sensors are driven and monitored by optical fibers, they provide great immunity to RF exposure. However, insufficient sensitivity makes these probes unsuitable for SAR evaluation in the range of the safety limits.

Thermistor probes were already applied in the late seventies for exposure assessments of handheld communications devices inside human phantoms [12, 13]. However, the minimum measurement time per point and the requirement of thermal equilibrium prior to each measurement make the technique unsuitable for standardized dosimetric evaluations, which usually involve several hundred measurement points. Today, temperature probes in the context of dosimetry are applied for temperature transfer calibrations of E-field probes as well as for specialized dosimetric tasks [11, 14, 15].

A more time efficient technique is *Radiation thermometry*, which was already used in the early seventies to assess the temperature rise due to absorption of electromagnetic energy [16]. The technique can be easily applied without any interference and provides valuable information about the temperature profile on surfaces. More recently, it has also been applied in the context of human exposure from mobile phones using dry ceramic phantoms [17, 18] or dry phantoms composed of resin and graphite [19]. For utilization during standardized compliance testing the technique has severe limitations with respect to sensitivity and requirements for ambient control [20], especially since volume scans are practically impossible due to the restrictions to certain cross-sections.

3.2 Dosimetric Field Probes

Miniaturized, isotropic three-dimensional diode loaded E-field probes were first presented in [21]. Since than, they have been used for various dosimetric investigations [22, 23]. The probes consist of three small orthogonally arranged dipoles which are directly loaded with Schottky diodes. The theory of this approach has been described in detail in [24].

Originally, transmission lines with relatively low resistance ($<50\,k\Omega$) were used, resulting in secondary modes of reception in particular at lower frequencies. More recently, improved performance over a broader frequency range ($10\,MHz$-$3\,GHz$) was achieved by employing high-ohmic lines and distributed filters just behind the sensor [25, 26]. Performance evaluations of several probe designs have revealed that probes with orthogonally arranged sensors result in spherical deviations from isotropic response of between ±0.6 and $\pm3.5\,dB$. These strong deviations are caused by diffraction inside the dielectric material of the probe which is significantly different than the surrounding media. Significantly improved performance was achieved by geometrically correcting the alignment of the sensor dipoles such that their receiving pattern is orthogonal with respect to the incident field [27, 26]. Probes with deviations from spherical isotropy of better than $\pm0.3\,dB$ in tissue simulating material have been realized [28, 26] (see Figure 2). More recently, miniaturized E-field probes have been developed for special tasks such as the characterization of larger probes (e.g., boundary effects, spatial resolution, etc.), dosimetric measurements inside small structures (e.g., small animals, in vitro dishes, etc.) and special physical phenomena. The one-sensor probe shown in Figure 4 has an outer tip diameter of only 1 mm. The orientation of the dipole sensor of 0.8 mm

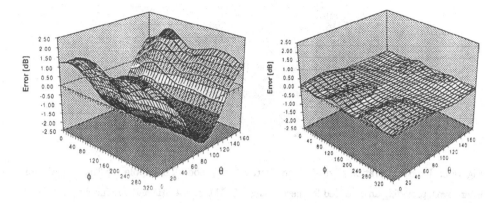

Figure 2: Spherical receiving patterns determined for brain-tissue simulating liquid evaluated for two different probes. Left: classic probe design in which the three sensor dipoles are orthogonally arranged; right: new design in which the spatial orientation of the sensors depart from the orthogonal arrangement in order to compensate for the field distortion inside the probe caused by the dielectric probe material (core, protective shell, etc.).

length inside the tip has been optimized to enable isotropic evaluation (deviation $<\pm0.2\,$dB) by rotating the probe around its axis and sequentially taking three measurements in 120° steps. Rotation at the spot ($< \pm0.1\,$mm) can be realized by using a high precision robot in combination with a light beam probe alignment device. The probe provides about the same spatial resolution as temperature probes but with significantly higher sensitivity ($<0.05\,$mW/g) and independence from temperature fluctuations.

Several groups are currently working on developing fiberoptical E-field sensors for various kinds of EMC applications utilizing electro-optical effects, e.g., the Pockels effect [29]. Advantages of this technique are the non-metallic construction, broad-band applicability and foremost the ability to measure in the time domain. However, the sensitivity and spatial dimensions of the suggested implementations do not meet current requirements for dosimetric evaluations.

The classic calibration technique for E-field probes in lossy media is transfer calibration using temperature probes [14, 15]. As an alternative, a waveguide technique has been suggested in [30]. A new approach also based on waveguides has recently been introduced providing improved precision and traceability of the procedure to power measurements, as well as great robustness and efficiency [31]. The standard uncertainty of this technique was assessed to be better than $\pm3.5\%$ but requires several waveguide setups to cover the frequency range between 800 MHz - 2.5 GHz. Due to the increasing size of waveguides with larger wavelengths, at lower frequencies calibration

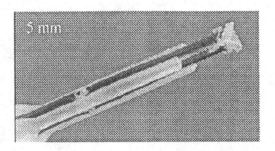

Figure 3: Latest generation of dosimetric electric field probe. The dipoles of 3 mm length are interleaved, providing an extended frequency range of 30 MHz to 4 GHz and a reduced tip diameter including protective shell of only 4 mm. The dynamic range is 0.001 mW/g to 100 mW/g. The deviation from spherical isotropy in brain simulating liquid at 900 MHz is < ±0.3 dB (axial isotropy < ±0.1 dB).

is best performed with transfer calibration using temperature probes. Another proposed technique is based on computer simulation. This method requires a detailed model of the probe and provides lower precision than experimental calibration techniques [15].

In addition to calibration techniques, procedures have been developed to fully characterize the performance of probes in tissue simulating liquids. This includes assessment of the uncertainties due to field disturbances, deviation from spherical isotropy, spatial resolution, boundary effects and secondary modes of reception [26].

3.3 Free-Space Field Probes

Free-space scans of the electric and magnetic near-field of RF transmitters provide important information about the quality of the RF design of the antenna and device.

For example, poor design of handheld wireless communication devices may result in significantly impaired radiation performance due to large increases of power loss inside the device and through absorption if operated in the closest vicinity of lossy structures. Even malfunction might occur in special near-field environments. Hence, free-space near-field evaluations can constitute a powerful design tool.

In special cases, the magnetic and electric near-field scans can also be used to demonstrate compliance with safety limits, e.g., indoor base stations or other indoor antennas (computer mounted antennas, etc.).

For these applications, near-field probes optimized for measurement in air have been developed.

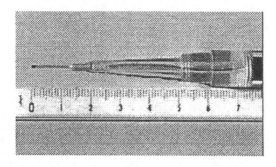

Figure 4: Miniaturized E-field probe designed and optimized for special tasks (e.g., study of boundary effects, dosimetry in small animals, etc.). The deviation from spherical isotropy in brain simulating liquid between 900 MHz and 1.8 GHz is < ±0.2 dB.

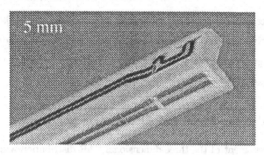

Figure 5: Free space electric field probe. The frequency range is from 100 MHz to 6 GHz with a dynamic range of 2 to >1000 V/m. The length of the dipole is 3 mm. The tip diameter including protective shell is 8 mm. The deviation from spherical isotropy is < ±0.4 dB.

In order to obtain maximum information, it is often advantageous to align the three sensors to a specific coordinate system with respect to the antenna. To easily enable this, the isotropic electric field probe shown in Figure 5 has one sensor aligned to the probe axis and the other two sensors normal to it. Equivalent to the optimization procedure employed in the design of the dosimetric probes, the spatial orientation of the sensors has been adjusted in order to compensate for the field distortion inside the probe caused by the core, substrate and protective shell. The achieved spherical isotropy is better than ±0.4 dB.

Figure 6 shows an isotropic near-field H-field probe optimized for the mobile frequency range [28]. This recently introduced probe consists of three orthogonally and concentrically arranged loops (Figure 6). The 3.8 mm loops with resistively loaded detectors were designed to achieve optimal sensitivity in the desired frequency range of 300 MHz to 2.5 GHz (deviation from isotropy:

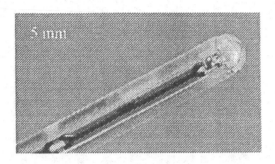

Figure 6: Isotropic magnetic field probe optimized for the frequency range from 300 MHz to 2.5 GHz with a dynamic range of 0.01 A/m to 2 A/m at 900 MHz. The diameter of the loops is 3.8 mm; the tip diameter including protective shell is 5 mm. The deviation from spherical isotropy is $< \pm 0.2$ dB.

< 0.2 dB). In free space, probes are usually calibrated in the far-field, e.g., [32]. Due to the small size of these near-field probes and their RF transparency, calibration in waveguides has been proposed for the frequency range between 800 MHz - 2.5 GHz and TEM-cells below 1 GHz [15], providing a precision of better than $\pm 5\%$.

3.4 Automated Near-Field and Dosimetric Scanners

The experimental assessment of the three-dimensional SAR distribution within a phantom easily involves measurements at several hundred points. Especially at higher frequencies, the locations of these points with respect to the phantom must be known with the greatest precision in order to obtain repeatable measurements in the presence of rapid spatial attenuation and field variations. High precision is especially required to accurately evaluate the SAR close to the surface, since the physical dimensions and boundary effects prevent measurement directly at the surface, i.e., the values at the surface must be obtained by careful extrapolations with appropriate functions. Due to the considerably large number of measurement points and extensive data processing required, it is obvious that the measurement process must be highly automated if such measurements must be performed routinely.

Several automated scanning systems based on E-field probes have been implemented. Systems developed in the 80s range from one-dimensional positioners [33] to three-axis scanners [34] and on to six-axis robots [35]. The first version of the latter goes back to the early 80s and has been continually improved and extended since then.

Systems explicitly designed for evaluation of handheld or body mounted transmitters are described in [25], [35] and [36].

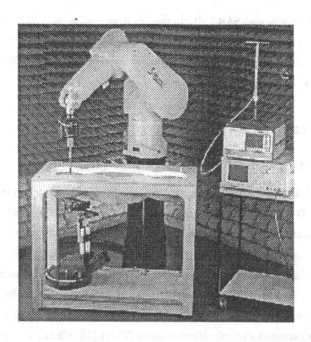

Figure 7: DASY3 near-field measurement system with generic twin phantom.

Figure 7 shows the recently introduced third generation of such a scanner. It incorporates a high precision robot (working range greater than 0.9 m and a position repeatability of better than ±0.02 mm), advanced dosimetric E-field probes (see previous sections), an optical proximity sensor for automated positioning of the probe with respect to the phantom surface (precision better than ±0.2 mm) and sophisticated software for data processing and measurement control. The generic twin phantom (enables testing of left-hand, right-hand and body mounted devices without requiring a change of phantoms) was developed with the objective of covering the maximum exposure occurring in over 80% of the entire user population [36]. A test procedure based on a calibrated dipole has been integrated which enables verification of system operation within its specifications. This check also enables assuring that all laboratories working with the above systems are assessing the same SAR values within ±12 % (k=2).

The analysis of the uncertainty budget performed in compliance with the NIS81 [37] and the NIST1297 [38] documents has shown that a precision for spatial peak SAR evaluation of better than ±12 % (k=1) is achieved, which is excellent when compared to the common uncertainty of far-field measurements. The uncertainty for the 95% confidence level (k=2) of < ±24 % is acceptable for compliance testing.

In addition to the spatial peak SAR limit, the same instrumentation also enables determination of the total energy absorbed in the user, which is another important measure for device performance.

Several other systems based on similar concepts have been developed at different universities [39, 40] or commercialized [41, 42].

4 Numerical Tools

4.1 Finite-Difference Time-Domain Technique

Although many different numerical techniques have been developed and applied to dosimetric research in the last 30 years [43], the *Finite-Difference Time-Domain* (FDTD) technique is currently clearly the dominant technique for exposure assessments. It was initially introduced by Yee [44] and basically involves discretization of the entire computational domain into voxels of homogeneous material and calculation of the electromagnetic fields in the time-domain following an explicit leapfrog algorithm starting from initial field conditions. Initial problems with open domain boundaries were subject of extensive research in the eighties. A boundary condition which has gained much popularity is the Berengers *Perfectly Matched Layer* (PML) [45], offering an increased dynamic range of up to 70 dB for numerical computations. State-of-the-art FDTD modeling comprises a variety of approaches such as *Contour-Path* modeling [46], *Finite-Volume* formulations [47], graded grids [48], partially filled cells [49], subgridding [50] and CAD data import [51].

FDTD has been widely used to perform near-field evaluation of RF transmitters and to obtain dosimetric information for non-homogeneous lossy bodies. In particular it has been used to assess a variety of RF transmitter related exposure problems, e.g., in [51], [52], and near-field evaluation of novel antennas for RF transmitters, e.g., in [53], [54]. Many of the exposure related studies have been based on human head models developed from clinical MRI data. Due to straight forward implementation of the algorithm, simple modeling of complexly shaped structures and available computational power, FDTD has become the first choice for RF transmitter related numerical exposure assessments. Major advantages are the direct modeling within rectilinear grids and the robustness of the technique, enabling the implementation of easy-to-use tools. The main limitation is the intrinsic problem of simulation techniques in assessing the result uncertainty with respect to actual conditions (solution validation). Additional restrictions are seen in the rather difficult departure from the commonly used rectilinear grid and in cell size limitations which prevent the modeling of very detailed handset structures which might be essential to fully characterize the RF.

4.2 Other Techniques

Approximation formulas based on analytical considerations are often well suited for assessing worst-case exposure at larger distances from the antenna. The near-field of complex antenna structures can best be numerically evaluated using boundary techniques such as MoM, GMT, etc. Several codes specifically implemented for antenna design purposes are available [55], [56].

Other techniques, lacking the flexibility of FDTD, are still used to obtain reference solutions for canonical problems or investigating basic interaction mechanisms, e.g., GMT [14], [57].

Another technique promising flexibility for non-homogeneous modeling is the Finite-Element (FE) method, which was made popular in electromagnetics by Silvester and Ferrari [58]. With FE the whole computational space is discretized, allocating the unknown field values originally in the nodes of the finite-element mesh. Linear or polynomial expansion functions are associated within each element. A variational method or a method of weighted residuals is used to obtain a system of equations with a sparse matrix. Original problems with spurious modes and open domains encouraged work related to hybrid schemes paired with boundary element methods and edge-element techniques [59]. Recent advances have been made toward adopting Berenger-style boundary conditions [60]. Early work on human tissue interaction with electromagnetic energy using the Finite Element method was reported in [61] for microwave absorption by a cranial structure. The thermal response of a coarse block model of a human in the near-zone of a resonant thin-wire antenna from 45-200 MHz using MoM and the FE method was calculated in [62]. More recently, the technique has been applied to study absorption during MRI diagnostics [63].

Although FE allows the modeling of more arbitrarily shaped bodies, it has not been used as extensively for RF transmitter related absorption studies as FDTD, since the problem of generating robust 3D meshes of arbitrary complex 3D bodies has not yet been solved.

Different hybrid approaches combining various methods and realized in explicit and implicit schemes have been proposed, developed and applied to the exposure of dielectric bodies, e.g., GMT and FE in [64] and GMT and MoM in [65]. However, the intrinsic problem of hybrid approaches, an implementation in which modeling of realistic setups is straight forward and robust and which does not require much insight of the user, has not yet been realized.

5 Special Considerations for Compliance Testing

The availability of reliable instrumentation or simulation techniques for precise assessment of SAR distribution inside tissue simulating materials is the primary condition for the implementation of compliance testing standards. However, in order to enable generalization of dosimetric evaluations

172

conducted with a particular setup to the exposure occurring in a specific cross-section of the user population, extensive information about the dependence of the absorption mechanism on various parameters is necessary. This includes:

- the dominant absorption mechanisms, in order to derive the parameters which most affect absorption;

- the effects of anatomical variability, such as head shape, tissue distributions and tissue parameters, etc.;

- the effect of the hand;

- the effects of accessories such as metallic frames of optical glasses, jewelry, medical implants, etc.;

- the effect of the variability of the position of the phone with respect to the head;

- the requirements for handset modeling, in case the instrumentation or procedure does not allow testing of the actual physical device.

A number of studies investigating these issues have been conducted during recent years, the results of which are summarized in the following.

5.1 General Absorption Mechanism in the Near-Field

The energy absorption mechanism in the close near-field of dipole antennas operating above 300 MHz was investigated in [14]. The dependence of the absorption on the frequency, electrical dipole length, distance between dipole and tissue, and shape of the lossy structures was investigated. The results revealed that the dominant interaction mechanism for all configurations is inductive coupling, i.e., surface/eddy currents induced by the incident magnetic field. The strong dominance of inductive coupling over capacitive coupling enabled the proposition of an approximation formula, the precision of which was assessed to be better than ±3 dB independent of frequency, distance, tissue parameters, antenna length, etc. These findings were essentially confirmed by [66]. In [67], it was further shown that this approximation also results in very accurate results for helix antennas. In order to allow for estimations of the spatial peak SAR, the scope of the approximation was extended in [43].

The main conclusion of these findings is that both absorption as well as the spatial peak SAR values are proportional to the square of the magnitude of the magnetic field (i.e., ≈current/distance) and not to the input power. In other words a matched antenna with a feedpoint impedance of

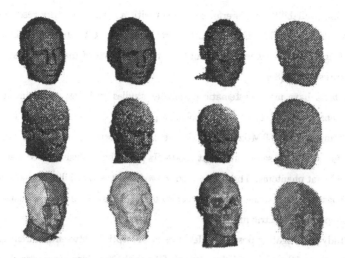

Figure 8: Numerical and experimental head phantoms of varying anatomical complexity, shape, origin and age. The voxel sizes of the numerical head phantoms vary between $10.5\,mm^3$ and $0.125\,mm^3$. In the third row, two experimental shell phantoms, a 5 tissue phantom described in [68] and its numerical representation are shown.

10 Ohms has the potential of inducing five times higher absorption when compared to a 50 Ohm antenna for the same antenna input power.

5.2 Modeling of the Human Head

Many different phantoms have been used in the past to assess RF transmitter related exposure, ranging from simply shaped bodies such as spheres, boxes and cylinders to more realistic complex phantoms including MRI and CT derived models with resolutions down to $0.125\,\text{mm}^3$.

Whereas early numerical assessments were largely limited to simple geometries due to computational restrictions, i.e., spheres [69, 70], prolate spheroids [71, 72], multilayered slabs [73] or brick phantoms [74, 75], complex experimental phantoms were used from very early on. Anatomically shaped homogeneous phantoms filled with tissue simulating liquids or gels were used in [13, 23, 76] as well as dry ceramic phantoms [17, 18, 19, 77]. In addition, studies have been performed using multi-tissue phantoms [78, 33] and phantoms based on real human skulls packed with different tissues [12, 33]. A realistic phantom of the whole body of a man including simulated bone, brain, muscle and lungs was used in [79]. Most recently a 5-tissue head phantom consisting of skin, muscle, bone, eye and brain tissue has been constructed and made commercially available [68, 78].

The increasing availability of powerful computers with adequate memory resources from the early 90s onwards has enabled the simulation of realistic MRI, CT, etc. based head phantoms using FDTD (Figure 8). Today, resolutions of $0.125\,mm^3$ for parts of the head are state-of-the-art on high-end workstations [80].

These new tools have made systematic parameter studies with respect to internal anatomy possible. The absorption in several high-resolution phantoms derived from MRI data of various persons was compared at 900 MHz for exactly the same exposure conditions in [81]. A similar study was performed by the same authors at 1800 MHz [82]. The analysis showed a small variation between the different phantoms. The studies also showed that the variability of the tissue structure can be encompassed by replacing the non-homogenous structure by a homogeneous phantom of appropriate shape and dielectric material.

An early study investigating possible differences in absorption between children and adults by scaling the phantom of the adult male by a factor of 0.7 to the head size of a 1 year old infant found comparable or lower values for the smaller head model [83]. Another study on possible differences in absorption between children and adults was reported in [84] by scaling the phantom used in [85] down to the size of a child according to body size and weight. Significantly higher 1 g averaged SAR at 835 MHz and a greater penetration depth was found in the case of the children's heads. These findings could not be confirmed by [86] using head models of MRI data of 3 and 7 year old children as well as differently scaled heads. No significant increase in averaged SAR values or penetration depth was found when the dipole source was kept at the same distance from the head. The spatial SAR values for children were within the variations found between different adults, suggesting that the maximum exposure occurring in the user group including adults and children can be assessed by using a single phantom of appropriate shape and composition. An explanation for part of the differences found in [85] might be the closer proximity of the source for the scaled heads as compared to adult heads.

Although the head shape does not significantly alter the absorption, it is nevertheless of great importance, since it defines the distance between the source of the magnetic field and the tissue for a given position of the RF transmitter with respect to the head. The compliance testing requirement to cover the highest absorption occurring in real life can be satisfied if the distance between the phantom and phone parts in the test setup is not larger than the minimum distance occurring within the user group between the phone and the skin. In order to develop a phantom satisfying this requirement, data on the head shape in the ear region of a total of 52 adult volunteers (male and female) was measured, from which a generic phantom head was derived. This was constructed so that the distance between the phone structure and phantom surface was always smaller than

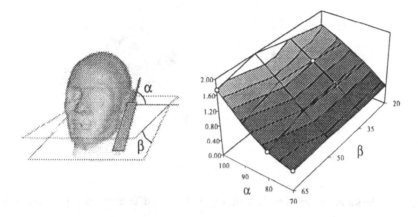

Figure 9: The spatial peak SAR value (averaged over a cube of 1 g tissue) induced by a commercially available GSM phone (carrier frequency: 900 MHz; nominal antenna input power: 0.25 W) as a function of the phone's position. The angles α and β were varied (the phone's earpiece was kept fixed with respect to the entrance of the auditory canal). The dots represent the four positions defined in the CENELEC document prES59005.

for 90% of the investigated people [36]. The thickness of the compressed ear was also evaluated, resulting in an average thickness of about 6 mm and a 10% percentile thickness of 4 mm [36]. So far, no investigation on the thickness of the compressed ear of children has been conducted.

5.3 Effect of the Hand

In [87] numerical research was focused on the effect of the head and hand on the performance parameters of monopole and PIFA antennas mounted on a simplified radio. It was reported that the antenna impedance in the case of the monopole was largely unaffected by both: the presence of the hand at the investigated distance, and the complexity of head modeling. However, radiation patterns were affected to a greater extent.

Authors in [88] report that the varying hand positions scarcely affect maximum spatial SAR values as long as the hand does not shade the antenna. [89] and [83] report that modeling a hand around the device results in lower spatial peak SAR values than the same configuration without the hand at 900 MHz. At 1800 MHz the opposite result was found in [83].

In the case of a sleeved dipole antenna the maximum spatial SAR values were unaffected [90] at 900 MHz whereas decreased values in the range of 30% were found for a whip antenna at the same frequency.

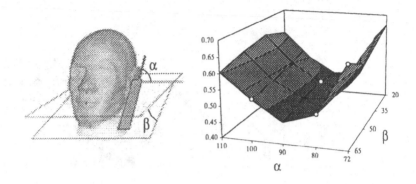

Figure 10: The same evaluation as shown in Figure 9 with a different commercially available GSM phone.

The problem inherent in these numerical studies is that the hand was simulated using simple block models which can only poorly represent the complex anatomy of a real hand. Due to the difficulties of modeling a realistic hand holding the phone in various positions, the effect of the hand was experimentally studied in [36] using a real hand. The studies were performed using commercially avaliable cellular phones. The frequency bands investigated were 450, 900 and 1800 MHz. Each phone was grasped in three different ways. The highest spatial peak SAR values of these three ways of grasping/holding the phone resulted in no or only a slight reduction of SAR values as compared to the values assessed in the absence of the hand. This suggests that the upper exposure range occurring under real-life conditions is best described by neglecting the effect of the hand.

5.4   Metallic Accessories & Environmental Effects

The effects of metallic accessories, such as spectacle frames and jewelry, have been investigated in [78] and summarized in [36] based on worst-case scenarios. The authors concluded that it is unlikely that metallic accessories significantly increase absorption.

Possible enhancement in SAR due to medical implants was investigated in [91], simulating resonant wire and disc structures inside and outside a tissue-simulating spherical body. The excitation was a resonant dipole. Enhancement of up to a factor of about 40 were found for the local peak SAR and a factor of 1.6 for the spatial peak SAR value averaged over 1 g. Similar results were reported in [78] for various tissue types in which enhancement factors of several hundred times were reported. Such discrepancies are not surprising since the local peak SAR strongly depends on the sharpness of the metallic structures.

5.5   Handset Position

Figures 9 and 10 show the strong dependence of the spatial peak SAR on the position of the phone at the head. Even small changes of the position can change the spatial peak SAR value significantly. Such strong dependences were expected on the basis of the absorption mechanism. Figures 9 and 10 also demonstrate that the variations of the exposure in function of the angles α and β are strongly dependent on the design of the phone. This underlines the importance of precisely defining the shape of the phantom and test positions in order to achieve reproducible results in various laboratories. It also emphasizes the importance of phone design. Another conclusion from these results is that the average user exposure cannot be determined by evaluating only one position.

5.6   Handset Modeling

Since the RF current flowing on antenna and phone (distribution and magnitude) is determined by the RF design and the distance of these currents from the the tissue by the geometric design of the phone, the spatial peak SAR value greatly varies between different phone designs (see Figure 11). In order to obtain an idea of the variations within the same type of phone, three samples of each of the 18 phones were evaluated. Large variations of up to 30% were found even though the output power at the adapter terminal was within ±0.1 dB.

Another study underlined the high sensitivity of the current distribution on absorption [52]. The comparison of numerical and experimental procedures showed that the spatial peak SAR differed by more than 50% for some phone positions, even though the field magnitude and current distribution differed only slightly.

In view of the fact that the current distribution can be altered by minor design details in the phone (e.g., matching network, capacitive coupling between various parts of the device, etc.), compliance testing requiring modeling of the phone is not practical due to the fundamental problems involved in assessing uncertainties across all differences arising from modeling the phone and the actual phones including manufacturing tolerances. Experimentally, this problem can be effectively handled by evaluating the spatial peak SAR of various samples of the mass production and using appropriate statistical methods to assess the uncertainty source.

Studies in which the position and phone can be precisely described show good agreement between measurements and computations [92, 93, 94].

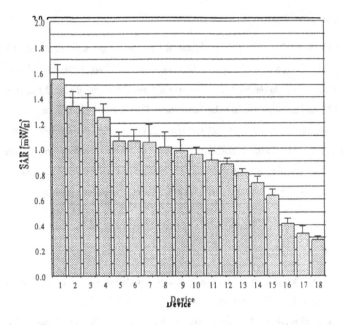

Figure 11: Eighteen different GSM devices were tested according to the CENELEC prES59005 document (Channel 62 = 902.4 MHz; Nominal time-averaged antenna input power (Class 4) = 0.25 W). Three samples of each phone were tested in the position which resulted in the greatest spatial peak SAR out of the four positions defined in the standard. The SAR values shown are the assessed spatial peak SAR values averaged over 10 g and the standard deviation among the different sample units.

6 Conclusion

Considerable progress has been made in experimental and numerical dosimetry during the last few years. Today a broad range of tools is available enabling precise dosimetric analysis for the frequency range of mobile communications.

These new techniques and tools have been applied to investigate the sensitivity of the effects of various parameters on absorption in the user's body caused by handheld RF transmitters. This information was essential in enabling the development of measurement setups for compliance testing. Several of these setups have been made commercially available, so that dosimetric evaluations of handheld RF transmitters have today become routine.

Optimal setups for experimental analysis of exposure setups for biological experiments are best evaluated by means of numerical tools. However, due to the large uncertainties introduced by

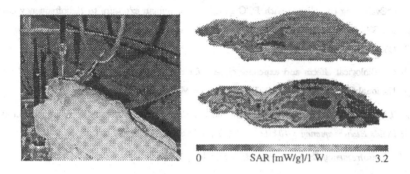

0   SAR [mW/g]/1 W   3.2

Figure 12: RF safety research requires specialized tools to analyze, evaluate and optimize the exposure setup for specific applications. Left: Experimental verification of the numerical dosimetry at specific locations; Right: Results of the numerical analysis.

modeling, any setup should also be experimentally validated prior to the biological experiment using specialized tools (see Figure 12).

Further improvements in experimental and numerical techniques for the evaluation of the near-field of RF transmitters can be expected in the next few years. The greatest potential for improvement is seen in tools providing comprehensive information for analysis and optimization of antennas which must operate efficiently in complex environments. The future will also see a new area of near-field evaluation which is closely related to exposure assessment, namely, the electromagnetic interference and compatibility problems that arise in wireless medical and life support systems. These problems will require new tools, such as more advanced scanners providing not only accurate amplitude but also phase information and enhanced spatial resolution (sub-millimeter).

References

[1] ICNIRP, *CLC/TC211: Human Exposure to Electromagnetic Fields*, Mar. 1998.

[2] ANSI/IEEE, *C95. 1-1992, IEEE Standard for Safety Levels with Respect to Human Exposure to Radio Frequency Electromagnetic Fields, 3 kHz to 300 GHz*, New York, NY 10017, 1992.

[3] CENELEC, *prES 59005, CLC/TC211 (SEC) 17, Considerations for evaluation of human exposure to Electromagnetic Fields (EMFs) from Mobile Telecommunication Equipment (MTE) in the frequency range 30 MHz - 6 GHz*, Brussels, Mar. 1998.

[4] ARIB, *STD-T56, Specific Absorption Rate (SAR) Estimation for Cellular Phone*, Jan. 1998.

180

[5] FCC, "Evaluating compliance with FCC guidelines for human exposure to radiofrequency electromagnetic fields", Tech. Rep. OET Bulletin 65, Federal Communications Commission, Washington, D.C. 20554, 1997.

[6] NCRP, "Biological effects and exposure criteria for radiofrequency electromagnetic fields", Tech. Rep., National Council on Radiation Protection and Measurement , Report No. 86, 1986.

[7] CENELEC CLC/TC111B, European Prestandard (prENV 50166-2), Human Exposure to Electromagnetic Fields High-Frequency : 10 kHz - 300 GHz, CENELEC, Brussels, Jan. 1995.

[8] ARIB, Radiofrequency Exposure Protection Standard, 1993.

[9] R.R. Bowman, "A probe for measuring temperature in radio-frequency-heated material", IEEE Transactions on Microwave Theory and Techniques, vol. 24, no. 1, pp. 43–45, Jan. 1976.

[10] R. G. Olsen and R. R. Bowman, "Simple nonperturbing temperature probe for microwave/radio frequency dosimetry", Bioelectromagnetics, vol. 10, pp. 209–213, 1989.

[11] M. Burkhardt, K. Pokovic, M. Gnos, T. Schmid, and N. Kuster, "Numerical and experimental dosimetry of petri dish exposure setups", Bioelectromagnetics, vol. 17, pp. 483–493, 1996.

[12] Q. Balzano, O. Garay, and F. R. Steel, "Energy deposition in simulated human operators of 800-MHz portable transmitters", IEEE Transactions on Vehicular Technology, vol. 27, no. 2, pp. 174–181, Nov. 1978.

[13] Q. Balzano, O. Garay, and F. R. Steel, "Heating of biological tissue in the in the induction field of VHF portable radio transmitters", IEEE Transactions on Vehicular Technology, vol. 27, no. 2, pp. 51–56, May 1978.

[14] N. Kuster and Q. Balzano, "Energy absorption mechanism by biological bodies in the near field of dipole antennas above 300 MHz", IEEE Transactions on Vehicular Technology, vol. 41, no. 1, pp. 17–23, Feb. 1992.

[15] K. Meier, M. Burkhardt, T. Schmid, and N. Kuster, "Broadband calibration of E-field probes in lossy media".

[16] A. W. Guy, "Analyse of electromagnetic fields induced in biological tissues by thermographic studies on equivalent phantom models", IEEE Transactions on Microwave Theory and Techniques, vol. 19, no. 2, pp. 205–215, Feb. 1971.

[17] T. Kobayashi, T. Nojima, K. Yamada, and S. Uebayashi, "Dry phantom composed of ceramics and its application to SAR estimation", IEEE Transactions on Microwave Theory and Techniques, vol. 41, no. 1, pp. 136–140, 41 1993.

[18] T. Nojima, S. Nishiki, and T. Kobayashi, "An experimental SAR estimation of human head exposure to UHF near fields using dry-phantom models and a thermograph", IEICE Transactions on Communications, vol. 77, no. 6, pp. 708–713, June 1994.

[19] A. Antolini and A. Leoni, "Thermographic method for the determination of SAR caused by cellular phones", *CSELT*, vol. 25, no. 1, pp. 131–137, Feb. 1997.

[20] T. C. Cetas, "Practical thermometry with a thermographic camera – calibration, and emittance measurements", *Rev. Sci. Instrum.*, vol. 49, no. 2, pp. 245–254, 1978.

[21] H. I. Bassen, M. Swicord, and J. Abita, "A miniature broad-band electric field probe", *Annals New York Academy of Science*, vol. 20, no. 5, pp. 481–493, 1975.

[22] G. H. Wong, S. S. Stuchly, A. Kraszewski, and M. A. Stuchly, "Probing electromagntec fields in lossy spheres and cylinders", *IEEE Transactions on Microwave Theory and Techniques*, vol. 32, no. 8, pp. 824–828, Aug. 1984.

[23] Indira Chatterjee, Yong-Gong Gu, and Om P. Gandhi, "Quantification of electromagnetic absorption in humans from body-mounted communication transceivers", *Transactions on Vehicular Technology*, vol. 34, no. 2, pp. 55–63, May 1985.

[24] H. I. Bassen and G. S. Smith, "Electric field probes - a review", *IEEE Transactions on Microwave Theory and Techniques*, vol. 31, no. 5, pp. 710–718, May 1983.

[25] T. Schmid, O. Egger, and N. Kuster, "Automated E-field scanning system for dosimetric assessments", *IEEE Transactions on Microwave Theory and Techniques*, vol. 44, no. 1, pp. 105–113, Jan. 1996.

[26] Katja Poković, Thomas Schmid, and Niels Kuster, "Design and characterization of E-field probes for lossy media", 1999, submitted.

[27] K. Pokovic, T. Schmid, and N. Kuster, "E-field probe with improved isotropy in brain simulating liquids", in *ELMAR'96*, Zadar Croatia, June 1996, pp. 172–175.

[28] Katja Pokovic, Thomas Schmid, Oliver Egger, and Niels Kuster, "High precision near-field scanner for analysis of handheld transmitters", in *USNC/URSI National Radio Science Meeting*, Atlanta, USA, June 1997, p. 196.

[29] M. Schwerdt, J. Berger, B. Schueppert, and K. Petermann, "Integrated optical E-field sensors with a balanced detection scheme", *IEEE Transactions on Electromagnetic Compatibility*, vol. 39, no. 4, pp. 386–390, Nov. 1997.

[30] D. Hill, "Waveguide technique for the calibration of miniature implantable electric-field probes for use in microwave-bioeffects studies", *IEEE Transactions on Microwave Theory and Techniques*, vol. 30, pp. 92–99, 1982.

[31] Katja Poković, Thomas Schmid, and Niels Kuster, "Robust setup for precise calibration of E-field probes in tissue simulating liquids at mobile communications frequencies", in *ICECOM'97*, Dubrovnik, Croatia, October 15–17, 1997, pp. 120–124.

[32] N. S. Nahman, M. Kanda, E. B. Larsen, and M. L. Crawford, "Methodology for standard electromagnetic field measurements", *IEEE Transactions on Instrumentation and Measurement*, vol. 34, no. 4, pp. 490–503, Dec. 1985.

[33] Robert F. Cleveland, Jr., and T. Whit Athey, "Specific absorption rate (SAR) in models of the human head exposed to hand-held UHF portable radios", *Bioelectromagnetics*, pp. 173–186, 1989.

[34] S. S. Stuchly, M. Barski, and B. Tam et. al., "Computer-based scanning system for electromagnetic scanning", *Review on Scientific Intstrumentation*, vol. 54, no. 11, pp. 1547–1550, 1983.

[35] Quirino Balzano, Oscar Garay, and Thomas J. Manning, "Electromagnetic energy exposure of simulated users of portable cellular telephones", *IEEE Transactions on Vehicular Technology*, , no. 3, pp. 390–403, Aug. 1995.

[36] N. Kuster, R. Kästle, and T. Schmid, "Dosimetric evaluation of handheld mobile communications equipement with known precision", *IEICE Transactions on Communications*, vol. 80, no. 5, pp. 645–652, May 1997.

[37] NIS81 NAMAS, "The treatment of uncertainty in EMC measurement", *Tech. Rep., NAMAS Executive*, 1994.

[38] Barry N. Taylor and Christ E. Kuyatt, "Guidelines for evaluating and expressing the uncertainty of NIST measurement results", *Tech. Rep.*, 1994.

[39] Q. Yu, M. Aronsson, D. Wu, and O. P. Gandhi, "Automated SAR measurements for compliance testing of cellular telephones", in *IEEE Antennas and Propagation Symposium*, Atlanta, USA, June 1997, pp. 1980–1983.

[40] K. Haelvoet, S. Criel, F. Dobbelaere, and L. Martens, "Near-field scanner for the accurate characterization of electromagnetic fields in the close vicinity of electronic devices and systems", *IEEE Instrumentation and Measurement Technology Conference*, pp. 1119–1123, June 1996.

[41] IDX, "Near field measurement systems", Tech. Rep., IDX Systems Inc., 20 NE Granger Ave Bldg. B. , Coryallis OR 97330, USA, 1996.

[42] Ilsan, "3D near field scanner", Tech. Rep., Ilsan America Inc., 1997.

[43] N. Kuster, Q. Balzano, and J. C. Lin, *Mobile Communications Safety*, Chapman & Hall, London, 1997.

[44] Kane S. Yee, "Numerical solution of initial boundary value problems involving Maxwell's equations in isotropic media", *IEEE Transactions on Antennas and Propagation*, vol. 14, no. 3, pp. 302–307, May 1966.

[45] J. Berenger, "A perfectly matched layer for the absorption of electromagnetic waves", *Journal of Computational Physics*, vol. 114, pp. 185–200, 1994.

[46] Thomas Jurgens, Allen Taflove, Korada Umashankar, and Thomas G. Moore, "Finite-difference time-domain modeling of curved surfaces", *IEEE Transactions on Antennas and Propagation*, vol. 40, no. 4, pp. 357–365, Apr. 1992.

[47] Vijaya Shankar and Alireza H. Mohammadian, "A time-domain, finite-volume treatment for the Maxwell equations", *Electromagnetics*, vol. 10, pp. 127–145, 1990.

[48] Z. M. Liu, A. S. Mohan, and T. A. Aubrey, "Em scattering using nonuniform mesh FDTD, PML and Mur's ABC", *Electromagnetics*, vol. 16, no. 4, pp. 341–358, July 1996.

[49] Thomas Weiland, "Verlustbehaftete Wellenleiter mit beliebiger Randkontur und Materialbelegung", *Electronics and Communication*, vol. 33, no. 4, pp. 170–174, 1979.

[50] S. S. Zivanovic, K. S. Yee, and K. K. Mei, "A subgridding method for the time-domain finite-difference method to solve Maxwell's equations", *IEEE Transactions on Microwave Theory and Techniques*, vol. 39, no. 3, pp. 471–479, Mar. 1991.

[51] A. D. Tinniswood, C. M. Furse, and O. P. Gandhi, "Computations of SAR distributions for two anatomically based models of the human head using cad files of commercial telephones and the parallelized FDTD code", *Transactions on Antennas Propagation*, vol. 46, no. 6, pp. 829–833, 1998.

[52] Achim Bahr, Sheng Gen Pan, Thomas Beck, Ralf Kästle, Thomas Schmid, and Niels Kuster, "Comparison between numerical and experimental near-field evaluation of a DCS1800 mobile telephone", *Radio Science*, vol. 33, no. 6, pp. 1553–1563, Nov. 1998.

[53] K. L. Virga and Y. Rahamat-Samii, "Low-profile enhanced-bandwidth PIFA antennas for wireless communcations packaging", *IEEE Transactions on Microwaves Theory and Techniques*, vol. 45, no. 10, pp. 1879–1888, Oct. 1997.

[54] Z. D. Liu, P. S. Hall, and D. Wake, "Dual-frequency planar inverted-F antenna", *IEEE Transactions on Antennas and Propagation*, vol. 45, no. 10, pp. 1451–1457, Oct. 1997.

[55] Ansoft Corporation, "Maxwell strata, product information", Tech. Rep., Four Station Square, Suite 660, Pittsburgh, PA 15219-1119, USA, 1994.

[56] Inc. Boulder Microwave Technologies, "Ensemble, product information", Tech. Rep., 2336 Canyon Blvd, Suite 102, Boulder, Colorado 80302, USA, 1994.

[57] Roger Yew-Siow Tay, Quirino Balzano, and Niels Kuster, "Dipole configurations with strongly improved radiation efficiency for hand-held transceivers", *IEEE Transactions on Antennas and Propagation*, vol. 46, no. 6, pp. 798–806, 1998.

[58] Peter P. Silvester and Ronald F. Ferrari, *Finite Elements for Electrical Engineers*, Cambridge University Press, Cambridge UK, 1983.

[59] A. Bossavit, "A rationale for edge elements in 3-D fields computations", *IEEE Transactions on Magnetics*, vol. 24, pp. 74–79, 1988.

[60] U. Pekel and R. Mittra, "A finite-element-method frequency domain application of the perfectly matched layer (PML) concept", *Microwave and Optical Technology Letters*, pp. 117–122, June 1995.

[61] Michael A. Morgan, "Finite element calculation of microwave absorption by the cranial structure", *IEEE Transactions on Biomedical Engineering*, vol. 28, no. 10, pp. 687–695, Oct. 1981.

[62] R. J. Spiegel, "The thermal response of a human in the near-zone of a resonant thin-wire antenna", *IEEE Transactions on Microwave Theory and Techniques*, vol. 30, no. 2, pp. 177–185, Feb. 1982.

[63] Dina Šimunić, Paul Wach, Werner Renhart, and Rudolf Stollberger, "Spatial distribution of high-frequency electromagnetic energy in human head during MRI: Numerical results and measurements", *IEEE Transactions on Biomedical Engineering*, vol. 43, no. 1, pp. 88–94, Jan. 1996.

[64] Lars H. Bomholt, "Coupling of the generalized multipole technique and the Finite Element method", *ACES*, vol. 9, no. 3, pp. 63–68, 1994.

[65] H. O. Ruoss, U. Jakobus, and F. M. Landsdorfer, "Iterative coupling of MoM and MMP for the analysis of metallic structures radiating in the presence of dielectric bodies", *Applied Computational Electromagnetics Society*, 1998.

[66] R. W. P. King, "Electromagnetic field generated in model of human head by simplified telephone transceiver", *Radio Science*, vol. 30, no. 1, pp. 267–281, Jan. 1995.

[67] N. Kuster, "Multiple multipole method for simulating EM problems involving biological bodies", *IEEE Transactions on Biomedical Engineering*, vol. 40, no. 7, pp. 611–620, July 1993.

[68] Camelia Gabriel, "Phantom models for antenna design and exposure assessment", in *IEE Colloqium on Design of Mobile Antennas for Optimal Performance in the Prescens of Biological Tissue* , Jan. 1997.

[69] A. Hizal and Y. K. Baykal, "Heat potential distribution in an inhomogeneous spherical model of a cranial structure exposed to microwaves due to loop or dipole antennas", *IEEE Transactions on Microwave Theory and Techniques*, vol. 26, no. 8, pp. 607–558, Aug. 1978.

[70] Y. Amemiya and S. Uebayashi, "Distribution of absorbed power inside a sphere simulating the human head in the near field of a $\lambda/2$ dipole antenna", *Electronics and Communications in Japan*, vol. 66, no. 9, pp. 64–72, Sept. 1983.

[71] Magdy F. Iskander, Peter W. Barber, Carl H. Durney, and Habib Massoudi, "Irradiation of prolate spheroidal models of humans in the near-field of a short electric dipole", *IEEE Transactions on Microwave Theory and Techniques*, vol. 28, no. 7, pp. 801–807, July 1980.

[72] Akhlesh Lakhtakia, Magdy F. Iskander, Carl H. Durney, and Habib Massoudi, "Near-field absorption in prolate spheroidal models of human exposed to a small loop antenna of arbitrary orientation", *IEEE Transactions on Microwave Theory and Techniques*, vol. 29, no. 6, pp. 588–594, June 1981.

[73] I. Chatterjee, M. J. Hagmann, and O. P. Gandhi, "Electromagnetic absorption in a multilayered slab model of tissue under near-field exposure conditions", *Bioelectromagnetics*, vol. 1, pp. 379–388, 1980.

[74] D. P. Nyquist, K. M. Chen, and B. S. Guru, "Coupling between small thin-wire antennas and a biological body", *IEEE Transactions on Antennas and Propagation*, vol. 25, no. 6, pp. 863–866, Nov. 1977.

[75] K. Karimullah, K. M. Chen, and D. P. Nyquist, "Electromagnetic coupling between a thin-wire antenna and a neighboring biological body: Theory and experiment", *IEEE Transactions on Microwave Theory and Techniques*, vol. 28, no. 11, pp. 1218–1225, Nov. 1980.

[76] Arthur W. Guy and Chung-Kwang Chou, "Specific absorption rates of energy in man models exposed to cellular UHF mobile-antenna fields", *IEEE Transactions on Microwave Theory and Techniques*, , no. 6, pp. 671–680, June 1986.

[77] H. Tamura, Y. Ishikawa, T. Kobayashi, and T. Nojima, "A dry phantom material composed of ceramic and graphite powder", *IEEE Transactions on Electromagnetic Compatibility*, vol. 39, no. 2, pp. 132–137, May 1997.

[78] Klaus Meier, *Scientific Bases for Dosimetric Assessments in Compliance Tests*, PhD thesis, Diss. ETH Nr. 11722, Zurich, 1996.

[79] Maria A. Stuchly, Andrzej Kraszewski, Stanislaw S. Stuchly, George W. Hartsgrove, and Ronald J. Spiegel, "RF energy deposition in a heterogeneous model of man: Near-field exposures", *IEEE Transactions on Biomedical Engineering*, , no. 12, pp. 944–949, Dec. 1987.

[80] Michael Burkhardt and Niels Kuster, "Appropriate modeling of the ear for compliance testing of handheld MTE with safety limits", in *Twentieth Annual Meeting of the Bioelectromagnetics Society*, St. Pete Beach, Florida, USA, June 1998, p. 79.

[81] Volker Hombach, Klaus Meier, Michael Burkhardt, Eberhard Kühn, and Niels Kuster, "The dependence of EM energy absorption upon human head modeling at 900 MHz", *IEEE Transactions on Microwave Theory and Techniques*, vol. 44, no. 10, pp. 1855–1863, Oct. 1996.

[82] K. Meier, V. Hombach, R. Kästle, R. Y.-S. Tay, and N. Kuster, "The dependence of electromagnetic energy absorption upon human-head modeling at 1800 MHz", *IEEE Transactions on Microwave Theory and Techniques*, vol. 45, no. 11, pp. 2058–2062, Nov. 1997.

[83] P. J. Dimbylow and S. M. Mann, "SAR calculations in an anatomically realistic model of the head for mobile communication transceivers at 900 MHz and 1. 8 GHz", *Physics in Medicine and Biology*, vol. 39, pp. 1537–1553, 1994.

[84] Om P. Gandhi, Gianluca Lazzi, and Cynthia M. Furse, "Electromagnetic absorption in the human head and neck for mobile telephones at 835 and 1900 MHz", *IEEE Transactions on Microwave Theory and Techniques*, vol. 44, no. 10, pp. 1884–1897, Oct. 1996.

[85] Om P. Gandhi and Jin Yuan Chen, "Electromagnetic absorption in the human head from experimental 6-GHz handheld tranceivers", *IEEE Transactions on Electromagnetic Compatibility*, vol. 37, pp. 547–858, 1995.

[86] F. Schoenborn, M. Burkhardt, and N. Kuster, "Differences in energy absorption between heads of adults and children in the near field of sources", *Health Physics*, 1998.

[87] Michael A. Jensen and Yahya Rahmat-Samii, "EM interaction of handset antennas and a human in personal communications", in *Proceedings of the IEEE*, 1995, vol. 83, pp. 7–17.

[88] Soichi Watanabe, Masao Taki, Toshio Nojima, and Osamu Fujiwara, "Characteristics of the SAR distributions in a head exposed to electromagnetic fields radiated by hand-held portable radio", *IEEE Transactions on Microwave Theory and Techniques*, vol. 44, no. 10, pp. 1874–1883, Oct. 1996.

[89] Michal Okoniewski and Maria A. Stuchly, "A study of the handset antennas and human body inter-action", *IEEE Transactions on Microwave Theory and Techniques*, vol. 44, no. 10, pp. 1855–1863, Oct. 1996.

[90] Paolo Bernardi, Marta Cavagnaro, and Stefano Pisa, "Evaluation of the SAR distribution in the human head for cellular phones used in a partially closed environment", *IEEE Transactions on Electromagnetic Compatibility*, vol. 38, no. 3, pp. 357–366, Aug. 1996.

[91] J. Cooper and V. Hombach, "Increase in specific absorption rate in human heads arising from im-plantations", *Electronics Letters*, vol. 32, no. 24, pp. 2217–2219, 1996.

[92] S. Mazur, D. Martensson, and C. Toernevik, "Comparisons of measurements and FDTD calcula-tions of electromagnetic near-fields and SAR distributions", in *Eighteenth Annual Meeting of the Bioelectromagnetics Society*, Victoria, Canada, June 1996, p. 123.

[93] M. Siegbahn, S. Mazur, and C. Törnevik, "Comparisons of measurements and FDTD calculationsof mobile phone electromagnetic far-fields and near-fields", in *IEEE Antennas and Propagation Sympo-sium*, Montreal, Canada, July 1997, pp. 978–981.

[94] M. Burkhardt, N. Chavannes, K. Pokovic, T. Schmid, and N. Kuster, "Study on the FDTD perfor-mance for transmitters in complex environments", in *ICECOM'97*, Dubrovnik, Croatia, Oct. 1997, pp. 83–86.

HIGH-RESOLUTION MICROWAVE DOSIMETRY IN LOSSY MEDIA

A. G. PAKHOMOV[1,2], S. P. MATHUR[1], Y. AKYEL[1], J. L. KIEL[2] and
M. R. MURPHY[2]
[1]McKesson BioServices, US Army Medical Research Detachment and
[2]Directed Energy Bioeffects Division, Human Effectiveness Directorate,
Air Force Research Laboratory, Brooks Air Force Base, San Antonio,
Texas, 78235-5324, USA

1. Introduction

Measurement of the local specific absorption rate (SAR) in microwave-exposed specimens is a necessary task in a variety of bioelectromagnetic studies. Numerical techniques, such as finite-difference time-domain (FDTD) modeling, are widely used to evaluate local SAR and its distribution [1]. This modeling is laborious and expensive, and involves certain approximations of the spatial organization of exposed subjects and their dielectric properties. Since these approximations introduce an error into SAR predictions, the accuracy of the method may be difficult to assess.

A precise instrumental technique of local SAR measurement in absorptive media could be an alternative to numerical modeling and could also be employed to verify FDTD predictions. While measurement of local heating dynamics under microwave exposure is a standard approach to establish SAR, common thermometry techniques have substantial limitations [2]. The accuracy of thermometry relies on several conditions, including possible field distortion by a probe, the probe's size, heat capacity, and response time.

A few non-field-perturbing devices available on the market (devices with non-metallic probes, such as fluoroptic thermometers) tend to be slow and noisy. In a Luxtron 850 fluoroptic thermometer, for example, the measurement rate is limited to 30 samples/sec, and, at this rate, sequential measurements of a constant temperature may vary by as much as 0.5-1 °C. Averaging of sequential measurements reduces the "noise", but also reduces the temporal resolution, so a fast heating cannot be recorded correctly. Besides, these thermometers cannot be used to detect temperature fluctuations over small distances of 1-2 mm, which are comparable with the size of the measuring probe itself.

The possibility of using thermometers with metal-containing probes (e.g., thermistors and thermocouples) for microwave dosimetry remains in question. Such probes may cause field distortion, and electric currents induced in the probe and wires may result in erroneous temperature readings and excessive heating of the probe itself. However, these artifacts should decrease with decreasing the dimensions of the probe and wires;

187

B.J. Klauenberg and D. Miklavcic (eds.), Radio Frequency Radiation Dosimetry, 187-197.

at some point, the artifacts may become negligible. Other advantages of miniaturization are decreasing of the probe's heat capacity and improvement of its spatial and temporal resolution.

In this study, we have analyzed the performance of a microthermocouple (MTC) under most "unfavorable" exposure conditions (very steep field gradients produced by extremely-high peak power microwave pulses). It was shown that the dynamics of microwave heating recorded by the MTC is qualitatively different from the dynamics recorded by larger probes. Due to its exceptionally fast response and high sensitivity, MTC was capable of recording a temperature plateau after a microwave pulse or a brief train of pulses. The level of this plateau was linearly proportional to the local specific absorbed dose (SAD). Calculation of SAR from the initial slope of the heating curve was not possible because of its contamination with recording artifacts; at the same time, SAR could be precisely calculated from SAD and known duration of microwave pulse(s). The recording artifacts and possible field distortion by the MTC were shown to have no effect on SAD or SAR measurements.

2. Exposure Setup

Square microwave pulses (9.5 GHz, 1 μs width, 75-115 kW) were produced by an EPSCO Model MH300 system with an MF-IM65-01 RF plug-in into a WR90 waveguide (22.86 x 10.16 mm). Incident and reflected powers in the waveguide were measured via directional couplers by a dual-channel HP 438A power meter with HP 8481A power sensors. Microwave pulses or pulse trains were triggered externally from a Grass Instruments S8800 stimulator. The shape of microwave pulses was monitored via an HP 432 detector on a TEK 2430A digital oscilloscope.

The design of the exposure cell was similar to the one proposed by Chou and Guy [3]. A vertical section at the end of the waveguide was separated by a quarter-wave matching plate and filled with 6 ml of a physiological solution. The waveguide walls in the cell were covered with a lacquer to prevent its electrical contact with the solution. Complex dielectric constant of this solution at 9.5 GHz and at various temperatures was calculated by equations given by Stogryn [4]. With the solution normality of 0.12-0.14, its relative permittivity (ε_r) and conductivity (σ) at 25 °C were calculated as 63.4 and 15.9 S/m, respectively. To verify the calculations, the dielectric properties of the solution were also obtained experimentally using an HP 8510 measurement system (see [5] for detailed procedures). Measured values ($\varepsilon_r = 63$, $\sigma = 16$ S/m) were virtually the same as the calculated ones. The respective linear loss coefficient of the solution was 3.71 Np/cm.

SAR in the solution along the axis of the waveguide decreased exponentially with increasing the distance above the matching plate, and could be calculated according to [3] as:

$$ SAR = (1/\rho) \frac{2 (P_i - P_r)}{S} 2\alpha e^{-2\alpha z} \quad (W/g) \qquad (1) $$

where ρ is the saline density (g/cm^3). P_i and P_r are the incident and reflected power values (W; their difference will be regarded below as a "transmitted power"), S is the cross-section of the waveguide (2.32 cm^2), α is the linear loss coefficient in the saline (Np/cm), and z is the distance above the matching plate (cm). Based on this equation, a power of 100 kW transmitted to the exposure cell would produce SAR values of 610, 304, 145, and 69 kW/g at the distances of at 0, 1, 2, and 3 mm above the plate, respectively (at 9.5 GHz and 25 °C). In other words, SAR in the saline would decrease by approximately twofold per millimeter. Obviously, measurement of local SAR from heating dynamics in such a steep gradient makes it imperative to use temperature probes much smaller than 1 mm.

In some experiments, we also used relatively low power, longer microwave pulses (10-100 ms, 5-10 W, 9.5 GHz). These pulses were produced by an HP 8690A Sweep Oscillator connected via a Hughes 8020H Traveling Wave Tube Amplifier to the same waveguide exposure system.

3. Microthermocouple and Temperature Recording

MTCs made of 25-μm diameter bare copper and constantan wires was purchased from Omega Engineering, Inc. For convenient handling, MTC was fixed in a custom-made holder. Bare wires ending with the thermocouple junction extended from the holder by about 3 mm; they were covered with a lacquer for electrical insulation from saline. The lacquer coating was made as thin as technically achievable, under a visual control using a microscope.

The MTC was precisely positioned in the exposure cell by means of a micromanipulator. The accuracy of the micromanipulator movement was 0.05 mm in any orthogonal direction.

MTC signal was recorded by a BIOPAC MP100 data acquisition system via a universal DC amplifier (Gould 5900 frame). The higher cutoff frequency of the amplifier in most experiments was set at 1 or 10 kHz. The entire system (MTC + amplifier + MP100) was calibrated against a precision mercury thermometer. Within studied range of temperatures and achievable accuracy, the system response was linearly proportional to the temperature.

In most experiments, acquisition of the signal from MTC began shortly (10-50 ms) before the microwave pulse train, and ended some period after the train was over. Records of temperature changes produced by repeated pulses or pulse trains could be averaged, that substantially decreased the noise. The accuracy of temperature readings was about 0.05 °C without signal averaging and better than 0.01 °C after averaging. This accuracy could be further increased by measuring mean temperature over 5-50 ms periods (when the temperature of the medium was constant).

One should note that the above numbers apply only to "relative" temperatures, or temperature changes within relatively short periods of time (10^{-2}-10^3 s). The absolute temperature of the medium could only be measured using two thermocouples, one of them rendered as a reference and kept at a known constant temperature [6]. For the purpose of SAR measurement, knowing the exact absolute temperature was not essential, so the reference thermocouple was not used in our experiments.

In some experiments, temperature in the exposure cell was simultaneously recorded by the MTC and a Luxtron 850 Multichannel Fluoroptic Thermometer.

4. Results and Discussion

In the first series of experiments, we obtained simultaneous records of microwave heating by MTC and by a "recognized" artifact-free thermometer, Luxtron 850. MTC junction was put either in contact with the fluoroptic probe, or 0.2-0.4 mm apart. A sample record is shown in the Figure 1: At any transmitted power (which, in the shown case, was proportional to the pulse repetition rate), both the probes recorded the same heating dynamics. These data established the general possibility of using MTC under microwave irradiation, and also confirmed MTC calibration.

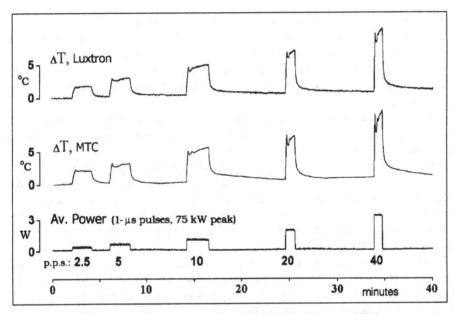

Figure 1. Parallel recording of microwave heating by a fluoroptic probe (Luxtron 850 thermometer, upper trace) and by a microthermocouple (MTC, middle trace). The probes are in contact with each other at the bottom of the exposure cell. The lower trace is the average transmitted power, which is proportional to the pulse repetition rate (p.p.s.).

Such similarity of MTC and Luxtron recordings as shown in Figure 1 was only possible at slow recording rates and long exposures (minutes) at a relatively low average transmitted power. Slow recording conceals much faster response time of the

MTC, and, at a low average power, allows enough time for temperature equilibration between neighboring spots with different local SAR. Additional peaks on the heating curves (particularly noticeable at 20 and 40 p.p.s.) resulted from liquid convection in the exposure cell. Such peaks disappeared entirely when the saline was solidified with a 1% agar-agar, and heating curves attained a "classic" appearance.

If the presence of the MTC in the exposed medium caused field distortion and altered microwave heating, then the Luxtron thermometer readings would depend on the presence of the MTC next to the fluoroptic probe. Experiments showed that this was not the case: Luxtron-recorded heating curves in response to identical exposures were the same with MTC positioned next to the fluoroptic probe, MTC removed from the solution, and also after returning MTC back into the original position.

However, the fluoroptic probe might be too slow and bulky to detect fine alterations of the field and heating pattern introduced by the MTC. Therefore, we performed more accurate experiments, in which the MTC-recorded heating curves served as "self-control" (Figure 2). The micromanipulator that we used for MTC positioning had an option to lift the MTC and then slide it momentarily down to the exact original position. The upper heating curve in Figure 2 was recorded when the MTC was permanently kept at the bottom of the exposure cell (*Position 1*). The lower curve (the same exposure parameters) started when the MTC was elevated into *Position 2*. After the exposure was over, the MTC was quickly slid into the *Position 1*.

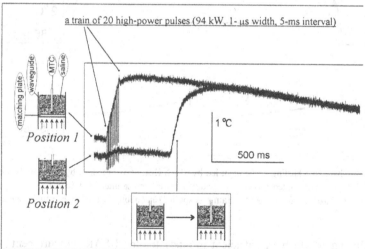

Figure 2. Microwave heating recorded by microthermocouple (MTC). Upper trace: MTC was at the bottom of the exposure cell (*Position 1*) during and after exposure (20 high-power pulses). Lower trace: MTC was at the top of the (*Position 2*) during the exposure, and was slid into the *Position 1* after the exposure was over. See text for further explanation.

Solution in the cell was exposed to repeated trains of microwave pulses (1 train/10 sec), which resulted in a temperature difference between two MTC positions (*Position 2* was farther away from the matching plate and therefore was cooler). Microwave pulses produced large artifacts and pronounced heating in the *Position 1*, and smaller artifacts and only subtle heating in the *Position 2*. Due to the high field attenuation by the saline (about 40 times per 5 mm), the presence of the MTC in the *Position 2* could have no effect on microwave heating at the bottom of the cell.

Hence, the temperature recorded on the lower curve after sliding the MTC down is the real temperature of the medium in the *Position 1*, not affected by the presence of the MTC there during exposure. This temperature was exactly the same as the one recorded with the MTC being in the *Position 1* throughout the exposure (the upper curve). This experiment was repeated numerous times, with various MTC positions and exposure parameters, and with the same result as illustrated. Thus, the presence of MTC in the medium during exposure produced no measurable field distortion and did not affect heating of the medium.

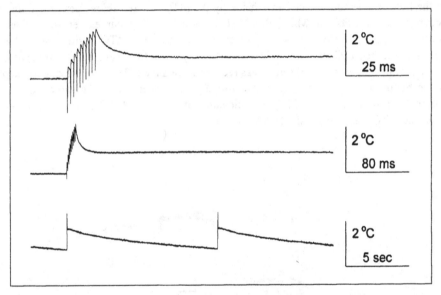

Figure 3. The shape of a heating curve recorded by MTC. Heating is produced by trains of 10 high-power pulses (110 kW peak transmitted power, 1 μs width, 1 ms interpulse interval, 1 train/12 s). Note recording artifacts (they also mark the time when the train was applied) and a long-lasting temperature plateau after the train.

This finding would be a sufficient grounds for local SAR measurement with the MTC unless the initial slope of the heating curve was contaminated by recording artifacts. These artifacts can be seen even better in Figure 3, which shows a heating curve recorded at three different time scales. The presence of microwave pulse artifacts makes it difficult or impossible to measure the initial slope of the heating curve. Hence, the "standard" approach to SAR measurement (Figure 4) may not be used in this case.

At the same time, there is an important difference between heating curves recorded by MTC and by larger conventional probes. This difference is the extended temperature plateau after a short microwave pulse or a brief train of pulses (Figure 3). It should be emphasized that this is an actual plateau, and not a part of the declining slope that was artificially expanded to make it appear flat.

The existence of this plateau can be explained as follows. In our set-up, brief and intensive microwave exposure creates a significant SAR and temperature gradient in the exposed medium. After the exposure is over, the heat dissipates from warmer to cooler areas of the exposed medium and further into the environment. Thus, the MTC submerged into the medium is simultaneously warmed up by the heat flow from the warmer areas and cooled down by the heat flow to the cooler areas. The heat flow to cooler areas will obviously prevail, and the entire exposed medium will eventually cool down to the ambient temperature. However, for a very brief time interval after exposure and within an adequately small medium volume (like the volume in the immediate vicinity of the MTC), the warming and cooling heat flows will be virtually equal. As a result, the MTC will record a temperature plateau, as if no heat dissipation takes place.

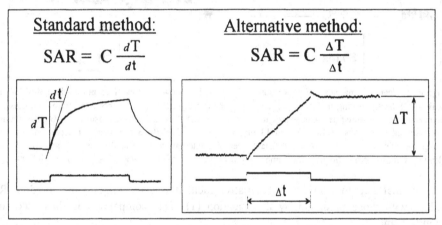

Figure 4. "Standard" and "alternative" methods of specific absorption rate (SAR) measurement from heating dynamics. Shown are the microwave pulse duration (lower traces) and sample heating curves (upper traces), The standard method uses the initial slope of the heating curve (C is the specific heat capacity of medium). The alternative method calculates SAR from the level of the after-pulse plateau and the duration of the pulse. Both methods can be used with the same result when the plateau is present and no artifacts are recorded.

If this is true, the plateau level (i.e., the difference between the temperature before the exposure and the plateau) will be linearly proportional to the delivered energy (e.g., the number of pulses in a train) and, within certain limits, will not depend on the pattern of the energy delivery (e.g., interpulse interval). Figure 5 shows that this was the case indeed. Hence, the locally absorbed dose (SAD) can be found as a product of the plateau level and specific heat capacity of the medium.

Once we know the SAD, we do not need the initial slope of the heating curve to calculate SAR. Instead, SAR can be calculated as a ratio of SAD and known duration of microwave pulse(s), as shown in Figure 4 as an "alternative method".

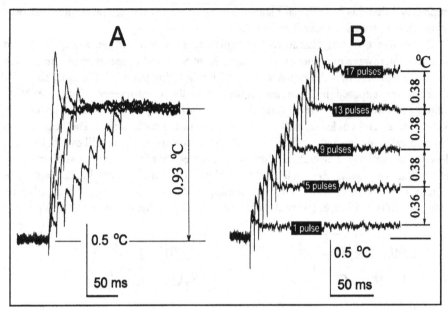

Figure 5. Heating dynamics under exposure to trains of high-power pulses (1-μs pulse width, 400 kW/g peak SAR). Spikes during the ascending portion of the curves (artifacts) correspond to microwave pulses. Temperature is measured as mean values for 40-ms periods before and after exposure. A: 10-pulse trains with interpulse intervals of 1, 3, 5, and 13 ms. The plateau level (0.93 °C) does not depend on of the interpulse interval. B: Trains with a constant interpulse interval (5 ms) and different number of pulses. Heating is linearly proportional to the number of pulses per train. Heating by a single pulse is about 0.09 °C.

The final step to justify this alternative method was to compare measured SAR values with those predicted by the equation (1). This comparison is illustrated in Figures 6 and 7.

In Figure 6, A, heating of 0.1 °C was produced by a single high-power pulse. Calculated and measured SARs were 585 and 420 kW/g, respectively. As stated by Bassen and Babij [2], "*an absolute accuracy of ±3 dB is the best case measurement uncertainty that can be achieved when attempting to determine the maximum and minimum SARs within an irradiated biological body.*" In our setup and with the use of MTC, the difference between calculated and measured SARs was only 1.4 dB.

The match was even better for exposure to low-power, longer microwave pulses (Figure 6, B). Measured values were only 0.4-0.5 dB apart from the calculated ones. This small error could easily be attributed to a combined inaccuracy of power meters, field probes and other devices, or even to some imperfection of the analytical equation itself (the equation is developed for TE_{10} wave propagation mode, while various other modes also arise in the waveguide after the matching plate). Figure 6 also demonstrates that the presence of a temperature plateau after a microwave pulse is not a specific attribute of exposure to extremely high-power pulses. The plateau is as well pronounced with reasonably low amplitude pulses, and the proposed method of SAR measurement with MTC can be used as long as the plateau can be detected.

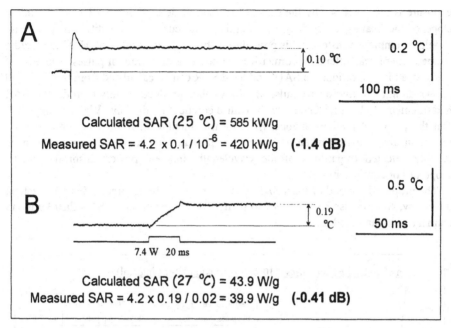

Figure 6. Comparison of calculated and microthermocouple-measured local SAR values. A, a high-power pulse (96 kW transmitted power, 1 µs width); the moment of exposure coincides with the artifact on the heating curve. B, a moderate power pulse (7.4 W, 20 ms width, indicated on the lower trace). The temperature before the pulse and during the plateau was measured as a mean of a 25-ms period. Calculated SAR values were obtained using the equation (1), see text. The difference of the measured and calculated SAR was -1.4 dB (A) and -0.41 dB (B).

Figure 7 gives an example of the field mapping with MTC and illustrates its spatial resolution. Heating curves recorded in spots as close as 0.5 mm to each other were substantially different, and measured SAR values were always close to the calculated ones.

MTC was also used for horizontal field mapping in a cross-section of the waveguide. Horizontal field distribution was calculated according to standard waveguide equations [7] for TE_{10} mode. Again, measured and calculated values were within ±1.5 dB apart.

5. Conclusions

The experiments established that MTC is an appropriate tool to measure local SAR in an absorptive medium. Small size, negligible heat capacity, fast response time, and high sensitivity make MTC particularly useful in high-gradient fields with complicated patterns of SAR distribution and also at high SARs, when larger probes may be too slow.

However, the presence of recording artifacts when the radiation is on prompts that, in general, the temperature of the medium can be measured accurately only when the exposure is off. That is why the standard method of determining SAR from the initial slope of the heating curve may be essentially inaccurate, and a different procedure should be employed instead. This procedure is based on the ability of MTC to record a temperature plateau after a short microwave pulse or a brief train of pulses. The level of this plateau is proportional to SAD, and local SAR can be calculated from the SAD and known duration of microwave pulse(s). This method produced highly reproducible SAR measurements, which matched closely with theoretical predictions. We can conjecture that the proposed method of local SAR measurement with MTC will produce valid results in any situation when a temperature plateau after a pulsed exposure can be reliably detected (regardless of the wavelength, incident power, absorptive media properties or configuration).

The proposed method of local SAR measurement can be recommended for practical dosimetry, and can also be employed to verify FDTD predictions of SAR distribution in animal and human models.

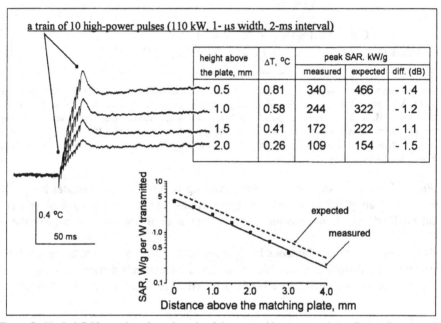

Figure 7. Vertical field mapping along the axis of the waveguide: A comparison of microthermocouple-measured and expected local SAR values. Heating curves (left) produced by a train of 10 high-power pulses were recorded at the heights from 0 to 3 mm above the matching plate (only four curves are shown). The "expected" local SAR was calculated by the equation (1). The table and the graph compare measured and expected local SAR values.

6. Acknowledgments

The work was supported in part by the U.S. Army Medical Research and Materiel Command and the U.S. Force Research Laboratory under US Army contract DAMD17-94-C-4069 awarded to McKesson BioServices. The views expressed are those of the authors and should not be construed as reflecting the official policy or position of the Department of the Army, Department of the Air Force, or the United States Government.

7. References

1. Gandhi, O.P. (1990) Numerical methods for specific absorption rate calculations, in O.P. Gandhi (ed.), *Biological Effects and Medical Applications of Electromagnetic Energy*, Prentice Hall, Englewood Cliffs, New Jersey, pp. 113-140.
2. Bassen, H.I. and Babij, T. M. (1990) Experimental techniques and instrumentation, *Ibid.*, pp. 141-173.
3. Chou, C-K. and Guy, A. W. (1978) Effects of electromagnetic fields on isolated nerve and muscle preparations, *IEEE Trans. Microwave Theory Tech.* 26, 141-147.
4. Stogryn, A. (1971) Equations for calculating the dielectric constant of saline water, *IEEE Trans. Microwave Theory Tech.* 19, 733-736.
5. Bao, J. Z., Swicord, M., and Davis, C. Microwave dielectric characterization of binary mixtures of water, methanol, and ethanol. *J. Chem. Phys.* 104, pp. 4441-4450.
6. Benedict R. P., ed. (1983) *Manual on the Use of Thermocouples in Temperature Measurement*, ASTM Special Technical Publication 470B, ASTM, Baltimore, MD, p. 258.
7. Paris, D.T. and Hurd, F.K. (1969) *Basic Electromagnetic Theory*, McGraw Hill Book Company, NY, pp. 398-451.

INFRARED THERMOGRAPHY IN EXPERIMENTAL DOSIMETRY OF RADIO FREQUENCY AND MILLIMETER WAVELENGTH RADIATION EXPOSURE

E. P. KHIZHNYAK and M. C. ZISKIN
Institute of Cell Biophysics, Russian Academy of Sciences,
Pushchino, Moscow region, 142292 RUSSIA;
Richard J. Fox Center for Biomedical Physics, Temple
University Medical School, Philadelphia PA, 19140, USA

Abstract - Distribution of 800-1500 MHz and 37-78 GHz microwave absorption in biological tissue phantoms and real biological objects was studied using method of Infrared Thermography. Multicomponent and 0.1-0.2 mm thin-layer phantoms were irradiated in near and far field of radiating antennas. SAR distribution was calculated as the function of initial heating rate. Temperature sensitivity was better than 0.02 K at up to 60 frames per second sampling rate. It was found that nonuniform SAR distribution could appear in near field area due to a geometrical resonance resulting from a secondary wave-mode interaction between an irradiated object and the corresponding critical cross-section of the horn antennas, and due to the biological heterogeneity. Local SAR in hot-spots can significantly exceed the spatially averaged values. These findings provide an explanation for a number of frequency-dependent and for so-called non-thermal effects of microwaves.

1. Introduction

One of the main mechanisms responsible for biological effects of radio frequency (RF) and millimeter wavelength (mm-wave) electromagnetic irradiation is heating due to absorption of microwave energy in biological structures. Assuming that the most part of the absorbed energy is converted into the thermal form, SAR could be calculated as a function of initial heating rate. Such approach requires the measurement of the heating rate pattern in biological objects during exposure.

It is relatively easy to calculate the SAR pattern in far-field and in homogeneous objects. However, the problem of SAR calculation and experimental measurements became much more difficult when the heterogeneous object with non-flat surface is exposed in near field area of irradiating antenna. Also, experimental proof of any theoretical calculation of SAR distribution is necessary.

B.J. Klauenberg and D. Miklavcic (eds.), Radio Frequency Radiation Dosimetry, 199-206.
© 2000 *Kluwer Academic Publishers. Printed in the Netherlands.*

It is difficult to accurately measure the heating rate pattern during irradiation using temperature sensors because the presence of these sensors may lead to distortions of the field pattern in the area of measurement [10, 11].

Infrared (IR) thermography [3, 10, 11] is ideal for measuring the rate of heating caused by millimeter wave's absorption in biological tissue, because the skin layer for 3-5 mkm IR-irradiation is close to the penetration depths of the mm-waves. At the same time, the differences in frequency between IR and mm-waves are too great to produce interference between these two types of radiation. Remote temperature sensing, based on measuring the IR emission irradiated by the object, entirely rules out the possibility of the above-mentioned artifacts occurring when any sensors are used.

Another problem in SAR determination is liquid evaporation from the surface of an exposed object, especially in the presence of sharp gradients of temperature rise. Evaporation creates an additional non-linearity in the dynamics of heating and makes the relationship between the initial heating rate and the steady-state value of overheating more complex [11]. The role of evaporation during microwave exposure needs to be considered when calculating the balance between absorbed and dissipated energy, and the relationship between SAR and steady-state overheating.

2. Methods

2.1. GENERATORS

The main requirement for a generator is it's output power. Because the measurement of the minimal heating rate is limited by the sensitivity of the IR camera, the output power of generators should be high enough to perform the reliable measurements in terms of acceptable errors.

Two type of generators were used in our study:

1. In 800-1500 MHz frequency range we used a generator based on the LB-7 oscillator with output power of up to 500 W;

2. In millimeter wave range we used two generators based on backward-wave oscillators with output power of up to 50 mW, one for the 37.5-53.57 GHz and the other for the 53.57-78.33 GHz frequency range.

All generators were equipped with isolators to prevent the influence of reflected waves on output parameters and with variable attenuators to control output power.

2.2. RADIATING ANTENNAS

In RF frequency range phantoms and real biological objects were exposed from the open side of the rectangular waveguide or placed inside it.

Different horn-type antennas were used for mm-wave exposure: a round antenna with a 17 mm horn aperture for 39-52 GHz frequency range, as well as rectangular horn antennas with 10x20 mm and 26x38 mm apertures for the 38-53 GHz frequency range and 10x20mm and 17x26 mm for the 53-78 GHz frequency range. Phantoms and real biological objects were exposed in far and near field areas of these horn antennas.

2.3. PHANTOMS

Since most of the mm-wave energy in the 35-78 GHz frequency range is absorbed within 0.1-0.2 mm layer of physiological solution [4], [8], the distribution of SAR in millimeter wavelength was studied using thin-layer film-type phantoms [11]. These phantoms were prepared using 0.1-0.2 mm thick dielectric filters possessing a 0.2 mkm pore diameter and saturated with a saline solution (100 mM NaCl, 10 mM tris, pH=7.2). A saline-saturated filter was placed on a 5 mkm thick water-resistant and IR-transparent film stretched over a rectangular holder. Constructed in this manner, the phantom is practically a flat absorbing medium, thick from the point of view of microwave energy absorption, yet thin from a thermodynamic point of view. The combination of these two features made it possible to record heating patterns during irradiation from the opposite side of the irradiated surface. Preliminary testing of such phantoms showed that the difference in the heating rates distribution occurring at the front and back surfaces and caused by microwave irradiation did not exceed 5%. Detail description of experimental set-up and the method of measurements using such phantoms was published in [11].

To assess the effect of evaporation, the kinetics of heating was measured using two types of phantoms: an evaporation-disabled and an evaporation-enabled phantom. The evaporation-disabled phantom was completely covered on both sides with a water-resistant film. The evaporation-enabled phantom was completely covered only on the non-irradiated side. On the irradiated side there was a window in the water-resistant film to permit evaporation during microwave exposures.

Multi-component gel-type phantoms (prepared using 1-5% Agar) consisting of high conductive cylindrical heterogeneous structures located in flat layers with lower conductivity were used in RF studies. Such phantoms were irradiated from the open side of the rectangular waveguide or placed inside it. Surface heating-rate patterns were recorded during exposure. To access the heating pattern inside of the phantoms, they could be opened at different cross-sections. The opening time varied from 0.3 to 1 second depending on the size of phantoms and the shape of the opening profile. Computer reconstruction of the heating rate dynamics was performed using the sequence of heating patterns obtained at different time intervals of exposure.

2.4. IR CAMERA

Two IR cameras with 3-5 mkm spectral windows of sensitivity were used for recording the heating patterns:
- AGA model 780/SW (AGA-Infrared Systems, Sweden) with a spatial resolution of 128 X128 pixels per frame and 0.1 K temperature sensitivity;
- AMBER model 4256 (Amber Engineering, Inc., USA) with a spatial resolution of 256 X 256 pixels per frame, 0.02 K temperature sensitivity and up to 60 frames per second sampling rate.

Previously developed computer image analysis techniques [3, 10, 14] employing thermal image filtration and averaging procedures made it possible to calculate differential thermograms, and to improve temperature resolution to 0.05 K when using the AGA-780 IR camera. A reference frame recorded prior to the start of each exposure was subtracted from each of the heating patterns recorded during irradiation. This minimized measurement errors caused by the non-uniform sensitivity of the IR camera.

The duration of the time interval from the start of irradiation used for calculation of SAR-patterns was a compromise between the following two conditions:

1. It should be long enough to provide an acceptable ratio between the irradiation-induced temperature elevation and the thermal noise-equivalent of the IR camera; and

2. It should be short enough to ensure a linear temperature rise in the maximum heating area, and the absence of significant distortions due to heat conduction.

The latter condition was tested by comparing differential thermograms calculated at different time intervals from the starting point of exposure, and observing the spread of the heating pattern due to thermal conduction [11]. Because the maximal value of overheating in our measuring was usually less than 5 K, the heat capacity of phantoms was regarded as a constant. Therefore, the recorded heating patterns closely reflected the SAR distribution in the irradiated object.

3. Results and Discussion

The experiments carried out have shown that a mm-wave horn antennas produce a non uniform SAR distribution on the surface of exposed objects when these objects are irradiated in the antenna's near-field . The spatial parameters of heating patterns, such as the number, size, and location of the SAR maxima in the surface layer of irradiated objects strongly depend on frequency, geometrical parameters of the specific horn antenna, distance between the horn aperture and the irradiated object, and electrical properties of the object [11].

Single and double hot-spots heating patterns were produced by 17 mm round horn antenna in 44-52 GHz frequency range. A cyclical repetition of the heating rate patterns was observed at 5.5 GHz frequency intervals. At 20 mW irradiated powers, local SAR values were as high as 5 kW/kg in extremely non-uniform heating patterns. Rectangular horn antennas produced much more complex heating patterns with up to 4 local hot-spots in case of 10x20 mm aperture, and up to 6 local hot-spots in case of 26x38 mm aperture. Multiple hot-spot patterns with four or more local overheating maxima observed in 44-52 GHz frequency range using 26x38 mm horn antenna and in 57-78 GHz frequency range with 17x26 mm horn antenna, are highly dependent on irradiation frequency. Sharp hot-spots were effectively flattened when frequency was shifted for 0.004 of the value corresponded to the most sharp peak. Relative heating rates at the same point on the irradiated surface at different frequencies may vary more than a factor of ten. The same significant changes in heating patterns could be caused by 0.1-0.2 mm changes in the distance between the horn aperture and the surface of irradiated objects [11].

The mechanism responsible for the effects obtained is a geometrical' resonance resulting from secondary wave-mode interaction between an irradiated surface and the corresponding critical cross-section of the radiating horn antenna [11]. A combination consisting of a horn antenna and an absorbing layer with a non-zero reflectivity located in the near field, can act as a very complex resonator in which different wave-modes can exist simultaneously. Consequently, the distribution of incident power density at the irradiated surface will change with the changes in the frequency and the geometrical parameters of the resonator system.

A numerical analysis of possible power density profiles for a relatively simple case of an H-plane 2-D resonator model [1] showed that a relative frequency change of 0.0046 can lead to a qualitative change in the field's profile from unimodal to bimodal with more than a 6-fold change of power density in the central area of the horn antenna. This means that even in such a simple case it is possible to observe what seems to be a sharp resonance-type biological effect with an equivalent Q-factor of approximately 500. Obviously, it is necessary to characterize the heating patterns when frequency-dependent biological effects of microwaves are studied.

Horn antennas with apertures in both the E-and the H- planes frequently used in biological experiments and EHF-therapy [4], [6], [7], [9] can produce more complex frequency-dependent multi hot-spot heating patterns in irradiated objects, and the equivalent Q-factor of biological effects can exceed 500 when such antennas are used.

All such multi hot-spot heating patterns were observed using flat homogeneous phantoms. The problem of SAR-distribution in real heterogeneous biological objects with non-flat shape becomes much more difficult.

It is almost impossible to predict the SAR-distribution on the surface of biological objects because:

    1. The numerical analysis of the near-field structure in the E- and H- planes of a horn antenna is very complex when a surface with a non-zero reflection is placed at a short distance (about 1-2 wavelengths) from the horn aperture; and

    2. The distribution of the phases of reflected waves from the irradiated surface needs to be taken into account, but cannot be measured for heterogeneous biological objects with non-flat surfaces.

In such cases, SAR-distribution can not be calculated using the average value of incident power density, and it is only possible to empirically measure the distribution of microwave absorption in the irradiated objects.

Significant SAR increase was observed in experiments using multi-component phantoms consisting of high conductive cylindrical heterogeneous structures located in flat layers with lower conductivity. More than 100 times SAR increase was observed during RF and mm-wave exposure of such phantoms when the axis of cylindrical structures was parallel to the E-vector of the field, and there was no concentration of SAR when cylindrical axis was perpendicular to the E-vector. Cylindrical form of heterogeneity was chosen because such structures simulate well a real biological heterogeneity, such as sweat pore channels located in low conductive epidermis, blood vessels, as well as the heterogeneity due to the presence of the temperature sensors.

Two problems related to the presence of temperature sensors in the area of exposure was revealed in our studies: distortion of the field absorption patterns around the sensors and difference in heating rates measured using these sensors and IR equipment [11]. Maximum field distortions were observed when thermistor was placed in the area of local SAR concentration. The presence of thermistor can lead to 50-60% concentration or decrease of the field in the area of thermistor together with the change of the shape of SAR distribution within a radius of 2 mm from the boundary of the thermistor. The same problems related to the presence of some field sensors in area of exposure were revealed in our studies.

The reality of the local field and SAR distortion in area of heterogeneous biological structures was experimentally demonstrated in the studies of dynamics of the process by which sweat pores in human skin open and close [13]. More than two time increase in heating rate was observed in the areas of sweat pores during 40-72 GHz mm-wave exposure. At the same time, the additional increase of steady-state overheating was very low due to the increase of evaporation.

A low spatial-averaged steady-state temperature increase can not be used as the sole criterion for claiming that microwave induced bio-effects are non-thermal, because the SAR values in a thin surface layer can be high enough to produce significant changes in biological structures. Furthermore, some biological processes depend not only on the steady-state temperature increment, but also on the heating rate [2, 5] and temperature gradients [12]. Therefore, the data pertaining to both heating rate and steady-state overheating patterns are necessary.

Significant difference in kinetics of heating and steady-state temperature increment was observed between evaporation-enabled and evaporation-disabled phantoms under identical exposure conditions [11]. The linear part of the temperature rise after the start of exposure vas 2-4 times longer and the steady-state temperature increment was up to 4 times greater in evaporation-disabled phantoms. Such differences in steady-state temperature increments show that most of the absorbed microwave energy (up to 75%) can be dissipated by evaporation. The relationship between the initial heating rate and steady-state overheating in objects with unpredictable evaporation became ambiguous.

Evaporation processes can play a significant role in the mechanism of biological and therapeutic effects of mm-waves. Dehydration in surface layers of human skin can occur without a considerable temperature increase. This will promote a deeper penetration of mm-waves into the skin, and will promote sweating to achieve thermal homeostasis.

4. Conclusion

Our study has demonstrated that mm-waves can produce a nonuniform heating pattern in irradiated objects, especially in the near field area. The same is true for RF frequency range with corresponding correction of scale depending on the frequency. Parameters of hot-spots are strongly dependent on the frequency and coupling conditions between the irradiating antenna and the irradiated surface; heterogeneity and surface geometry of biological objects. Because the pattern of electrodynamic heterogeneity in real biological object is usually unpredictable, and continuously changing due to different dynamic processes in living systems, such as opening and closing of sweat pores, the heating pattern will be also unpredictable and changing. Therefore, the incident power density is not a sufficient parameter for dosimetry and it is the heating rate pattern which is to be measured during radiation for full characterization of microwave absorption in irradiated objects.

5. References

1. Balantsev, V.N., Lebedev, A.M., Permjakov, V.A., Plotnikov, S.A., Sevastjanov, V.V., and Kuznetsov, A.N. (1991) Numerical investigation of SAR distribution in two-dimensional models of horn antennas with biological objects, *Dig. of the Int. Symposium on Millimeter Waves of Non-Thermal Intensity in Medicine, Moscow, USSR*, part 3, 660-664.
2. Barnes, F.S. (1984) Cell membrane temperature rate sensitivity as predicted from the Nerst equation, *Bioelectromagnetics* 5, 113-115.
3. Betsky, O.V., Petrov, I.Yu., Tyazhelov, V.V., Khizhniak, E.P., and Yaremenko, Yu.G. (1989) The distribution of electromagnetic fields of the millimeter wavelength range in phantoms and biological tissues in the near field area of the irradiators, *Doklady Akademii Nauk SSSR* 309, 230-233.
4. Betsky, O.V. (1991) Mechanisms of biological effects of mm waves interaction with living organisms, *Dig. of the Int. Symposium on Millimeter Waves of Non-Thermal Intensity on Medicine, Moscow, USSR*, part 3, 521-528.
5. Bolshakov, M.A., and Alekseev, S.I. (1986) A heating-rate dependent change in the electrical activity of pace-maker pond snail neurons, *Biofizika* 3, 521-523.
6. Deviatkov, N. D. (1973) Influence of millimeter-band electromagnetic radiation on biological objects, *Usp. Fiz Nauk* 110, 452-455, Transl: in *Sov. Phys. Usp.*, 16, 568-569, 1974.
7. Deviatkov, N.D., Golant, M.B., and Betsky, O.V. (1991) Millimeter waves and their role in vital activity, (in Russian) *Radia i Svyaz*, 1-169.
8. Furia, L., Hill, D.W., and Gandhi, O.P. (1986) Effect of millimeter-wave irradiation on growth of Saccharamyces cerevisiae, *IEEE Trans. Biomed. Eng.* **BME-33**, 993-999.
9. Golant, M.B., Grinberg, K.N., and Zdanovich, O.F. (1987) Determining the components of the millimeter wavelength range exciting electromagnetic oscillations in cells and a selection of an optimal shape of the irradiator, (In Russian) *Electoronnaya Tekhnika, seriya Electronica SVCh* 8, 52-54.
10. Khizhnyak, E.P., Betski, O.V., Voronov, V.N., Tyzhelov,V.V. and Yaremenko, Yu.G. (1991) Role of the distribution of microwave absorbtion in bioeffects of the EHF-irradiation, *Dig. of the Int. Symposium on Millimeter Waves of Non-Thermal Intensity in Medicine, Moscow, USSR*, part 3, 630-634.
11. Khizhnyak, E.P., and Ziskin, M.C. (1994) Heating patterns in biological tissue phantoms caused by millimeter wave electromagnetic irradiation, *IEEE Trans. Biomed. Eng.* **BME 41**, 865-873.
12. Khizhnyak, E.P., and Ziskin, M.C. (1996) Temperature oscillations in liquid media caused by continuous (nonmodulated) millimeter wavelength electromagnetic irradiation, *Bioelectromagnetics* 17, 223-229.

13. Khizhnyak, E.P., and Ziskin, M.C. (1997) Millimeter waves as a factor synchronizing biological processes in irradiated objects, *Abstract book of Second World Congress for Electricity and Magnetism in Biology and Medicine, Bologna, Italy*, 82-83.
14. Pashovkin, T.N., Khizhnyak, E.P., and Sarvazyan, A.P. (1984) Thermographic investigation of ultrasonically induced temperature distribution in biological tissues and tissue-equivalent phantoms, *Archives of Acoustics*, **9**, 15-21.

A COMPARISON OF SAR VALUES DETERMINED EMPIRICALLY AND BY FD-TD MODELING

T. J. WALTERS[1], P. A. MASON[1], K. L. RYAN[2], D. A. NELSON[3], and W. D. HURT
Air Force Research Laboratory, Human Effectiveness Directorate, Directed Energy Bioeffects Division, Brooks AFB, TX, 78235, [1]Veridian, Inc., San Antonio, TX, 78216, [2]Trinity University, San Antonio, TX, [3]Michigan Technological University, Houghton, MI, 49931

1. Introduction

Specific absorption rate (SAR) is defined by the National Council on Radiation Protection and Measurements as "...the time derivative of the incremental energy absorbed by (dissipated in) an incremental mass contained in a volume of a given density" (NRCP, 1981). The whole-body and partial-body SAR form the basis of permissible exposure limits for radio frequency radiation (RFR). The whole-body SAR provides very useful information regarding the influence of frequency, polarization, and orientation on RFR absorption [1,2]. However, RFR is not absorbed uniformly throughout a biological system. Numerous factors contribute to the heterogeneity of SAR values in biological system, including differences in electrical properties of different tissues, impedance mismatches at tissue boundaries, and the complex geometry of individual organs and structures [1,2]. Due to these factors, local SAR must be used to reveal the distribution of RFR absorption within the animal. Without this information, bioeffects data obtained from one species cannot be meaningfully extrapolated to another.

2. Experimental Estimation of Local SAR

2.1. BASIS OF THE METHOD

Local SARs are most commonly determined from thermal data obtained using non-perturbing temperature probes inserted into a region of interest. This method involves making a linear extrapolation from the rate of temperature change during exposure using the following relationship:

$$SAR = (\Delta T \cdot c) \cdot t^{-1} \qquad (1)$$

B.J. Klauenberg and D. Miklavcic (eds.), Radio Frequency Radiation Dosimetry, 207-216.

208

where SAR is in W/kg, T is temperature (°C), t is the time (sec) of the sampling period, and c is the specific heat of the tissue of interest (J/°C per kg) [2]. The specific heat can vary significantly between tissue types; e.g., some accepted values are 1,500 J/kg • °C for bone, 3,800 J/°C per kg for gray matter, and 3,500 500 J/°C per kg for white matter [3]. Thus, the use of a single specific heat value may not be appropriate to heterogeneous systems. The specific heat of brain tissue is 3,700 J/°C per kg. This method assumes: 1) homogenous tissue mass; and, 2) negligible heat loss via thermal conduction, convection, blood flow, and other thermoregulatory mechanisms. At the same time, the induced increase in temperature must not exceed 10°C, due to non-linearities that may result from the influence of temperature on the properties of the tissue. In order to satisfy this assumption, the system must be exposed to an incident power density capable of producing an SAR of at least 20 W/kg, with a typical exposure period of a maximum of 30 seconds [1]. In addition to these considerations, there are several other theoretical sources of error in using thermal data to estimate SAR that will be discussed subsequently.

2.2. REGIONAL SAR IN THE RAT BRAIN AS A FUNCTION OF ORIENTATION

Other investigators [4,5], as well as ourselves [6], have previously examined the influence of orientation relative to the RFR source on regional brain heating (and SAR). In these reports [1,4,5,6], data obtained using different experimental procedures and

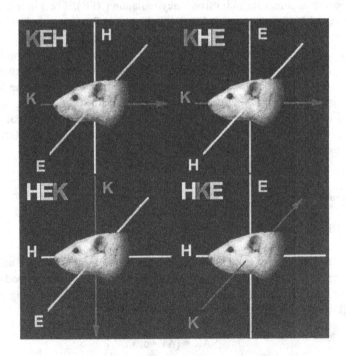

Figure 1. Representations of the four orientations examined.

equipment were combined in order to ascertain orientation effects. In the studies reported herein, however, a systematic series of experiments was performed to examine the influence of orientation on regional brain heating and SAR. Our studies concentrated on four orientations (khe, keh, hke, and hek (Figure 1)) representative of those that might be encountered by a freely moving rat.

2.3. EXPERIMENTAL METHOD

Recognizing the influence of body mass on SAR, all rats used in this study were food-restricted and maintained within a weight range of 365-375 g (n=8). Rats were anesthetized and stereotaxically implanted with a guide cannula in each of the four brain regions to be examined. These regions included the olfactory bulb, hypothalamus, cerebral cortex, and the brainstem. The cannulae were held in place with dental acrylic. Prior to RFR exposure, a Vitek temperature probe was placed in each guide cannula. The output of each Vitek, along with that of the power meter, was digitally recorded using a PC-based data acquisition program. All exposures took place in the far field, at a frequency of 2.06 GHz, and at an incident power of 1.0 W/cm². In addition to determining local SAR, whole-body SAR's were determined for all of the orientations according to the method described by Padilla and Bixby [7].

2.4. ORIENTATION AFFECTS LOCAL SAR IN A GIVEN BRAIN REGION

The orientation of the rat relative to the RFR source had a profound impact on the heating rate among individual regions during the same exposure (Fig. 2; Table 1). Figure 2 shows the influence of orientation on heating rate in a selected brain region as a function of orientation. Note the 4-fold difference in heating rate between the khe and keh orientations.

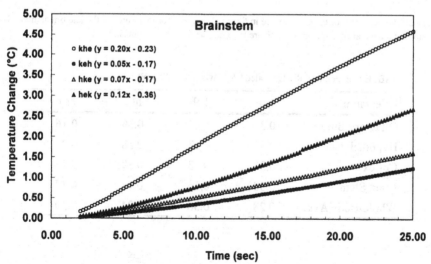

Figure 2. The mean rates of temperature rise in the brainstem during exposure in four different orientations. Lines (not shown) were fitted to the time interval between 2 and 25 seconds. The slope of each line (equation shown in the figure legend) was then used to calculate the SAR.

2.5. DIFFERENCES BETWEEN REGIONAL SAR IN A GIVEN ORIENTATION

Notable differences were also found among heating rates as a function of location. Figure 3 shows the differences between heating rates among the brain regions measured during exposure in the khe orientation. The estimated SAR values of the four brain regions for the four orientations examined are contained in Table 1. Note that for all orientations, a two-fold range of SAR values exists. Comparison of the whole-body average SAR with the regional SAR values shows that both "hot" and "cold" spots exist in the brain as a function of location and orientation.

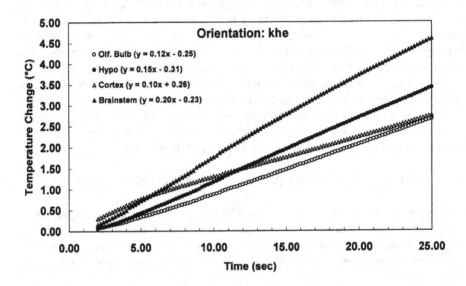

Figure 3. The mean rate of temperature rise in four brain regions for exposures in the khe orientation. The equation for the linear fit is shown in the figure legend. The fitted line is not shown.

TABLE 1. Experimentally Estimated SAR (W/kg per mW/cm²)

Orientation	khe	keh	hke	hek
Olfactory Bulb	0.35	0.23	0.39	0.16
Hypothalamus	0.45	0.13	0.38	0.28
Cortex	0.31	0.13	0.39	0.24
Brain Stem	0.58	0.16	0.21	0.37
Whole-Body Avg.	0.28	0.38	0.24	0.30

3. Comparison of Experimental Results with FD-TD Predicted SAR Values

The experimental data (above) clearly demonstrate the need to account for regional SAR in order to perform meaningful bioeffects experiments. However, it is not possible to empirically determine the regional SAR values for all tissues, under all exposure conditions and orientations. With the development of finite-difference time-domain (FD-TD) codes, it is possible to model SAR values under a wide range of exposure conditions in a number of laboratory animals, as well as humans. While these models have gained acceptance within the RFR community, little, if any, direct comparison of experimental data with SAR values predicted by FD-TD modeling exists. We were therefore interested in comparing our experimental data with FD-TD predictions performed in our laboratory. The brain offers an ideal organ for comparison, due to the reliability of stereotaxic probe placement, as well as the ability to easily confirm probe placements in frozen tissue sections following experimentation.

3.1. FD-TD OUTPUT

The process of obtaining SAR values from the FD-TD output has been described in this book by Mason et al.[8]. The anatomical model of the rat used in this investigation has been described by Mason et al. [9].

3.2. WHOLE-BODY SAR

We first compared the whole-body SAR determined calorimetrically with that predicted by the FD-TD for each orientation (Figure 4). The predictions from the FD-TD are slightly lower than those determined calorimetrically. However, the relative influence of orientation on whole-body SAR is similar for both methods.

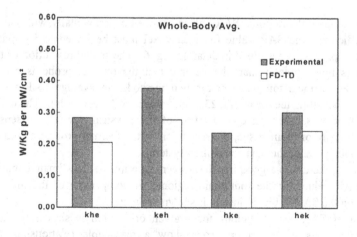

Figure 4. A comparison of the whole-body average SAR determined calorimetrically with the value predicted by FD-TD modeling as a function of orientation.

212

3.3. AVERAGE WHOLE-BRAIN SAR

The next level of comparison we performed was between the average whole-brain SAR's obtained with the two methods. The experimentally derived whole-brain SAR was estimated from the mean of the four brain regions at each of the four orientations. The results of this comparison are shown in Figure 5. The agreement between the two methods is good, and certainly within the error of the experimental method. The greater apparent value for the FD-TD prediction, compared with the experimental estimation, in the khe orientation will be discussed below.

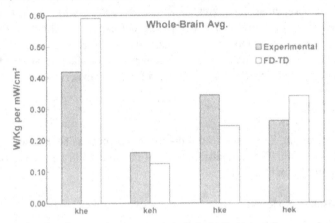

Figure 5. A comparison between the average whole-brain SAR obtained from experimental methods and the FD-TD prediction.

3.4. REGIONAL BRAIN SAR

In order to compare the regional SAR values obtained experimentally with those from FD-TD predictions, the SAR value for each voxel must be linked to its anatomical location. The process is described in detail in Fig. 6. The spatial resolution of the FD-TD output is much greater than the volume that the thermal probe is capable of resolving. For comparison between the two methods, an average SAR value was obtained by calculating the mean of a 23.5 mm^3 area (7 x 7 x 3 voxels). The use of this sample volume was based on consideration of the physical size of the sensor, the thermal diffusivity of brain tissue, the resolution of the thermal probe, sampling time, and the sampling rate of the data acquisition system.

For most cases, there was a good match between experimentally estimated and FD-TD predicted SAR values for the four brain regions. A comparison of the influence of orientation on the SAR in the brainstem is shown in Figure 7.

Comparison of SAR across regions for a given orientation is shown in Figure 8. Overall, the results of the comparisons show a reasonable relationship between

213

methods. As would be expected, the match is closest when the sampling volume around the temperature probe contains uniform SAR values that are close to the whole-body SAR. Mismatches between methods occurred when the FD-TD predicted relatively high or low SAR. In these cases, the extremes in SAR were not reflected in the thermal data, i.e., high and low FD-TD predicted SARs corresponded to under- or overestimation of experimentally estimated SARs, respectively. The olfactory bulb provides the best example of this situation.

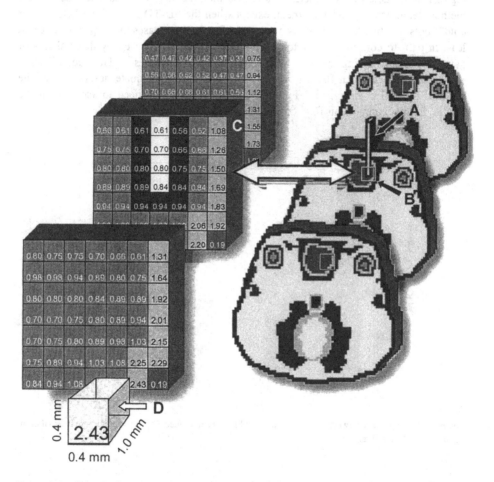

Figure 6. A schematic diagram illustrating the procedure used to locate the appropriate sampling area. In this example the area corresponds to the olfactory bulb. The location of each thermal probe was first mapped on the corresponding anatomical image (A), as can be seen in the middle slice. The numeric output from the FD-TD was then combined with the graphic image containing the anatomical information. The portion of this procedure corresponding to the sample area (B) is shown in the color-coded table (C). The sample area for each slice consisted of a 7 voxel x 7 voxel (x,y) area with the tip of the probe in the center of the area. Each voxel was 0.4 mm x 0.4 mm x 1.0 mm (x,y,z) (D). A similar size region was sampled in the adjacent slices, thus the volume of the sample area was 23.5 mm³ area (7 x 7 x 3 voxels).

4. Conclusions

The orientation, relative to the RF source, has a profound influence on the regional SAR, and must be accounted for in order to validly execute and interpret any bioeffects experiment. It is not practical to empirically determine regional SAR for all experimental conditions. The FD-TD method for determining regional SAR offers an attractive alternative to laborious experimental methods. Initial comparisons between regional SAR determined experimentally, via thermal methods, and with the FD-TD method demonstrate good agreement, except when the FD-TD method predicts "hot or cold" spots (relative to whole-body average SAR). In these cases the disparity is due, at least in part, to confounding factors in the thermal method caused by thermal loss or gain from surrounding regions with dissimilar SAR values. This complication emphasizes the fact that SAR alone does not provide an adequate description of the regional thermal environment. Future computer models will need to account for the role of thermal components (e.g. conduction, convection, blood flow).

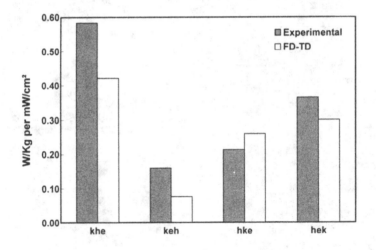

Figure 7. A comparison between experimentally and FD-TD determined SAR as a function of orientation in a single brain region (brainstem).

5. Notes

Experimental procedures were approved by the Animal Care and Use Committee at the Air Force Research Laboratory, Brooks AFB. The animals used in this study were procured, maintained and used in accordance with the Animal Welfare Act and the

"Guide for the Care and Use of Laboratory Animals" prepared by the Committee on Care and Use of Laboratory Animals of the Institute of Laboratory Animal Resources - National Research Council. Views presented are those of the authors and do not reflect the official policy or position of the Department of the Air Force, Department of Defense, or U.S. Government.

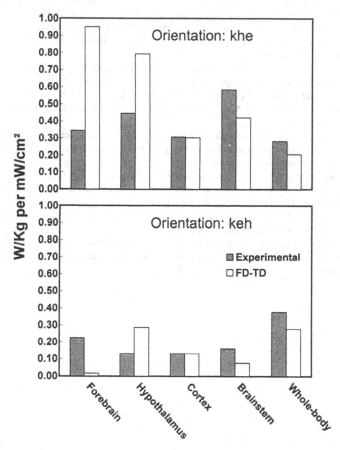

Figure 8. SAR across brain regions for the khe and keh orientations.

6. Acknowledgments

Research was funded by U.S. Air Force Contracts F33615-90-C-0604, F33615-90-D-0606, HQ Human Systems Center, Air Force Materiel Command, Brooks AFB, TX, 78235-5000. The authors would like to extend their gratitude to the following individuals for their invaluable technical support: George Lantrip; Heather Lehnert; Kevin Kosub; Kavita Mahajan; and SSgt. Luther M. Tate.

216

7. References

1. Chou, C.K., Bassen, H., Osepchuck, J., Balzano, Q., Peterson, R. Meltz, M. Cleveland, R, Lin, J.C., and Heynick, L. (1996) Radio frequency electromagnetic exposure: Tutorial review on experimental dosimetry, *Bioelectromagnetics* **17**, 341-353.
2. Durney, C.H., Massoudi, H., and Iskander, M.F. (1986) *Radiofrequency Radiation Dosimetry Handbook. USAF School of Aerospace Medicine Report,* **SAM-TR-85-73**, Brooks AFB, TX.
3. Olsen, R.W. (1985) *Temperature Distributions During Induced Deep hypothermia and Subsequent Circulatory Arrest: An Experimental and Numerical Study.* Ph.D. dissertation, The University of Texas Health Science Center at Dallas, p. 188.
4. Chou, C.K., Guy, A.W., McDougall, J.A., and Lai, H. (1985) Specific absorption rate in rats exposed to 2,450-MHz microwaves under seven exposure conditions, *Bioelectromagnetics* **6**, 341-353.
5. Ward, T.R., Svensgaard, D.J., Spiegel, R.J., Puckett, E.T., Long, M.D., and Kinn, J.B. (1986) Brain temperature measurements in rats: A comparison of microwave and ambient temperature exposure, *Bioelectromagnetics* **7**, 243-258.
6. Walters, T.J., Ryan, K.L., Belcher, J.C., Doyle, J.M., Tehrany, M.R., and Mason, P.A. (1998) Regional brain heating during microwave exposure (2.06 GHz), warm-water immersion, environmental heating and exercise, *Bioelectromagnetics* **19**, 341-353.
7. Padilla, J.P., and Bixby, R. (1986) *Using Dewar-Flask Calorimetry and Rectal Temperatures to Determine the Specific Absorption Rates of Small Rodents. USAF School of Aerospace Medicine Report*, USAFSAM-TP-86-3, Brooks AFB, TX.
8. Mason, P.A., Ziriax, J.M., Hurt, W.D., Walters, T.J., Ryan, K.L., Nelson, P.A., and D'Andrea, J.A. (1999) Recent advances in dosimetry and modeling, *This Volume.*
9. Mason, P.A., Walters, T.J., Fanton, J.W., Erwin, D.N., Gao, J.H., Roby, J.W., Kane, J.L., Lott, K.A., Lott, L.E., and Blystone, R.V. (1995) Database created from magnetic resonance images of a Sprague-Dawley rat, rhesus monkey, and pigmy goat, *FASEB J.* **9**, 434-440.

EXPOSURE AND DOSIMETRY OF LOW FREQUENCY RF FIELDS

P. VECCHIA and A. POLICHETTI
Physics Laboratory, National Institute of Health
Viale Regina Elena 299, 00161, Rome, Italy

1. Introduction

While extensive literature has been produced over the last fifty years on the biological and health effects in the higher frequency range of radiofrequency (RF) and microwave (MW) electromagnetic fields (EMF), only sparse data exist on the effects of fields of frequency below a few megahertz, i.e. in the lower range of RF.

In effect, the lower border of the RF spectrum is not clearly defined: in most standards, it is more or less implicitly assumed somewhere between 1 kHz and 100 kHz. However, within the scope of the International EMF Project of the World Health Organization (WHO) the RF range is defined as the interval 300 Hz - 300 GHz.

The scarce attention paid until a few years ago to the low-frequency RF fields may be explained by the limited number of technologies involved. However, the population potentially exposed is very large, motivating increasing interest and research. A workshop on "Intermediate frequency range EMF: 3 kHz - 3 MHz" was held in April 25-26, 1998, in the frame of the COST Action 244bis. An international seminar on the same topic is scheduled in 1999 by the International EMF Project.

The main scope of this paper is the evaluation of exposure and dosimetry of low frequency RF fields. For a more general overview, a brief review is also given of the literature on biological effects, and of current safety standards.

2. Biological Effects of Low Frequency RF Fields

A recent review of the literature by Juutilainen [1] has indicated that the research in this area is limited and sparse: different biological endpoints have been investigated at different frequencies, and a coherent picture of the findings is difficult to draw.

The number and the dimension of the studies is also insufficient to draw firm conclusions on any biological or health effect of low frequency RF fields. However, although many studies are negative, some interesting findings have been reported, even at low field intensities. These findings must be independently replicated before they form the basis for any protection measure, but clearly indicate the need for further research.

B.J. Klauenberg and D. Miklavcic (eds.), Radio Frequency Radiation Dosimetry, 217-226.

Some of the reported effects are quite similar to those from ELF fields, suggesting the possibility of some common mechanism of action, possibly related to electromagnetic induction. On the other hand, induced electric voltages and currents linearly increase with increasing frequency. Therefore it has been suggested [1] that any related biological effect should be more apparent at higher frequencies. On the contrary, in the aforementioned review it is remarked that no study reports stronger effects at higher frequencies.

In conclusion, there is at present no established mechanism accounting for the sparse and inconclusive reported effects. However, on theoretical grounds, the induction of electric currents is the most relevant interaction mechanism of electric, and mainly magnetic, fields with the body, up to at least 100 kHz [2, 3].

3. Safety Standards

All the international exposure guidelines consider the electric current density as the appropriate dosimetric quantity in the frequency range below about 100 kHz [3-6].

Based on the experimental findings, all the standards assume that the effectiveness for biological effects decreases with increasing frequency, and basic exposure limits increase accordingly. Basic limits recommended by the international organizations are reported in Table 1.

TABLE 1. Basic exposure limits on current density recommended by international standards

Organization	Category of exposure	Frequency range (kHz)	Basic limit (mA/m²)
IEEE	Controlled environment	3-100	350 f
CENELEC	Workers	1-10,000	10 f
ICNIRP	Workers	1-10,000	10 f
IEEE	Uncontrolled environment	3-100	157 f
CENELEC	General public	1-10,000	4 f
ICNIRP	General public	1-10,000	2 f

f is the frequency in kilohertz

The numerical values are quite different among standards, indicating some divergence in the evaluation of effects reported in the literature, and in safety factors.

Following a well established procedure, reference levels of more practical use for compliance are deduced from the basic limits, through appropriate dosimetric models.

All the standards are consistent in setting reference levels which decrease with increasing frequency in a transitional frequency range, to take into account the resonant absorption of electromagnetic energy by the body. Further discrepancies appear in the reference levels, which are due to the particular choice of numerical factors in the mathematical curves (in particular the limits of the transition frequency range). Such arbitrariness reflects the lack of a unique, widely accepted dosimetric model.

4. Sources of Exposure

The technologies, which make use of low frequency fields, though limited, are progressively increasing. Most of them exploit magnetic fields through inductive processes, and for this reason the term "inductive frequencies" is sometimes used to designate this portion of the electromagnetic spectrum.

Because of the long wavelength (3 km at 100 kHz), the electric and magnetic fields are actually decoupled and should be considered separately for most applications.

The electric field is of importance in radio broadcasting. Low frequencies are used for long and medium wave AM transmission over long distances.

Magnetic fields, on the contrary, are exploited in a number of processes which include:
- industrial heating and melting of metal;
- surface hardening;
- magnetic shaping;
- metal detection;
- non destructive tests.

A recent review by Gaspard [7] lists a number of such applications, with typical operating frequencies in the range of tens to hundreds kilohertz, and powers reaching even some megawatts.

Besides industrial technologies, a few applications are of special importance because of the great proportion of the potentially exposed population. These include in particular:
- domestic inductive cookers;
- electronic surveillance devices;
- personal identification systems.

According to Gaspard [7] the penetration of inductive cookers is still relatively low (of the order of 5,000 units per years in Europe), but might increase substantially if the market expands from the present area essentially limited to restaurants, to the consumers.

Electronic Article Surveillance (EAS) systems are anti-theft devices which are commonly installed in supermarkets, department stores, and retail shops to prevent the removal of items. From the point of view of the exposure, they are similar to proximity readers, a term used to indicate systems which allow the identification of personnel by the at-a-distance reading of magnetic badges.

Finally, low frequency (around 20 kHz) magnetic fields are used to drive the electronic beam inside cathode tubes of VDTs and TV sets. Although this application has raised great attention (much of the scientific research on bioeffects has been performed with 20 kHz fields), the exposure levels actually experienced by people are very low.

In contrast, there is some merit in considering in more detail the exposure conditions in the case of proximity readers and EAS devices, given the large dissemination of these sources and the social and economic impact of any controversy about health risks possibly related.

5. Characteristics of Proximity Readers

Proximity readers are devices which allow a fluid and easy control of personnel crossing fixed passages, e.g. at the entrance/exit at workplaces or for access to restricted areas. The person wears a "magnetic" badge (in a pocket, in a purse, etc.), without any need to insert it in a slot. When passing in front of a magnetic antenna, the badge is activated and sends a coded signal detected by a receiver and processed by suitable software. In some cases, an electric battery is inserted inside the badge, but in general the electronic circuit in the card is feeded by a coil sealed in the badge itself.

The coil couples with the magnetic field generated by the antenna, storing enough magnetic energy to activate the circuit. To make the system reliable, enough energy must be transferred even in the case of the worst coupling (i.e. the most unfavourable orientation of the badge to the field), and that requires an adequate intensity of the magnetic field. "Active" badges (i.e. badges energized by an internal battery) obviously require lower fields, but some intensity is however needed to trigger the circuit which normally stands-by to save the battery.

To ascertain possible health risks, as well as compliance with exposure restrictions, an extensive analysis of these devices has been carried out in Italy by Polichetti and Vecchia [8]. The results of experimental measurements show that the electric field strength is well below any recommended limit, as expected due to the fact that the devices are based on magnetic coupling. As regards the magnetic field, representative values measured inside the corridors are listed in Table 2, together with the main characteristics of the sources, produced by different manufactures.

TABLE 2. Magnetic field strength H (A/m) inside the corridor of proximity readers(maximum values)

Frequency (kHz)	Waveform	Badge type	H (A/m) 10 cm from wall	H (A/m) at centre
117	CW Sinusoidal	Passive	10	6
132	Sinusoidal pulses	Active	6	0.7
120	CW Sinusoidal	Passive	5	0.8
125	AM Sinusoidal	Passive	7	4

To test the compliance with exposure standards, we compared the above values with recommended reference levels at 120 kHz (the typical frequency of operation of proximity readers) listed in Table 3.

Comparison of Tabs. 2 and 3 immediately shows that some reference levels are exceeded at some locations. However, according to basic principles of protection against non ionizing radiation [3,9], reference levels may be exceeded, provided basic exposure limits (in this case, limits on the current density) are complied with. Therefore, appropriate dosimetric models and theoretical calculations are needed to evaluate current densities, which cannot be directly measured.

TABLE 3. Reference levels for exposure to magnetic fields (A/m) at 120 kHz

	Workers[1]	General public[1]
CENELEC	13.33	5.83
ICNIRP	13.33	5
ICNIRP (peak value)	22.6	8.5
IEEE-ANSI	135.8	135.8

(1) In the IEEE-ANSI standard, a distinction is made between controlled environments and uncontrolled environments, rather than between workers and the general population.

This fundamental step of the exposure assessment will be discussed in the next section, with special reference to proximity readers. It is important however to note that problems of the same kind, but of greater severity, are likely to arise from EAS systems at supermarkets, department stores, and retail shops. Whereas no extensive survey on this type of device has been performed, some data made available by the manufacturers seem to indicate that reference levels are exceeded more than in the case of proximity readers.

6. Evaluation of Current Densities

Deriving current densities from external fields is complicated by both the complex structure of the human body and the highly inhomogeneous distribution of the fields. The problem is discussed here with special reference to specific sources, but is quite general for low frequency fields, where exposures always occur in the near field.

There are basically two possible approaches, consisting of i) a rough evaluation based on a simplified model of the body and a simplified distribution of the fields, or ii) a careful calculation based on refined body modelling, detailed field mapping and sophisticated numerical codes.

The second one is obviously of importance from a scientific point of view, but is not affordable in most practical cases of exposure assessment. On the other hand, approximations could lead to unreliable results. An estimate of the error margins for different approximations helps to determine if simplified dosimetry is satisfactory for any given purpose.

We performed analytical calculations on a simple model of the body (or part of it, namely the head and trunk), approximated by an ellipsoid of one single tissue. In Fig. 1, the distribution of current density is shown, with the maximum at the periphery of the body on the axis Y.

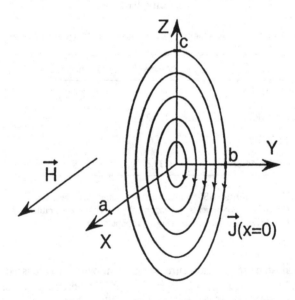

Figure 1. Distribution of induced currents at the mid-section of the ellipsoidal model of the body (or the head and trunk)

The maximum value of current density is given by:

$$J_{max} = 2 \pi f \mu H \sigma \frac{b}{1 + \eta^2} \qquad (1)$$

where f is the frequency, H the magnetic field strength, μ and σ the magnetic permeability and the conductivity of the tissue, respectively, and $\eta = b/c$ is an anisotropy factor taking into account the body shape. It is immediately seen that for a given width of the body (i.e. for a given value of b), J_{max} varies within a factor of two, namely between:

$$J_{max} = \pi f \mu H \sigma b \qquad (2)$$

in the limit case b = c, i.e. circular loops, and:

$$J_{max} = 2 \pi f \mu H \sigma b \qquad (3)$$

in the limit case $\eta = 0$, i.e. an indefinite cylinder.

The sensitivity of the model to the electrical characteristics of the tissue is shown by data in Table 4, where induced currents in the indefinite cylindrical model of the whole body are reported, corresponding to different values of the dielectric constant ε_r and conductivity σ found in the literature for the muscle.

In the last row, the most recent data by Gabriel [10] are reported; the corresponding value of the current density may be considered the most reliable with regard to the electrical characteristics, and as an upper limit with regard to the body size.

TABLE 4. Dependence of the maximum induced current
on the dielectric characteristics of the tissue (muscle)

ε_r	σ (S/m)	$|\sigma^*|$ (S/m)	J_{max} (mA/m^2)
$2 \cdot 10^4$	0.40	0.42	800
$2 \cdot 10^4$	0.59	0.60	1140
$3 \cdot 10^4$	0.40	0.45	850
$3 \cdot 10^4$	0.59	0.62	1180
$9 \cdot 10^3$	0.40	0.40	760

Notes: 1) Exposure: H = 10 A/m, homogeneous; f = 120 kHz
2) Model: homogeneous indefinite cylinder (radius = 20 cm)
3) $\sigma^* = \sigma + j 2 \pi f \varepsilon_r \varepsilon_o$

Data in Table 4 also indicate that capacitive current gives only a minor contribution to the total induced current.

To evaluate the dependence on the model orientation and size, calculations were performed for different values of b and c: results are reported in Table 5. The data correspond to a homogeneous magnetic field H = 10 A/m at 120 kHz frequency, and to an ellipsoid of muscle (σ = 0.40 S/m).

TABLE 5. Induced current densities through different mid-sections of the human body (standard man). Homogeneous magnetic field H = 10 A/m

Section	b (cm)	c (cm)	J_{max} (mA/m^2)
1) Whole body, front	20	85	720
2) Whole body, lateral	15	85	550
3) Head and trunk, lateral	15	55	530
4) Reference section[1]	20	20	380

(1) The reference section, which is given here for comparison, is used for dosimetric calculations in international standards

To evaluate the degree of approximation, a comparison can be made with data extrapolated from numerical calculations performed by Andreuccetti et al. [11] on a cell model of the human body using the impedance method.

For the same exposure conditions as above (H = 10 A/m, f = 120 kHz), and for a realistically shaped, single tissue model, the maximum current density in the torso is 610 mA/m^2. Therefore, the geometrical approximation (the ellipse 3 in Table 5 instead of the real shape) alone leads to an underestimation of induced currents by about 13%. On the other hand, calculations on the realistic, multiple-tissue model indicate that the maximum current density occurs inside the body (not at the periphery) and equals 330 mA/m^2. It may therefore be concluded that the ellipsoid single-tissue model overestimates the induced current by a substantial factor (about 1.6 for the exposure conditions under consideration).

It is also interesting to note that the maximum current density at the periphery of the body is much lower, due to the dielectric characteristics of the skin, which are very different from the muscle.

Experimental and theoretical procedures described above with reference to proximity readers can be applied to other similar sources, in particular to EAS systems. To our knowledge, no extensive survey has been performed on this kind of device. Sparse data available from the manufactures indicate that various frequencies are used, typically of the order of some kHz. As a realistic example, we can consider a 5 kHz magnetic field whose strength within the control area varies between 30 and 40 A/m. Analytical calculations based on the ellipsoidal model 1 in Table 5 give, for the limit cases of homogeneous fields equal to the lower and upper value, current densities of about 80 and 110 mA/m^2, respectively. No data are available for the realistic cell-model of the body. However, since the dielectric constants only slightly decrease with decreasing frequency from 120 to 5 kHz [10], a reduction of the order of 40%, as in the case of proximity readers, can be expected.

7. Health Risk Considerations

A comparison of Tables 2 and 3 indicates that local values of the magnetic field inside proximity readers may exceed reference levels provided by CENELEC and ICNIRP standards, whereas they are well below those recommended by IEEE-ANSI. Given the fact that all the standards are based on the same literature, and all include conservative safety factors, compliance with only one of them is sufficient to exclude at least major health risks. However, non-compliance with one standard may be a reason for concern and opposition by the public. Therefore, a more detailed analysis on current densities is appropriate. According to all the standards, in fact, reference levels may be exceeded provided basic limits are complied with.

Values of current density in Table 5 are still above basic limits recommended for the general public by CENELEC (480 mA/m^2 at 120 kHz) and ICNIRP (240 mA/m^2 at 120 kHz), whereas values calculated from a realistic body model are about 40% lower, complying with the CENELEC standard but not with the ICNIRP one. However, calculations were performed under the conservative assumption of a uniform field equal to the maximum experimentally measured. Taking field inhomogeneity into account, it may be concluded that proximity readers are very likely to comply with even the most restrictive standards.

The case of EAS devices is more complex because current densities calculated even for the lowest field intensity (of the order of 80 mA/m^2 in the example considered) is substantially higher than the limits recommended e.g. by ICNIRP (50 mA/m^2 at 5 kHz for the exposure of workers, 10 mA/m^2 at 5 kHz for the exposure of population). Also in this case, whereas compliance with the IEEE-ANSI standard (785 mA/m^2 for uncontrolled environments at 5 kHz) is sufficient to exclude major risks, the discrepancies between different regulations might create problems of acceptance for this kind of device.

8. Conclusions

Low frequency RF fields have so far been considered quite superficially, both in the research and in the development of safety standards. The rapid development of technologies involving the exposure of a large proportion of the public requires however a much better understanding of exposure characteristics and biological effects.

In particular, good dosimetry is needed to carefully evaluate current densities for different devices and exposure conditions. Whereas this paper reports numerical estimates only for identification systems and anti-theft devices, preliminary data indicate that exposures to several other sources, including domestic cookers and industrial heaters, may exceed limits recommended by at least a part of existing standards. The relevance of localized exposures and of the exposure time for biological effects should also be clarified. Finally, a harmonization of exposure standards is needed, to avoid controversies which could have a great social impact.

9. References

1. Juutilainen, J. (1998) Biological effects of electromagnetic fields at the kHz and low MHz frequency range, in L. Miro and R. De Seze (eds.), *Proceedings of the 3rd COST 244bis Workshop on "Intermediate Frequency Range"*, Paris, April 25-26, 1998, pp. 88-93.
2. WHO (1993) *Environmental Health Criteria 137: Electromagnetic fields (300 Hz to 300 GHz)*, World Health Organization, Geneva.
3. ICNIRP (1998) Guidelines for limiting exposure to time-varying electric, magnetic, and electromagnetic fields (up to 300 GHz), *Health Phys.* **74**, 494-522.
4. IEEE-ANSI (1992) *IEEE Standard for safety levels with respect to human exposure to radio frequency electromagnetic fields. 3 kHz to 300 GHz. (Standard IEEE C95.1-1991. Revision of ANSI C95.1-1982)*, Institute of Electrical and Electronics Engineers, New York.
5. CENELEC (1995) *Human exposure to electromagnetic fields. Low frequency (0 - 10 kHz). Experimental European Standard ENV 50166-1*, European Committee for Electrotechnical Standardization, Paris.
6. CENELEC (1995) *Human exposure to electromagnetic fields. High frequency (10 kHz - 300 GHz). Experimental European Standard ENV 50166-2*, European Committee for Electrotechnical Standardization, Paris.
7. Gaspard, J.Y. (1998) Assessment of the use of intermediate frequency range, in L. Miro and R. De Seze (eds.), *Proceedings of the 3rd COST 244bis Workshop on "Intermediate Frequency Range"*, Paris, April 25-26, 1998, pp. 4-20.
8. Polichetti, A. and Vecchia, P. (1998) Exposure of the general public to low- and medium-frequency electromagnetic fields, in L. Miro and R. De Seze (eds.), *Proceedings of the 3rd COST 244bis Workshop on "Intermediate Frequency Range"*, Paris, April 25-26, 1998, pp. 21-30.
9. IRPA/INIRC (1985), Review of concepts, quantities, units and terminology for non-ionizing radiation protection, *Health Phys.* **49**, 1329-1362.
10. Gabriel, C. (1996) *Compilation of the Dielectric Properties of Body Tissues at RF and Microwave Frequencies*, Report AL/OE-TR-1996-0037, NTIS, Springfield, VA.
11. Andreuccetti, D., Fossi, R. and Petrucci, C. (1998) *Esposizione della popolazione ai varchi magnetici: analisi di un caso tipo (Population exposure to proximity readers: A case-study analysis)*, Report No. TR/ICEMM/2.98, Institute for Research on Electromagnetic Waves, Florence (in Italian).

MOLECULAR DOSIMETRY

J. L. KIEL
Air Force Research Laboratory
8308 Hawks Rd.
Brooks Air Force Base, Texas 78235 U.S.A.

1. Introduction

The ideal dosimeter to support health and safety monitoring and epidemiological studies to determine if any long term chronic effects of non-ionizing electromagnetic radiation exist would be a personal one. Such a dosimeter would measure the actual dose received by an individual at risk. It would have to take into account whole body average and local exposures and in itself minimally perturb the exposure being assessed. Furthermore, the dosimeter would have to measure an effective dose in a way that was very sensitive to potential biological effect mechanisms. Preferably, the dosimeter would have to be responsive without restrictions of geometry, that is, it would have to respond at the molecular level as if it were in an infinite volume. Its response would be determined by its location on the person being assessed, that is, the human's geometry rather than the dosimeter's.

Molecular "biodosimeters" that have been suggested and whose selection has been supported by theoretical calculations include DNA, enzymes, lipid bilayers, and hemoglobin [1, 2, 3, 4]. However, none of these have stood up to scrutiny in respect to their demonstrating a particular resonance response to microwave and radio frequency radiation or a broad enough frequency response to be useful [5, 6]. Another approach that has been successfully pursued is the designing of an artificial polymer. A substance with several specific characteristics was sought out as a candidate for a synthetic molecular dosimeter. First, it had to participate in a thermally-sensitive chemical reaction. Second, the limiting reaction had to demonstrate steady-state kinetics (zero-order rate law). Third, the substance and/or its reactants had to show electric and/or magnetic field susceptibility. Fourth, energetically degenerate long-lived states, that are interconvertible but restricted by an energy barrier between them, had to be involved in the reactions of the substance. A material with these properties, diazoluminomelanin (DALM), was synthesized by co-polymerization, either synthetically or bio-synthetically, of 3-amino-L-tyrosine and luminol (5-amino-2,3-dihydrophthalazine-1,4-dione) induced by nitrite [7, 8, 9]. Although this polymer's structure has yet to be fully resolved, its reaction kinetics, influences of various states, thermodynamics, electromagnetic susceptibility, and photochemistry fit the profile described above and can be modeled. These properties are discussed here, and the predicted and observed responses compared. The kinetic, thermodynamic, and quantum mechanical models displayed are general enough so as to be applicable to other biochemical and chemical reactions beyond this special case.

227

B.J. Klauenberg and D. Miklavcic (eds.), Radio Frequency Radiation Dosimetry, 227-237.

2. Experimental and Computational Approaches

Diazoluminomelanin (DALM) when added to a solution containing hydrogen peroxide and sodium bicarbonate or carbonate luminesces [8]. It does this at a level that corresponds to the temperature of the solution. This steady-state luminescence can persist for hours and can be cycled from one temperature to another and back with a comparable cyclic change in the intensity of the luminescence. The response of the luminescence of this peroxidizing reaction can be described by the following Arrhenius-like equation:

$$I = Ae^{-Ea/kT} \qquad (1)$$

Where I = intensity of the luminescence, A is a pre-exponential factor, E_a is the apparent activation energy of the process, k = Boltzmann's constant, and T = the absolute temperature in degrees Kelvin. If $\Delta H - \Delta ST$, where ΔH = the enthalpy change and ΔS = the entropy change, is substituted for E_a, k_o is substituted for I/A, and the equation is converted to its logarithmic form and rearranged, then it becomes the following:

$$T = \Delta H/(\Delta S - k\ln k_o) \qquad (2)$$

The kinetic constant k_o is the zero-order rate constant which is equal to the steady-state luminescence. Expression (2) for temperature, in the form of enthalpy and entropy change and steady-state luminescence, can be substituted for absolute temperature in the following expression for Gibbs free energy change of an equilibrium reaction:

$$|\Delta G| = NkT\ln K_{eq} \qquad (3)$$

The Gibbs free energy is given as an absolute value because the reaction can switch between exergonic and endergonic forms. The equilibrium constant K_{eq} is assumed to be the same as the steady-state luminescence rate, in other words, equal to k_o. N is the number of molecules. The Gibbs free energy is considered equal to the steady-state input of energy, that is, the specific absorption rate (SAR) of all microenvironmental energy. In the case of microwave irradiation, SAR is the absorbed dose rate of the microwave energy added to the ambient rate of energy input. Equation (3) with substitution of equation (2) for T becomes the following:

$$SAR = \Delta HNk\ln k_o/(\Delta S - Nk\ln k_o) \qquad (4)$$

Because entropy is so difficult to determine in biochemical reactions, the above equation needed to be simplified. One approach is to quantize heat capacity, which leads to quantization of entropy. This process is described in the following equations:

$$S = \int_0^T C_p \, d\ln T = \int_0^T (\partial H / \partial T)_p \, d\ln T = Nk\ln q \qquad (5)$$

$$\therefore \Delta S = Nk\ln(q'/q_o) \tag{6}$$

In (5) and (6), C_p = heat capacity at constant pressure, q = the molecular partition function which is proportional to the number of kinds of states, and q' and q_o represent the molecular partition functions of the initial and final conditions, respectively. Next, let $kN\ln(q'/q_o)$ approach $kN\ln k_o$ in value. Then, ΔS is substituted for $Nk\ln k_o$ in equation (4). The result is as follows:

$$SAR = \Delta H\Delta S/(\Delta S\text{-}Nk\ln k_o) \tag{7}$$

Equation (7) is a hyperbolic function of the following form:

$$y = ac/(a-x) \tag{8}$$

In this equation, y = SAR, a = ΔS, c = ΔH, and x = $\ln k_o$, where k_o is the molar rate constant equal to photons per second per mole.

3. Comparison of Computational and Experimental Results

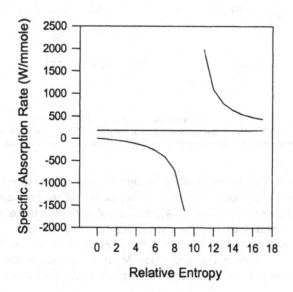

Figure 1. Computational results of the hyperbolic function for specific absorption rate (SAR) when the luminescence rate is held constant. The straight line is the sum of the partial differentials of entropy and the natural logarithm of the luminescence rate.

230

Figure 1 shows a plot of this equation, when $\ln k_o$ is arbitrarily set equal to 10 and the entropy is changed. The straight line is equal to the activation energy. Figure 2 shows the results when the entropy is held

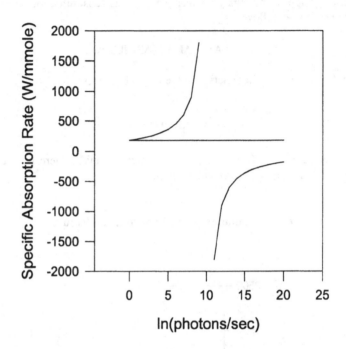

Figure 2. Computational results of the hyperbolic function for SAR. Entropy is held constant and the luminescence rate is varied. The straight line is the sum of the partial differentials for entropy and the natural logarithm of the luminescence rate constant, that is the Gibbs free energy of the steady-state reaction..

constant and the natural logarithm of the luminescence is varied. In Fig. 2, the straight line is also the activation energy, which is, in fact the sum of the two partial differentials of entropy and enthalpy, respectively. When the absolute SAR (ambient plus microwave energy input) is plotted against the natural logarithm of the luminescence, then a saturation curve is generated as shown in Figure 3. Figure 4 shows actual data plotted of input wattage of microwave radiation (2450 MHz, continuous wave) versus relative luminescence intensity of peroxidizing DALM made in HL-60 human leukemia cells [10].

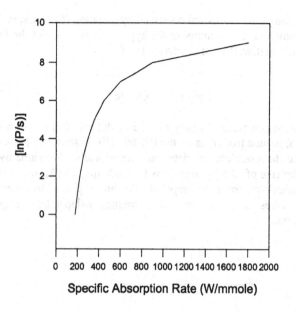

Figure 3. Computed luminescence saturation curve when SAR (the total energy input of environmental and microwave energy) is applied as an independent variable.

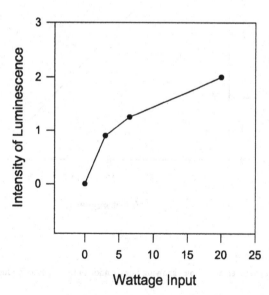

Figure 4. Experimental thermochemiluminescence results of the exposure of HL-60 cells loaded with peroxidizing DALM to 2450-MHz microwave radiation.

Therefore, the semi-empirical model described by equation (8) seems to fit experimental data. To overcome the discontinuity in the hyperbolic function of the DALM response model, the whole equation (8) can be inverted as follows:

$$1/y = (-1/ac)x + 1/c \qquad (9)$$

This transformation generates a linear plot. The model results are shown in Figure 5, and the actual results, plotted from data on the HL-60 cells, is shown in Figure 6. This linear plot allows one to calculate or determine graphically the enthalpy and entropy. Therefore, the inverse of the y-intercept is the enthalpy, and knowing this, the entropy can be easily calculated from the slope of the linear plot. However, none of these representations takes into account any frequency-dependent responses of the biochemical system.

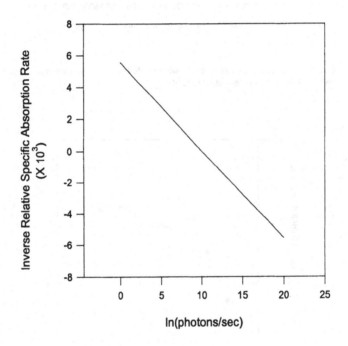

Figure 5. Calculated inverse SAR plot showing linearization of the hyperbolic function for SAR.

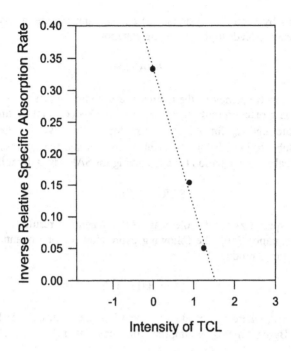

Figure 6. Experimental data of HL-60 human leukemia thermochemiluminescence fitted to a linear regression inverse plot of SAR.

4. Determining Frequency Dependency of Biochemical Reactions

Biochemical reactions, where the hyperbolic model describes the reaction thermodynamics, can be thought of as capable of switching between exothermic and endothermic reaction pathways. The smallest distance between the limbs of the hyperbolic function plot represents a "forbidden" transition. Heating or cooling the reaction will only make each energy pathway limb reach a maximum reaction rate or entropy, respectively. The point of the discontinuity in the function, that is, infinite energy absorption or release is where the above classical approach breaks down. This point is where the entropy and energy of the zero-order reaction rate are nearly equal. However, in both paths of the hyperbolic function, when the reaction rate is held constant, cooling either increases entropy toward a maximum or toward infinity. Therefore, the reaction is entropy "driven", or "dominated". Calculating the energy gap between the limbs by vector analysis (sums) and treating the energy as a quantum of electromagnetic energy leads to the following equation:

$$\nu = 2(SAR)/Nh \qquad (10)$$

In equation (10), ν = frequency of the microwave or radio-frequency radiation, N = the number of absorbing reacting molecules, and h = Planck's constant. Using equation (2), one can substitute $kNlnk_o$ for ΔS, and lnk_o for T to solve for ΔH in terms of luminescence. Substituting $Nk(lnk_o)^2 - k(lnk_o)^2$ for ΔH and the luminescence intensity equivalent of enthalpy into equation (7) and solving for SAR gives the following:

$$SAR = Nk(lnk_o)^2 \qquad (11)$$

This result is the quantization of molecular SAR. When the result of equation (11) is substituted into equation (10), the following expression for one quantum of absorbed microwave radiation is produced:

$$\nu = 2k(lnk_o)^2/h \qquad (12)$$

Because $SAR = \sigma E^2$, where σ = conductivity and E = the electric field, then by analogy $E^2 = k(lnk_o)^2$. Based on this assumption, one can calculate conductivity from the following equation:

$$\sigma = SAR/k(lnk_o)^2 \qquad (13)$$

Therefore, N is equivalent to conductivity, the number of electrons transported through a given length or through a given cross-sectional area.

5. Quantum Effect Results

Figure 7 shows the conductivity (imaginary part of the permittivitty) calculated using equation (13) compared to actual reported measurements of a 5% (weight/volume) solution of DALM in water [11]. The divergence of the measured values from the calculated curve results from the frequencies approaching the relaxation frequency of water where conductivity increases. If the reaction rate reaches the point in the discontinuity of the hyperbolic function where the limbs of the hyperbole are closest and is pulsed with microwave photons equivalent to the energy of the gap, a very strong non-linearity in absorption and emission (thermal release) are likely to occur. This oscillation is from the rapid switching between endothermic and exothermic pathways. This predicted result has been experimentally observed for 1.25 GHz-radiation with 6 μs pulses at 10 pps and with a peak power of 2 MW [12]. Sound production, apparent cavitation, bubble collapse, and plasma release with accompanying electrical discharge have all been observed under these conditions with peroxidizing DALM. Also, in the HL-60 cell experiments with intracellular peroxidizing DALM, heating, during the microwave exposures, was greater than expected for the ionic strength of the suspension.

Another strange property of DALM can be explained with the hyperbolic function model. When oxidized DALM is immobilized on epoxy polymer and exposed to 366-nm-wavelength light and microwave radiation or heat simultaneously and then rapidly cooled, the fluorescent properties change [13]. When re-exposed to the UVA light, a yellow afterglow (thermally-activated delayed fluorescence; TDF)

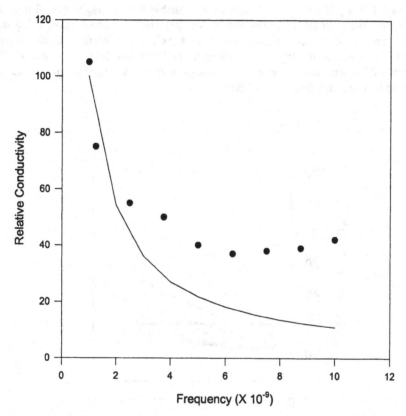

Figure 7. Calculated (line) compared to measured (points) conductivity of DALM. The divergence is related to the imposed conductivity of water.

occurs that is proportional to the length of UVA/microwave constant-intensity exposure. The TDF response is retained for long periods after the original exposure, especially if the surface is protected from oxygen (air). According to Figures 1 and 2, this happens because the emissive states (determining luminsecence and the endothermic reaction) should continue to increase to a maximum, if the surface is heated ,and the non-emissive states (determining entropy and the exothermic reaction) should be retained at a maximum level, if the surface is rapidly cooled. However, during the heating (or microwave exposure) phase, TDF is not apparent but re-appears after cooling. This result is in accord with Figures 1 and 2 because heating the emissive states leads to an anti-Arrhenius response by driving the non-emissive states (entropy) to zero. These non-emissive states are the reservoir for generating emissive states. Upon cooling, the non-emissive states accumulate again. In practical terms, these properties form the basis for an integrating microwave and radio frequency radiation "film badge" for personal dosimetry.

Finally, Figure 8 shows that by making small changes in the DALM concentration in these reactions, different maxima for steady-state luminescence can be reached for different microwave frequencies absorbed. The SAR used to make the calculations in Fig.8 was 0.4 W/kg, the standard maximum permissible limit averaged over 6 minutes [14]. These results strongly suggest that dosimeters could be constructed, using this reaction type, that could distinguish frequency as well as the intensity of the energy absorbed. As noted in the figure, by setting an arbitrary threshold for detection, the frequency differences could be easily distinguished by appearance of successive fluorescent sections on a dosimeter surface.

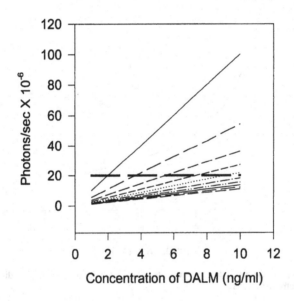

Figure 8. Calculated maximum thermochemiluminescence of peroxidizing DALM (various concentrations) at frequencies of 1-10 GHz (each line from top to bottom represents successive increases in frequency: n+1 GHz). The long-dashed line is an arbitrarily set threshold of detection. The SAR was held constant at 0.4W/kg.

6. Conclusions

The behavior of DALM reactions in microwave and radio frequency radiation fields is explicable by well-established laws of chemical kinetics and thermodynamics.

Frequency sensitivity is based on the statistical mechanical treatment of heat capacity. Finally, because the modification of the DALM polymer structure should lead to changes in its intrinsic responses (zero-order rate constant, frequency sensitivities, and luminescence/fluorescence emission spectra), the model can serve as a basis for designing sensors that will respond to different chemical as well as different physical microenvironments. The luminescence and fluorescence emission spectra dependence on these factors remains to be modeled and is an area of intensive on-going research in our laboratory.

7. Acknowledgments

This work was sponsored, in part, by the Air Force Office of Scientific Research. The opinions expressed herein are those solely of the author and do not reflect opinions or policy of the United States Air Force or any other Federal Agency.

8. References

1. Edwards, G., Davis, C., Saffer, J., and Swicord, M. (1985) Microwave-field-driven acoustic modes in DNA, *Biophysical Journal* **47**, 799-807.
2. Balcer-Kubiczek, E. and Harrison, G. (1991) Neoplastic transformation of C3H/10T1/2 cells following exposure to 120-Hz modulated 2.45-GHz microwaves and phorbol ester tumor promoter, *Radiation Res.* **126**, 65-72.
3. Merritt, J., Kiel, J., and Hurt, W. (1995) Considerations for human exposure standards for fast-rise-time high-peak-power electromagnetic pulses, *Aviat Space Environ Med* **66**, 586-589.
4. Kondepudi, D. (1982) Possible effects of 10^{11} Hz radiation on the oxygen affinity of hemoglobin, *Bioelectromagnetics* **3**, 349-361.
5. Gabriel, C., Grant, E., Tata, R., Brown, P., Gestblom, B., and Noreland, E. (1987) Microwave absorption in aqueous solutions of DNA, *Nature* **328**, 145-146.
6. Meltz, M., Walker, K., and Erwin, D. (1987) Radiofrequency (microwave) radiation exposure of mammalian cells during UV-induced DNA repair synthesis, *Radiation Res.* **110**, 255-266.
7. Kiel, J., O'Brien, G., Dillon, J., and Wright, J. (1990) Diazoluminomelanin: A synthetic luminescent biopolymer, *Free Rad.. Res. Comms.***8**, 115-121.
8. Kiel, J. and O'Brien, G. (1991) *Diazoluminomelanin and a method for preparing same*, U.S. Patent 5,003,050.
9. Bruno, J. and Kiel, J. (1993) Effect of radio-frequency radiation (RFR) and diazoluminomelanin (DALM) on the growth potential of bacilli, in M. Blank (ed.), *Electricity and Magnetism in Biology and Medicine*, San Francisco Press, San Francisco, pp. 231-233.
10. Bruno, J. and Kiel, J. (1994) Synthesis of diazoluminomelanin (DALM) in HL-60 cells for possible use as a cellular-level microwave dosimeter, *Bioelectromagnetics* **15**, 315-328.
11. Kiel, J., Gabriel, C., Simmons, D., Erwin, D., and Grant, E. (1990) Diazoluminomelanin: A conductive polymer with microwave and radiowave absorptive porperties, in P. Pedersen and B. Onaral (eds.), *Proceedings of the Twelfth Annual International Conference of the IEEE Engineering in Medicine and Biology Society*, vol. 12, no. 4, IEEE, Philadelphia, pp. 1689-1690.
12. Kiel, J., Seaman, R., Mathur, S., Parker, J., Wright, J., Alls, J., and Morales, P. (1998) Pulsed microwave induced light, sound, and electrical discharge enhanced by a biopolymer, *Bioelectromagnetics*, in press.
13. Holwitt, E., Kiel, J., and Erwin, D. (1997) *Microwave-sensitive Article*, U.S. Patent 5,658,673.
14. (1991) *IEEE Standard for Safety Levels with Respect to Human Exposure to Radio Frequency Electromagnetic Fields, 3 kHz to 300 GHz*, C95.1-1991, Institute of Electrical and Electronics Engineers, New York.

DOSIMETRY OF MAGNETIC FIELDS IN THE RADIOFREQUENCY RANGE

M. S. MARKOV
EMF Therapeutic, 4 Square Bus Center, Ste. 160
Chattanooga, TN 37405 USA

1. Introduction

Two therapeutic applications of time varying pulsed electromagnetic fields (EMF) became popular in the last three decades. Both provide a non-invasive no-touch means of applying EMF signals that are detectable at the cell/tissue level [1, 2]. The EMF currently used in orthopedics consist of low frequency signals, with maximum induced electric fields in the mV/cm and the induced current density in the range of several $\mu A/cm^2$. Therefore, the induced secondary magnetic field has a negligible effect on the value of the induced electric field and the resulting magnetic field is relatively independent of the presence of a cell/tissue load. The second therapeutic EMF modality utilizes a short wave pulsed radio frequency (PRF) signal of 27.12 MHz sinusoidal waves. This classical diathermy signal is known to produce heat when applied in continuous mode. In recent applications for reduction of pain and edema the 27.12 MHz sinewave signal is pulsed as 65 μs pulse burst with a repetition rate of 80-600 pps and a peak magnetic field of 2 Gauss. The low duty cycle (less than 4%) assures the minimal elevation of temperature ($< 1°C$ for 30 min treatment period). Therefore, the accumulated heat in the target tissue is negligible and no dosimetry related to temperature measurements can be applied. Pulsed radiofrequency (PRF) electromagnetic fields (EMF) have been increasingly used clinically to treat a variety of conditions including soft tissue injuries, wounds and burns.

2. Magnetic Field Dosimetry of High Frequency Signals

The interaction of EMF with different tissues of the human body is a very complex problem and depends on a variety of factors including the electrical and intrinsic properties of the target tissue and the physiological characteristics of the individual. Since the interactions of PRF with living tissues are a complex function of numerous parameters [3] the electromagnetic signal should not only satisfy the dielectric properties of the target tissue but should also induce sufficient voltage and current within that tissue. The inhomogeneity of the dielectric properties of living tissues and the complexity of their shape create serious difficulties in dosimetry, especially in the

239

B.J. Klauenberg and D. Miklavcic (eds.), Radio Frequency Radiation Dosimetry, 239-245.

case of high frequency EMF because the secondary field is large enough to perturb the primary field. The induced internal fields strongly depend on the parameters of the incident field as well as on the size, shape, and dielectric properties of the exposed body. The spatial configuration between the signal and the exposed body, as well as presence of other objects in close proximity, also has to be taken into account. The tissue dielectric properties are important for a prompt calculation of the internal electric fields resulting from exposure to radiofrequency EMF, in particular with respect to therapeutic applications of radiofrequency fields.

The majority of EMF devices in clinical use inductively couple the signal to the target site: the incident time varying magnetic field induces a time varying electric field via a coil applicator. This approach requires the target tissue to be placed at close proximity to the coil. In these devices the incident magnetic field is predominant when compared to the incident electric field. The electric field induced in the load is proportional to the range of change of current (dI/dt) in the coil applicator. The amplitude of the primary magnetic field is always perturbed by a tissue load due to secondary field from the induced current, which tends to reduce the primary magnetic field. The actual magnetic field configuration inside a target is a complicated function of the variations in the electromagnetic properties of different types of tissues [4-6]. Most PRF transmitters are designed to operate at 50 Ω for maximum efficiency. This means that maximum power is delivered to a load (applicator plus tissue) when its impedance at 27.12 MHz most closely approximates this value. Since the impedance of a given tissue load is a function of its composition, mass and volume, the PRF dosimetry depends on the total impedance presented to the transmitter at any treatment site.

3. Material and Methods

This study was designed to evaluate the magnetic field, tissue impedance and standing wave ratio (SWR). The method developed by us provides a simple approach in evaluating important biophysical parameters, which can be applied to examine the interaction of EMF with intended body targets. To effectively administer PRF treatment for most conditions, it is essential to take into consideration the intended target and its interactions with different body targets. The clinical importance of this study revolves around the conclusion that the targeted tissue must be taken into consideration in order to achieve optimum treatment. Therefore, treatment must be geared differently for each pathology. The model developed is also a valuable tool to test next generation applicators and products.

This study was performed using a commercially available shortwave device sofPulse. The signal consists of 27.12 MHz sinusoidal waves, delivered as 65 μs pulse bursts with a repetition rate of 80-600 pps and a peak magnetic field of 2.5 Gauss. In this device, the predominant field is the incident magnetic field and not the incident electric field.

A model was developed to simulate this PRF signal with a signal which is only a fraction of the normal output. Fifteen different body targets were used to evaluate the response to applied PRF fields by measuring the magnetic field, voltage standing wave ratio (VSWR) and the impedance. VSWR is the ratio of the magnitude of the voltage across an unknown load to that across a characteristic impedance load. A VSWR of 1 denotes a perfect match and is the optimum value. Because the sofPulse applicator is designed to operate at 50 Ω for maximum efficiency, the maximum coupling to the load (applicator plus tissue target) is achieved when the impedance is closest to 50 Ω.

Figure 1 shows the system setup including a HP 3577A Network Analyzer, a magnetic field probe (a 5 mm diameter Fischer Custom Communication circular loop) and a PC. All measurements were done

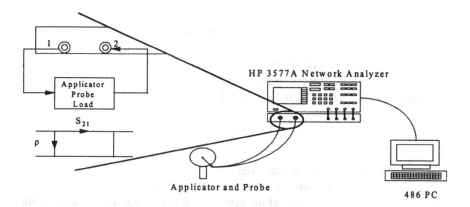

Figure 1 System set-up

using the Network Analyzer, applicator and the magnetic field probe held in place with a custom-made plastic apparatus on the exterior of the applicator. Using the output signal from the Network Analyzer, the reflection coefficient (ρ) and insertion loss (S_{21}) were applied to calculate VSWR and magnetic field. The reflection coefficient was used to determine the VSWR and a value of zero for the reflection coefficient denotes no loss and a perfect match (VSWR=1). S_{21} represents the difference (loss) between the input signal and the output signal. This value, in addition to a scaling factor based on the output voltage of the generator at the maximum setting, was used to calculate the magnetic field. Impedance values (real and imaginary) were obtained directly from the Network Analyzer. Software was developed using Visual Basic to automate the collection of data.

4. Method of Calculation

The magnetic field values are calculated through a series of steps. After determining S_{21}, the insertion loss between ports 1 and 2, the power (P) at port 2 is calculated using Equation 1. After converting the power from dBW to Watts using Equation 2, the voltage is determined by substituting into Equation 3.

$$P_2 = P_1 + S_{21} \tag{1}$$

Where P_2 is the power input to port 2
P_1 is the power output from port 1

$$P = \log^{-1}\left(\frac{P_2}{10}\right) \tag{2}$$

Where P is the power in Watts

$$V = \sqrt{PR} \tag{3}$$

Where V is the voltage
R is the resistance = 50 Ω
The magnetic field is calculated using Equation 4.

$$B = 0.66cV \tag{4}$$

Where **B** is the magnetic field,
0.66 is the Fischer probe calibration factor,
c is the ratio of the generator output voltage and the output from port 1 of the network analyzer,
V is voltage detected by Fisher probe
VSWR is calculated using Equation 5:

$$VSWR = \frac{1 + \rho}{1 - \rho} \tag{5}$$

Where ρ is the reflection coefficient

The **impedance** values were obtained directly from the Network Analyzer. The magnitude was calculated using the square root of the sum of squares of the real and imaginary parts.

The magnetic field probe was placed on the geometrical center of the applicator with a minimal air gap at the applicator/tissue interface. Figure 2 shows the measured magnetic field **B** as a superposition of the incident magnetic field **B**$_0$ and the secondary magnetic field **B**$_1$ which is a result of induced electric field in the body target and is always oppositely oriented toward the initial field: **B** = **B**$_0$ - **B**$_1$. This approach provides important possibilities regarding tissue characteristics in their connection with the resulting magnetic field. The measured magnetic field does not directly reflect the actual magnetic field delivered under load mismatch conditions. This accounts for the higher measured magnetic field values for the targets containing bone and cartilage because of their tissue composition and non-homogeneity. Such a defined difference between body targets containing predominantly soft tissues (calf, buttock, hip) versus bone/cartilage containing tissues, such as ankle, fingers, wrist is due not only to the difference in composition, but also to the degree of homogeneity of tissues.

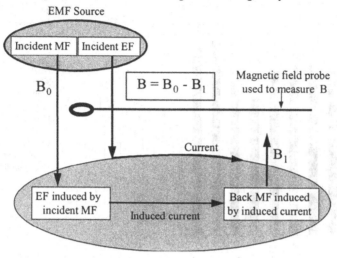

Figure 2. Magnetic field measured by any instrument will be a superposition of incident magnetic field and induced back magnetic field

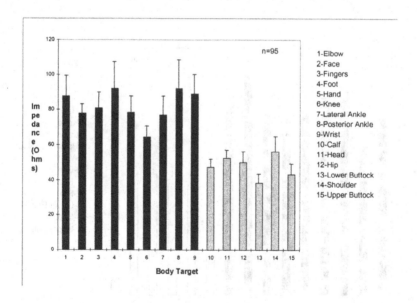

Figure 3. Magnetic field as a function of body targets

244

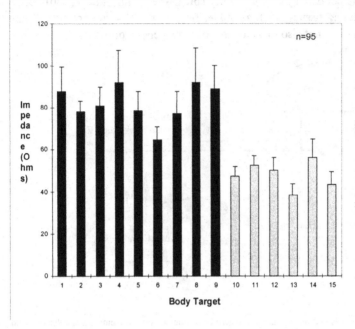

Figure 4. Impedance as a function of body targets

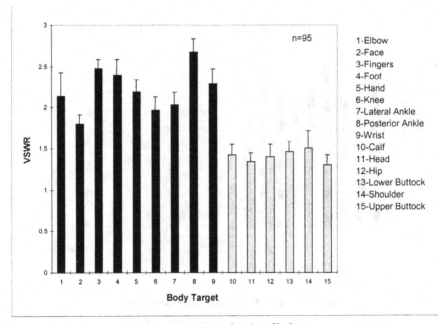

Figure 5. VSWR as a function of body targets

Figure 3-5 illustrates the clear separation into two categories of the body targets depending on their soft tissue versus bone content. The composition and homogeneity of the treatment areas affect the induced current pathways. The homogeneous tissues provide better pathways for the induced electric current; therefore, the resulting induced back magnetic field will have higher value. Thus, in turn the measured magnetic field will have lesser values than in bone/cartilage tissues.

5. Discussion

The technique developed for this study provides a simple method of evaluating important biophysical and technical parameters which can be applied to examine the interaction of EMF-delivering devices with intended body targets. The scaling factor method designed here may be easily used to calculate the magnetic field at different signal configurations. Variations in the measured parameters were large between targets, but the response of the targets was repeatable from person to person.

The most important observation was that the measured magnetic field is a superposition of the incident magnetic field and the secondary, oppositely-oriented magnetic field (Figure 2). This accounts for the higher measured magnetic field for the targets containing more bone because of their tissue composition and non-homogeneity. There is a definite difference between the predominantly soft tissue containing group and the mostly bone/cartilage containing group. This difference is due not only to the difference in composition of the two groups but also to the degree of homogeneity of the area. Tissue dielectric properties need to be evaluated because they play a crucial role in the coupling of EMF with living tissues.

6. Acknowledgement

It is my pleasure to thank S. Otano-Lata and V. Iyer for their assistance in this project.

7. References

1. Pilla, A.A. (1993) State of the art in electromagnetic therapeutics, in M. Blank (ed.) *Electricity and Magnetism in Biology and Medicine*, San Francisco Press Inc., San Francisco pp. 17-22.
2. Markov, M.S. (in press) Magnetic field dosimetry – Biophysical and clinical aspects, in F. Berzani (ed.) *Electricity and Magnetism in Biology and Medicine*, Plenum Press, New York.
3. Markov, M.S. and Pilla, A.A. (1995) Electromagnetic field stimulation of soft tissue: Pulsed radiofrequency treatment of post operative pain and edema, *Wounds* 7, 143-151.
4. Stuchly, M.S. and Stuchly, S.S. (1996) Experimental ratio and microwave dosimetry, in C. Polk and E. Postow (eds.) *Handbook of Biological Effects of Electromagnetic Fields*, CRC Press, Boca Raton pp. 295-336.
5. Polk, C. Bioelectromagnetic dosimetry, in M. Blank (ed.) *Electromagnetic Fields: Biological Interactions and mechanisms*, Amer. Chem. Soc., Washington DC, pp. 57-75.
6. Massoudi, H. Durney, C.H. and Iskander, M.F. (1980) Long-wavelength analysis of near-field irradiation of prolate spherical models of man and animals, *Electronics Letters* 16, # 3:99.

CURRENT STATUS OF DOSIMETRY IN EVALUATING WIRELESS COMMUNICATIONS TECHNOLOGIES

P. BERNARDI, M. CAVAGNARO, G. D'INZEO and S. PISA
Department of Electronic Engineering
University of Rome "La Sapienza"
V. Eudossiana, 18
00184 Rome - Italy

1. Introduction

The diffusion of wireless communication is a key target in the growth of economy, both for high technology societies and for developing countries. The population concern about health risks associated to the exposure to electromagnetic fields can influence the expansion of these new technologies. The answer to public concern must came from independent studies performed by the scientific community, and from the identification and application of safety standards for new equipment.

In this work an evaluation of the studies undergoing in the world to standardise the exposure of humans to emerging wireless communication systems is performed.

Theoretical and experimental studies related to the exposure of humans to the electromagnetic fields emitted by portable cellular phones will be presented and critically analysed considering the current state of the art (§2).

Among several initiatives developing in the world on such arguments, two particular on-going projects will be pointed out. These projects, namely the COST244 (§3) and the CEPHOS project (§4), are related to the European activity in the field of dosimetry, and deeply involve the Authors.

Growing interest is now emerging on the diffusion in the work environments of communication systems based on wireless local area networks (WLANs), that allow easier and cheaper high velocity connection among different workstations and computers. These systems need specific and accurate analysis to test their safety

B.J. Klauenberg and D. Miklavcic (eds.), Radio Frequency Radiation Dosimetry, 247-255.
© 2000 *Kluwer Academic Publishers. Printed in the Netherlands.*

2. Dosimetric Studies Related to Portable Cellular Phones

At the beginning of 70s, due to the rising concern about the fields generated by radio-frequency equipment, the first studies on electromagnetic field absorption in exposed bodies were developed. These works were based on analytical and numerical techniques in conjunction with simplified models of the human body, or on experimental techniques in conjunction with homogeneous phantoms. Excellent reviews can be found in [1, 2].

After a period of limited interest, starting from '92 new attention has been devoted to dosimetry, due to the great increase of mobile personal communications. As with the older works, also the newer ones are based both on numerical and experimental techniques. The most recently published theoretical results take advantage of the development of new numerical techniques and from the improvement of computer speed and memory, while the experimental results are based on the introduction of new sensors.

2.1. NUMERICAL DOSIMETRY

The numerical evaluation of the SAR (power absorbed per unit mass) distribution inside the head of a cellular phone user requires the use of detailed anatomical models of the human head [3-12]. These models have been obtained either from NMR images, or from the anatomical cross section atlas. Cell dimensions as small as 1 millimetre have been used, and up to 26 different tissues have been identified. Particular attention has been recently devoted to the realisation of accurate models of the ear [13, 14].

Once the anatomical model of the head has been obtained, it is necessary to electrically characterise the different head tissues at the frequencies adopted by cellular phone technologies (around 800/900 and 1800/1900 MHz). Nevertheless, it is very difficult to measure the electrical characteristics of living tissues; as a consequence the permittivity and conductivity values used by different authors show variations from 20 to 80 % [3-12]. Recently, new values have been obtained in particular for fat, bone and cartilage from measurements performed on freshly excised animals [15]. Different authors noted that these new values produce an increase in the maximum SAR, as averaged over 1g of mass, of about 30%, which grows up to 100% if an homogeneous model of the head is considered [8, 10, 11].

With reference to the source modelling, the first theoretical works were based on simplified models of the phone antenna [3, 4]. In these works the emitting element was simulated as a $\lambda/2$ dipole or as a $\lambda/4$ monopole on a metallic box. In [7] other kinds of antennas, corresponding to more actual devices, have been considered: the sleeve dipole antenna, and the whip antenna. The first behaves very close to a half-wavelength dipole, while the second one is constituted of a 5/8-wavelength monopole with a coil at its base. When the monopole is collapsed, the coil forms together with the metal in the radio case a RF radiator. In order to allow the operation of hand-held communication apparatuses in multipath fading environments, a planar inverted F antenna has been suggested [5] and analysed [12]. Finally, cellular phones more close to commercial systems and equipped with helical antennas have also been studied [16, 17].

In Figure 1 the maximum SAR, as averaged over 1g of mass, obtained in the head of a cellular phone user by different authors is shown as a function of the distance between head and phone. The phones considered in Figure 1 are those modelled as a monopole on a conducting box or those equipped with the whip antenna, since they have similar radiating properties. The radiated power has been assumed to be 600 mW.

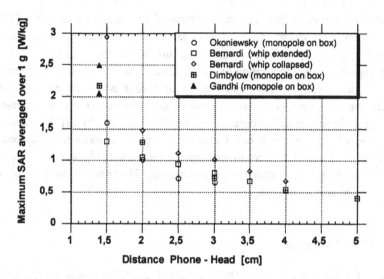

Figure 1. SAR averaged over 1g as a function of the phone-head distance

It is interesting to note that the reported values are below the IEEE/ANSI exposure limit of 1.6 W/kg when the phone-head distance is greater than 1.5 cm. The previous limit is exceeded only when the antenna is in close contact with the ear, or when the whip antenna is used without extracting the monopole [7]. It must be pointed out that new technologies (such as GSM or DCS) irradiate an average power of lower than half the power used in Figure 1.

Finally, in [7] the influence on the SAR distribution of perfectly conducting walls placed close to the head, to simulate an exposure in partially closed environments (e.g. the interior of a car) has been studied. It has been found that a horizontal reflecting wall above the head smoothes the SAR distribution by increasing the lower values and by reducing the higher ones. Instead, a vertical wall parallel to the phone and close to it increases the maximum SAR up to 100%.

2.2. EXPERIMENTAL DOSIMETRY

The experimental evaluation of the SAR distribution inside a phantom head exposed to a cellular phone requires the use of head models both with reference to the anatomy and to the tissue dielectric properties [18-27]. The most recent phantom realisations are addressed toward non-homogeneous models [22, 25, 27]. These models are better than the homogeneous ones in simulating the head anatomy, but they suffer from the difficulties in inserting the probe inside them. On the contrary, homogeneous models are less close to reality but simpler to use and seem to give a worst case approximation of the SAR values [27].

Various kinds of probes have been proposed for measuring the SAR [26]; they mainly measure the electric field (or its square) or the temperature. This last kind of probes is very simple since it measures the temperature increase due to the electromagnetic energy absorption by the material. The problem is to minimise heath diffusion: short exposure times and high instantaneous power are necessary. On the contrary, the test of personal phones involves low power measures, as a consequence the temperature probes lack accuracy and sensitivity [23, 24, 26-29]. The electric field probes are able to perform a direct measure of the field inside the phantom. Small probes have to be used in order to measure the field accurately without perturbations: the drawback is that they suffer from low sensitivity. Nevertheless, SAR values up to 1 mW/kg can be appreciated [28].

3. The COST244 Project

In 1992 a European project called 'Biomedical Effects of Electromagnetic Fields' began, also referred to as COST244, with a foreseen duration of 4 years. In 1996, due to the relevance of the work carried out, the project was renewed under the name of COST244bis. Objective of the COST244 project is to obtain a general insight into the state of the art in the area of electromagnetic field effects and safety standards. To this aim, COST244 co-ordinates the European research activity, promotes national research projects and encourages collaboration among experts in different disciplines.

Inside the COST244 project, 3 working groups (WG) have been established. Since its constitution the WG3 (Systems Application & Engineering), co-ordinated by one of the Authors, has been supporting the researches on mobile communication aspects. In particular, from 1994 the WG3 activity has been focused on the comparison of all the results obtained in the evaluation of cellular phone hazards, both numerically and experimentally. So, in July 1994, a *physical canonical problem* and a *numerical canonical problem* have been proposed to all the interested researchers [29].

The numerical problem consists in the illumination of a phantom (cube or sphere) by an antenna ($\lambda/2$ dipole or $\lambda/4$ monopole on a conducting box). In Figure 2 an example of the proposed geometry is shown. The phantom could be both homogeneous and a two layers one, and the frequencies considered were 900 and 1800 MHz, close to those used in the GSM and DCS technologies. SAR values in some particular points of the phantom and radiation pattern values were asked [30].

Figure 2. Example of the geometry for the numerical canonical problem

The results of the canonical exercise showed good agreement in the normalised data, while great differences were found in the absolute data, e.g. in the maximum SAR. These rather confusing results induced a deep analysis of all the possible error sources, which led to the conclusion that the great differences could be explained with the different choices and approximations done by the research groups in developing the numerical solution, e.g. in modelling the radiating antenna. To overcome all of these confounding factors an exercise (Hertzian dipole in the presence of a sphere) with the corresponding analytical solution was circulated [31].

Finally, at the COST244bis Workshop held in Trento in December 1997, a new proposal was issued. It consists of a helical antenna on a box illuminating a sphere or a cube, and the proposed comparison is among numerical and experimental data [32]. This time the objective is to analyse the accuracy of numerical techniques in modelling cellular phones, taking into account the most recent mobile antennas. Information and results of the COST244 activity related to the canonical numerical problems are available in the sites: *http://www.radio.fer.hr/cost244* and *http://serverest.itc.it/*~pontalti/ *canonical_cases.htm*

4. The CHEPHOS Project

The CEPHOS project, co-ordinated by one of the Authors, has started on October 1997 and will be completed by October 1999. Purposes of the project are to support the formulation of European standards on electromagnetic exposure through a scientific assessment of the power absorbed in humans in exposure conditions related to hand-held terminals. The technical objectives of the project are:

- development and maintenance of a data-base on literature and standards referring to dosimetry and exposure problems;
- design and study of canonical situations to be analyzed both numerically and experimentally in order to assess the limits and uncertainties of the two approaches;
- development of numerical and experimental phantoms in order to assess the human exposure to actual mobile terminals in realistic conditions;
- definition of a procedure to test the mobile phones in order to assess their compliance with a given safety guideline.

The consortium that is carrying out the project represents 15 organisations from six European Countries (Denmark, Finland, France, Greece, Italy, and UK). The team is composed of seven universities, four laboratories and four industrial partners.

During the first months of the project out the collection of published work on numerical and experimental dosimetry related to cellular phones and on possible related health effects was carried. Objective of this activity was to define the state of the art related to the subject as the basis for the successive work to be carried out within the project. The gathered literature will be organized within a database which will be accessible to the scientific community. This database, still under construction, will be accessible at the site *http://www.cephos.finsiel.it*. In the meantime, the collection and critical analysis of the existing national and international safety guidelines concerning the frequencies of mobile cellular phones was performed. This is an important issue for outlining and analyzing the different requirements from the regulatory bodies. Similarly, a critical analysis of the published work dealing with the numerical evaluation of the field absorbed by a user was carried out as a starting point for developing the comparison among the different numerical techniques used in dosimetric problems. Finally, an activity devoted to the numerical analysis of simple canonical cases is being developed. These cases will be used for comparison among the different numerical codes used in dosimetry, for comparison among the different experimental set-ups, and, finally, between numerical and experimental techniques. This part of the CEPHOS project is being carried out taking into account the experience gained within the COST244 projects as illustrated in the previous section.

5. Dosimetry Studies Related to Wireless Local Area Networks.

Wireless Local Area Networks (WLAN) are used in services requiring high-speed data transmission over small areas such as, for example, the interior of a building [33, 34]. To transmit data, wireless systems use the field emitted at microwave or at millimetre-wave frequencies by the antennas placed at the mobile terminals and by the base station antenna usually located on the ceiling.

In this arrangement, the user can find himself in close proximity to the radiating mobile antenna, exposed to an electromagnetic field composed by the waves directly coming from the antennas and by the waves reflected and scattered from the objects present into the area.

In the frequency range of WLAN systems, the eye seems to be the most hazardously exposed organ. In fact, at these frequencies the absorption takes place mainly in the superficial layers of the skin, and the eye, at least when the eyelids are open, is not protected by the skin layer.

The oldest studies related to WLAN dosimetry were performed by using geometrical optics [1]. In this technique the wavelength is supposed to be very short compared to the dimensions of the exposed body (a sphere or a prolate spheroid), so that the incident radiation can be described by rays.

This approach seems to be not particularly suitable for studying the exposure of the human eye due to its anatomical and physiological properties.

The first numerical study on the interaction between microwave fields and the human eye was performed by Umashankar and Taflove [35] by using the FDTD method. Further analytical considerations can be found in [36]. In [37-38] a complete numerical study has been performed by using a high-resolution (0.5 mm) model of the eye and surrounding tissues. This model takes into account 12 tissues whose electrical properties have been extrapolated from experimental data by using the Debye's equation. The SAR evaluation has been performed on three-dimensional open and closed eye models. The obtained results show that in the closed eye model the absorption takes place mainly in the superficial skin region, while in the open eye model the lens, the cornea, and the humour regions are affected by sensible absorption. Recently a study at the frequency of 77 GHz has been also presented [39].

Finally, in [40] a thermal analysis has been also conducted. On the basis of the results obtained, it appears that the use of the exposure limits proposed by the regulatory bodies gives rise to temperature increases far from the induction of lens opacification. However, it has been shown that the same SAR values, as averaged over the whole eye (about 10 g), can produce different heating when different frequencies are considered. This is due to the fact that heating is influenced not only by the mean power absorbed into a volume, but also by the way in which the SAR is distributed into that volume.

6. References

1. Durney, C.H. (1985) "The Physical Interactions of Radiofrequency Radiation Fields and Biological Systems", AGARD Lecture Series, no. 138, pp. 2.1-2.19.
2. Lin, J.C. Computer method for field intensity prediction (1986) in C. Polk and E. Postow (eds.) *Handbook of Biological Effects of Electromagnetic Fields*, CRC Press.

3. Toftgard, J., Hornsleth, S. N., and Andersen, J. B. (1993) Effects on portable antennas of the presence of a person, *IEEE Trans. Antennas Propagat.* **41**, 739.

4. Dimbylow, P. J. and Mann, S. M. (1994) SAR calculations in an anatomically realistic model of the head for mobile communication transceivers at 900 MHz and 1.8 GHz, *Phys. Med. Biol.* **39**, 1537.

5. Jensen, M. A. and Rahmat-Samii, Y. (1995) EM interaction of handset antennas and a human in personal communications, *IEEE Proc.* **83**, 7.

6. Martens, L. De Moerloose, J. De Zutter, D., De Poorter, J., and De Wagter, C. (1995) Calculation of the electromagnetic fields induced in the head of an operator of a cordless telephone, *Radio Science* **30**, 283.

7. Bernardi, P. Cavagnaro, M., and Pisa, S. (1996) Evaluation of the SAR distribution in the human head for cellular phones used in a partially closed environment, *IEEE Trans. Electromag. Compat.* **38**, 357.

8. Gandhi, O. P. Lazzi, G., and Furse, C. M. (1996) Electromagnetic absorption in the human head and neck for mobile telephones at 835 and 1900 MHz, *IEEE Trans. Microwave Theory Tech.* **44**, 1884.

9. Okoniewski, M. and Stuchly, M. A. (1996) A study of the handset antenna and human body interaction, *IEEE Trans. Microwave Theory Tech.* **44**, 1855.

10. Hombach, V., Meier, K., Burkhardt, M., Kuhn, E., and Kuster, N. (1996) The dependence of EM energy absorption upon human head modeling at 900 MHz, *IEEE Trans. Microwave Theory Tech.* **44**, 1865.

11. Watanabe, S., Taki, M., Nojima, T., and Fujiwara, O. (1996) Characteristics of the SAR distributions in a head exposed to electromagnetic fields radiated by a hand-held portable radio, *IEEE Trans. Microwave Theory Tech.*, **44**, 1874.

12. Bernardi, P., Cavagnaro, M., Pisa, S., and Piuzzi, E. (1997) Temperature elevation induced in the head of a cellular phone user, *International Scientific Meeting on Electromagnetics in Medicine Abstracts*, Chicago, pp. 76, 1997.

13. Hamada, T., Watanabe, S., and Taki, M. (1998) Effect of pinna on the local SARs in human head exposed to microwave by a cellular telephone, *20th Annual Meeting of the Bioelectromagnetics Society Abstracts*, St. Petersburg, pp. 99-100.

14. Burkhardt, M., and Kuster, N. (1998) Appropriate modeling of the ear for compliance testing of hand held MTE with SAR safety limits, *20th Annual Meeting of the Bioelectromagnetics Society Abstracts*, St. Petersburg, pp. 79-80, 1998.

15. Gabriel, S., Lau, R. W., and Gabriel, C. (1996) The dielectric properties of biological tissues: III. parametric models for the dielectric spectrum of tissues, *Physics in Medicine and Biology* **41**, 2271-2293.

16. Cavagnaro, M. and Pisa, S. (1996) Simulation of cellular phone antennas by using inductive lumped elements in the 3D-FDTD algorithm, *Microwave Opt. Technol. Lett.* **13**, 324-327.

17. Lazzi, G. and Gandhi, O. P. (1998) On modeling and personal dosimetry of cellular telephone helical antennas with the FDTD code, *IEEE Trans. on Antennas and Propagation* **46**, 525-530.

18. Balzano, O., Garay, F., and Steel, R. (1978) Heating of biological tissue in the induction field of VHF portable radio transmitter, *IEEE Trans. Veh. Tech.* **27**, 51-56.

19. Chatterjee, I. and Gandhi, O. P. (1985) Quantification of electromagnetic absorption in humans from body-mounted communication transceivers, *IEEE Trans. Veh. Tech.* **34**, 55-62.

20. Stuchly, S. S., Kraszewski, A., Stuchly, M. A., Hartsgrove, G., Adamski, D. (1985) Energy deposition in a model of man in the near field, *Bioelectromagnetics* **6**, 115-129.

21. Guy, A. and Chou, C. K. (1986) Specific absorption rates of energy in man models exposed to cellular UHF mobile-antenna fields, *IEEE Trans Veh. Tech.* **34**, 671-680.

22. Cleveland R. F. Jr., and Athey, T. W. (1989) Specific absorption rate (SAR) in models of human head exposed to hand-held UHF portable radios, *Bioelectromagnetics* **10**, 173-186.

23. Lovisolo, G. A., Raganella, L., Nocentini, S., Bardati, F. Gerardino, Tognolatti, P. (1993) Hand-held portable telephones: Electromagnetic radiation absorption studies", *Proc. COST 244 Meeting On Mobile Comm. and ELF Fields*, Bled, pp. 47-52.

24. Handerson, V., and Joyner, K. H. (1995) Specific absorption rate levels measured in a phantom head exposed to radio frequency transmission from analog hand-held mobile phones, *Bioelectromagnetics* **16**, 60-69.

25. Schmid, T., Egger, O., and Kuster, N. (1996) Automated E-field scanning system for dosimetric assessment, *IEEE Transactions on Microwave Theory and Tech.*, **44**, 1.

26. Chou, C. K., Bassen, H., Osepchuck, J., Balzano, Q., Petersen, R., Meltz, M., Cleveland, R., Lin, J. C., and Heynick, L. (1996) Radiofrequency electromagnetic exposure: Tutorial review on experimental dosimetry, *Bioelectromagnetics* **17**, 195-208.

27. Bertotto, P., Schiavoni, A., Richiardi, G., Bielli, P., Gabriel, C. (1998) Comparison between measured SAR distributions into homogeneous and not homogeneous phantoms generated by cellular phones, *20th Annual Meeting of the Bioelectromagnetics Society Abstracts*, St. Petersburg, pp. 255.

28. Pokovic, K., Schmid, T. and Kuster, N. (1998) New generation of miniature, isotropic and broadband near-field probes, , *20th Annual Meeting of the Bioelectromagnetics Society Abstracts*, St. Petersburg, pp. 59.

29. Simunic D. (ed.) (1994) Proc. of the COST244 Meeting. "*Reference Models for Bioelectromagnetic Test of Mobile Communication Systems*", Roma, Italy.

30. Bach-Andersen J. et al, co-ordinated by G. d'Inzeo (1994) Proposal for numerical canonical models in mobile communication, *COST244 WG3 Numerical Program*, July 1994.

31. Hornsleth, S. et al, co-ordinated by G. d'Inzeo (1996) Proposal for hertzian dipole exercise, *COST244 WG3 Numerical Program*, Feb. 1996.

32. Wiart, J. (1998) Helix antenna close to tissues: Modelling, analysis, measurement and comparisons in near field at 900 MHz", *COST244 WG3 Numerical Program*, April 1998.

33. Pahlavan, K. and Levesque, A. H. (1994) Wireless data communications, *Proc. IEEE* **82**, 9.

34. Ali, F., and Horton, J. B. (1995) Introduction to special issue on emerging commercial and consumer circuits, systems, and their applications, *IEEE Trans. on Microwave Theory and Techn.* **43**, 1633-1637.

35. Umashankar, K., and Taflove, A. (1982) A novel method to analyze electromagnetic scattering of complex objects, *IEEE Trans. on Electromag. Compat.* **24**, 397-405.

36. Gandhi O. P., and Riazi, A. (1986) Absorption of millimeter waves by human beings and its biological implications, *IEEE Trans. on Microwave Theory and Techn.*, **34**, 228-235.

37. Bernardi, P., Cavagnaro, M., and Pisa, S. (1996) Evaluation of the power absorbed in human eyes exposed to millimeter waves, *Proc. Int. Symp. on Electromagnetic Compatibility*, 194-199, Rome, Italy.

38. Bernardi, P., Cavagnaro, M., and Pisa, S. (1997) Assessment of the potential risk for humans exposed to millimeter-wave wireless LANs: The power absorbed in the eye, *Wireless Journal*, **3**, 511-517.

39. Bahr, A., Gustrau, F., Pan, S. G., Kullnick, U., and Wolff, I. (1998) SAR analysis and heating patterns of cm/mm wave electromagnetic fields in the human skin and eye, *20th Annual Meeting of the Bioelectromagnetics Society Abstracts*, St. Petersburg, 158-159.

40. Bernardi, P., Cavagnaro, M., Pisa, S., and Piuzzi, E. (1999) SAR distribution and temperature increase in an anatomical model of the human eye exposed to the field radiated by the user antenna in a wireless LAN, *Trans. on Microwave Theory and Techn.*, in press.

SUMMARY OF SESSION E: CONTACT AND INDUCED CURRENTS

M. ISRAEL
National Centre of Hygiene
Medical Ecology and Nutrition
Department of Physical Factors
15, Dimiter Nestorov Str.
1431 Sofia, Bulgaria

Most of the existing standards and guidelines for electromagnetic exposure on human body for the frequency range up to 10 MHz use the parameters of contact and induced currents, as basic limits. There are several numerical methods to calculate the reference levels - easy to be measured. But most of the real exposures occur in the near-field, in non uniform fields. It is also difficult, to evaluate local, partial body exposures, such from reradiating objects, etc.

Exposure assessment needs implementation of standard measuring procedures, accurate methods for different kinds of sources, working places, residential areas.

This session provides answers to a number of questions connected with dosimetry and its use for standardization.

The need of comparability of the results of exposure assessment for their use in epidemiological studies, also for practical dosimetry is discussed by Dr. M. Israel and P. Gajsek. Here, Dr. M. Israel presents several standard methods for exposure assessment used in Bulgaria, also the problems connected with the measuring equipment, education, standards and legislation, databases, parameters for evaluating the exposures.

Information for the measuring procedures used in Slovenia, and results of practical measurements in different occupations, residential areas, and around sources of radiation exists in the manuscript of P.Gajsek.

A review of mechanisms responsible for excitation of nerve, muscle and synapse, gives to Dr. J.Reilly the possibility to propose criteria for exposure limits based on such responses. Here, parameters for exposure criteria are discussed, also. For example, the in-situ electric field is the fundamental force, not the current density, for the membrane polarization effects. Also, the stimulus energy is not a pertinent descriptor of excitation thresholds. Dr. J.Reilly proposes thresholds for short-term reactions and the comparison with some standards (C95.1 and ICNIRP guidelines) shows that only the phosphene's curve crosses the standard limit at 100 Hz.

Dr. S.Tofany calculates the current distributions as a function of frequency, human height and grounding conditions, corresponding to the SAR distribution. The field strength limits proposed to protect against local SARs of 20 W/kg, have to be as low as

257

B.J. Klauenberg and D. Miklavcic (eds.), Radio Frequency Radiation Dosimetry, 257-258.
© 2000 *Kluwer Academic Publishers. Printed in the Netherlands.*

16 V/m. The value of local SAR=20 W/kg is in agreement with the thresholds for a sensation of warmth.

A detailed review of the existing standards developed on a criteria of induced currents based on SAR measurements is made by Dr. J. Leonowich. A recommendation for a voltage criteria is forwarded. A discussion on the averaging time shows that for local SARs there is not a good rationale for averaging over 1 s for the range of 100 kHz to 100 MHz.

Dr. P.Chadwick connects the induced current densities and SARs from the near-field exposures to RF magnetic fields. The analysis permits to use the calculated levels with the existing UK exposure guidelines. For time-averaging exposures to magnetic fields between 100 kHz and 10 MHz, investigation and ceiling levels have been developed for plane wave exposures.

The findings of peak near-field exposures shown by R.Tell illustrates the need for a better dosimetric understanding of near-field exposures to allow more accurate hazard assessment.

Many questions for further discussion arise at the time of the session as:

- Existing standards are good based on the thermal effect criteria.What criteria should be used both for neurophysiological responses and for long term exposures?
- What kinds of rationales for standards should be used for near-field exposures, for non uniform fields, for partial body exposures; which local temperature increment is acceptable in local exposures: 1 °C or less?
- Parameters connected with the basic limits are difficult to measure in human body. How to use them through the reference levels for direct exposure assessment? What is the need of basic limits in the standards if we can monitor the derived levels only?
- There is a need for developing of exposure assessment procedures for different occupations, residential environment, sources of radiation, etc. How to collect such standard information for databases?
- What kinds of probes give more information in measurements - isotropic or anysotropic ones?
- What means and what is the need of using the calculated power densities (S) for near-field exposures?
- Recommendations for some changes in the exposure standards criteria: voltage criteria, averaging time, the physiological point of view, why "current density"?
- What kind of measuring procedures should be taken for occupations such as maintenance of medical sources of radiation, also for working near broadcast antennas, or other high powered transmitting equipment?

Many questions connected with the partial body dosimetry, the near-field dosimetry, SARs and field's orientation, exposures from reradiation objects, etc. may be read at the conclusions of R.Tell's manuscript.

EXPOSURE ASSESSMENT OF EXTREMELY LOW FREQUENCY, RADIOFREQUENCY AND MICROWAVE RADIATION: METHODS AND STANDARDS FOR DATABASES

M. ISRAEL
Ass. Prof. Ph.D.
National Centre of Hygiene,
Medical Ecology and Nutrition
Department of Physical Factors
15, Dimiter Nestorov Str.
1431 Sofia, Bulgaria

1. Introduction

One of the main problems in electromagnetic biology, occupational and environmental hygiene and epidemiology is metrology. Some of the priorities of the WHO Electromagnetic Fields (EMF) Project are connected with human exposure and risk assessment by EMF with different frequencies in occupational and residential area. The idea of this project to "incorporate research results into WHO Environmental Health Criteria monographs where health risk assessment of exposure to EMF will be made" is impossible without any standardization of the measuring and exposure assessing methods. The priorities of research in this field are connected with experiments and epidemiological studies with exact dosimetry, full characterization of the exposure techniques and the EMF characteristics. Data collection and analysis have to be made to ensure the right conclusions for the biological results; standard protocols responding to quality assurance (QA) statistics, and to good laboratory practice (GLP) procedure have to be used in every study.

The exposure assessment of hazardous electromagnetic radiation (EMR) is one of the most difficult tasks in occupational and environmental research. The difficulty is due to the nature of EMR, also to the specificity of its propagation.

Many new sources of radiation with diverse characteristics have been produced in the last years: equipment for medical use, new types of communication systems, especially mobile, different kinds of distribution networks and household devices.

Some years ago we were speaking only about the big errors of the measuring equipment, but not about the procedure of measuring, nor for the additional errors as a result of the lack of standards in evaluation of the exposures. Now we can say that some of the problems of the EMR measurements are solved, but there are many new difficulties on the international level.

259

B.J. Klauenberg and D. Miklavcic (eds.), Radio Frequency Radiation Dosimetry, 259-269.
© 2000 *Kluwer Academic Publishers. Printed in the Netherlands.*

One of the main problems in EMR evaluation remains the need to make "near-field" measurements in the extremely low frequency (ELF), and in the radio frequency (RF) ranges.

Another problem is the diversity of measuring devices used by different laboratories and having different level of calibration, detection, construction of probes, quality, time constants, frequency response, etc.

There are few standards in the world dealing with the measuring methods with any kind of description of the procedure of measuring. Methods and procedures for exposure assessment are not recommended in national or in international standards.

There are neither recommendations nor guides for exposure assessment of EMR by different or specific emitters, for working places, residential areas, dwellings, except the standards for microwave ovens, video display units (VDUs).

Many of the articles published in the last 10 years do not present an evaluation of the error of the method of exposure assessment.

Data collecting for epidemiological studies is possible only by using standard methods of measurement, of exposure evaluation, unified for different sources of EMR, for places of human exposure, for different quality of equipment, etc. In statistics these are the homogeneous exposure groups (HEG).

Here, we will try to present some of the requirements for "good" measurement and exposure assessment of EMF around different kinds of sources of radiation, workplaces, residential areas, also standard methods and databases for EMR sources used in Bulgaria.

2. Requirements for Good Laboratory Practice and Quality Assurance

Measurements have to be made with reliability and documentation in relation to the international standards dealing with quality assessment and assurance (QA) [1, 2, 3]. The main requirements for QA of the methods for exposure assessment include the following: calibration, linearity, accuracy, repeatability, reproducibility, uncertainty, absolute error, specificity, limits of measurement, comparison's method.

The requirements for good laboratory practice (GLP) of the measuring laboratories are described in the European standards series EN 45000.

Here we don't intend to discuss these requirements, only to mention them and to direct our attention to the need of their application in practice.

3. Parameters for Exposure Assessment. General Methods

Most of the parameters for evaluating the EMR exposures as incident characteristics of radiation, parameters of the biological object, also of the absorbed electromagnetic energy are discussed in different publications [4, 5, 6].

Currently, the exposure assessment parameters depend on the exposure standards. Using contemporary criteria for standardization two kinds of exposure assessment characteristics have to be evaluated: these connected with the basic limits, and the other ones - for the derived/reference levels [7, 8]. These parameters are:

3.1. BASIC LIMITS

- Induced current - I, mA (root mean square - RMS) for: current through each foot (both without and with contact to metallic bodies) - ANSI; body to ground for f < 10 MHz - IRPA/ICNIRP; touching conductive receiving bodies - VDE.
- Current density - J, A/cm^2 (RMS) for: averaging area 1 cm^2, and for averaging time 1 s - ANSI, ICNIRP and VDE.
- Whole body average specific absorption rate SAR, W/kg for different periods of time: 6 min - ICNIRP, NRPB and VDE; depending on the frequency - ANSI.
- Local SAR, W/kg for averaging mass: 0.001 kg - ANSI and NRPB; 0.1 kg - ICNIRP and VDE.
- Local SAR, W/kg for special cases: hands, wrists, feet, ankles - ANSI and ICNIRP; hands, wrists, ankles - VDE; limbs - NRPB.
- Specific absorption SA, mJ/kg: for averaging - Specific absorption SA, mJ/kg: for averaging mass 0.01 kg - VDE.
- Power flux density, W/m^2: for averaging time depends on the frequency - in NRPB, ICNIRP.

The ICNIRP Guidelines also recommend for the low-frequency range basic restrictions for current densities induced by transient or very short-term peak fields.

3.2. DERIVED LEVELS/ REFERENCE LEVELS

- Electric field strength ; also RMS value - E, V/m (every standard for near field);
- Magnetic field strength, also RMS value - H, A/m (every one for near field);
- Magnetic flux density - B, T or G;
- Derived levels (far fields) of the power flux density, S, W/m^2 (every one);
- Limit values (dose) for time < 6 min: E^2.t [(V/m)^2.min]; H^2.t [(A/m)^2.min]; S.t [W/m^2.min] - VDE standard;
- Peak limit values of E, V/m; H, A/m and S, W/m^2 - VDE;
- Equivalent parameters E, H and S - calculated by measurements in far zone or in near-field zone for the other - every one standard;
- Threshold voltage U, V for pacemaker users - VDE.

Two standards in Bulgaria [9,10] propose the use of "energetic loading" (dose parameters - E^2.t [(V/m)^2.min]; H^2.t [(A/m)^2.min]; S.t [W/m^2.min]) of the organism as additional parameters for exposure assessment, similar to the ones used in VDE Standard (time duration is in hours, not in minutes).

The aim of the real exposure assessment is to receive information by measurement, calculation or evaluation of the exposure time, the dose levels, or different parameters easy to compare with the derived levels or better, with the basic limits used in the standards.

Our opinion, in general, is that the exposure assessment can be made using two groups of methods of measurement:

spot methods - based on single measurements of the incident EMF parameters at the assessed work place or at men's residences. These methods can be applied if the values of EMR are comparatively stationary during the time of exposure - work shift or twenty four hours.

dosimetric methods - they comprise personal dosimetry during the work shift, evaluation of the time duration for an individual or for a group with homogeneous exposure, direct dose measurements, etc. Some of the indirect methods of dosimetry are described as follows:

- scenario method - spot measurements with assessment of exposure based on the professional duties;
- individual dosimetry - during a work shift or in 24-hour periods in different professional groups, mainly for assessment of the magnetic component of the field and for the determination of maximal and mean values of electric and magnetic fields;
- assessment of the electromagnetic effects in retrospective period - for the entire length of service or for shorter periods (months, years).

The results from the measurement of the EMF parameters can be processed statistically to assess time weighted average (TWA) values of radiation per work day, week or length of service. There are another methods for assessment of EMR with measurement or calculation of the time duration of exposure, mean values of the exposure parameters, peak values, special methods for evaluation of intermittent exposures, non-homogeneous fields, etc.

Direct dosimetry includes methods for assessment of the energetic loading of the organism, induced currents, whole-body or localized SAR, etc.

4. State of Exposure Assessment in Bulgaria

4.1. LEVEL OF LEGISLATION. MEASURING LABORATORIES

Legislation in Bulgaria permits different kind of organizations, companies, laboratories, private or governmental, to do measurements and exposure assessment of EMR. In the last 2 years, the process of this legislation is going very fast, and many quite new measuring laboratories are entering in the field of control of EMR.

Most of the control laboratories belong to the Ministry of Health, Ministry of Environment, and to the Ministry of Labor and Social Policy. There are more than 20 inspections owning measuring equipment for different ranges of EMR, and with qualified industrial hygienists, medical physicists, and technicians. More than 15 private laboratories have been notified up to now by the Ministry of Health to assess the working conditions. Unfortunately, only 4 of them have any kind of devices for EMF measuring. There are, also, some new occupational health services, having measuring equipment.

The biggest laboratory in Bulgaria doing measurement, evaluation, standardization, control and research work in the field of exposure and risk assessment of EMR is the Department of Physical Factors in the National Centre of Hygiene, Medical Ecology and Nutrition (NCHMEN) in the Ministry of Health.

4.2. MEASURING EQUIPMENT

The measuring equipment for EMR used in Bulgaria are shown on Table 1.

TABLE 1. List of the measuring equipment, used in Bulgaria

DEVICES	MEASURING PARAMETERS	FREQUENCY RANGE	DYNAMIC RANGE	ERROR
NFM – 1	E [V/m]	50 Hz	$2 \div 40$ kV/m	± 20 %
(Germany)		60 kHz - 350 MHz	$2 \div 1500$ V/m	
	H [A/m]	60 kHz - 10 MHz	$0.1 \div 10$ A/m	
EFA-1, 2, 3	B [T]	5 Hz - 30 kHz	$0.1 \div 10$ mT	± 5 %
EMR 200, 300	E [V/m]	3 MHz - 26 GHz	$1 \div 1000$ V/m	± 3 dB
(Germany)	H [A/m]	10 MHz - 1 GHz	$0.03 \div 16$ A/m	
HI 3604	E [V/m]	20 - 2000 Hz	1 V/m $\div 199$ kV/m	± 30 %
(USA)	B [T]	20 - 1000 Hz	0.1 µT $\div 2$ mT	
HI 3603	E [V/m]	2 - 300 kHz	$1 \div 1999$ V/m	± 30 %
(USA)	H [A/m]	8 - 300 kHz	$1 \div 1999$ mA/m	
HI 1501 (USA)	S [W/m²]	2450 MHz	$0 \div 100$ mW/cm²	± 20 %
HI 1600 (USA)	S [W/m²]	915 MHz	$0 \div 20$ mW/cm²	± 20 %
		2450 MHz	$0 \div 10$ mW/cm²	
RAHAM 495	S [W/m²]	200 kHz - 40 GHz	2 µW/cm² $\div 20$ mW/cm²	± 40 %
(USA)				
PO-1 (Russia)	S [W/m²]	300 MHz - 16,7 GHz	$0,1$ µW/cm² $\div 1$ W/cm²	± 30 %
PO-9 (Russia)	S [W/m²]	300 MHz - 16,7 GHz	0.1 µW/cm² $\div 1$ W/cm²	± 30 %
IPE-39	E [V/m]	500 Hz -	$2 \div 8000$ V/m	± 20 %
(Bulgaria)	H [A/m]	4,7 MHz	$0.005 \div 1000$ A/m	
IPE-D93	E [V/m]	60 kHz -	$1 \div 1000$ V/m	± 20 %
(Bulgaria)	W [(V/m)².h]	300 MHz	$1 \div 99999$ (V/m)² h	
Narda 8532	B [T]	12 Hz - 50 kHz	$0.001 \div 0.2$ µT	± 1 %
(USA)				
Narda 8718	E [V/m]	3 kHz - 40 GHz	$0.5 \div 800$ V/m	< 3 dB
(USA)	H [A/m]	0.3 - 300 MHz	$0.02 \div 7.3$ A/m	
	S [W/m²]		0.1 µW/cm² $\div 2$ W/cm²	

Many problems exist using different kind of measuring equipment. Good exposure assessment can be made only by using equipment having many characteristics, the combination of which may not be possible in a single instrument.

- the instrument should measure in terms of every parameter defined in the standards depending on the frequency range (far or near fields);
- the frequency response of the probe must be linear in the frequency range measured or corresponding to used standard;
- the sensor of the field probe should be much smaller than the shortest wavelength of the fields to be measured;
- the probe should not cause significant scattering of the field;
- the probe should be dependent or independent of its angular orientation in the field, depending on the requirements;
- the instrument should be capable of reading either peak, RMS or average values for complicated waveforms;
- it should have a dynamic range of at least 30 dB without having to change probes;
- the instrument should be direct reading, without any need of additional calculations;
- it should give, in addition, an information about the exposure time, or the dose, or calculation of $E^2.T$, $H^2.T$ or $S.T$, where is needed.

In addition to the above characteristics, the instrument should be stable, rugged, lightweight, battery operated, etc.

Here we don't intend to criticize the production of different companies for EMR devices, but we have to mention that some of the devices cited above have serious imperfections. Some examples are:

- some are calibrated only for far-field measurements, even though the instructions give the possibility for using them in near-fields;
- most of them measure indirectly the parameters of EMR (there is a need of additional calculations);
- there are difficulties in measuring of complex fields (with many frequencies, intermittent, non-homogeneous);
- most of them have low level of overload of detectors;
- some probes are large;
- some isotropic diagrams of the probes are not tested correctly;
- they use many probes for limited frequency ranges;
- some devices measure very few parameters;
- it is impossible to use them for direct exposure assessment (most of them are not feasible to measure energy parameters);
- most of the existing devices measure RMS fields, but not peak values, nor intermittent fields;

- big errors - about ± 20% and more; etc.

Here we do not discuss the additional errors connected with the procedure of measuring using concrete device.

Now there are some companies producing equipment with many improvements in the field of measuring parameters (also induced currents, SAR, etc.), having automatic calibration, isotropic testing, with possibility for evaluation of the exposure duration, measuring the "dose" characteristics of the exposure, etc. Some of the devices produced in the last years have shaped frequency response to different standards, giving a possibility for direct data collecting, also with some frequency analysis.

4.3. STANDARDS

TABLE 2. The standards for EMR in Bulgaria

Standard	Frequency range	Parameters	Purpose
Ordinance No. 41/1995	0 Hz-60 kHz	E [kV/m], B [mT]	Occupational
BNS 12.1.002-78	50 Hz	E [kV/m]	Occupational
BNS 14525-90	60 kHz- 300 MHz	E [V/m], H [A/m], dose values W_E, W_H	Occupational
BNS 17137-90	300MHz-300 GHz	power density S [μW/cm^2], dose value W_S	Occupational
Ordinance No. 9/1991	30 kHz-300 MHz	E [V/m]	Population
	300 MHz-30 GHz	S [μW/cm^2]	Population
Ordinance No.8/1996	0-300MHz	E [V/m], B [T]	VDUs-Occupational
Ordinance No.9/1994	20 Hz - 400 kHz	E [V/m], B[T]	VDUs-children
Ordinance No.7/1996	50 Hz	E [V/m], Safety zones	Population

The Regulation No.41/1995 is an adaptation of the European Directive 391/89. There the American Conference of Governmental Industrial Hygienists TLVs [7] for static and ELF electric and magnetic fields till 60 kHz is adapted. Above this frequency, two standards for RF and microwave radiation are in use since 1990. [9,10]

The European pre-standards ENV 50166-1 and ENV 50166-2 are in process of translation and introduction now [11, 12].

We need new standards for EMR exposure to the population in residential areas and in dwellings. The present regulation used in Bulgaria is based only on assessing the electric field strength in the RF region. The exposure assessment is based on calculation of safety zones around the RF and microwave sources, also on the measuring values after the installation of the aerials.

266

4.4. EDUCATION

There are different forms of education in the field of measurement and exposure assessment of EMR. Some courses with short and long duration are organized by the Technical University, Ministry of Health, Ministry of Labor and Social Policy. Sofia and Shumen Universities are training medical physicists and specialists in metrology.

4.5. DATABASES FOR SOURCES OF EMR

We have developed a data base system of sources of electromagnetic radiation since 1988. It consists of 3 main parts depending on the source frequency: ELF, RF and microwaves - MW. All investigated sources are included in the data base and the new information is immediately updated after performing measurements. The system possesses the opportunities to: search in dependence of the source parameters (frequency, power, etc.); evaluate the dose compared to the current hygienic standards for working or living environment. Data collection is improved by using contemporary software ("Access"). We intend to use the data base system for epidemiological studies according to the requirements to the second level of information collecting.

4.6. METHODS FOR EXPOSURE ASSESSMENT

We use both spot measuring method and dosimetry - indirect and direct ones. The following methods for measurement and exposure assessment are developed and standardized in Bulgaria:

- Method for studying the EMR of microwave ovens for household and similar use;
- Methods for studying and assessment of the EMR of VDUs - two methods: one for evaluating the emission, second for the working place of the operator;
- Method for determination of hygienic safety zones around objects, emitting EMF in residential territories;
- Method for studying and assessment of EMR at work places in physiotherapy and rehabilitation cabinets;
- Method for studying and assessment of electromagnetic radiation at work places in high voltage substations;
- Method for studying and hygienic assessment of a constant magnetic field in the working environment.

The items, common for the standardized methods, are a part of the standard measurement protocol. They include the requirements for obtaining precise results, following the obligations for QA and GLP.

The differences between the methods cover the following:

- they concern different users: EMR sources, work places, residential areas;

- they have different purposes: some concern control of EMR sources and others the hygienic assessment of possible human irradiation by EMR;
- different current standards applied;
- specific features of the measurement equipment;
- the way of presentation of the results depending on the requirements of different standards.

Principally the methods for assessment of the sources of electromagnetic radiation differ from those for their measurement at the work places by the selection of points for measurement and by the procedure. Naturally, the standards used for the assessment are also different. The major differences are contained in the measurement procedures, the latter corresponding to the particular measurement equipment, sources, and standards. For example, the method for investigation of microwave ovens includes the standard procedure referred by IEC 335-2-25 [13] while the assessment of VDUs involves the general requirements of Swedish standards [14].

The methods have been practically standardized in the whole country.

Here we are presenting a standard measurement protocol containing the common requirements of all above cited methods for exposure assessment:

4.7. MINIMUM DATA FOR A STANDARD PROTOCOL

Date/Specific No (depending on the source)...
Testing Laboratory; Address ....
Document for Notification/Accreditation...
Company..
Technical characteristics of the source (frequency, power, etc.).....................................
Description of the emitting elements (antennae, inductor, etc.).......................................
Description of the working place.(location, protection, etc..
Number and sex of the workers..
Working shift, duration of exposure...
Data from medical surveillance...
Technological process (broadcasting, welding, etc.)...
Need/purpose of measurement (for shielding, risk evaluation, etc.)...............................
Current Standards...
Measuring data..
Data of exposure duration (measured or calculated)..
Measuring equipment: type, No...
Measuring/exposure assessment method...
Measuring procedure (depends on the source, standard, etc.)..
Additional conditions of measurement (other hazards, special procedure, etc)..........
Measuring operators (names)...
Conclusions (risk assessment, hazard evaluation, etc.)...
Recommendations (safety measures, medical surveillance, etc.)....................................
Other information (required for the exposure assessment or for safety measures).......

268

5. Further Activities

The further activities for creation and unification of methods for measurement and assessment of electromagnetic radiation should be oriented towards comprising more typical sources and work places. We consider the development of standard methods like:

- *for emissions* - emitters for high frequency processing of materials (plastics, metals); electric medical devices, nuclear-magnetic resonance equipment, machines for induction heating of metals, for electric erosion, for thermal processing of dielectrics, electric welding machines, sources of electrostatic fields
- *for work places* - in communication systems, in electric vehicles and at work places with the listed sources.

The main goals for creating and unification of the methods for measurement and assessment of electromagnetic radiation are:

- comparability of the results from studying sources of similar type and possibilities for their use in epidemiological studies;
- use of the collected data from measurements for creation of information systems and data bases and their further application for assessment of the risk and impact of EMF;
- possibility for checking the hygienic normative criteria;
- receiving results corresponding to the requirements of QA and following the GLP procedure.

6. References

1. ISO 5725 (1986) Draft (1991) *Accuracy (Trueness and Precision) of Measurement Methods and Results, Part 1: General Principle*, International Organization for Standardization.
2. ISO/TAG 4WG3 (1992) *Guide to the Expression of Uncertainty in Measurement*, June 1992, International Organization for Standardization.
3. European Standard EN 485. *General Requirements for the Performance of Procedures for Workplace Measurements*. Draft 1991.
4. Israel, M.S. (1994) Electromagnetic radiation - parameters for risk assessment, *Reviews on Environmental Health* 10, 85-93.
5. Environmental Health Criteria 137 (1993), *Electromagnetic Fields (300 Hz to 300 GHz)*, World Health Organization, Geneva.
6. Tech. 3278-E (1995) *Radiofrequency Radiation Hazards. Exposure Limits and Their Implications for Broadcasters*, European Broadcasting Union, G.T. Waters (ed.), Geneva.
7. TLVs and BEIs (1998) *Threshold Limit Values for Chemical Substances and Physical Agents*, American Conference of Governmental Industrial Hygienists (ACGIH).
8. ICNIRP (1998) Guidelines for limiting exposure to time-varying electric, magnetic, and electromagnetic fields (up to 300 GHz), *Health Physics* 74, 493-522.
9. Bulgarian National Standard BNS 14525-90 (1990) *Labor Protection. Radiofrequency Electromagnetic Fields. Permissible Levels and Requirements for Control*, Committee of Quality, Sofia, Bulgaria.

10. Bulgarian National Standard BNS 17137-90 (1990) *Labor Protection. Microwaves. Permissible Levels and Requirements for Control*, Committee of Quality, Sofia, Bulgaria.

11. CENELEC ENV 50166-1 (1995) *Human Exposure to Electromagnetic Fields Low-Frequency (0 Hz to 10 kHz)*, European Committee for Electrotechnical Standardization.

12. CENELEC ENV 50166-2 (1995) *Human Exposure to Electromagnetic Fields High-Frequency (10 kHz to 300 GHz)*, European Committee for Electrotechnical Standardization.

13. IEC 335-2-25 (1988) *Safety of Household and Similar Electrical Appliances, Part 2: Particular Requirements for Microwave Ovens*.

14. Svensk Standard SS 436 1490 (1995) *Computers and Office Machines - Measuring Methods for Electric and Magnetic Near Fields*, Svenska Elektriska Kommissionen, SEK.

BIOPHYSICAL BASIS FOR ELECTRICAL STIMULATION OF EXCITABLE TISSUE: APPLICATION TO LOW FREQUENCY EXPOSURE STANDARDS

J. P. REILLY
Metatec Associates
12516 Davan Drive
Silver Spring, MD 20904

1. Introduction

Nerve and muscle are specialized to respond to electrical stimuli by becoming "excited', that is the cellular membrane undergoes a marked nonlinear change of conductivity if it is sufficiently depolarized from its normal resting potential. Excitation of a nerve cell leads to a propagating nerve action potential, which in an efferent nerve normally proceeds from a sensory receptor to the central nervous system (CNS). In an afferent nerve cell, an action potential is initiated in the CNS, from whence it propagates to muscle connections called *motor end plates*. Communication from one nerve cell to another or at the motor end plate takes place across synapses by means of neurotransmitters.

These normal electrical processes can be activated or modified by means of electrical forces artificially introduced into the body through applied contact currents, or through electromagnetic induction. If controlled, applied electrical forces can be used for therapeutic or diagnostic purposes. If uncontrolled, the same forces can be detrimental.

Excitable tissue effects are typically observed within a short duration after the application of the stimulus, often within milliseconds to seconds. These "acute" effects stand in contrast to responses to chronic electromagnetic exposure effects that many investigators have studied at much lower exposure levels for possible implications on human health.

Although there are many questions remaining about short term electrical effects, we largely understand the underlying mechanisms, can verify theoretical mechanisms in humans and animals, the experimental results are robust, and we can define biological end points in the intact human. It is therefore valuable to define limits to human exposure based on our understanding of acute excitable tissue effects.

The purpose of this paper is to review the mechanisms that lead to responses to short-term electrical exposure of nerve, muscle, and synapse. We will develop thresholds of human response based on such responses, and propose limits to

271

B.J. Klauenberg and D. Miklavcic (eds.), Radio Frequency Radiation Dosimetry, 271-291.
© 2000 *Kluwer Academic Publishers. Printed in the Netherlands.*

electromagnetic exposure. Such limits should be understood as upper limits on acceptable human exposure.

2. Principles of Electrical Effects on Excitable Cells

2.1. CELLULAR POLARIZATION

Biological cells normally maintain an interior potential V_r which is negative with respect to its exterior; for nerve and muscle cells, typical values of V_r are -65 and -90 mV respectively (potential of cell interior with respect to exterior). The electrical forces on the cellular membrane are enormous. Considering the membrane potential ($\approx$ 0.1 V) and thickness ($\approx 10^{-8}$ m), the electric field developed across the resting membrane is around 10^7 V/m. Conductivity properties of the excitable membrane are intimately tied to the electric field developed across it. Disturbances from the resting condition can lead to profound changes in the membrane's electrical properties, which ultimately initiate and sustain the functional responses of nerve and muscle.

The membrane potential of elongated cells are most affected by an electric field in the surrounding medium. Figure 1 illustrates the distribution of current flow around an elongated cell that is placed in a medium having a uniform electric field (i.e., uniform current density). The cell is presumed to be oriented parallel to the undisturbed field. The flux lines suggest that the current through the membrane, and hence the disturbance of membrane polarization, is greatest at the ends of the fiber. The anode-facing end of the cell will be hyperpolarized, and the cathode-facing end will be depolarized. In an alternating field, the sites of hyperpolarization and depolarization will alternate every one-half cycle of the field oscillation.

The potential disturbances of the elongated cell can be analyzed using the theory of electrical cables, which was originated by Oliver Heaviside in 1876 in connection with the analysis of the first Trans-Atlantic telegraphy cable. Consider a cable of length $2L$ in a longitudinal static field of strength E. The steady-state solution for membrane voltage is given by [1]

$$V_m(X) = -E\lambda\left(\frac{\sinh X}{\cosh L/\lambda}\right)$$

(1)

where $X = x/\lambda$, x is the distance from the center of the cell, λ is the space constant, and $2L$ is the length of the cell. Note that $X = 0$ is taken as the center of the cell, and the ends are at $\pm L$. The space constant λ, also known as the *electrotonic distance* of the membrane, defines the distance along the membrane that a steady-state voltage disturbance due to point current injection will decay to e^{-1} of the value at the disturbed location. Space constants for invertebrate nerve are in the range 0.23 - 0.65 cm[2].

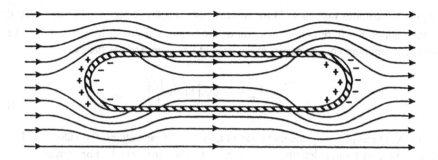

Figure 1. Representation of current flow around elongated cell placed in medium having a uniform electric field. The membrane is assumed to be semipermeable to current flow.

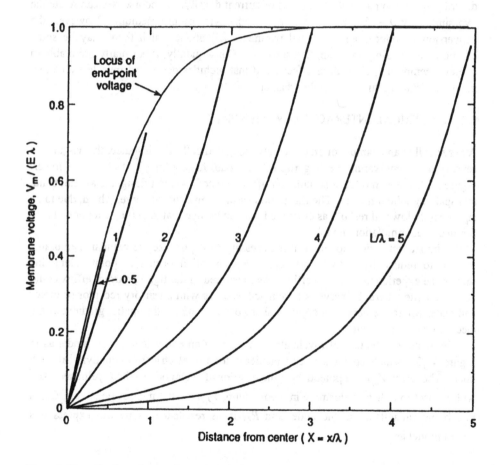

Figure 2. Normalized membrane voltage of a finite cable immersed in a static field of strength E.

Figure 2 illustrates Equation (1) for several cable lengths. Since V_m has odd-valued symmetry about $X = 0$, only one quadrant of the function needs to be illustrated. The maximum membrane voltage occurs at the ends of the fiber, and has the value

$$V_m = - E\lambda\left(\tanh\frac{L}{\lambda}\right)$$

(2)

For very long cells ($L \to \infty$), the membrane's terminus voltage attains the value $E\lambda$. But even for fibers of modest length, that value is closely approached. For instance, with $L/\lambda = 2$ (total length $= 4\lambda$), the membrane voltage at the ends is $\pm 0.964E\lambda$.

The fundamental force for membrane polarization effects is the in-situ electric field (E), rather than current density (J). Many investigators, however, have preferred to describe excitation parameters in terms of current density. Of course, we can relate the two simply by $J = E\sigma$, where σ is the conductivity of the medium. However, the conversion introduces an additional parameter (σ) about which there may be some additional uncertainty in an applied situation. Consequently, it is usually preferable to express membrane polarization effects, and that includes nerve and muscle excitation, in terms of the in-situ E-field rather than current density.

2.2. ELECTRICAL INTERACTION WITH NERVE

A nerve cell is an example of an extremely elongated cell. For instance, the length of a sensory nerve enervating the fingertip or toe would have a length of about one meter. Figure 3 illustrates modes of stimulation of a nerve cell, which I designate as end, bend, and spatial gradient modes. The illustration shows a myelinated nerve, which, due to its significantly lower threshold as compared with an unmyelinated nerve, is a good choice for electrical stimulation models.

To initiate an action potential, it is necessary to depolarize the cellular membrane, that is, to reduce its resting potential. Depolarization occurs at points along the membrane experiencing current efflux. As illustrated in the figure, current efflux could occur at a site where the nerve is terminated, such as with a sensory receptor or motor end plate, where the nerve undergoes a sharp bend, or where the spatial gradient of the electric field is maximum.

We model stimulation of myelinated nerve using an equivalent circuit model as in Figure 4 [3], which contains circuit elements for the electrical conductivity at each node. The terms $V_{e,n}$ are potentials at the exterior of each node of the myelinated axon with respect to a distant electrode in the medium, $V_{i,n}$ are the interior potentials, C_m is membrane capacitance at the node, and R_m is its resistance. One can express this circuit model as

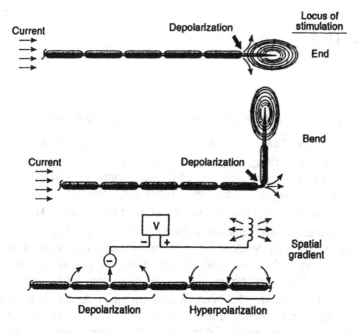

Figure 3. Modes of neural stimulation. Excitation is initiated at points of maximal current efflux across neural membrane. Potential excitation sites consist of fiber terminals, sharp bends, and maximal gradient of E-field.

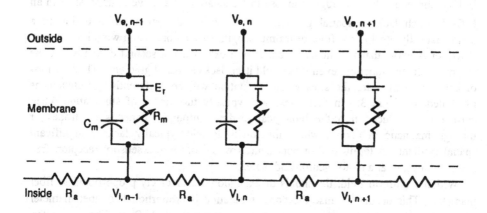

Figure 4. Equivalent circuit models for excitable membranes. The response near the excitation threshold requires that the membrane conductance be described by a set of nonlinear differential equations.

$$\frac{dV_n}{dt} = \frac{1}{C_m}\left[G_a\left(V_{n-1} - 2V_n + V_{n+1} + V_{e,n-1} - 2V_{e,n} + V_{e,n+1}\right) - I_{i,n}\right]$$

$$(3)$$

which is a discrete form of the cable equation, where V_n is the membrane potential at node n. The term $V_{e,n-1} - 2V_{e,n} + V_{e,n+1}$, which is the driving function for membrane polarization change, is a second difference (i.e., derivative) of the spatial potential measured along the long axis of the nerve fiber, or, equivalently, the first derivative of the longitudinal electric field. The term $I_{i,n}$ representing the ionic current flowing across the membrane (i.e., through the element R_m in Figure 4) is governed by a set of nonlinear differential equations applying to the myelinated nerve membrane [4].

Equation (3) has been developed as a computer model consisting of an arbitrary number nodes, and with a threshold criterion based on propagation of an action potential [5, 6]. The model titrates the magnitude of stimulating potentials ($V_{e,n}$) between threshold and no-threshold conditions to determine the threshold of excitation within 1%. The model is designed to accommodate an arbitrary spatial and temporal variation distribution of the potential disturbances, $V_{e,n}$. The computer implementation of the myelinated nerve model has been identified as the Spatially Extended Nonlinear Node (SENN) model.

Equation (3) requires a finite gradient of the longitudinal E-field to cause a change in V_n. Consequently, one might suppose that excitation of a nerve is impossible in an E-field which lacks a spatial gradient. If we were to restrict our attention to a mathematically ideal nerve fiber of infinite length, such a conclusion would be correct. However, a nerve that terminates or bends will experience a second derivative of the external potential function, even if the field itself lacks a spatial gradient. The locations of ends and bends are the sites where excitation will be preferentially initiated, as illustrated in Figure 3. In fact, these are typically the modes of stimulation when current is introduced into the biological medium through cutaneous electrodes or through magnetic induction, where the in-situ E-field typically lacks a significant spatial gradient. A terminated or bent neuron would occur at a sensory receptor, free nerve endings, or at a motor neuron end plate.

Within a uniform field, thresholds of excitation are inversely proportional to fiber diameter. This occurs because the nodal separation is proportional to fiber diameter such that $d = 100\,D$, where d is nodal separation, and D is fiber diameter. Consequently, the voltage difference from node to node is directly proportional to fiber diameter. For example, a 10-μm fiber would have an internode spacing of 1 mm. The membrane potential change at the terminal node is approximately equal $V_m = Ed$. Compare this result with the cable relationship mentioned above, for which $V_m = E\lambda$.

The distribution of myelinated fiber diameters in peripheral nerve effectively covers the range 2 - 20 µm. Since the lowest thresholds correspond to the largest fibers, we use a 20 µm fiber to model response thresholds for peripheral nerve stimulation. In the case of brain neurons, the diameter distribution is shifted to smaller diameters, and we use a 10 µm fiber for a conservative analysis of brain stimulation.

2.3. STRENGTH-DURATION LAW OF EXCITATION

Figure 5 shows thresholds of excitation based on the SENN model for a point electrode 2 mm distant from a 20-µm nerve fiber. Threshold curves are shown for a monophasic square wave, and for a biphasic square wave and a single cycle of a sine wave. The left axis shows the threshold in terms of charge units (µC); the right axis is in terms of peak current (mA). Thresholds for the monophasic stimulus fall to a minimum plateau called *rheobase* as the pulse width is increased. For short duration stimuli, thresholds converge to a minimum charge threshold. Stimulus energy is not a pertinent descriptor of excitation thresholds, as some have erroneously supposed.

Thresholds for monophasic stimulus conform to the empirical relationships

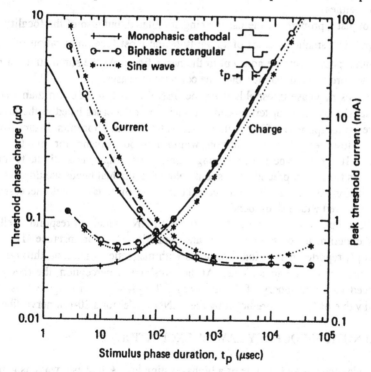

Figure 5. Strength-duration relationships derived from the myelinated nerve model: current thresholds and charge thresholds for single-pulse monophasic and for single-cycle biphasic stimuli with initial cathodic phase, point electrode 2 mm distant from 20-µm fiber. Threshold current refers to the peak of the stimulus waveform. Charge refers to a single phase for biphasic stimuli.

$$\frac{I_T}{I_o} = \frac{1}{1 - e^{-t/\tau_e}} \tag{4}$$

$$\frac{Q_T}{Q_o} = \frac{t/\tau_e}{1 - e^{-t/\tau_e}} \tag{5}$$

where I_T and Q_T are thresholds for peak current and charge, respectively, I_o and Q_o are the rheobase values, and τ_e is the strength-duration time constant. Equations (4) and (5) are redundant in that one can be derived from the other. Note that two parameters completely describe the strength-duration law of the excitable tissue: the rheobase and the strength-duration time constant.

The S-D relationships discussed here are a universal property of excitable tissue. However, τ_e varies widely with the type of tissue being stimulated. The smallest values of τ_e are found in nerve tissue; muscle tissue exhibits values approximately 10 times greater, and synaptic processes about 100 greater than nerve tissue values. These variations have great significance in defining thresholds of excitation in various frequency regimes.

τ_e is not just a property of the tissue being stimulated, but also of the focality of the stimulation. The smallest values of τ_e apply to the most focal application of current, such as with a point electrode adjacent to the nerve fiber. As the stimulation current is applied in a more gradual manner, the time constant increases.

Table 1 lists rheobase thresholds from the SENN model of a 20-μm diameter fiber that is either sharply bent, or terminated at a node. For the cases listed in the table, the lowest thresholds apply to a fiber with a sharp 180° bend at a location distant from the terminus. However, such a condition would not be realistic for practical fiber trajectories. If we examine the remaining cases, we see that the straight, terminated fiber provides the lowest practical threshold. Note that gradual bends would necessitate higher excitation thresholds, since the second derivative of voltage would necessarily be lower as compared with a sharp bend.

The theoretical rheobase indicated by the SENN model correspond well with numerous experiments of both humans and animals [6]. For instance, Havel and colleagues [7] reported perception thresholds with magnetic stimulation through a coil encircling the arm of human subjects. At the threshold of perception, the rheobase E-field induced in the periphery of the arm was 5.9 V/m -- which is quite close to the theoretical value of 6.2 V/m predicted by the SENN model for a 20-μm nerve fiber.

2.4. STRENGTH-FREQUENCY LAW OF EXCITATION

The sinusoidal wave is an example of a biphasic stimulus. A biphasic wave is generally less effective for excitation than a monophasic wave of the same magnitude and duration. This occurs because the biphasic phase reversal will tend to reverse a developing action potential that was initiated on the initial phase. The result is higher thresholds of excitation for biphasic stimuli.

TABLE 1. Excitation requirements for end and bend modes of stimulation.

Bend angle (deg.)	Bend node (#)	E threshold (V/m)	τ_e (μs)
0	1	6.21	128.2
90	2	8.55	126.0
90	4	9.84	114.3
90	6	9.96	112.9
90	8	9.96	112.9
180	2	6.56	101.4
180	4	5.45	105.0
180	6	5.10	110.3
180	8	5.04	111.6

Thresholds apply to 20-μm nerve fiber within constant E-field that is oriented parallel to the nerve beyond the bend point.

We can express sinusoidal thresholds as a strength-frequency (S-F) curve, as in Figure 6. The solid curves in this figure have been derived from the SENN model with a point electrode as in Figure 5. Separate theoretical curves are shown for a single cycle stimulus, and for a continuous stimulus.

Figure 7 illustrates results from the SENN model showing the variation of thresholds with number of cycles of sinusoidal stimulation, where the waveforms starts at a zero crossing. The threshold at one-half cycle of stimulation is relatively low because the waveshape is monophasic. At one cycle of stimulation, the waveshape is a charge-balanced biphasic wave, which requires a relatively high threshold. Thresholds alternate with half-cycle increments in a sawtooth pattern, gradually falling until reaching a minimum plateau when the total duration of stimulation is about one millisecond, which is effectively equivalent to continuous stimulation. The minimum threshold for continuous stimuli converges to a value equal to the threshold of a monophasic wave having a duration equal to the phase duration (half-cycle time) of the sinusoidal wave. Similar waveform sensitivity effects have been demonstrated experimentally in both human [8] and bovine [9] subjects.

The strength-frequency curve is a U-shaped function with a minimum plateau at mid frequencies, and an upturn at both low and high frequencies. The low frequency upturn

280

occurs with sinusoidal wave shapes, and does not occur with square-wave biphasic waves, as indicated in Figure 5 for long phase durations. This upturn is a consequence

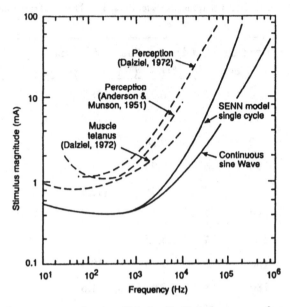

Figure 6. Strength-frequency curves for sinusoidal stimuli. Dashed curves are from experimental data. Solid curves apply to SENN model. Experimental curves have been shifted vertically to facilitate comparisons.

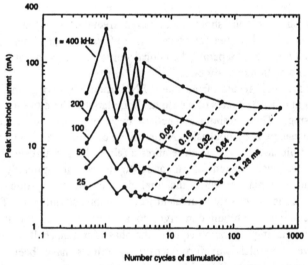

Figure 7. Excitation thresholds as a function of number of cycles of sinusoidal stimulation. Stimulus duration stepped in half-cycle increments out to four cycles, and full cycle increments beyond that. Dashed lines indicate duration of stimulus. Point electrode 2 mm distant from 20 μm myelinated nerve fiber.

of the slow rate of rise of a low-frequency sinusoidal wave, and is known as *accommodation*. If the stimulus waveform consisted of an oscillating square wave, rather than a sinusoid, thresholds would not have the low-frequency upturn, although the high frequency behavior would be similar to that for sinusoidal waveforms.

An empirical fit to S-F data from the myelinated nerve model is given by

$$\frac{I_T}{I_o} = \left[1 - \exp\left(-\frac{f_e}{f}\right)\right]^{-a} \left[1 - \exp\left(-\frac{f}{f_o}\right)\right]^{-b}$$

(6)

where f_e and f_o are constants that determine the points of upturn in the S-F curve at high and low frequencies, respectively. Data from the SENN model provide the values $b = 0.8$; below 80 kHz $a = 1.45$ for single cycle stimulation, and $a = 0.9$ for continuous cycle stimulation; from 80 kHz to 400 kHz, $a = 1.7$ for single cycle, and $a = 1.0$ for continuous stimulation [6]. Experimentally derived values of f_e encompasses a wide range, with values around 1000 Hz for electrocutaneous nerve stimulation, 100 Hz for cardiac muscle stimulation; f_o lies in the range 10 - 50 Hz for nerve stimulation, and 10 Hz for cardiac stimulation.

The frequency limits for which Equation (6) is valid are not known. With continuous sinusoidal stimulation, human electrocutaneous perception data follow Equation (6) up to 100 kHz, above which thresholds reach a maximum plateau as a result of thermal perception due to tissue heating [10, 11]. For pulsed stimuli, the frequency above which thermal perception will dominate electrical thresholds depends greatly on the fraction of on-time (duty factor), since the heating capacity of electrical current (i.e., its rms value) is proportional to the square root of the duty factor. Consequently, one can extend the frequency where thermal perception dominates electrical perception by using pulsed stimuli of low duty factor. One could infer experimental verification of Equation (6) up to perhaps 10 MHz based on human sensory thresholds with pulsed stimuli as short as 0.1 µs, the shortest duration tested [6]. Experiments with sinusoidal stimulation of rats show reasonable correspondence up to 1 MHz, the highest frequency tested [12].

2.5. ELECTRICAL STIMULATION OF MUSCLE AND CARDIAC TISSUE

Muscle tissue may be electrically stimulated directly, or indirectly through stimulation of motor neurons. In general, skeletal muscle tissue is more readily stimulated via a motor neuron, considering the fact that S-D time constants for stimulation of muscle tissue are approximately a factor of 10 greater than for nerve tissue. With transcutaneous stimulation of normal muscle tissue, time constants for motor reaction of approximately 0.1 ms are observed. Such a time constant is typical of nerve stimulation since the muscle tissue is being excited indirectly through the enervating motor neurons. However, if the same tissue is denervated, time constants of several ms are observed [13]. In this case, the muscle is stimulated directly.

The heart is a special case of muscle tissue. The heart is most sensitive to electrical excitation (a premature beat) during the diastolic (relaxed) period. It is most sensitive to fibrillation during the partial recovery period within the systolic (contractile) phase. Since excitation thresholds are much lower than fibrillation thresholds, the former is appropriate for electrical safety considerations.

Experimental data from several sources suggests that the rheobase for the heart is about 12.0 V/m at the median of a statistical distribution among healthy canine subjects [6]. At the one-percentile rank, the rheobase is approximately 6.0 V/m -- a factor of two smaller.

Excitation thresholds of the heart exhibit an S-D law, much like that for nerve excitation, except that the time constants are shifted to much larger values, similar to those observed with skeletal muscle stimulation. Like nerve tissue, time constants for cardiac excitation increase as the stimulus current is applied less focally to the stimulated tissue. Typical values of τ_e for cardiac tissue with large area stimulation is about 3 ms -- a factor of ten or more greater than τ_e for nerve stimulation [6].

2.6. STIMULATION OF NEURAL SYNAPSES AND PHOSPHENES

Whereas the nerve cell requires membrane depolarization of approximately 15 mV to initiate an action potential, synaptic processes can be affected with alteration of the presynaptic membrane potential by less than 1 mV, and possibly as little as 60 µV with electrical stimulation of the synaptic processes in the retina [14, 15] - a factor 250 times lower than neural excitation thresholds. Consequently, we should consider the synapse as a potentially sensitive mode of neural interaction with applied electrical stimulation. The fact that the central nervous system (CNS) is rich in synapses makes this mode of stimulation especially interesting.

While properties of the neural synapse have been widely studied as a branch of neurophysiology, the role of synaptic effects in electrical stimulation has been little studied. An important property of the synapse is that a relatively small change in presynaptic potential can have a much larger percentage change in post synaptic potentials [16]. Considering that the postsynaptic cell sums the presynaptic inputs from several cells, the resulting effect on the membrane potential at the postsynaptic cell can be considerable.

Polarization of presynaptic processes due to an in-situ electric field can result in enhancement or inhibition of postsynaptic action potentials. An example of this effect is attributed to the phenomenon of electro- and magnetophosphenes, which are visual effects resulting from electric currents or magnetic fields applied to the head [17, 18, 19, 20, 21, 22, 23, 24, 25, 26]. Experimental evidence suggests that phosphenes are generated through modification of synaptic potentials in the receptors and neurons of the retina. The retina is rich in synaptic junctions of photoreceptors and neurons that comprise a visual processing system. Phosphenes are most sensitive to current or an in-

283

Using data from magnetophosphenes [24, 25] we can calculate the corresponding induced E-fields using principles of magnetic induction (see below). The maximum E-field in the head is calculated to be 0.079 V/m for stimulation at the most sensitive frequency tested (20 Hz). At the location within the retina, where the electrical interaction is thought to take place, the calculated field is 0.053 V/m-rms, which is consistent with the current density threshold of 0.008 A/m^2 at the retina determined for electro-phosphenes [25], assuming the conductivity of the brain is 0.15 S/m.

Experimental data show that the S-D time constant for phosphenes using electrodes on the temples is about 14 ms [20, 21]. Figure 8 illustrates curves for electrically

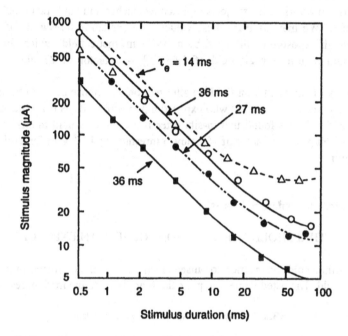

Figure 8. Strength-duration curves for electrically evoked potentials in the retina of the frog's eye. Curves represent various experimental procedures. (Adapted from Knighton, 1974).

evoked potentials in the frog's eye, showing τ_e in the range 14 - 36 ms [14]. These values are consistent with the phosphene data described above, but are about 100 times greater than corresponding values for peripheral nerve.

Phosphene thresholds are quite unlike those for nerve excitation. For one thing, the internal E-field corresponding to phosphene perception at the optimum frequency is a factor of 100 or so below rheobase thresholds for neural stimulation. Furthermore, the

S-D time constants for phosphenes are approximately 100 times greater than for nerve stimulation.

Although photoreceptors are not found in the CNS, the brain and spinal column are rich in neuro synaptic junctions. It is logical to inquire whether CNS interactions are possible at the low phosphene thresholds pertaining to retinal stimulation. While we lack a clear answer to this question, experimental evidence shows that CNS interactions are indeed possible with magnetic stimulation of the brain at intensities well below levels necessary to excite neurons. For instance, [26] exposed the human head to sinusoidal magnetic fields and recorded visual evoked potentials (VEP) on the surface of the scalp in response to a visual stimulus. With a 50 Hz stimulus at a flux density of 60 mT, Silny observed significantly altered VEP patterns in 12 of 15 test subjects. It is remarkable that these alterations persisted for as long as 70 minutes after cessation of the stimulus. In addition, his subjects reported headaches and "indisposition" above 60 mT exposure. We may compare these values with excitation thresholds of a cortical neuron through exposure of the head by a 50-Hz magnetic field, which for a 10-μm neuron would require a peak field of 1.4T [6] -- a factor of 23 above the VEP thresholds.

The ability of sub excitation fields to alter neuronal response has also been reported by Bawin and associates [27, 28] who exposed hippocampal slices from the rat brain to magnetic fields. It was found that in-situ effective induced E-field intensities were as low as 0.75 V/m peak -- a factor of 16 below a threshold of 12.3 V/m for excitation of a 10-μm neuron.

3. Applications to Exposure Limits

3.1. IN-SITU THRESHOLDS OF REACTION FOR HUMAN EXPOSURE.

Table 2 lists thresholds of reaction to pulsed electrical stimulation, listed in the order of rheobase E-field. The listed values are peak fields applying to the median response of a

TABLE 2. Reaction Thresholds for Pulsed Stimulation

Responding Tissue	Rheobase E-field (V/m-pk)	S-D time constant (ms)
Retinal synapse	0.075	25.00
20-μm nerve fiber	6.2	0.12
10-μm nerve fiber	12.3	0.12
Cardiac muscle	12.0	3.00

Median response; peak E-field.

statistical distribution. The rheobase for photoreceptor synapse stimulation derives from human magnetophosphene data of Lövsund and is consistent with rheobase thresholds determined by others mentioned above. The time constant applies to S-D data developed for human phosphenes [20, 21], and frog retina stimulation [14, 15]. The listed time constant is also consistent with phosphene frequency sensitivity data of Lövsund, as will be explained below. Nerve excitation rheobase thresholds and time constants are derived from the SENN model. The data for a 20-μm fiber apply to large peripheral nerve fiber; the 10-μm data apply to a large brain neuron. The cardiac data applies to the median response for healthy hearts.

3.2. THRESHOLDS OF REACTION WITH MAGNETIC EXPOSURE

A time-varying magnetic field will induce an electric field within a conducting material, including biological materials. A simple, but effective model for calculating induction effects in the human body with large area exposure treats the torso or head as an ellipse with homogeneous conductivity [6]. For an elliptical cross section in the y-z plane, and an incident magnetic field in the x-direction, the induced E-field is given by

$$E = -\frac{dB_x}{dt}\left(\frac{a^2 y a_z - b^2 z a_y}{a^2 + b^2}\right)$$

(7)

where dB_x/dt is the rate of change of the magnetic flux density, a and b are the semi-major and semi-minor axes of the ellipse, and a_y and a_z are unit vectors in the y- and z-directions respectively. Equation (7) is a long wavelength solution, in which the wavelength of the magnetic field is much greater than the maximum dimensions of the body. For a circular cross section, $r = (x^2 + y^2)$ along a circular path of radius r, and Equation (7) reduces to

$$E = -\frac{r}{2}\frac{dB}{dt}$$

(8)

which is the familiar expression of Faraday's Law. According to Equation (8), the maximum induced E-field occurs with the periphery of the largest circular cross section. With a elliptical cross-section, one can demonstrate with Equation (7) that the maximum induced field occurs at the outermost point on the minor axis of the ellipse.

To develop magnetic exposure criteria, we represent the torso of a large adult human with ellipses as follows: sagittal cross section: $a = 0.4$, $b = 0.17$ m; frontal (coronal) cross section: $a = 0.4$, $b = 0.2$; longitudinal cross section: $a = 0.2$, $b = 0.17$. We model the human head as an ellipse with $a = 0.13$ and $b = 0.1$ m. The cortex of the brain is approximately 1.5 cm below the surface of the scalp.

We calculate magnetic thresholds using the excitation criteria of Table 2, and the induction model of Equation (7). The result is tabulated in Table 3, and illustrated in Figure 9.

The pulsed field limits of Figure 9 can be converted to sinusoidal thresholds by applying principles and assumptions developed previously [6]. One assumption is that at the excitation threshold, the peak of a pulsed stimulus is equivalent to the peak of a sinusoidal stimulus of many cycles. Another assumption is that the strength-duration time constant τ_e and the strength-frequency constant f_e are related by $f_e = 1/(2\tau_e)$. A further relationship is that the peak flux density, B, and the peak time rate of change, dB/dt, are related by

$$B_o = \frac{1}{\pi}\dot{B}\,t_p$$

(9)

where $\dot{B}_o \equiv dB/dt$, and t_p = phase duration. A final assumption is that the rms value of a sinusoid is its peak value divided by $\sqrt{2}$. The result is illustrated by curves (a)-(d) in Figure 10.

TABLE 3. Reaction Thresholds for Sinusoidal Stimulation

Responding Tissue	Rheobase E-field (V/m-rms)	Upper transition frequency (Hz)
Retinal synapse	0.053	20.0
20-μm nerve fiber	6.2	0.12
10-μm nerve fiber	12.3	0.12
Cardiac muscle	12.0	3.0

Median response; rms E-field.

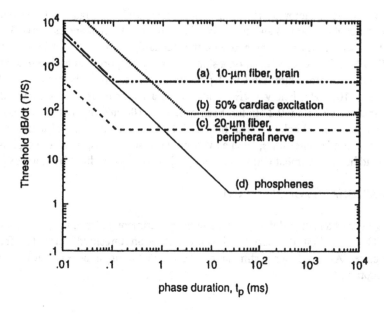

Figure 9. Strength-duration threshold criteria for monophasic *dB/dt* pulses. Separate curves are shown for nerve, heart, and retinal synapse.

3.3. LIMIT CRITERIA

The lower limit of the thresholds in Figure 10 may be used to derive acceptability limits for whole-body human exposure to magnetic fields. The thresholds are intended to represent approximate median values among a population of healthy individuals. To derive protective standards from thresholds, it is customary to apply an acceptability factor to account for particularly sensitive individuals, those in a pathological state, and for uncertainties in the methodology of determining thresholds. Although we do not know the statistical distribution of thresholds for all the reactions shown in Figure 10, it would not be unreasonable to assume a log-normal distribution, similar to the threshold distribution for electrocutaneous perception, or for cardiac excitation of fibrillation [6]. Accordingly, a reasonable estimate is that magnetic thresholds at the one-percentile is a factor of 2 or 3 below median values. Further allowances should be made for individuals in a pathological state, which can lead to greater population variance. Considering these factors, an acceptability factor of 10 will be applied to the median thresholds of Figure 10.

Note that the acceptability factor of 10 applied here is a conservative one when applied to a field magnitude, rather than a specific absorption rate (SAR) metric, as in existing standards adopted for higher frequencies in the IEEE standards [30] and others. This is true because the SAR value is proportional to the square of the field. Consequently, a factor of 10 applied to SAR is equivalent to a factor $\sqrt{10}$ in the associated electric or magneticfield under free-field propagation conditions.

If we apply a factor of 10 to the lowest thresholds in Figure 10, we obtain the criteria defined by Table 4 and curve (e) of Figure 10. At frequencies above 3 kHz, the acceptance curve has been merged with the standards of IEEE C95.1 by extrapolating the C95.1 low frequency plateau on a slope inversely proportional to frequency below 3 kHz. With this procedure, the C95.1 plateau is a factor of 5.0 below curve (c) in the region 3 to 100 kHz; below 3 kHz, curve (e) is a factor of 10 below curve (d). The acceptance curve below 0.2 Hz has been capped at 100 mT-rms (141 mT-peak). This limit is a factor of 10 below a peak field of 1.41 T, in consideration of human reactions reported at similar intensities, such as vertigo, taste sensations, nausea and phosphenes with rapid head movement within the field [31], as well as cardiac rate changes [32].

3.4. RATIONALE FOR PROTECTIVE LIMITS

The purpose of a standard it to protect against a detrimental effect, not just a perceptible one. Consider first the peripheral nerve excitation threshold curve (c) from that perspective. Although nerve stimulation is not a disturbing experience at the threshold of perception,

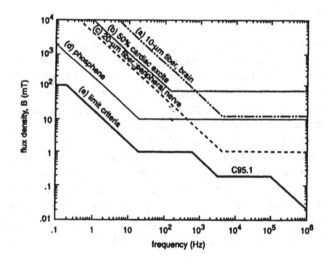

Figure 10. Thresholds for short-term reactions to sinusoidal magnetic fields with whole body exposure of a large adult person. Curves (a)-(d): human reaction thresholds; (e): derived limit criteria with acceptance factor.

TABLE 4. Magnetic Field Exposure Limits Based on
Mechanisms of Short-Term Reactions

Frequency range (Hz)	B_0 (mT-rms)	H (A/m-rms)
<0.2	100	7.95×10^4
0.2 – 20	20/f	$1.59 \times 10^4/f$
20 - 615	1	795
615 - 3×10^3	615/f	$4.89 \times 10^5/f$
3×10^3 - 10^5	0.205	163
10^5 - 10^6	$2.05 \times 10^4/f$	$1.63 \times 10^7/f$

Criteria above 3 kHz based on existing IEEE/ANSI C95.1.
Criteria below 3 kHz based on threshold reactions, with
acceptability factor of 10.

unpleasant or painful sensations are experienced with magnetic stimuli that exceed the peripheral nerve perception threshold by only 40% or so [8, 33]. Consequently, curve (e) at frequencies above 615 Hz is approximately a factor of 7 below a clearly detrimental effect. Curve (d) is based on phosphene perception. While this phenomenon has not been reported to be disturbing in a laboratory setting, it is not clear whether this would be the case for affected individuals in an uncontrolled environment. Furthermore, CNS reactions at levels somewhat above the phosphene threshold has been indicated in some experiments. Until more definitive studies have been performed, induction of phosphenes will be assumed to be a situation that should be avoided in an uncontrolled environment.

The acceptance curves developed here are derived from established bioelectric mechanisms involving acute effects of excitable tissue. This in not the only basis to be used in standards. Other information to be considered includes epidemiological data, and laboratory studies that are deemed reliable, but for which a mechanistic explanation may not presently be established. Such information must be judged as to its reliability, application to an intact human, and evaluated risk. Consequently, curve (e) should be considered as an upper limit on acceptable exposure to magnetic fields in uncontrolled environments.

For particular applications in controlled situations, it may be acceptable to allow greater exposure than indicated by curve (e). For instance, in magnetic resonance imaging procedures, higher exposure levels are permitted by existing guidelines [34, 35, 36].

290

4. REFERENCES

1. Sten-Knudsen, O. (1960) Is muscle contraction initiated by internal current flow? *J. Physiol.* **151**, 363-384.
2. Rall, W. (1977) Core conductor theory and cable properties of neurons, in *Handbook of Physiology: A Critical, Comprehensive Presentation of Physiological Knowledge and Concepts*, Vol. 1. American Physiological Society, Bethesda, MD, pp. 39-97.
3. McNeal, D.R. (1976) Analysis of a model for excitation of myelinated nerve, *IEEE Trans. Biomed. Eng.* **BME-23**, 329-337.
4. Frankenhaeuser, B. and Huxley, A.F. (1964) The action potential in the myelinated nerve fiber of Xenopus Laevis as computed on the basis of voltage clamp data, *J. Physiol.* **171**, 302-315.
5. Reilly, J.P. (1988) Electrical models for neural excitation studies, *John Hopkins APL Tech. Digest*,9, 44-59.
6. Reilly, J.P. (1998) *Applied Bioelectricity: From Electrical Stimulation to Electropathology*, Springer-Verlag, Berlin.
7. Havel, W.J., Nyenhuis, J.A., Bourland, J.D., Foster, K.S., Geddes, L.A., Graber, G.P., Waniger, M.S., and Schaefer, K.J. (1997) Comparison of rectangular and damped sinusoidal dB/dt waveforms in magnetic stimulation, *IEEE Trans. Magnetics*, Sept.
8. Budinger, T.F., Fischer, H., Hentschel, D., Reinfelder, H., and Schmitt, F. (1991) Physiological effects of fast oscillating magnetic field gradients, *J. Computer Assisted Tomography* **15**, 904-909.
9. Reinemann, D.J., Stetson, L.W., Reilly, J.P., Laughlin, N.K., McGuirk, S., and LeMire, S.D. (1966) Dairy cow sensitivity and aversion to short duration transient currents. Paper 963087, *American Society Agricultural Engineers International Meeting*, Phoenix, AZ.
10. Dalziel, C.F. and Mansfield, T.H. (1950) Effects of frequency on perception currents, *AIEE Trans.* **69**, 1162-1168.
11. Chatterjee, I., Wu, D., and Gandhi, O.P. (1986) Human body impedance and threshold currents for perception and pain for contact hazard analysis in the VLF-MF band, *IEEE Trans. Biomed. Eng.* **BME-33**, 486-494.
12. LaCourse, J.R., Miller, W.T., Bogt, M., and Selibowitz, S.M. (1985) Effect of high-frequency current on nerve and muscle tissue, *IEEE Trans. Biomed. Eng.* **BME-32**, 82-86.
13. Sunderland, S. (1978) *Nerves and Nerve Injuries*, Churchill Livingstone, New York.
14. Knighton, R.W. (1975a) An electrically evoked slow potential of the frog's retina, I: Properties of response, *J. Neurophysiol.* **38**, 185-197.
15. Knighton, R.W. (1975b) An electrically evoked slow potential of the frog's retina, II: Identification with PII component of electroretinogram, *J. Neurophysiol.* **38**, 198-209.
16. Katz, B. and Miledi, R. (1967) The study of synaptic transmission in the absence of nerve impulses, *J. Physiol. (London)* **192**, 407-436.
17. Adrian, D. (1977) Auditory and visual sensations stimulated by low frequency currents, *Radio Sci.* **12**, 243-250.
18. Barlow, H.B. Kohn, H.I., and Walsh, E.G. (1947a) Visual sensations aroused by magnetic fields, *Am. J. Physiol.* **148**, 372-375.
19. Barlow, H.B., Kohn, H.I., and Walsh, W.G. (1946b) The effect of dark adaptation and of light upon the electric threshold of the human eye, *Am. J. Physiol. 148*, 376-381.
20. Baumgart, E. (1951) Sur le seuil phosphène électrique. Quantité liminaire it pseudo-chronaxie. *Comptes Rendeu Soc. Biol.* (Nov. 24), 1654-1657.
21. Bergeron, J. Hart, M.R., Mallick, J.A., and String, L.H. (1995) Strength-duration curve for human electro- and magnetophosphenes, *Proc. Bioelectromagnetics Soc. Annual Meeting*, Boston.
22. Budinger, T.F., Cullander, C., and Bordow, R. (1984) Switched magnetic field thresholds for the induction of magnetophosphenes, *Proc. Annual Meet. Mag. Res. Med.*, New York, p.18.
23. Carstensen, E.L. (1985) Sensitivity of the human eye to power frequency electric fields, *IEEE Trans. Biomed. Eng.* **BME-32**, 561-565.

24. Lövsund, P., Öberg, P.A., Nilson, S.E., and Reuter, T. (1980a) Magnetophosphenes: A quantitative analysis of thresholds, *Med. Biol. Eng. Comput.* **18**, 326-334.

25. Lövsund, P., Öberg, P.A., and Nilson, S.E. (1980b) Magneto- and electrophosphenes: A comparative study, *Med. Biol. Eng. Comput.* **18**, 758-764.

26. Silny, J. (1986) The influence of threshold of the time-varying magnetic field in the human organism in J. H. Bernhardt (ed.), *Biological Effects of Static and Extremely Low Frequency Magnetic Fields*, MMV Medzin Verlag, Munchen, Germany.

27. Bawin, S.M., Sheppard, A.R., Mahoney, M.D., and Adey, W.R. (1984) Influences of sinusoidal electric fields on excitability in the rat hippocampal slice. *Brain Res.* **323**, 227-237.

28. Bawin, S.M., Sheppard, A.R., Mahoney, M.D., Abu-Assal, M., and Adey, W.R. (1986) Comparison between the effects of extracellular direct and sinusoidal currents on the excitability in hippocampal slices, *Brain Res.* **362**, 350-354

29. Sienkiewicz, Z.J., Haylock, R.G., and Saunders, R.D. (1998) Deficits in spatial learning after exposure of mice to a 50 Hz magnetic field, *Bioelectromagnetics* **19**, 79-84.

30. IEEE, (1992) *IEEE standard for Safety Levels with Respect to Human Exposure to Radio Frequency Electromagnetic Fields, 3 kHz to 300 GHz.* Document IEEE C95.1-1991, published by Institute of Electrical and Electronics Engineers, New York.

31. Schenk, J.F., Kumoulin, C.L., Redington, C.L., Kressel, R.W., Elliot, H.Y., and McDougall, I.L. (1992) Human exposure to 4.0 Tesla magnetic fields in a whole-body scanner, *Med. Phys.* **19**, 1089-1098.

32. Jehenson, P., Duboc, D., Lavergne, T., Guize, L., Guerin, F., Degeorges, M., and Syrota, A. (1988) Change in human cardiac rhythm induced by a 2-T static magnetic field, *Radiology* **166**, 227-230.

33. Bourland, J.D., Nyenhuis, J.A., Foster, K.S., and Geddes, L.A. (1997) Threshold and pain strength-duration curves for MRI gradient fields, Proc. Soc. Mag. Res. Med., 5th Ann. Meeting, Vancouver, Apr. 12-18, 1974.

34. FDA, (1992) FDA safety parameter action levels, in R. L. Bagin, R. L. Liburdy, and B. Persson (eds.), *Biological Effects and Safety Aspects of Nuclear Magnetic Resonance Imaging*, New York Academy of Sciences, New York, pp. 399-400.

35. FDA, (1995) *MRI guidance update. Draft statement*, R. A. Phillips (Chief), Computed Imaging Devices Branch, ODE/CDRH, Rockville, MD (Nov. 11, 1995).

36. IEC, (1995) *Medical Electrical Equipment -- Part 2: Particular Requirements for the Safety of Magnetic Resonance Equipment for Medical Diagnosis.* International Electrotechnical Commission Publication 601-2-33, Geneva, Switzerland.

DOSIMETRY OF RF INDUCED BODY TO GROUND CURRENT AND IMPLICATIONS ON SAFETY STANDARDS

S. TOFANI
Health Physics Department
Ivrea Hospital - Azienda USL 9
10015 - IVREA, ITALY

1. Introduction

The current flowing to the extremity, induced by exposure to radio frequency (RF) electromagnetic (EM) fields, is an important parameter in the assessment of RF hazards because of its relationship with EM energy deposition (Specific Absorption Rate or SAR). This energy deposition is greatest around the ankles where the current density is relatively large because of the narrow conductance path afforded by the small cross-sectional area of the wet tissue at the ankle. This paper reports on current distribution as a function of frequency, human height and grounding conditions, as well as on the corresponding SAR distribution. Thermal implications will be also reported as a base for exposure restriction on safety standards regulation.

2. Induced Current Magnitude

The magnitude of induced current flow generated by an ambient electric field has been studied for some time. Deno [1] published and validated a simple empirical formula based on a quasi-static analysis for calculating the magnitude of the current flow induced in a person from high voltage transmission lines at 60 Hz.

For a freestanding person in a uniform vertically polarized electric field, E, of angular frequency, ω, Deno proposed that the induced current, I, is:

$$I = j\omega\varepsilon_0 E S \tag{1}$$

where ε_0 is the permittivity of free space, S is the effective current collecting area of the body.

Equation 1 has been recast by Gandhi et al. [2] by formulating the variables into convenient units and noting that $\omega = 2\pi f$, $\varepsilon_0 = 8.85*10^{-12}$ Fm^{-1}, and making S = 1,936h^2

$$I = 0.108 \, h^2 f \, E \tag{2}$$

293

B.J. Klauenberg and D. Miklavcic (eds.), Radio Frequency Radiation Dosimetry, 293-299.
© *2000 Kluwer Academic Publishers. Printed in the Netherlands.*

where I is the magnitude of the ankle current (mA), h is the height of the subject (m), f is the frequency of the field (MHz), and E is the magnitude of the vertical component of the E-field (V/m).

Despite the quasi-static assumption, Equation 2 appears to hold well for frequencies close to 40 MHz, i.e., the resonance whole body frequency for a grounded adult. It should be noted however, that Gandhi et al. have factored S to be about 20% higher than the corresponding value used by Deno in order to obtain a better fit to their experimental data in the RF frequency range [2]. Also, it should be noted that equation 2 applies to subjects in bare feet.

For frequencies around resonance Gandhi et al. [3] postulate the following formula for a 1.75m tall man:

$$I = 11.0(f/f_r)\sin(\pi/2 * f/f_r) E \qquad (3)$$

where f_r is the resonant frequency of the body (MHz), corresponding to 32 MHz for a 1.75m tall man.

In formulating equation 3 Gandhi et al. [3] made use of an experimental observation (up to 50 MHz) that the distribution of current in a standing human being is fairly similar to the current distribution on a metallic monopole antenna in that the current distribution may be approximated by a fraction of a sine wave with the maximum value always at the base (i.e. the feet) when the frequency is less than or equal to f_r. At frequencies larger than f_r, as anticipated from the standing wave current variation on a monopole antenna, the maximum current would no longer be at the base, but rather at a point increasingly higher up the monopole. A comparison between theoretical, equation (3), and experimental data of Gandhi et al. for a 1.75m tall man is reported in Table 1. The parameters Fv reported in Table 1 is computed as the ratio between the induced current I and the corresponding vertical component of the electric field E (Fv = I/E). Data from a grounded semispherical model of the human body [4] agree fairly well with the Gandhi formula.

TABLE 1. Comparison of Experimental and Theoretical Values of Fv (I/E) for a grounded (bare-foot) human subjects exposed to a vertically polarized electric field between 25 and 50 MHz (data from [3])

Frequency MHz	Fv mA/(V/m) Experimental	Theoretical
25	-	8.0
27	8.7	9.2
30	10.1	-
35	11.4	11.7
40	12.6	-
45	12.0	12.3
50	10.3	10.9

Measured induced current passing through the feet of nine subjects exposed to vertically polarized electric fields for nearby antennas, transmitting at frequencies between 90 and 104 MHz has also been reported by Tofani et al. [5, 6].

The 9 subjects exposed to RF fields were modeled using the anatomically based model of the human body described in detail in several earlier publications [7-11]. Induced electric fields components and current densities were calculated using the FDTD method for 90-110 MHz plane wave exposure conditions [6].

Table 2 reports Fv parameter experimentally and theoretically evaluated for a vertically polarized incident field E = 1 V/m for subjects having two different heights (175 cm and 191 cm tall people respectively). The agreement between experimental and numerical values of Fv is excellent.

TABLE 2. Comparison of Experimental and Numerical Values of Fv for a grounded (bare-foot) human subjects of different height, exposed to a vertically polarized electric field between 90 to 110 MHz (data from [6])

Frequency MHz	Fv mA/(V/m) Experimental		Numerical	
	h= 1.75 m	1.91 m	1.75 m	1.91 m
90		5.42	4.38	5.11
95	4.46	-	4.10	4.90
100	-	4.87	3.97	4.64
104	3.90	4.45	3.85	4.35
105	3.45	-	3.84	4.28
110	-	-	3.78	3.88

For the tallest subject of height 191 cm the foot current is 20% higher than the 175 cm tall average person, showing that the foot current increases as (height)^2 also at higher frequency according to the equation 2 previously observed at frequencies lower than 50 MHz.

Wearing shoes does not substantially modify the above situation since measurements of foot currents in subjects wearing rubber and leather-soled shoes have shown a reduction of approximately 19.1% with rubber-soled shoes and 10.5% with leather-soled shoes respectively [6]. These results are in agreement with the trend of data reported in Gandhi et al. [3].

3. Energy Deposition

The current which flows through the feet results in high to local SARs in the ankles, which from numerical calculations has also been shown to give the highest local SAR in a human exposed to plane waves [6,8,11,12,13].

The reason for this can be found by the equation

$$SAR_{ankle} = \frac{j^2}{\sigma\rho} = \frac{I^2}{4A_e^2 \sigma_m \rho} \qquad (4)$$

where I is the average vertical current in the muscle of both ankles which can be assumed equal to foot current, Ae is the effective ankle cross sectional area through which the current flows, σ_m is the muscle conductivity and ρ is the tissue mass density. The effective cross sectional area A_e is given

$$A_e = \sum_i \frac{\sigma_i}{\sigma_m} A_i \qquad (5)$$

where σ, and Ai are tissue conductivity and areas.

Because of the small conductivity of bone and cartilage tissues, the effective area is about 10 cm^2 corresponding to about 1/5 of the physical area [3,12].

Following the above procedure Gandhi et al [2, 3], have first shown that SAR at ankle section of grounded free-standing humans being exposed to vertical polarized fields of 1 mW/cm^2 (61.4 V/m) would be as high as 243 W/kg, that using electromagnetic scaling concepts would became 534 W/kg for five, year old children.

This data were in agreement with the results of SARs calculated using the FDTD method in a 5628 cells anatomically based model of a human for plane-wave exposure from 20-100 MHz [8]. These SAR calculations are also in agreement with corresponding SAR experimentally measured in humans exposed to RF fields up to 50 MHz [14].

4. Thermal Implications

Since the human body is lossy, heat is generated from the electrical resistive losses in the body tissues. These losses are greatest around the ankles where the current flow is relatively large and the local resistance is high because of the narrow conductance path afforded by the small cross-sectional area of wet tissues at the ankle.

The thermal implications of high SAR's in the body extremities have been studied by Chen and Gandhi [15] and by Hoque and Gandhi [16].

Referring to ankle surface, the relation connecting temperature increment and SAR was experimentally obtained [15] as:

$$\Delta T = 0,0048 \times SAR \quad [°C/min] \tag{6}$$

Equilibrium rates of heat transfer were obtained within 10 minutes.

In a more defined FDTD thermal model where inhomogeneous volume-averaged tissue properties, blood-flow, metabolism, thermal conductivity and specific heat were used for assessing SAR with an impedance method [16], temperature distribution in human legs was obtained showing hot spots in the internal region of the leg section with a corresponding temperature increment of + 2 °C higher than the corresponding surface temperature increment.

Exposure conditions corresponding to SAR_{ankle} of 217 W/kg would result in an internal temperature elevation of 8,5 °C. Temperature distributions were experimentally validated with surface temperature measurements. The equilibrium rates of heat transfer were found within 12 minutes. This value was in agreement with the 10-15 minutes of equilibration rates of heat transfer measured in 8 volunteers whose hands and wrists were immersed in warm water [17].

Considering the above reported data a limit of 20 W/kg should be set for the extremities in order to avoid a local temperature increment of more that 1°C.The values of 1°C has been suggested as an upper limit of local deep body temperature increase that avoids detrimental health effects [18].

The 20 W/kg limit are also in agreement with the sensation of warmth in the hand or wrist regions experienced by human volunteers when current passed through their wrist in a grasping contact with an estimated SAR of the order of 20 W/kg [19].

Setting an SAR of 20 W/kg from equation 4 we can estimate a maximum foot current of 283 mA in the 90-110 frequency range (σ_m= 1 S/m) and 220 mA at 40 MHz (σ_m= 0.6 S/m).

For these reasons, the induced foot current (body to ground) limits adopted by the most recent Safety Standards [20,21,22] are 200 mA.

From Tables 1 and 2 we can estimate that in order not to exceed 200 mA the electric field should not be higher than 16 V/m at 40 MHz and than 40 V/m at 100 MHz.

5. Conclusions

For grounded subject that are standing in vertically polarized fields, ankle SAR can be intolerably high. Field-strength limits that would protect against local SARs in excess of 20 W kg^{-1} would have to be as low as 16 V m^{-1}. This would on the other hand be unnecessarily restrictive for exposure situations where either no good ground contact exists or for exposure in horizontally polarized fields or for partial body exposures. An alternate safety measure would be to introduce induced current limits for ankles and wrists.

6. Acknowledgements

The author wishes to express his gratitude to Emanuela Noascone for her assistance in the preparation of the manuscript.

7. References

1. Deno, D.W. (1977) Currents induced in the human body by high voltage transmission line electric field - measurement and calculation of distribution and dose, *IEEE Trans. Power App. Syst.* **PAS-96**, 1517-1527.
2. Gandhi, O.P., Chaterjee, I., Wu, D. & Gu. Y. (1985) Likelihood of high rates of energy deposition in the human legs at the ANSI recommended 3-30 MHz RF safety levels, *Proc. IEEE* **73**, 1145-1147.
3. Gandhi, O.P., Chen, J.Y. & Riazi, A. (1986) Currents induced in a human being for plane-wave exposure conditions 0-50 MHz and for RF sealers, *IEEE Trans.Biomed. Eng.* **BME-33**, 757-767.
4. Jokela, K., Puranen, L. and Gandhi, O.P. (1994) Radio frequency currents induced in the human body for medium-frequency/high-frequency broadcast antennas, *Health Physics* **66**, 237-244.
5. Tofani, S., d'Amore., G., Fiandino, G., Benedetto, A. and Gandhi, O.P. (1993) Evaluation of induced foot-currents in humans exposed to radio-telecommunication EM fields, *Proceedings of the 2nd International Scientific Meeting: Microwave in Medicine* 11-14 October 1993, Rome.
6. Tofani, S., d'Amore, G., Fiandino, G., Benedetto, A., Gandhi O.P. and Chen., J.Y. (1995) Induced foot-currents in humans exposed to VHF radio-frequency EM fields", *IEEE Transactions on Electromagnetic Compatibility*, **37**.
7. Sullivan, D.M., Borup, D.T. and Gandhi, O.P. (1987) Use of finite difference time-domain method in calculating EM absorption in human tissues, *IEEE Transactions on BME* **34**, 148-157.
8. Chen, J.Y. and Gandhi, O.P. (1989) RF currents induced in an anatomically-based model of a human for plane wave exposures (20-100 MHz), *Health Physics* **57**, 89-98.
9. Chen, J.Y. and Gandhi, O.P. (1989) Electromagnetic deposition in an anatomically based model of man for leakage fields of a parallel-plate dielectric heater, *IEEE Trans. Microwave Theory Tech* **37**, 174-180.
10. Chen, J.Y., Gandhi, O.P., Conover, D.L. (1991). SAR and induced current distributions for operator exposure to RF dielectric sealers, *IEEE Transactions on Electromagnetic Compatibility* **33**, 252-261.

11. Gandhi, O.P., Gu, Y., Chen, J.Y. and Bassen, H.I. (1992) Specific absorption rates and induced current distributions in an anatomically based human model for plane-wave exposures, *Health Physics* **63**, 281-290.

12. Dimbylow, P.J. (1991) Finite difference time-domain calculations of absorbed power in the ankle for 10-100 MHz plane wave exposure, *IEEE Transactions on BME* **38**, 423-428.

13. Gam, H. and Gabriel, C. (1995) Present knowledge about specific absorption rates inside a human body exposed to radiofrequency electromagnetic fields, *Health Physics* **68**, 147-156.

14. Hill, D.A. and Walsh, J.A. (1985) Radio-frequency current through the feet of a grounded human, *IEEE Transactions On Electromagnetic Compatibility* **EMC 27**, 18-22.

15. Chen, J.Y. and Gandhi, O.P. (1988) Thermal implications of high SAR's in the body extremities at the ANSI-recommended MF-VHF safety levels, *IEEE Transactions on Biomedical Engineering* **35**, 435-441.

16. Hoque, M. and Gandhi, O.P. (1988) Temperature distributions in the human leg for VLF-VHF exposures at the ANSI-recommended safety levels, *IEEE Transactions on Biomedical Engineering* **35**, 442-449.

17. Sienkiewicz, Z.J., O'Hagan, J.B., Muirhead, C.R. and Pearson, A.J. (1989) Relationship between local temperature and heat transfer through the hand and wrist,. *Bioelectromagnetics* 10 **77**.

18. World Health Organization (1993) Electromagnetic fields (300 Hz to 300 GHz), *Environmental Health Criteria* **137**, WHO Geneva, 176-177.

19. Chatterjee, I, Wu, D., and Gandhi, O.P. (1986) Human body impedance and threshold currents for perception and pain for contact hazard analysis in the VLF-MF band, *IEEE Transactions on Biomedical Engineering* **BME 33**, 486-494.

20. ANSI/IEEE C95.1. (1992) *American National Standard Safety Levels with Respect to Human Exposure to Radio Frequency Electromagnetic Fields (3 kHz - 300 GHz) IEEE*, Inc, Piscataway, NJ.

21. CENELEC. (1995) Human exposure to electromagnetic fields: High frequency (10 KHz to 300 GHz), *European Prestandard, ENV 50166-2*, European Committee for Electrotechnical Standardization.

22. ICNIRP - International Commission on Non-Ionizing Radiation Protection (1998) Guidelines for limiting exposure to time-varying electric, magnetic, and electromagnetic fields (1 Hz - 300 GHz), *Health Physics* **74**, 494-522.

DEVELOPMENT OF INDUCED AND CONTACT CURRENT (ICC) LIMITS IN THE HF AND VHF REGIONS

J. A. LEONOWICH
Institute for Environmental, Safety
and Occupational Health Risk Analysis
2402 E Street
Brooks AFB, TX 78235-5324

1. Introduction - The Need for Induced and Contact Current Limits

Radio frequency (RF) exposure limits have existed in one form or another since the early 1950's. These limits were initially given in terms of either electric field strength in volts per meter (V/m), magnetic field strength in amperes per meter (A/m), or equivalent plane wave power density in watts per square meter (W/m^2). These field parameters are relatively easy to measure or calculate using commercially available broadband instrumentation. However, from 100 kHz to 10 GHz, the basic RF dosimetric quantity is specific absorption rate (SAR) in watts per kilogram (W/kg). Hurt [1], has published an excellent overview of SAR based RF dosimetry. SAR is a difficult parameter to measure outside of the laboratory, and the purposes of field exposure limits are to insure that SAR limits will never be exceeded, thus preventing either whole body or localized heating in tissue. Another consequence of exposure to RF fields is that currents can be induced in the body, which can also cause localized heating, particularly in areas with a small cross-sectional area, such as the wrist or ankles. It is therefore necessary to limit the magnitude of these currents induced in the body. This paper will review the historical development of these induced and contact current (ICC) limits, their current status, and the need for future refinement and development of these limits, as well as techniques to measure them accurately.

2. The IEEE C95.1 Standard - (1991)

The most important RF safety standard in the United States is the IEEE C95.1 standard, which is a consensus standard published by the transnational IEEE Standards Coordinating Committee (SCC) 28 in 1991, and adopted by the American National Standards Institute (ANSI) in 1992 [2]. C95.1 has existed since it was chartered as an ANSI committee sponsored by IEEE and the U.S. Navy in 1959. The recommendations of C95.1 have continued to be refined and have become increasingly sophisticated since the first frequency independent limit of 10 mW/cm^2 was issued as a standard in 1966

B.J. Klauenberg and D. Miklavcic (eds.), Radio Frequency Radiation Dosimetry, 301-307.
© 2000 *Kluwer Academic Publishers. Printed in the Netherlands.*

and reaffirmed in 1974. Based on the availability of better dosimetric data, a frequency dependent standard was released by the C95.1 committee in 1982 [3]. Following the release of this document, Gandhi et al, [4,5] calculated that ankle currents could exceed the SAR limits specified by this standard, even if field parameters such as the E field limits were not exceeded. The IEEE SCC28 committee therefore recommended ICC limits in the revision of the 1982 standard released in 1991. These limits are shown in Table 1. The IEEE rationale was based on limiting SAR's in extremities below 20 W/kg, as averaged over 10 grams of tissue for controlled areas, which by theoretical calculations equates to an induced current of 100 mA. For uncontrolled areas, where individuals have no knowledge of their exposure, the IEEE reduced the SAR limits by a factor of 5, thus the current limits are reduced by the *square root* of 5. There was no good rationale for the 1-second averaging time at frequencies above 100 kHz, where SAR (i.e., tissue heating) is assumed to be the predominant dosimetric quantity of importance. Below 100 kHz, where electrostimulation of nerve tissue is of concern, a 1-second averaging time is appropriate. Another important point about the IEEE ICC limits for contact currents is that it is assumed that there is a full hand grasping contact with the conducting object. The IEEE ICC limits therefore provide protection against shock and burn only under conditions of direct contact and do not protect against momentary spark discharge phenomenon associated with making or breaking contact with conducting objects!

TABLE 1. IEEE ICC Limits Per Foot or Hand Contact – Averaged Over 1 Second (f in MHz)

Exposure Characteristics	Frequency Range	Maximum Contact Current (mA)
Controlled Area	Up to 3.0 kHz	Not covered
	3.0 kHz - 100 kHz	1000 f
	100 kHz - 100 MHz	100
Uncontrolled Areas	Up to 3.0 kHz	Not covered
	3.0 - 100 kHz	450 f
	100 kHz - 100 MHz	45

C95.1 is a living document that is always under active revision. After considerable review, Subcommittee 4 of SCC 28 proposed increasing the averaging time for ICC exposures above 100 kHz to 6 minutes, which is consistent with the averaging time used when SAR is the dosimetric quantity of primary importance. It was also proposed to place a ceiling value of 500 mA as the maximum value of the current allowed. Based on the ceiling value of 500 mA, an exposure to this level of current would be allowed for 14.4 seconds per six-minute averaging period. Assuming a rise of surface skin temperature of $2°$ C per minute, this equates to an incremental temperature increase of about $0.5°$ C. This suggests that even *higher* ceiling values of may be appropriate in the next revision of C95.1. Recent work by Olsen [6], measuring SAR in rhesus monkey ankles, also suggests that the 100 mA limit is overly conservative in humans. The proposed changes to the present C95.1 ICC limits shown in Table 2 were approved by the IEEE Standards Board in December 1998 and will be published as a supplement to the C95.1 standard in 1999. In addition to the change in averaging time, the supplement will include *de minimis* E-field conditions, below which ICC measurements will not be required.

TABLE 2. IEEE ICC Limits Per Foot or Hand Contact – Adopted December 1998 (f in MHz)

Exposure Characteristics	Frequency Range	Maximum Contact Current (mA)	Averaging Time
Controlled Area	Up to 3.0 kHz	Not covered	N/A
	3.0 kHz - 100 kHz	1000 f	1 second
	100 kHz - 100 MHz	100, with 500 mA ceiling	6 minutes
Uncontrolled Areas	Up to 3.0 kHz	Not covered	
	3.0 - 100 kHz	450 f	1 second
	100 kHz - 100 MHz	45, with 220 mA ceiling	6 minutes

3. The Guidelines of the International Commission on Non-Ionizing Radiation Protection (ICNIRP) – (1998)

The ICNIRP is an independent scientific organization, which investigates the hazards associated with non-ionizing radiation. ICNIRP released revised guidelines covering time-varying electromagnetic fields, including ICCs, in 1998 [7]. ICNIRP defines underlying dosimetric quantities such as SAR as so-called *basic restrictions*, and field limits as *reference levels*. The ICNIRP reference levels for contact currents are shown in Table 3.

TABLE 3. ICNIRP Reference Levels for Contact Currents from Conductive Objects (f in kHz)

Exposure Characteristics	Frequency Range	Maximum Contact Current (mA)
Occupational Exposure	up to 2.5 kHz	1.0
	2.5 - 100 kHz	0.4 f
	100 kHz - 110 MHz	40
General Public Exposure	up to 2.5 kHz	0.5
	2.5 - 100 kHz	0.2 f
	100 kHz - 110 MHz	20

There really is no analogous limit in IEEE C95.1 to these contact current limits. ICNIRP *believes* that these limits are protective against RF shock and burns, even though no voltage criteria are specified. IEEE C95.1 assumes grasping contact, which may not be protective against startle reaction, pain, or RF burns due to a spark discharge caused by making or breaking contact with a conducting object. The difference between the ICNIRP and IEEE contact current limits can partially be explained by the fact that ICNIRP assumes a single finger point contact, versus the IEEE grasping hand contact criteria. ICNIRP also does not specify an averaging time for contact currents, rather it is assumed that any interactions will occur essentially instantaneously.

The ICNIRP reference levels for current induced in any limb is show in Table 4. These are analogous to the IEEE induced current limits, but cover a different frequency range. IEEE C 95.1 induced current limits end at 100 MHz, which is approximately the middle of the FM radio band, which covers the frequencies from 88 to 108 MHz. By extending their reference levels to 110 MHz, the whole FM band is covered by ICNIRP. The ICNIRP reference levels also end in the first third of the HF band. The ICNIRP guidelines assume that SAR is the only basic restriction above 10 MHz, and that from 100 kHz to 10 MHz, both SAR and current density, J, in amperes per square meter (A/m^2) are potentially important. IEEE C95.1 assumes that the transition to a SAR based phenomenon occurs rapidly at frequencies greater than 100 kHz, and that current density is not an important parameter at MF and HF frequencies. The ICNIRP guidelines for induced currents are therefore more difficult to apply for field personnel than the IEEE limits. One other interesting difference between IEEE and ICNIRP is that there is no ceiling value for the limb current, so currents greater than 500 mA are allowed, albeit for a very short exposure duration per each 6-minute averaging time period. The ICNIRP reference levels for induced current are similar to the investigation levels for induced current found in the United Kingdom's National Radiological Protection Board's (NRPB) restrictions for time-varying electromagnetic fields. A recent paper by Chadwick [8, 9], of the NRPB, proposes a method for dealing with the transition between current density and SAR based phenomenon between 100 kHz and 10 MHz.

TABLE 4. ICNIRP Reference Levels for Current Induced in any Limb (averaged over 6 minutes)

Exposure Characteristics	Frequency Range	Maximum Current (mA)
Occupational Exposure	10 MHz - 110 MHz	100
General Public Exposure	10 MHz - 110 MHz	45

4. Protection Against RF Shock and Burns

Spark discharges, which can cause RF shock and burns, may occur when a person contacts a conducting object in a high electric field. Neither IEEE C95.1 nor the ICNIRP guidelines address protection against RF shock and burns explicitly. Reilly [10] has reviewed the complicated physics associated with the generation of spark discharges and arcs. It is clear from his analysis that the induction of arcing or spark discharges is a function of both open circuit voltage (V_{oc}), and amperage. The U.S. Navy has used a V_{oc} of 140 volts, as measured with a high impedance "burngun", as a predictor of situations where RF shock and burns may be encountered. Data taken at a joint U.S. Air Force/Navy ICC workshop held at the Naval Surface Weapons Center, Dahlgren, Virginia, as shown in Table 5, indicates that the 140 volt criteria is useful in predicting an RF shock/burn hazard.

TABLE 5. Burn Threshold Data Taken Around F-18 Aircraft At the Dahlgren Ground Plane

Location	Frequency (MHz)	E (V/m)	V_{oc} HP-410C	V_{oc} Hi-Z Gun	V_{oc} Lo-Z Gun	I_{wrist} Phantom	Arc ?
Wing Tip	6.4	50	165	160	192	392	Yes
Pitot Tube	6.4	25	170	165	130	260	Yes
Tow Bar	6.4	50	195	200	140	295	Yes
Pitot Tube	18.036	100	71	62	68	175	No
Pitot Tube	18.036	150	125	140	124	254	Yes
Tail	26.88	200	213	202	197	350	Yes

Adoption of voltage/current criteria for protection against RF shock and burns is planned for the next revision of IEEE C95.1. In order to facilitate this, a systematic collection of data between MF and VHF frequencies, where most RF shock and burns occur, is needed.

5. Measurement Issues

In principle, the measurement of induced currents is reasonably straightforward, but in practice can be quite difficult. Following the release of the IEEE C95.1 ICC criteria in 1991, several manufacturers have produced devices to measure ICCs. The type of measurement instrument to use is still open to choice - stand-on "bathroom scale" versus transformer (ankle) clamps to measure induced currents. Comparison of the two basic types of instruments was performed at an ICC workshop held at Brooks AFB, Texas in May 1996. The main conclusion of the workshop were as follows:

(1) Transformer/ankle probes were generally in good agreement, however, large variations were found in the stand-on probes between manufacturers and different units of the *same* manufacturer.

306

(2) Transformer/ankle probes show less variation with changes in grounding and are less susceptible to radio frequency interference (RFI).
(3) Stand-on probes underestimate true ankle current if not firmly in contact with ground, most likely due to the flow of displacement current around the platform probe.

Because of these variations, there remains an urgent need for rigidly standardized measurement protocols when making ICC measurements. In addition, induced current measurement of individuals standing in a uniform RF field show markedly different results, which is a complex function of body height, girth, footwear, and tissue composition. One solution to this problem is to use a standardized phantom to make these measurements. At least one manufacturer is marketing a human equivalent antenna (HEA), which has impedance similar to a human being. The appropriateness of using an HEA as a surrogate for human exposure is still open to study.

6. Conclusion

The development of ICC limits is maturing. Part of the differences between the IEEE standard and ICNIRP guideline can be explained by the fact that they are measuring different parameters. The following issues still need to be resolved:

(1) The change in averaging time from 1 second to 360 seconds at 100 kHz in the IEEE standard should be justified - although it appears that the switch from electrostimulation to a SAR based phenomenon is fairly steep between 100 - 150 kHz. This is a major difference between ICNIRP and IEEE.
(2) Criteria for contact currents should be better specified – are the limits based on touch or grasping contact? Are the limits based on perception, annoyance, or the avoidance of injury? Since the perception of current is fairly well know, what level of safety margin is required? Is a safety margin less than 10 acceptable?
(3) Any ICC measurement protocols should be as simple as possible - industrial hygienists are baffled by these measurements.
(4) Are human surrogates such as the HEA or some other device preferable to using humans with ICC measurement devices? Individuals standing in RF fields show wide variability in induced currents generated through their ankles, also complicated by the response of different types of measurement devices – this makes a good case for an HEA standard.
(5) *De Minimis* conditions should be clearly pointed out - use of E field graphs such as the ones recently approved by the IEEE should be encouraged.
(6) Voltage criteria for protection against RF shock and burns should be included in any future contact current limits.

The recent efforts of World Health Organization International EMF project to harmonize world-wide RF standards provides an excellent venue for many of these important issues to be resolved.

7. Acknowledgements

This paper was sponsored as part of an Intergovernmental Personnel Agreement between Armstrong Laboratory (now the Air Force Research Laboratory), Brooks AFB, Texas, and Pacific Northwest National Laboratory, Richland, Washington. The views expressed in this paper are those of the author and do not reflect the official policy or position of the U.S. Department of the Air Force, U.S. Department of Defense, or the Pacific Northwest National Laboratory, operated for the U.S. Department of Energy by Battelle Memorial Institute.

8. References

1. Hurt, W.D. (1997) Dosimetry of radiofrequency (RF) fields, in K. Hardy, M. Meltz, and R. Glickman, (eds.) *Non-Ionizing Radiation: An Overview of the Physics and Biology* Health Physics Society, 1997 Summer School, Medical Physics Publishing, Madison, Wisconsin.
2. IEEE Standards Coordinating Committee 28 (1991) *IEEE Standard for Safety Levels with Respect to Human Exposure to Radio Frequency Electromagnetic Fields, 3 kHz to 300 GHz*, IEEE Std C95.1-1991, Institute of Electrical and Electronics Engineers, Inc., New York, New York.
3. American National Standards Institute (1982) *Safety Levels With Respect to Human Exposure to Radio Frequency Electromagnetic Fields, 300 kHz to 300 GHz*, Institute of Electrical and Electronics Engineering, New York, New York.
4. .Gandhi, O.P., Chatterjee, I., Wu, D., and Gu, Y.G. (1985) Likelihood of high rates of energy deposition in the human leg at the ANSI recommended 3-30 MHz safety levels, *Proc. IEEE.* **73**, 1145-47.
5. Chen, J.Y., and Gandhi, O.P. (1988) Thermal implications of high SAR's in the body extremities at the ANSI recommended MF-VHF safety levels, *IEEE Trans. Biomed. Eng.* **33**, 435-441.
6. Olsen, R.G. (1999) SAR measurements in the rhesus monkey ankle: implications for humans, in B. Klauenberg and D. Miklavčič, (eds.), *Radio Frequency Radiation Dosimetry and Its Relationship to the Biological Effects of Electromagnetic Fields*, Kluwer Academic Publishers B.V., Dordrecht, The Netherlands, pp371-378.
7. International Commission on Non-Ionizing Radiation Protection (1998) Guidelines for limiting exposure to time-varying electric, magnetic, and electromagnetic fields (Up to 300 GHz), *Health Phys.* **74**, 494 – 522.
8. Chadwick, P. (1998) Induced current densities and SARs from non-uniform, near-field exposure to radiofrequency magnetic fields, in B. Klauenberg and D. Miklavčič (eds.), *Radio Frequency Radiation Dosimetry and Its Relationship to the Biological Effects of Electromagnetic Fields*, Kluwer Academic Publishers B.V., Dordrecht, The Netherlands, pp. 325-338.
9. Chadwick, P. (1998) *Occupational Exposure to Electromagnetic Fields: Practical Application of NRPB Guidance, NRPB-R301*, NRPB, Chilton, United Kingdom.
10. Reilly, J. P. (1998) *Applied Bioelectricity*, Springer-Verlag, New York, New York.

RADIOFREQUENCY MEASUREMENTS AND SOURCES

P. GAJŠEK
Non-ionizing Radiation Department,
Institute of Public Health of the Republic of Slovenia,
Ljubljana, SLOVENIA

1. Introduction

Except for light and heat which can be perceived by exposure to very high RF fields, electromagnetic radiation is not generally detectable by human beings. The presence of the radiation must be measured by instruments or approximated by theoretical calculations.

This paper focuses on electromagnetic measurement procedures and radiation sources in the radiofrequency (RF) spectrum. RF energy is used in a wide variety of ways in the modern society, and new uses are being increasingly introduced. Important applications of RF radiation include radio and TV broadcasting, telecommunications, radar, medical applications and various applications in industry. Therefore, the identification and survey of RF exposure has become an increasingly important aspect of radiation protection.

2. Preliminary Considerations

There are several steps necessary for the accurate assessment of RF exposure [1]:
- the source and exposure situation must be characterized so that the most appropriate measurement technique and instrumentation can be selected,
- the correct use of this instrumentation requires knowledge of the quantity being measured and the limitations of the instrument used,
- knowledge of relevant exposure standards is essential.

Prior to the commencement of a survey of potentially hazardous RF fields, it is important to obtain as much information as possible about the known characteristics of:
- RF source,
- the exposure situation.

B.J. Klauenberg and D. Miklavcic (eds.), Radio Frequency Radiation Dosimetry, 309-319.
© 2000 *Kluwer Academic Publishers. Printed in the Netherlands.*

Information about the exposure situation may include:
- geometry - height and distance from source.
- existence of any scattering objects. Scattering can enhance the E-fields by a factor of 2 and hence power density by a factor of 4 [2].

This information is required for the estimation of the expected field strengths and the selection of the most appropriate survey instrumentation. Information about the RF source may include:
- frequencies present, including harmonics,
- power transmitted,
- antenna characteristics (type, gain, beamwidth and scan rate),
- polarization (orientation of E-field),
- modulation characteristics (peak and average values),
- duty cycle, pulse width and pulse repetition frequency.

2.1. NEAR-FIELD VERSUS FAR-FIELD CRITERIA

Determining whether exposure takes place in the near- or far-field of the source is perhaps one of the most confusing aspects of this work and it is instructive to spend some time in a discussion of the near- versus far-field criteria. The frequency of an operation, the dimensions of the source and the distance of the exposure position from the source, determine whether or not exposure takes place in the near- or the far-field. It is convenient at this stage to distinguish between the field regions of small radiators (dimensions of one wavelength or less) and large radiators (dimensions larger than one wavelength):
- The field regions of small radiators ($D \ll \lambda$, where D is the largest dimension /in meters/ of the antenna, λ is the wavelength /in meters/) are the electrostatic, the induction and the radiation or far-field regions. In the first two of these regions, the E-fields vary with the inverse cube and inverse square of the distance from the source respectively, whereas in the far-field region the E-fields vary with the inverse of the distance from the source. The radiation field region is commonly taken to begin at a distance of $\lambda/2\pi$ meters from the source.
- The field regions of large radiators ($D \gg \lambda$) are the reactive near-field region, the radiating near-field region, and the far-field region. In the latter two regions the radiating field predominates, i.e. the E-field varies with the inverse of the distance from the source, but in the case of the radiating near-field region the angular distribution of the E-field varies with distance from the antenna. The commonly accepted distance to the boundary of the reactive near-field is $\lambda/2\pi$ meters whereas the far-field region is commonly taken to exist at distances greater than $2D^2/\lambda$ meters.

The power flux density, S, is expressed in watts per square meter (W/m^2) and can be derived from the following simple relationships, provided the waves are plane waves (far field):

$$S = E.H \qquad \qquad (1)$$
$$S = E^2/377 \qquad \qquad (2)$$
$$S = 377H^2 \qquad \qquad (3)$$

where E is equivalent electric field-strength in V/m and H is equivalent magnetic field-strength in A/m. The value of 377 is the characteristic wave impedance of free space and is expressed in ohms.

In the reactive near-field the equivalent plane wave power flux density can only be calculated from equations (2) and (3) and not from the product of the measured E- and H-fields as per equation (1).

3. Measuring Instrumentation

Measuring instruments consist of three basic parts: **a probe, a lead and a monitor**. The probe is usually composed of two main components: a pickup and a detector.

All instruments use two, of the most widespread types of detectors: a diode or a thermocouple.

Advantages of the **diode detectors** are their extreme sensitivity, fast response time, and protection from overload. They are disadvantaged however, due to their limited range caused by the diodes' non-linear response. A diode acts as a square law detector only at low signal levels, tending to a linear detector at high signal levels, with an intermediate characteristic at moderate signal levels. In association with an R-C circuit, a diode detector tends to act as a peak detector. For measuring a short high-pulsed microwave signal, the compensation of the characteristics of the diode detector must be made. It has been proven that knowledge of the reaction of components of the measuring device and diode detector to a pulse signal can be used in making correction curves for a compensation of the instrument's non-ideal response [3]. Where there are two or more (in-band) signals of similar amplitude, the instrument will tend to read higher than true. It is preferable to measure each signal separately, either by turning off all except one signal, or by making narrow band measurements.

The **thermal detector** has the advantage of accurate measurements in pulse fields and a linear response with the incident power density. The readout is accomplished by the RF heating, so the instrument has a slow response and is susceptible to drift with environmental temperature (lower ranges).

There are two kinds of measurement techniques - pick ups of RF fields used so far:
- electric field measurements; for the E-field the pick up is usually a monopole or short dipole.

- magnetic field measurements; for the H-field the pick up is usually some kind of electrically small induction coil or loop.

To permit meaningful measurements the following instrumentation characteristics are required or desirable:
- The probe must respond to only one parameter and not have spurious responses (for example, respond to the E-field and not to the H-field and vice versa).
- The probe must not produce significant scattering.
- The leads from the probe to the monitor must not disturb the field at the probe significantly nor couple energy from the field.
- The frequency response of the probe must cover the range of frequencies present.
- The probe should be responsive to all polarization components of the field. This may be accomplished either by inherent isotropic response or by physical rotation of the probe through three orthogonal directions. An isotropic probe is a probe capable of measuring a radiation incident from all directions.
- If used in the reactive near-field, the dimensions of the probe sensor should preferably be less than a $\lambda/4$ at the highest frequency present.
- The instrument should indicate a true root mean square (RMS) value of the measured field parameter.
- The response time of the instrument should be known. It is desirable for the response time of the order to be 1 second or less, so that intermittent fields are easily detected.
- Good overload protection, battery operation, portability and rugged construction are other desirable characteristics.

In the near field, the E- and H- field are normally not in phase or related by wave impedance. Therefore, the magnitude and the direction of each E and H must be measured independently. The surveyor should be aware of the field parameter (E or H) to which the instrument responds and that exposure standards generally stipulate limits corresponding to both field parameters. Equivalent plane wave power flux density as calculated by equations (2) and (3) is certainly a convenient unit but in the reactive near-field both field components must be measured and both calculated values of S must be within the allowable limit.

In the far field (for planewave), apart from polarization and direction of propagation, only the incident power density has to be defined. However, no instrument actually measures average power flux density directly and it must be realized that this quantity is neither measurable nor is it the most meaningful, particularly in the reactive near-field of antennas. The instruments showing power density are actually measuring magnitude of the E- or H- fields.

4. Measurement methods

If the information on the RF source and exposure situation is well defined, then the surveyor, after making estimates of the expected field strengths and selecting the appropriate instrument, may proceed with the survey, using a high-power probe to avoid inadvertent probe burnout and a high sensitivity scale to avoid possible over-exposure of the surveyor.

In the reactive near-field of radiators operating at frequencies less than 300 MHz it is necessary to conduct separate measurements of the E and H-fields. An electrically small (largest dimension < 0.25λ) probe sensor is required due to large gradients in field components and the spatial resolution being critical (large probes will yield spatially averaged values). The use of an isotropic probe is strongly recommended.

Both occupied and accessible positions should be surveyed. The operator of the equipment under test and the surveyor should be as far away as possible from the test area. All objects normally present, which may reflect or absorb energy, must be in position.

Non-uniform field distributions result from reflections from natural and man-made structures. Peaks in the field distribution are separate by at least one-half of a wavelength with the maximum levels of E and H-fields occurring in different locations. Temporal (time) variations also occur as a result of scanning antennas, scanning radiation beams and changes in frequency. It is therefore imperative that any survey include a sufficiently large sampling of data points to preclude omission of hazardous combinations of conditions [1]. The surveyor should take precautions against RF burns and shock, particularly around high power systems. With a careful measurement technique and an appropriate choice of instrumentation, overall measurement uncertainties of ± 3 dB can be expected in practice.

Some other important points to note:
- it must be decided whether measurements should be carried out on frequency selective basis or on a broad-band basis
- the reference levels to be adopted will determine whether peak or RMS values of the filed-strength should be measured
- when determining RMS values, the integration period should be adjusted accordingly (a averaging time interval of 6 min is normally used)

5. Exposure in Living and Natural Environment

In 1996, the research project on human exposure to electromagnetic fields in Slovenia was completed. Its results used in this paper represent the basis for a general evaluation of public exposure in the living environment. We tested over 100 different power sources of RF output power from 100 W to several MW in the frequency range from several kHz to several 10 GHz [4].

314

There are only rare cases of general public exposure, caused by power RF transmitters, being greater than 0,1 W/m^2, because the public access to the immediate vicinity of the base of RF power sources is normally inaccessible and restricted by fence. Furthermore, the estimation shows that over 95% of general public is exposed below 50 $\mu W/m^2$.

5.1. LOW AND MEDIUM FREQUENCY RADIO TRANSMITTERS

Within 30 m of a 0.1 wavelength antenna transmitting 400 kW at 145 kHz, electric fields up to 600 V/m and magnetic fields up to 1 A/m could be measured. Measurements made so far at the AM medium-wave radio station (918 kHz, 300 kW) show that fields would be below 80 V/m at the main entrance (50 m), less than 25 V/m at distances greater than 100 m, and less than 5 V/m at distances greater than 300 m.

5.2. FM RADIO TRANSMITTERS

Usually such transmitters have a smaller coverage area than those at lower frequencies and therefore lower output powers. The antennas are mounted on towers at heights of several 10 m. Exposure in the main beam of the antennas can only occur at large distances. Measurements of FM stations (100 kW ERP) show that fields would be below 9 V/m at the main entrance (50 m), and less than 1 V/m at distances greater than 100 m.

5.3. TELEVISION TRANSMITTERS

The TV coverage is achieved by using a small number of high power transmitters mounted on very high towers, together with lower power transmitters which give local coverage in some areas. Due to the radiation pattern of the antenna, exposure to the main beam can occur only at distances around a kilometer or so from the tower. Measurements give the electric field strength at 1 km from a 400 kW ERP transmitter as 5 V/m. The grating lobe can give rise to a similar field exposure within only a several hundred meters of the antenna.

5.4. BASE STATIONS

Base stations are low power radio relay stations. Typical powers used in practice are not more than 40 W for GSM and 20 W for the analogue TACS. Base station antennas are either mounted on towers with typical heights in the range 15-50 m or on the roofs or sides of tall buildings. Due to the radiation pattern of the sector antenna, public exposure to main beams should not be possible at radial distances of less than 60 m [5]. Side lobes have power levels that are at least 20 dB below that of the main beam and therefore should not be considered.

Some preliminary results show that in worst case conditions, a predicted total electric field strength for GSM base station to which public can be exposed, is 3 V/m [5].

5.5. CITIZENS BAND (CB) RADIO

These devices usually operate within the frequency range from 27 to 450 MHz. Antennas are often mounted upon the bumpers of cars, on poles outside houses or on mobile handsets which are held close to the heads of users. Typical available RF powers are 4 W.

At close distances, the fields depend upon the precise length and structure of the antenna. Stronger electric fields are to be expected in the vicinity of the shorter antennas, up to 1000 V/m have been measured 2 cm from low power mobile antennas. Although such field strengths are exceeding guidelines, relevance of such localized reactive fields for radiation protection is limited.

A further band exists from 934 to 935 MHz where devices operate with output power of 8 W. In general, the antenna is mounted several meters above ground. Under these conditions, public exposure will not usually be possible at distances of less than 3 m, where the electric and magnetic field strengths will be less than 10 V/m and 0.05 A/m respectively. Public exposure to vehicle mounted transmitters may occur at distances of only a few centimeters, although this is only likely to occur for very short periods of time. Equipment with an integral antenna, such as a CB handset, is limited to an effective radiated power of 3 W. The exposure situation with a device such as this is similar to that described for a handset.

5.6. MICROWAVE LINKS

The frequencies used for microwave links are usually in the range 5 to 40 GHz and power levels range from less than 1 to 8 W. Highly directive dish antennas are used. However, they also have many side lobes which may be more significant in relation to public exposure, but the power is usually at least 20 dB below that in the main beam.

The antennas are mounted on towers or the tops of buildings with heights of at least 20 m, thus a typical main beam cannot intercept the ground at distances of less than 230 m. With a radiated power of 8 W and a gain of 50 dB, the power density would be near 2 W/m^2. Assuming a gain of 10 dB for a side lobe travelling directly downward, the power density at 20 m from an 8 W antenna will be below 0.08 W/m^2, under far-field conditions.

5.7. RADAR

These devices operate within a frequency range of 500 MHz up to 15 GHz, although there are some systems operating up to 100 GHz. The signals produced by radar are pulsed with very short duty cycles that give average powers relevant for radiation protection, which are several orders of magnitude less than the peak powers. The antennas used for radars produce main beams only a few degrees wide. In addition, many of the systems feature antennas of which direction is continuously varied by either rotating them in azimuth or varying their elevation by a nodding motion. Typically this rotation or nodding will reduce mean power by a factor of 100 and thus reduce RMS fields by a factor of 10. These considerations further reduce the likelihood of excessive exposure.

Air traffic control radars are scanning devices and therefore produce relatively low mean power densities in any direction. Measurements made in the vicinity of an air traffic radar operating at 2.88 GHz with a peak output power of 70 kW gave peak power densities with the rotating antenna of 2 W/m^2 at 15 m and 0.9 W/m^2 at 75 m. At the publicly accessible site (1000 m) there were power densities below 0.03 W/m^2.

Measurements made in the vicinity of a **meteorological radar** operating at 5.6 GHz with a peak output power of 250 kW gave peak power densities with the rotating antenna of 375 W/m^2 at 34 m; and 0.04 W/m^2 at 75 m.

Survey results of the **military radar** (3 GHz, peak output power 3 MW) indicate the peak power densities with the rotating antenna of 60 W/m^2 at 250 m.

Generally, the powers of **marine radars** are far lower than other radar systems with peak powers of up to 30 kW (mean power from 1 to 25 W). Under normal operating conditions with the antenna rotating, the average power density within a meter of the radar system is less than 10 W/m^2. The **traffic radars** have very low powers (0.5-100 mW) with operating frequency between 9-35 GHz. Measurements made so far show that power density would be below 250 mW/m^2 at 3 m, and less than 10 mW/m^2 at distances greater than 10 m.

5.8. MICROWAVE OVENS

A survey of microwave ovens in domestic use [6] showed that all ovens emitted less than 10 W/m^2 (all ovens meet the requirements easily) at 5 cm, and a statistical analysis indicated that 50% of ovens emit less than 1 W/m^2.

5.9. HANDSETS

Handsets are small compact transceivers with typical RMS powers of 0.6 W for TACS 0.44 W for GSM and 0.22 W for DCS. Higher power classes are possible, but unlikely to be used for handsets due to the restrictions imposed by the battery power supplies. Typically, the distance from the antenna to the head is only about 2 cm or less. Therefore, the user is in the near-field of the source and simple field measurements are not appropriate to assess exposure. Numerical calculations based upon coupling from handsets to an anatomically realistic numerical phantom of the head, have shown that during a normal operation, a radiated power of 1 W gives rise to a maximum Specific Absorption Rate (SAR) of 2.1 W/kg at 900 MHz, and 3.0 W/kg at 1800 MHz averaged over any 10 g of tissue [7]. Measurements show similar energy absorptions in a homogeneous head phantom when exposed to common handsets under normal operation conditions.

6. Occupational Exposure

Levels of occupational exposure vary considerably, and are strongly dependent upon the particular application. It covers whole radiofrequency spectrum from several kHz up to 300 GHz. In many cases, occupational exposure occurs in the near field of a source, and exposure assessment is therefore difficult.

6.1 INDUCTION HEATERS

These devices operate within one of three frequency ranges; 50 to 450 Hz, 1 to 10 kHz, and 200 kHz to 2 MHz with output powers between 6 kW and 1 MW. Survey results indicate that the exposure level increases as the frequency decreases and for those devices operating above 200 kHz operators can be exposed to H-fields in excess of current limits [8].

6.2. DIELECTRIC HEATING

These devices generally operate within the frequency range of 11 to 50 MHz and have available RF powers between 0.5 and 100 kW. The relatively high output power and the use of unshielded electrodes can produce relatively high fields around RF heaters. Surveys in Slovenia have indicated that 75% of the devices measured, exposed the operator to E fields greater than 250 V/m, and 35% to H fields greater than 0.7 A/m [9]. Similar results were obtained by [10]. The stray fields are localized in the immediate vicinity of the sealers, so that exposure of the body is highly inhomogeneous.

Hence, workers near some of these devices absorb RF energy at rates above the recommended limits. To reduce stray fields to acceptable levels, it is necessary to shield the electrode system, divert the RF energy, or reduce exposure by administrative measures [11].

6.3. COMMUNICATION

While most communication and radar workers are exposed to only relatively low-intensity fields, some can be exposed to high levels of RF. Surface fields on AM towers can reach values in excess of 1000 V/m. Commonly such towers are climbed for maintenance purposes while they are energized by the transmitter. The EM fields that may expose the worker in such situations are generally not well defined. Caution is warranted for such situations. The high voltages that exist on the towers may cause severe RF burns. At the base of FM towers and within distances of typically 30m, field strength may reach values of over 61 V/m. Workers climbing FM radio or TV broadcast towers may be exposed to electric fields up to 1000 V/m and magnetic fields up to 5 A/m [12].

6.4 RADAR

Radar systems produce strong EM fields on the axis of the antenna. However, in most systems, average field strengths are reduced typically by a factor of 100 to 1000, because of antenna rotation. With stationary antennas, which represent the worst case, peak power flux densities of 10 MW/m^2 may occur on the antenna axis up to a few meters from the source. In areas surrounding military radars, workers can be exposed to power flux densities of up to tens of kW/m^2 but normally those encountered are below 0.4 W/m^2 [13].

6.5. SHORT-WAVE DIATHERMY UNITS

These devices generally operate at 27.12 MHz with available RF powers up to 400 W. Only a part of a patient's body is exposed to RF energy and exposure duration is limited (10-20 minutes). However, exposure intensity is high and sufficient to cause a sustained increase in tissue temperature. Exposures to operators of short-wave diathermy devices may exceed 60 V/m and/or 0.16 A/m for operators standing in their normal positions (in front of the diathermy console) for some treatment regimes [14]. Stronger fields are encountered close to the electrodes and cables. In the "worst case", the exposure limits for the staff may be exceeded at distances lower than 1.5 - 2 m [15].

6.6. MICROWAVE DIATHERMY UNITS

These devices operate at one of three frequencies; 433 MHz, 915 Hz and 2450 MHz with available RF powers up to 400 W. Survey results indicate that stray or reflected radiation can expose the operator to levels in excess of current limits within 2 m of the patient [16].

6.7. VISUAL DISPLAY UNITS (VDU)

VDUs emit non-ionizing radiations such as visible light, together with very low levels of ultraviolet radiation. VDUs are not a source of X-rays nor of microwave radiation. In addition, low levels of radiofrequency, very low and extremely low electric and magnetic fields are normally found around VDUs. Depending on, for example, the humidity, electrostatic fields are also found.

The EMF, emitted by VDUs, are far below the current exposure limits [13].

7. Conclusion

In general, environmental RF fields routinely encountered by the public fall well below the recommended safety limits. The measurement study [4] has shown ambient field strength levels in residential areas are hundreds or thousands of times below present exposure recommendations. Hand held RF devices can create higher ambient fields in the immediate vicinity for short time periods during usage, but time averaged exposure comply with safety guidelines.

There have been some cases of excessive RF exposures from various emitters. These situations have usually involved occupational exposure where personnel routinely perform maintenance and other activities near broadcast stations, radars, or other high-powered transmitting equipment [17]. Protective actions must be taken and work procedures established to minimize such exposure.

8. References

1. EBU (1995) European Broadcast Union. *Radiofrequency Radiation Hazards - Exposure Limits and Their Implications for Broadcasters*, Tech. 3278-E
2. ANSI (1973) *American National Standards Institute. Techniques and Instrumentation for Measurement of Potentially Hazardous Electromagnetic Radiation at Microwave Frequencies*, IEEE, NY 10017
3. Šimunič, D. and Keller, H. (1998) A novel measurement method of a short high-pulsed electric field, BEMS Twentieth Annual Meeting, Florida, *Abstract book*, 261.
4. CRP (1996) General evaluation of human exposure to electromagnetic fields in living and natural environment in Slovenia (in Slovene language), Ministry of Environment of the Republic of Slovenia, Contract Nr. V2-6924-1538-96.
5. Petersen, R.C. and Testagrossa, P.A. (1992) Radiofrequency electromagnetic fields associated with cellular radio cell site antennas, *Bioelectromagnetics* 13, 527-542.
6. Matthes, R. (1992) Radiation emission from microwave ovens, *J. Radiol. Prot.* 12, 176-172.
7. Dimbylow , P.J. and Mann, S.M. (1994) SAR calculations in an anatomically realistic model of the head for mobile communication transceivers at 900 MHz and 1,88 GHz, *Phys. Med. Biol.* 39, 1537-1553.
8. Stuchly, M.A. and Lecuyer, D.L. (1985) Induction heating and operator exposure to EMF, *Health Physics*, 49, 693-700.
9. Gajšek, P. and Gajšek, J. (1992) Ten years of control on electromagnetic non-ionizing radiation in Republic of Slovenia (in Slovene language), *Electrotechnical Review*, 59, 219-222.
10. Stuchly, M. A.(1980) Radiation survey of dielectric heaters in Canada, *J. Microwave Power*, 15, 113-121.
11. Eriksonn, A. and Mild, K. H. (1985) RF electromagnetic leakage fields from plastic welding machines: Measurements and reducing measures, *Journal of Microwave Power*, 20, 95-107.
12. Mild, K. H. and Lovstrand, K. G. (1990) Environmental and professional encountered electromagnetic fields. in O. P. Gandhi (ed.) Biological Effects and Medical Applications of Electromagnetic Fields, *Prentice Hall, Inc.*, Engelwood Cliffs, New Jersey.
13. OEFZS (1988) Austria Forschungzentrum Siebersdorf Schutz vor Nichtionizierende Sthralung - Hochfrequenz und Mikrowellensthralung im Frequenzbereich 10 kHz-3000 GHz, *Report* 4436.
14. Gajšek, P. (1996) Review of occupational exposure in physiotherapy and protective measures with special fabric, BEMS Eighteenth Annual Meeting, Victoria, *Abstract book*: 242.
15. Stuchly, M. A., Repacholi, M. H., Lecuyer, D. W., and Mann, R. D. (1982) Exposure to the operator and patient during shortwave diathermy treatments, *Health Physics*, 42, 341-366.
16. Delpizzo, V. and Joyner, K. H. (1987) On the safe use of MW and shortwave diathermy units, *Australian J. Physiotherapy* 33, 152-162.
17. Gajšek, P. and Miklavčič, D. (1999) Biological Effects of Non-Ionizing Radiation, (in Slovene language), *Handbook*, Faculty of Electrical Engineering, University of Ljubljana.

INDUCED CURRENT DENSITIES AND SARS FROM NON-UNIFORM, NEAR-FIELD EXPOSURE TO RADIOFREQUENCY MAGNETIC FIELDS

P. CHADWICK
National Radiological Protection Board
Chilton
Didcot
Oxfordshire OX11 0RQ
UK

1. Introduction

Underlying most national and international standards on exposure to radio frequency magnetic fields are basic restrictions on induced current density and specific energy absorption rate (SAR) in the body. Derived levels for magnetic field strength are given, at or below which the basic restrictions will not be exceeded.

In the dosimetric models used to determine the investigation levels, it is usually assumed that exposure is uniform over the body and occurs under plane wave conditions. In these circumstances, it is the electric field component of the electromagnetic wave which couples maximally with the body. When exposure to a radiofrequency magnetic field occurs in the inductive near field, induced current density and SAR will be smaller than when exposure occurs under plane-wave conditions. In this situation, it may be appropriate to determine compliance with the basic restrictions on induced current density and SAR directly rather than compare measured magnetic field strengths with derived levels.

This paper describes a simple analytical technique for calculating induced current densities and SARs from magnetic fields under inductive near-field conditions and illustrates how this approach has been integrated with existing UK exposure guidelines.

2. Dosimetric Models for Magnetic Fields

2.1. LOW FREQUENCY MAGNETIC FIELDS

Guidelines which specify a dosimetric model for the derivation of the magnetic field reference levels, such as those of NRPB [1], generally assume a closed circular current path orthogonal to the incident magnetic flux density:

B.J. Klauenberg and D. Miklavcic (eds.), Radio Frequency Radiation Dosimetry, 321-334.

322

B: magnetic flux density normal to the body
J: induced current density

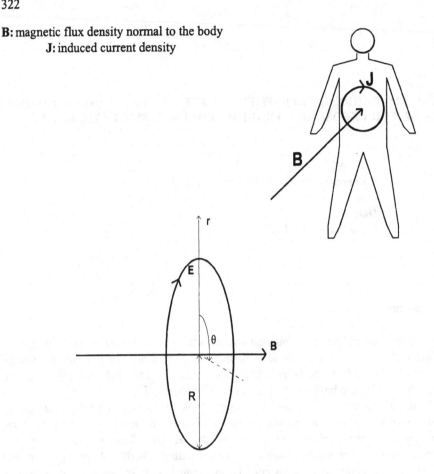

The integral of the electric field strength, **E**, along a closed circular current path normal to magnetic flux density **B** incident on the body will be given by

$$\oint \mathbf{E}.\mathbf{dl} = -\int_{r=0}^{R} \int_{\theta=0}^{2\pi} \left(-\frac{\partial |\mathbf{B}|}{\partial t} \right) r \, dr \, d\theta \qquad (1)$$

where r is the radial displacement vector and θ the circumferential displacement vector. R is the radius of the current path. If circumferential symmetry is assumed then the magnitude of the induced current density, **J**, along the current path is given by

$$|\mathbf{J}| = \frac{\sigma}{R} \frac{\partial}{\partial t} \left[\int_0^R (r\,|\mathbf{B}|)\,dr \right] \qquad (2)$$

where σ is the electrical conductivity of tissue.

For a single-frequency source

$$\left| \frac{\partial \mathbf{B}}{\partial t} \right| = \omega\,|\mathbf{B}| \qquad (3)$$

where ω is the angular frequency, and Equation 2 can be written as

$$|\mathbf{J}| = \frac{\sigma\omega}{R} \left[\int_0^R (r\,|\mathbf{B}|)\,dr \right] \qquad (4)$$

When the magnetic flux density is uniform across the body, Equation 4 becomes

$$|\mathbf{J}| = \frac{\sigma\,\omega\,R\,|\mathbf{B}|}{2} \qquad (5)$$

Equation 5 allows the derivation of an investigation level (in the case of NRPB guidelines) on magnetic flux density from consideration of the basic restriction on induced current density in the body.

2.2. RADIOFREQUENCY MAGNETIC FIELDS

At frequencies above 10 MHz, the magnetic field is generally considered in its interaction with the body as if it were a component of a plane electromagnetic wave. The dosimetric quantity is SAR. When non-uniform exposure occurs in the inductive near-field of a source, this approach is likely to overestimate SAR and it may be more appropriate in some circumstances to assess compliance with the basic restriction using a dosimetric model based on the interaction of the magnetic field with the human body. The model used in the derivation of the NRPB investigation levels below 100 kHz is

324

described above: in this paper, the approach has been extended to frequencies up to 50 MHz.

At frequencies between 100 kHz and 10 MHz, guidelines such as those of NRPB and ICNIRP[2] have basic restrictions on both induced current density and SAR; the eddy-current dosimetric model can be used to show compliance with the induced current density restrictions under conditions of non-plane-wave exposure. The analysis presented is based on a conduction path of radius 20 cm, appropriate for an adult oriented maximally to the incident field. It is likely to be conservative in that orientation may be less than maximal and conduction path radius may be less than 20 cm.

For uniform exposure of the trunk to an RF magnetic field, the induced current density at a given radius will be given by equation 5. Strictly, consideration should be given to permittivity as well as conductivity at radio frequencies and the quantity σ in equation 5 should be replaced by $\sigma + I\varepsilon\omega$, where ε is the permittivity of tissue at angular frequency ω and i is the square root of -1. In practice, this will make a negligible difference to the amplitude of the induced current density over the frequency range of interest.

Assuming an rms value for induced current density, SAR can be calculated from the relationship

$$SAR = \frac{|J|^2}{\sigma\rho} \qquad (6)$$

where ρ is the density of tissue. The average SAR over the exposure volume can be calculated from the SAR distribution across the trunk. From Equation 4, induced current density will be proportional to radius and from Equation 6, localised SAR will be proportional to the square of radius.

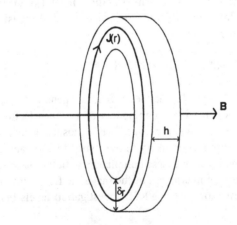

The annulus of tissue at radius r has volume $2\pi r\delta rh$ and mass $2\pi r\delta rh\rho$ where δr is the radial thickness of the annulus and h is its depth. The total power absorbed in the annulus will be its mass multiplied by the SAR at radius r, $SAR|_r$. It is assumed, pessimistically, that the SAR over the depth of the body, h, will be constant.

The total power absorbed in the disc of radius R and thickness h then is given by

$$\text{total power} = 2\pi h\rho \int_0^R (r\,SAR|_r)\,dr \qquad (7)$$

The total mass of the disc is $\pi R^2 h\rho$. The average SAR over the disc is given by the power absorbed in it divided by its mass:

$$\overline{SAR} = \frac{2}{R^2} \int_0^R (r\,SAR|_r)\,dr \qquad (8)$$

$SAR|_r$ is proportional to r^2:

$$SAR|_r = SAR|_R \frac{r^2}{R^2} \qquad (9)$$

Equations 8 and 9 can be combined to give

$$\overline{SAR} = \frac{2}{R^4} SAR|_R \int_0^R r^3\,dr \qquad (10)$$

Or

$$\overline{SAR} = \frac{SAR|_R}{2} \qquad (11)$$

$$\frac{\sigma\omega^2 R^2 |B|^2}{8\rho} < 0.4 \text{ W kg}^{-1} \qquad (12)$$

The NRPB basic restriction on whole-body SAR is 0.4 W kg^{-1}, as is the ICNIRP basic restriction for occupational exposure. If the mean SAR over the volume of tissue exposed in the trunk is below 0.4 W kg^{-1} then the whole-body average SAR will also be below the basic restriction. Equations 6 and 11 can be combined with equation 5 to derive the condition for compliance with this restriction:

Equation 12 can be rearranged with the numerical values used in this analysis substituted in to give the magnetic field strength H at which uniform exposure of the upper body will result in a whole-body average SAR below the basic restriction:

$$ H = \frac{51}{f} \; A \; m^{-1} \hspace{2cm} (13) $$

where f is the frequency in MHz. Equation 13 is also valid when the ICNIRP basic restriction for occupational exposure is considered. For the ICNIRP restriction on public exposure, a value of 23/f A m^{-1} rather than 51/f A m^{-1} is appropriate.

3. Integration with Existing UK Exposure Guidelines

3.1. EXISTING STRUCTURE

NRPB guidance for exposure to magnetic fields has basic restrictions on induced current density and SAR. Below 100 kHz, there is a basic restriction on induced current density only; exposures cannot be time-averaged for comparison with the guidelines. Above 10 MHz there is a basic restriction on SAR only and exposures can be time-averaged for comparison with the guidelines. Between 100 kHz and 10 MHz there are basic restrictions on induced current density and on SAR; exposures can be time-averaged if the basic restrictions on induced current density are not exceeded.

At frequencies between 1 kHz and 535 kHz, the NRPB investigation level for magnetic field strength is 64 A m^{-1}. This investigation level is based on the circulating eddy-current dosimetric model described above. At frequencies above 10 MHz, the investigation level is based on SAR considerations and the dosimetric model used for exposures assumes uniform exposure to a plane-wave electromagnetic field. Above 535 kHz the investigation level is given by the expression 18/f^2 where f is the frequency in MHz. At 10.6 MHz, this frequency-dependent investigation level intercepts the 0.16 A m^{-1} investigation level.

3.2. THE CEILING LEVEL

The analysis presented in this paper provides a framework for time-averaging exposure to magnetic fields at frequencies between 100 kHz and 10 MHz. Exposures can be averaged over time, subject to a "ceiling level" of 25.3 A m^{-1}. The ceiling level ensures compliance with the basic restriction on induced current density and corresponds to the lower frequency investigation level of 64 A m^{-1} after consideration has been given to the higher electrical conductivity of tissue at radiofrequencies.

- At frequencies below 10 MHz, the basic restriction on induced current density will be met if the magnetic field strength over the trunk does not exceed 25.3 A m^{-1}. This ceiling level cannot be time-averaged.

- Above 100 kHz there is a basic restriction on whole-body average SAR which will be met if the magnetic field strength over the trunk does not exceed 51/f A m^{-1}, where f is the frequency in MHz. This can be time-averaged on a power density or field strength squared basis.

- The magnetic field strength criteria for compliance with the two basic restrictions are equal at a frequency of 2 MHz.

The ceiling level of 25.3 A m^{-1} is equal to the existing investigation level at 847 kHz and equal to the SAR-driven magnetic field strength restriction at 2 MHz. Above 2 MHz, the SAR-driven magnetic field strength falls with frequency to 50 MHz while the current density-driven magnetic field strength is constant to 10 MHz, above which there is no basic restriction on induced current density. The frequency-dependencies of the investigation level and the ceiling level are shown in Figure 1.

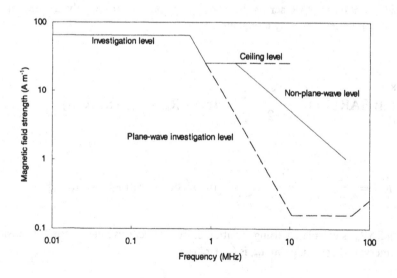

Figure 1. NRPB Investigation levels and time-averaging ceiling level for magnetic fields

328

4. Non-Uniform Magnetic Fields

When the magnetic field across the body is non-uniform, a solution must be found to equation 4. The simplest approach is to use a numerical integration technique and one of the simplest of these is the trapezoidal rule.

Application of the trapezoidal rule requires a series of equally spaced measurements of magnetic flux density to be made along a radius, such that the distance between each measurement point is δr. The induced current density over the Nth radial interval is given by

$$|\mathbf{J}|_N = \frac{2\sigma\omega}{r_N + r_{N+1}}\, \delta r \sum_{n=0}^{N}\left[\frac{r_n|\mathbf{B}|_n}{2} + \frac{r_{n+1}|\mathbf{B}|_{n+1}}{2}\right] \tag{14}$$

$$|\mathbf{J}|_N = \frac{2\sigma\omega}{2N+1}\, \delta r \sum_{n=0}^{N}\left[\, n|\mathbf{B}|_n + (n+1)|\mathbf{B}|_{n+1}\right] \tag{15}$$

Since $r_n = n\delta r$,

The current densities $|\mathbf{J}|_N$ are calculated for the interval r_n to r_{n+1} and for the purposes of further calculation, they are assigned to the midpoint of the interval, radius $(r_n + r_{n+1})/2$. Rather than using equation 11, calculation of mean SAR now must take into account the radial variation in SAR across the body. Equation 8 can be solved using numerical integration in the same way as equation 4:

$$\int_0^R (r\, SAR|_r)\, dr = \frac{\delta r}{2} \sum_{m=0}^{M} [r_m SAR_m + r_{(m+1)} SAR_{m+1}] \tag{16}$$

Or

$$\overline{SAR} = \frac{2}{R^2}\, \frac{(\delta r)^2}{2} \sum_{m=0}^{M} [m\, SAR_m + (m+1)\, SAR_{m+1}] \tag{17}$$

Each SAR_m is calculated, using equation 6, from the corresponding current density over the interval whose midpoint, r_m, is $(r_n + r_{n+1})/2$.

The solution of equations 15 and 17 can be carried out easily on a personal computer spreadsheet program.

5. Upper Frequency Limit of Analysis

The analysis presented in this Appendix assumes a circular current path and circumferential symmetry. With these assumptions, it should be valid as long as wavelength, λ, of the induced current is not comparable with the length of the current path. This criterion can be expressed as

$$\frac{\lambda}{4} > 2\pi R \qquad (18)$$

The wavelength in tissue will be less than the wavelength of electromagnetic radiation in free space, λ_0. The relationship between the two quantities is

where ε_r is the relative permittivity of tissue and $\tan\delta$ is the loss tangent, given by

$$\left(\frac{\lambda_0}{\lambda}\right)^2 = \frac{\varepsilon_r}{2}\left[\sqrt{1 + \tan^2\delta} + 1\right] \qquad (19)$$

$$\tan\delta = \frac{\sigma}{\varepsilon\omega} \qquad (20)$$

The highest loss tangent and shortest wavelength will be in muscle tissue. The relative permittivity of muscle varies smoothly from 150 at 10 MHz to 75 at 100 MHz. Its value at 50 MHz is approximately 100. The conductivity of muscle does not differ significantly from 0.5 S m^{-1} over this frequency range. Using these values in equations 19 and 20, it can be shown that the criterion in equation 18 is satisfied at 50 MHz but may not be at higher frequencies. It is recommended that the methods of SAR calculation outlined in this report are not used at frequencies above 50 MHz.

6. Example 1: 8 MHz Antenna

Figure 2 shows the distribution of the normal component of magnetic field strength at 2.5 cm above the surface of a square antenna, measured with a calibrated single-axis search coil and a spectrum analyser. Each side of the antenna was 36 cm long and measurements of magnetic field strength were made every 2 cm over a 52 cm square

330

grid, encompassing the antenna and extending 8 cm beyond it in each horizontal direction.

In the analysis presented in this paper, circumferential symmetry of magnetic field strength is assumed. The distributions of magnetic field strength along the four radii parallel to the antenna sides and along the four diagonal radii were determined and averaged to give the mean radial variation in magnetic field strength shown in Figure 3. Figure 3 also shows the corresponding radial variation in induced current density.

At 8 MHz, the investigation level for magnetic field strength is 0.28 A m^{-1} and the basic restriction on induced current density in the head, neck and trunk is 80 A m^{-2}. The maximum normal component of magnetic field strength from this antenna was 28 A m^{-1}, 100 times higher than the NRPB investigation level. The maximum induced current was calculated to be 33 A m^{-2}, below the basic restriction. From equation 14, the SAR corresponding to this current density was calculated as 2.2 W kg^{-1}, assuming a value of 1000 kg m^{-3} for the density of tissue. The magnetic field from this antenna was pulsed, with a pulse width of 250 μs and a repetition rate of 10 Hz. The time-averaged SAR is then 5.5 mW kg^{-1}, very much below the 0.4 W kg^{-1} basic restriction on whole-body average SAR.

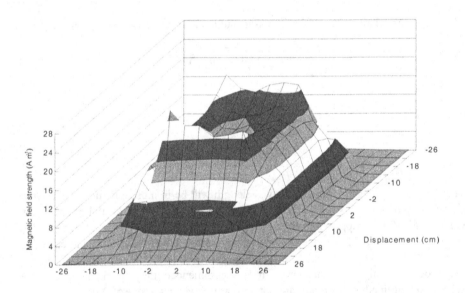

Figure 2. Normal component of magnetic field strength 2.5 cm above 8 MHz antenna

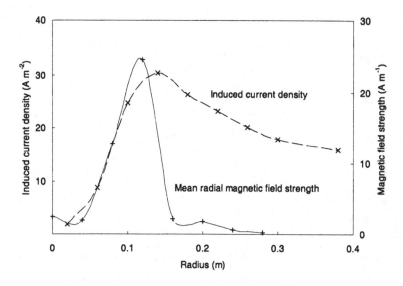

Figure 3. Mean radial variation in magnetic field strength from 8 MHz antenna and corresponding induced current density

7. Example 2: 27.12 MHz Wood Dryer

Figure 4 shows the vertical distribution of the component of magnetic field strength normal to the surface of the body of a person standing in front of a 27.12 MHz wood dryer. The dryer was shielded and the electric field strengths at the exposure position were below the investigation level. The horizontal distribution of this component of magnetic field strength was determined also.

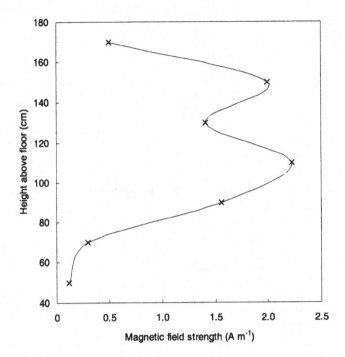

Figure 4. Vertical distribution of normal component of magnetic field strength at 15 cm from 27 MHz wood dryer

At this frequency there is no basic restriction on induced current density in the NRPB guidelines, but there is a basic restriction of 0.4 W kg^{-1} on whole-body average SAR. The appropriate investigation level for magnetic field strength is 0.16 A m^{-1}.

The condition for maximum SAR in the body would be for the spatial field strength maximum to occur over the centre of the trunk. The vertical and horizontal distributions of magnetic field strength in this region have been used to estimate the mean radial variation in magnetic field strength over the trunk as in example 1. Figure 5 shows this radial variation in magnetic field strength and the corresponding radial variation in SAR.

The localised maximum SAR in this exposure situation exceeds the 0.4 W kg^{-1} basic restriction on whole-body average SAR but not the 10 W kg^{-1} restriction on localised SAR in the trunk. It is appropriate to determine whether the exposure could result in a whole-body average SAR of greater than 0.4 W kg^{-1}. Equation 13 predicts that this will be the case - the spatial maximum magnetic field strength of 2.23 A m^{-1} is greater than the 1.89 A m^{-1} required to satisfy equation 12. The mean SAR over the trunk arising from a uniform exposure to a magnetic field strength of 2.23 A m^{-1} is 0.59 W kg^{-1}.

The calculated radial variation in SAR can be used with equations 6, 8, 15 and 17 to estimate the mean SAR over the trunk more accurately. The mean SAR over the trunk when the non-uniformity of the field is considered in this way is 0.35 W kg^{-1}.

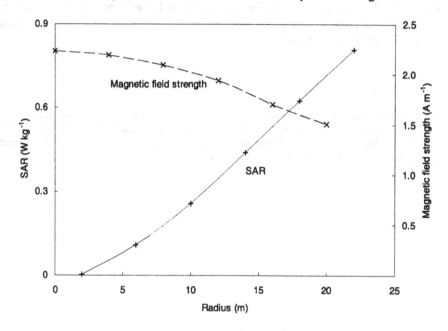

Figure 5. Mean radial variation in normal magnetic field strength and corresponding SAR from wood dryer

8. Conclusions

Investigation levels for radio frequency magnetic fields are calculated on the basis of uniform exposure under plane-wave conditions. For non-uniform exposures close to some sources, compliance with the basic restrictions can be demonstrated even when the investigation levels have been exceeded significantly. A simple analytical approach to the demonstration of compliance with exposure guidelines by calculation of induced current density and SAR for uniform and non-uniform exposures to inductive near-fields has been outlined in this report.

At frequencies below 10 MHz the NRPB basic restriction on induced current density will be met if uniform exposure over the trunk does not exceed 25.3 A m^{-1}. This has been used as a ceiling level in the practical application of NRPB guidelines. Above 100 kHz there is a basic restriction on whole-body average SAR which will be met if uniform exposure over the trunk does not exceed 51/f A m^{-1}, where f is the frequency in MHz.

334

Above 50 MHz, the assumptions underlying the dosimetric model outlined in this report may no longer be valid.

Magnetic fields at frequencies below 2 MHz cannot be time-averaged. Between 2 MHz and 10 MHz, exposures can be time averaged but only to a maximum of 25.3 A m^{-1}. Above 10 MHz, exposures can be time-averaged with no ceiling apart from specific absorption (SA) considerations.

9. References

1. NRPB (1993) Restrictions on exposure to static and time varying electromagnetic fields and radiation: Scientific basis and recommendations for the implementation of the Board's Statement, *Doc. NRPB* **4**.
2   ICNIRP (1998) Guidelines for limiting exposure to time-varying electric, magnetic and electromagnetic fields (up to 300 GHz), *Health Physics* **74**, 494-522.

EVALUATING PARTIAL-BODY RADIOFREQUENCY FIELD EXPOSURES:

The Need for Better Near-field Dosimetry

R. A. TELL
Richard Tell Associates, Inc.
8309 Garnet Canyon Lane
Las Vegas, Nevada 89129-4897
USA

Abstract

The potential for excessive radiofrequency (RF) exposure is generally greatest in the near-field of RF sources. Present-day RF exposure standards are primarily derived from dosimetry associated with uniform exposure of the entire body. This is particularly true for the limits on spatial peak specific absorption rate (SAR). Normally, however, the very intense and localized RF fields to which some personnel are exposed are clearly not uniform over the body and this raises the question of the appropriateness of partial-body relaxation provisions in various exposure standards. SAR measurement data obtained during the course of evaluating the effectiveness of RF protective clothing have demonstrated the existence of localized SARs that can substantially exceed recommended limits (e.g., 8 W/kg in any one gram of tissue), even though the local RF field incident on the body does not exceed maximum permissible exposure (MPE) limits for partial body exposures. These data indicate that body curvature plays a significant role in determining the peak SAR and may be most important for the head. The findings also suggest that the field polarization and source position relative to the body are critical determinants of the maximum, local SAR value. These results point to the need for a better understanding of how localized SAR is related to conditions of exposure and challenge the validity of peak SAR limits that are based on the assumption of uniform exposure.

1. Introduction

Present-day RF exposure standards are generally derived from an assumption of uniform field exposure of the entire body. While the fundamental basis of these standards is related to limiting the specific absorption rate (SAR) in the body, as averaged over the entire body mass, most standards, such as those of the IEEE [1], also

335

B.J. Klauenberg and D. Miklavcic (eds.), Radio Frequency Radiation Dosimetry, 335-342.
© *2000 Kluwer Academic Publishers. Printed in the Netherlands.*

contain Maximum Permissible Exposure (MPE) limits related to peak SAR that may occur at any point within the body, usually averaged over either one or ten grams of tissue. These spatial peak SAR limits are derived from observations of uniform exposure of laboratory animals and phantom models of both animals and humans.

In recognition of the non-uniform absorption of RF energy within the body, even with uniform field exposure over the body, the IEEE standard sets a limit on the spatial peak SAR at 20 times the whole-body average value. For example, while the whole-body average SAR is limited to 0.4 watts per kilogram (W/kg) in the whole body, a local SAR of 8 W/kg, averaged over one gram of tissue, is permitted. It is interesting to note that while specifications on local SAR were first included in the IEEE standard as early as 1982, even earlier laboratory work had already demonstrated the possibility of significantly elevated peak SARs, substantially greater than twenty times the whole-body average, for example, in the ankle region [2].

2. The Significance of the Near Field

Potentially hazardous RF exposures almost always occur in non-uniform, near-field environments. Work on broadcast antenna towers, at roof-top wireless communications sites or with RF heat sealers are all examples of practical, near-field exposure conditions that may result in excessive exposures. In each of these cases, personnel are normally located immediately near the active field source, whether it is an active antenna intended to radiate a signal or a non-intentional radiator, and are subject to highly non-uniform fields. In fact, real-world RF exposures are almost never uniform over the body. Nonetheless, the normal approach to evaluating these situations is, first, to measure, or calculate, the average field to which the person is exposed and then to compare this value to MPE limits based on uniform fields of the same magnitude.

To accommodate the presence of partial body exposure (highly non-uniform fields), some standards have specified MPEs for fields set to limit the spatial peak field to no more than, for example, 20 times the spatially averaged value [1]. This provision is based on the belief that RF fields of up to 20 times the basic MPE limit will not result in peak SARs that would exceed 20 times the whole-body average SAR.

There are two aspects to this particular evaluation approach that deserve comment: first, the resulting whole-body average and peak SARs in the exposed individual may not be comparable to the SAR limits in the standard based on the assumption of uniform field exposure and, second, it is not clear that local peak field exposures twenty times the permitted spatial average value, under near-field exposure conditions, will insure that the resulting peak SAR in the body will always comply with the standard's limits. The underlying assumption made in virtually all RF exposure standards is that the local SAR values at different points within the body that are associated with a given whole-body average SAR are of no special consequence and any particular biological responses that may be related to these RF hot-spots are automatically accounted for by limiting the whole-body average SAR to the prescribed values. A further philosophical question may be whether the above stated assumption is always true, especially under near-field exposure conditions.

A complicating factor in assessing RF hazards is the issue of partial body exposure and whether exposure standards sufficiently distinguish between partial body and non-uniform exposures. For example, in the present IEEE standard [1], partial body exposures with local fields up to twenty times the whole-body average value are permitted except when either the eyes or testes can be exposed. Partial body exposures are usually associated with a very localized application of RF energy to the body such as when placing a diathermy applicator or open-ended waveguide against the body. It is important to remember that the peak SAR limits in most exposure standards are based on the assumed twenty-fold increase in local SAR compared to the whole-body average value observed in whole-body, uniform exposure experiments. The present standards are generally not based on research of the biological effects of partial body exposures; in this context, one could argue that limits related to partial body exposures might permit greater peak SARs than those contained in our present standards.

A final aspect of the importance of developing better near-field dosimetry insights is related to the practical observation that instrumentation and measurement techniques for directly measuring SAR in humans is not currently available. Further, it is unlikely that such instrumentation will be developed in the near future. This lack of equipment that can be easily applied without undue intervention of the exposed subject argues for developing more accurate laboratory and analysis methods for quantifying SARs resulting from the typically nonuniform near-field exposures found in the work place.

3. Some Observations from Near-field Dosimetry

Some quantitative insight to the relationship between near-field exposures and resulting peak SARs can be gained by examining the results of laboratory measurements performed on full-sized human phantom models [3] and human subjects [4] as well as investigations of SARs resulting from exposures to near fields from reradiating structures [5].

In [3], near-field exposures of a full-sized human phantom to corner reflector antennas, were performed at 150, 450, 835 and 1,950 MHz for the purpose of evaluating the SAR attenuation effects of RF protective clothing fabricated from a material containing microscopic stainless steel fibers. Measurements of peak SARs were obtained at each frequency using a robotically controlled miniature, isotropic electric field probe. Figure 1 shows the SAR measurement setup. Vertically polarized near-field exposures, at a distance of 17 cm between the surface of the phantom and the radiating element of the various antennas, were performed at the thigh, the belly and the head, both without and with the protective garment on the phantom. Table 1 summarizes the net incident power densities, the peak SAR values obtained without the protective garment and the normalized SAR relative to the power density.

Inspection of Table 1 reveals normalized SARs in the range of 0.2-1.2 (W/kg)/(mW/cm^2). Of particular interest are the peak SARs found in the head where an

338

increasing efficiency of absorption is found with increasing frequency. Of even more interest is the finding that, for all frequencies greater than 150 MHz, power densities of 20 mW/cm^2 are related to peak SARs that exceed 8 W/kg, the peak SAR limit in the IEEE standard. In the case of 1,950 MHz exposure of the head, the peak SAR is over three times the underlying limit of 8 W/kg at the power density presumably associated with insuring that this limit would not be exceeded! This observation, alone, suggests the need for a better understanding of RF dosimetry associated with near-field exposures. For the same incident power densities, peak SARs in the chest region, much flatter than the head, were found to be five times less than in the head. Horizontal polarization at the head resulted in only one-half the peak SAR found with vertical polarization. These results suggest that body curvature may significantly affect the absorption of RF energy.

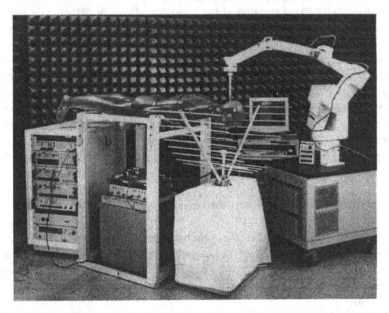

Figure 1. SAR measurement system for characterizing the SAR reduction properties of RF protective clothing in the near field at 150 MHz. Other corner reflector antennas were used at 450, 835 and 1,950 MHz.

Recently, in a study of thermophysiological responses of human volunteers exposed to 450 MHz far fields, Adair and colleagues measured normalized peak surface SARs in a human phantom of 0.33 (W/kg)/(mW/cm^2). This may be compared to a near-field measurement in Table 1 of 0.42 (W/kg)/(mW/cm^2), suggesting that near-field exposures to active radiating sources can result in higher peak SARs than far-field exposures. They estimated that a normalized peak SAR of 0.32 (W/kg)/(mW/cm^2) resulted when a living human was the target of the RF field.

TABLE 1. Summary of net incident power density, peak SAR and normalized SAR for near-field exposures.

Body area	Power density (mW/cm^2)	Peak SAR (W/kg)	Normalized SAR (W/kg)/(mW/cm^2)
	150 MHz		
Belly	8.96	2.80	0.31
Thigh	9.95	3.35	0.34
Head	14.27	4.91	0.34
	450 MHz		
Belly	19.01	7.94	0.42
Thigh	18.82	9.84	0.52
Head	18.86	8.97	0.48
	835 MHz		
Belly	19.18	3.75	0.20
Thigh	19.55	4.65	0.24
Head	19.17	17.0	0.88
	1,950 MHz		
Belly	3.0	0.72	0.24
Head	3.0	3.66	1.22

In a study of the capacity of reradiated fields to deliver high SARs [5], Kuster and Tell found that when compared to plane-waves, these fields resulted in substantially lower SARs in a simulated model of a human at 100 MHz. In this study, a uniform field having electric and magnetic field strengths essentially comparable to the plane-wave equivalent power densities of the IEEE standard [1] at 100 MHz (10 W/m^2) was assumed to illuminate a horizontally oriented conductive rod positioned 3 cm from the head of the subject. Peak SARs were calculated for the conditions of no reradiating rod present (free space conditions), with the head positioned near the center of the rod, half way between the center and end of the rod and at the end of the rod. When the peak SAR was normalized to the greater of the plane-wave equivalent power densities associated with either the electric or magnetic fields, the results demonstrated a very substantial reduction in absorption efficiency when compared to the plane-wave exposure case corresponding to far-field exposure.

TABLE 2. Summary of findings in study of peak SAR resulting from exposure to plane-wave, 100 MHz field and to reradiated fields of a conductive, horizontal resonant rod placed 3 cm from head of subject [5].

Condition	Maximum power density based on electric or magnetic field (mW/cm^2)	Normalized SAR (W/kg)/(mW/cm^2)	Field impedance (E/H) (ohms)
No rod	1 (E or H)	0.033	420
Center of rod	1700 (H)	0.00042	3.8
Middle	1100 (H)	0.00030	283
End of rod	1800 (E)	0.00002	1421

Table 2 summarizes these findings which support the hypothesis that high field strengths near the surface of reradiating objects are not good indicators of the SAR resulting from exposure. These data show that, for the case evaluated, the SAR in the head, and the body as a whole, are well below the SAR limits in the IEEE standard at 100 MHz, even though the local fields greatly exceed the MPE limits.

When contrasted with the peak near-field exposure SARs obtained with active antennas, it becomes clear that high RF fields, even those with magnitudes substantially greater than the MPEs of exposure standards, associated with reradiating objects do not necessarily have the capacity to deliver significant SAR. Hence, this finding again illustrates the need for a better dosimetric understanding of near-field exposures to allow more accurate hazard assessments. The field impedance is seen to be an apparently important factor in determining the efficiency of RF energy absorption, high impedance fields being particularly ineffective in causing significant SARs.

At higher frequencies, localized energy absorption efficiency in the near field can also be a function of the exact distance between the field source and the surface of the body. Figure 2 illustrates a finding [3] that suggests an impedance transformation effect in the distance range of 16-17 cm. It is interesting to note that the wavelength of the 1,950 MHz field is about 15.4 cm, closely corresponding to the point where the greatest normalized SAR was determined, being about twice the value either side of the observed peak.

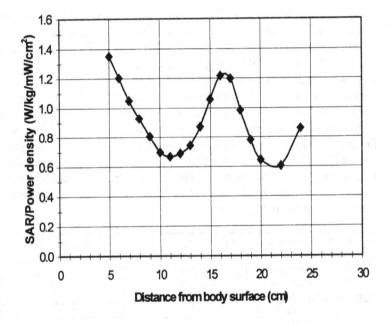

Figure 2. Normalized SAR in the head vs. exposure distance from an active 1,950 MHz antenna oriented for vertical polarization. The wavelength at this frequency is 15.4 cm, close to the distance correlated with the relative peak in SAR values suggesting an optimization in matching between the corner reflector antenna and the head of the phantom.

4. Questions, Insights, and Conclusions

Most of today's RF exposure standards are based, first, on controlling the whole-body averaged SAR and the electromagnetic field MPE limits corresponding to the selected SAR values are derived from the assumption of exposure to a uniform plane-wave field. This is not the case for the vast majority of practical exposures in which individuals may, indeed, be hazardously exposed. Several questions, insights and conclusions may be drawn from the above discussion:

1. Is it appropriate to make the assumption that the localized peak SARs associated with a given whole-body average SAR have no particular significance relative to the ultimate biological effect(s) that should be protected against across the electromagnetic spectrum and that controlling whole-body average SAR is sufficient?

2. More consideration should be given to dosimetric analysis of partial-body exposures and biological studies should concentrate on the effects of partial-body exposures, as opposed to simply nonuniform exposures, to provide a more meaningful basis for setting local SAR limits.

3. A greater emphasis on near-field exposure dosimetry is needed to determine whether present MPE limits are appropriate to apply in highly nonuniform exposure environments for controlling peak SARs to the given limits (e.g., 8 or 10 W/kg, depending on the standard).

4. Is the application of MPEs derived from <u>uniform field exposure conditions</u> unnecessarily conservative for typical work situations in which personnel may be subject to localized RF exposures?

5. Or, can such an application result in substantial underestimates of peak SARs with near-field exposures? Certainly some laboratory data indicate that peak SARs associated with recommended exposure limits can exceed the underlying peak SAR criteria [see, for example, 3].

6. Near-field dosimetry should be used to gain a better understanding of the role of body curvature in controlling peak SARs within the body. Data presented here indicate that curvature of the face and polarization can play a significant role in the resulting SAR. This means that the field source orientation relative to the face may have significant implications, e.g., frontal exposure vs. side exposure.

7. More near-field dosimetry research is needed to assess the significance of highly enhanced electric and magnetic fields produced by reradiating objects relative to SARs. Present data suggest that very high impedance electric fields are extremely ineffective in delivering significant SARs to absorptive tissue and, hence, measurements of such fields may lead to erroneous conclusions of non-compliance with applicable standards.

8. It would appear desirable to reevaluate the local SAR enhancement factor found in the body of both experimental research animals as well as humans when exposed to uniform fields. This finding has, to date, played a significant role in deriving limits for peak SARs; for example, the present assumption of a 20 to 1 ratio between local peak and whole-body average SARs dictates the local limit for exposures of all kinds, regardless of whether the exposure occurs in the far field or the near field.

9. How do near-field exposures compare with far-field exposures, for the same peak incident power density relative to spatial peak and whole-body average SARs? In one case the same power density exists over the entire body and in the other, only a limited portion of the body is subject to the maximum value.

10. What practical guidelines can be developed with our present knowledge to provide meaningful insight as to when a nonuniform exposure becomes a partial body exposure?

11. Any biological research performed with RF near-fields should give special consideration to the eyes and testes in the interest of assessing whether there are, in fact, any special aspects of these organs that merit special attention and protection in terms of peak SAR limits that would be any different from those for other parts of the body.

12. The need for more realistic exposure assessments, based on fundamental understandings of near-field dosimetry, is exacerbated by the lack of socially acceptable SAR probes that can be used in typical work near-field environments.

5. References

1. IEEE (1991) *IEEE Standard for Safety Levels with Respect to Human Exposure to Radio Frequency Electromagnetic Fields, 3 kHz to 300 GHz.* IEEE C95.1-1991, Institute of Electrical and Electronics Engineers, 345 East 47th Street, New York, NY 10017.

2. Guy, A. W., Webb, M. D., and Sorensen, C. C. (1976) Determination of power absorption in man exposed to high frequency electromagnetic fields by thermographic measurements on scale models, *IEEE Transactions on Biomedical Engineering* **BME-23**, 361-370.

3. Tell, R. A. (1997) *SAR Evaluation of the Naptex™ Suit for Use in the VHF and UHF Telecommunications Bands.* Technical report prepared for Motorola by Richard Tell Associates, Inc., 8309 Garnet Canyon Lane, Las Vegas, NV 89129.

4. Adair, E. R., Kelleher, S. A., Mack, G. W., and Morocco, T. S. (1998) Thermophysiological responses of human volunteers during controlled whole-body radio frequency exposure at 450 MHz, *Bioelectromagnetics* **19**, 232-245.

5. Kuster, N. and Tell, R. A. (1989) A dosimetric assessment of the significance of high intensity RF field exposure resulting from reradiating structures. Paper D-1-5 presented at the Eleventh Annual Meeting of the Bioelectromagnetics Society, Tucson, AZ, June 18-22, p. 24 (book of abstracts).

SESSION F: RESPONSES OF MAN AND ANIMALS I

E. R. ADAIR
Directed Energy Bioeffects Division
AFRL/HEDR
Brooks AFB, TX 78235

The papers presented during the first two days of this NATO ARW concerned various aspects of dosimetry, both theoretical and experimental. Whereas, in earlier years, such papers were concerned almost exclusively with the physical absorption of radio frequency energy in the tissues of the body, now we are beginning to hear mention of certain biological responses to that energy absorption. Theoretical dosimetry in particular has recognized the need to include such factors as tissue temperature, blood flow, metabolism, thermal conductivity, tissue inhomogeneity, and even evaporative capability. To the biologists attending the ARW in Slovenia, this change is like the dawn of a new day. This is so because we believe that in order to fully characterize the response of a whole organism, whether man or animal, to a RF exposure the physiological responses of that organism must be included. The present session, as well as Session G, provide some up-to-date information on the biological sequelae of RF energy absorption in humans and animals that will be important for expanding dosimetric modeling. Session F is concerned with thermal responses and begins with two theoretical papers. These are followed by a study of RF-induced tissue heating in the monkey ankle and two reports of the biological effects of high intensity fields.

B.J. Klauenberg and D. Miklavcic (eds.), Radio Frequency Radiation Dosimetry, 343.
© 2000 *Kluwer Academic Publishers. Printed in the Netherlands.*

THERMOREGULATION: ITS ROLE IN MICROWAVE EXPOSURE

E. R. ADAIR
Directed Energy Bioeffects Division
AFRL/HEDR
Brooks AFB, TX 78235

1. Introduction

The thermogenic properties of exposure to radio frequency (RF) and microwave fields are well recognized. Indeed, most current safety guidelines are based on the premise that significant increases in tissue temperature of RF-exposed human beings should not be permitted. While some investigators have argued that there are other responses that can be initiated at "non-thermal" levels of RF energy, concrete evidence to support this view is lacking. The term "non-thermal" is often defined in terms of the lack of a measurable increase in the temperature of exposed tissues. However, even when no changes can be measured in the deep or peripheral temperatures of the body, sensitive thermoregulatory mechanisms are mobilized to dissipate any heat generated in body tissues by the absorption of thermalizing energy from RF sources in the environment. When RF field strengths are significant, these mechanisms are so efficient that only modest increments in body temperatures may occur. However, when RF fields are very intense, as is the case with high-power microwave applications, regions of localized high specific absorption rate (SAR) may occur in which heat is generated faster than it can be dissipated. This condition poses a challenge for the thermoregulatory system of the exposed organism. In describing the role of thermoregulation in microwave exposure, this paper presents information on the organization and function of the controlling system for thermoregulation and some insights on the importance of both SAR and frequency to behavioral and autonomic thermoregulatory responses.

2. The Role of Frequency in RF Exposures

The consequences for thermoregulation of exposure to RF fields relate not only to the whole-body or localized SAR, which specifies the rate of energy absorption, but also to the frequency, which determines the depth to which the energy may penetrate below the surface of the skin. This is because, in mammals, there are two levels of neural receptors (thermoreceptors) that respond to changes in their own temperature by changes in firing rate, the initial neurophysiological event in the thermoregulatory

B.J. Klauenberg and D. Miklavcic (eds.), Radio Frequency Radiation Dosimetry, 345-356.
© 2000 *Kluwer Academic Publishers. Printed in the Netherlands.*

346

process. Some thermoreceptors are located within the first millimeter of the skin surface [1] and are primarily responsible for the generation of thermal sensations and the initiation of behavioral actions. The surface receptors are normally stimulated by conventional thermal stimuli in the environment (radiant, convective and conductive heating and cooling), but they can also be stimulated by RF radiation of high frequency that is superficially absorbed. Other thermoreceptors, which have similar operating characteristics, are located deep in the body, in the brain and spinal cord, and in the deep viscera. These receptors normally respond to changes in deep body temperature that occur during exercise, febrile episodes, slowly developing hypo- and hyperthermic states, etc. Under certain conditions, deeply penetrating RF energy may stimulate these receptors directly, almost to the exclusion of the peripheral receptors. Any deep stimulation that bypasses the surface receptors may also provide unusual thermal gradients within the body's tissues, at least during acute exposures when thermal equilibrium has not yet been attained. One would expect that the changes in thermoregulatory responses that may occur during RF exposure at frequencies near whole-body resonance will depend primarily on the configuration of stimulated thermoreceptors and ultimately by the organism's ability to dissipate heat generated locally in deep body tissues.

Classical thermal physiology has long held that both skin and core temperatures play a role in the generation of input signals to control heat production and heat loss in the body. Early studies by Benzinger [2] described threshold combinations of skin and core temperatures in cold-exposed humans at which an increase in heat production could be predicted. A basic linear pattern of responses, involving the independent deviation of core (T_c) and skin (T_{sk}) temperatures from set levels (T_{co}) and (T_{sko}), was confirmed in several species including the human [3], cat [4], dog [5], and non-human primate [6, 7]. For any given effector response R having a baseline value R_o, the equation for the basic interaction is:

$$(R - R_o) = a(T_c - T_{co}) + b(T_{sk} - T_{sko}). \tag{1}$$

The relative contribution of core *vs.* peripheral thermosensors to thermoregulation in assorted environments has been much debated over the years. Recent techniques by which these separate thermal inputs can be independently controlled (by means of implanted heat exchangers) have identified composite core and skin temperature signals as being of primary importance [8, 9]. In general, the experimental evidence points to peripheral temperature signals predominating when the organism is exposed to cold, while signals generated in the body core predominate when the organism is exposed to heat. These are important considerations because, during exposure to diverse RF and microwave fields, the loci of signal generation for thermoregulatory responses in any given thermal environment will be a complex function of the loci of RF energy deposition.

3. Dual Thermoregulatory Mechanisms

As has been well documented, two separate systems can accomplish thermoregulation in living endothermic mammals, autonomic (or physiological) and behavioral [10]. The former is automatic while the latter is under voluntary control. This paper describes the two thermoregulatory systems in some detail and how they may interact to regulate the temperature of a RF-exposed body most efficiently. As noted above, frequency is the most important attribute of RF energy to be considered, although polarization, field strength and local SAR also play very important roles in response mobilization. A comprehensive conceptual model will be introduced here as a context for our discussion. Various aspects of the role of thermoregulation in RF exposure will be illustrated by data collected on both animals and humans, as appropriate.

Thermoregulation is defined as the maintenance of the temperature or temperatures of a body within a restricted range under conditions involving variable internal and/or external heat loads. In terms of temperature regulation, human beings and many other mammals are classified as endotherms, organisms that regulate their body temperature by means of a controlled rate of metabolic heat production in the face of rather wide fluctuations in the thermal environment. In addition to heat production, autonomic mechanisms of heat loss, such as changes in peripheral vasomotor tone (vasodilation and constriction) and evaporative water loss (breathing, sweating), control the rate of heat exchange between the body and the environment and thereby accomplish thermoregulation. Adair [11] has described these autonomic mechanisms of heat production and heat loss in considerable detail.

Not all mammals have full endothermic capability, however. Some, e.g., rodents, are deficient in evaporative capability and must rely instead on a second mechanism, that of spreading saliva or urine over the skin to aid evaporative cooling. This is an example of behavioral thermoregulation, which involves complex response patterns of the skeletal musculature to heat and cold that modify the rates of heat production and/or heat loss. Examples include exercise, changes in body conformation, adjustment of thermal insulation (e.g., bedding, clothing), and the selection or production of an environment that reduces thermal stress and maximizes thermal comfort.

The comfortable environment for resting endotherms is often thermally neutral - - that is, it falls within the thermoneutral zone of vasomotor control [11]. Sometimes other, non-neutral environments feel comfortable, especially for humans. For example, after exercise, when the deep body temperature is elevated, a cool environment may feel comfortable. Conversely, after prolonged cold exposure, when the deep body temperature is depressed, a warm environment may feel comfortable. It is clear that the deep body temperature, as well as that of the skin, exerts a strong bias on sensations of thermal comfort. For this reason, the interplay between neural signals from peripheral and deep thermoreceptors is of great importance in determining the thermoregulatory response to any thermal stimulus, whether internal or environmental. We know that during RF exposure the depth to which the energy may penetrate is a function of frequency, especially in the resonance range for the organism in question.

348

4. Important Loci of Thermoreception

Figure 1 shows a schematic representation of the neural elements known to be involved in the thermoregulation of endotherms (both autonomic and behavioral) and some of the known and hypothetical links between them. This model, modified from Adair [12], is similar in many respects to other simple models [13,14] but, for clarity, does not contain the complexity of more contemporary models [15]. As shown in the figure, a change in the environment, whether a naturally occurring or behaviorally produced physical disturbance, alters the pattern of neural impulses arising in the thermosensitive structures of the skin (Ss). This neural traffic not only goes directly to the higher neural centers (giving rise to thermal sensation in humans), but also to a thermoregulatory integrating center. The integrating center must also receive neural impulses from many other

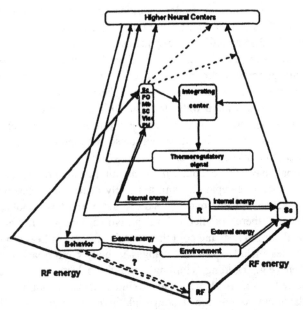

Figure 1. Schematic representation of the neural elements involved in thermoregulation, both behavioral and autonomic. Arrows denote direction of connecting links characterized as known neural (--------), hypothetical neural (- - - - -), and energy exchange (=====). Sc = sites located centrally; PO = preoptic area; Mb = midbrain reticular formation; SC = spinal cord: Visc = deep viscera; PH = posterior hypothalamus; Ss = thermosensitive skin structures; R = autonomic responses.

thermosensitive sites located centrally in the body (Sc). A substantial scientific literature describes the location and thermoregulatory significance of many of these sites. The most important is the medial preoptic/anterior hypothalamic area (PO) of the brainstem, considered by many to be the location of the "central thermostat." Other sites including the posterior hypothalamus (PH), midbrain reticular formation (Mb), spinal cord, (SC), and deep abdominal structures (Visc) have also been shown to influence thermoregulatory effector processes significantly. There is ample experimental evidence that many small afferent signals from diverse neural structures throughout the body can result in a strong, integrated effector command or signal that provokes a thermoregulatory response [9,16]. The thermoregulatory signal generated by the integrating center activates the forward controlling elements of the autonomic system, internally energized, in the form of physiological responses, to combat hyperthermia (vasodilation, sweating, decreased metabolic heat production, polypnea) or hypothermia (vasoconstriction, piloerection, shivering or nonshivering thermogenesis). The signals related to change in body temperature, the regulated variable, will then be fed back via the central and peripheral sensors to close the loop.

In the freely active organism, the behavioral regulatory system also operates, often to the exclusion of the autonomic system. Many neural elements are common to both systems. Neural pathways providing for the awareness of physiological responses and the sensation of peripheral stimuli, together with their modification by the central signals, are represented in Figure 1 as converging on the higher neural centers (e.g., the somatosensory cortex). Current theory holds that, at least in humans, thermal discomfort originates here and serves as the drive signal to behavioral responses. For very intense thermal stimuli, such as burning pain, a simple reflex may be operative. A pathway that carries the integrated thermoregulatory signal to the higher neural centers, bypassing the autonomic responses, is also necessary because thermal stimulation of the PH can alter behavioral, but not physiological, thermoregulatory responses [12]. No matter at what level the drive originates, it stimulates the behavioral responses by which the organism uses physical energy to modify the thermal environment.

5. Role of RF Energy Absorption

Exhaustive research has been conducted during the last 50 years to quantify the biological effects of mammalian exposure to various RF fields. Much recent work has dealt with the thermoregulatory consequences that accompany such exposure at thermogenic field strengths. Although extrapolation of animal data to human beings is still uncertain, recent studies of volunteers exposed to some of these fields have begun to provide a refined understanding of human response capabilities. Thus, it is appropriate to ask how absorbed RF energy may be introduced into our simple schematic diagram or model (Figure 1).

Some complex models [15] introduce "radiation" as an environmental variable, together with air temperature, humidity, and air movement; however, this variable usually designates radiant energy from the sun or nearby objects (λ = 1mm – 100nm). RF and microwave radiation (λ = 1km – 1mm) should properly appear as a separate entity, because of its unique property of absorption below the surface of the skin. When the long axis of the body is parallel to the electric field vector, the depth of penetration is maximal at 0.4 λ (resonance), and the potential for stimulation of central thermoreceptors is optimized. Limited studies of animals exposed at resonance have demonstrated clearly the selective mobilization of thermoregulatory mechanisms

350

(vasodilation, saliva spreading, sweating) designed specifically to eliminate heat from the body [17, 18, 19]. Such responses are not nearly as robust or effective as when the animals are exposed at frequencies above resonance.

Figure 2 shows some additions to the basic model depicted in Figure 1; these are designed to indicate some of the avenues by which absorbed RF energy may influence mammalian thermoregulation. A full range of frequencies, or range of penetration depths, cannot be easily shown in such a diagram; however the extreme cases of peripheral (skin) and core energy deposition are accounted for. Intermediate frequencies will, of necessity, involve combinations of these two extremes in direct relation to the absorbed frequency. The possibility of intermediate levels of receptors is not explored here either, although their existence can not be ruled out.

The source of RF energy is shown at the bottom of the model and heavy arrows depict the absorption of this energy near the skin sensors (Ss) and/or the central sensors (Sc). The number and location of stimulated sensors will always depend on the penetration depth, the field strength, and, especially, on the orientation of the organism in the field, as

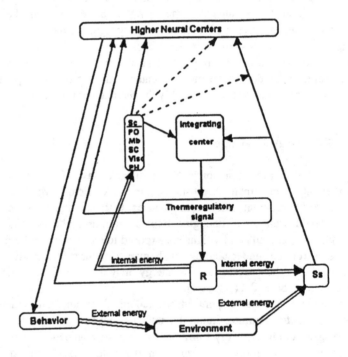

Figure 2. RF energy added to the schematic representation of neural elements depicted in Figure 1. The source of RF energy (RF) may stimulate (_____) either thermosensitive skin sites (Ss) or central sites (Sc), depending on the RF frequency. A questionable behavioral link (------) depends upon attribution.

shown by Durney, *et al.* [20]. Chou, *et al.* [21] studied local energy absorption in rats exposed to 2.45 GHz under seven different exposure conditions (restrained in near- and far-field, E and H polarization and unrestrained in circular waveguide and miniature anechoic chamber). Their general finding was that the local SARs were highest in the tail and hypothalamus, regions that are important for thermoregulation in the rat. Recent studies of local temperature increases in rat brain during exposure to high power microwaves [22, 23] reinforce the possibility of selective RF heating in sites that may be intimately involved in the control of thermoregulatory processes.

As shown in Figure 2, no additional neural pathways are posited for the internal distribution of signals generated by RF energy. However, a behavioral link has been added to the model that is based on the questionable attribution of thermal discomfort to the RF source in the environment. The matter of attribution has been raised and discussed by Justesen [24], who found that rats and mice were unable to escape or avoid a lethal multipath RF field unless a concomitant cue, such as a bright light or intense sound, was provided. Directional RF energy, such as that launched by an antenna, is easily discriminated, however [25]. Also, the autonomic responses that facilitate heat loss will be mobilized as required, even though no behavioral response, such as turning off the RF source, occurs. The thermoregulatory efficiency of the autonomic heat loss responses will depend on the strength of the RF signal, the physiological attributes of the species exposed, and whether the environment is conducive to heat loss.

6. Resonant Exposure as a Special Case

Current thinking about the thermoregulatory system includes the notion of redundancy of neural sites within the central nervous system [9], the purpose of which is to protect the organism from loss of function. Thus, the possibility that other sites in the CNS, which could take over the control or integrative functions of the medial preoptic/anterior hypothalamic (PO/AH) area, should that area not be the one being thermally stimulated, is particularly relevant to RF exposures. At frequencies close to either whole-body or head resonance for a particular endothermic species, significant energy may be deposited in deep neural structures that may contain thermosensitive cells. If the PO/AH is not one of these structures, whole-body hyperthermia might easily develop if it were not for the ability of other sites, such as the midbrain reticular formation or the medulla, to take over the thermoregulatory function of the PO/AH. Adair [26] demonstrated that changes in autonomic thermoregulatory responses of squirrel monkeys, during whole-body exposure to 2.45-GHz CW energy, depend on the integral of energy absorption by the whole body, not on the energy deposited in the PO/AH alone. Thus, redundancy of neural sites may help to ensure rapid distribution of heat generated by absorbed RF energy, particularly if the field strength is moderate.

During whole-body exposure at resonance, the efficiency of the thermoregulatory system will depend largely on the heat loss capabilities of the organism in question. Human beings have an extraordinary capability to lose body heat through the evaporation of sweat from the body surface [27]. Adequate hydration and a favorable environment (i.e., minimal clothing, low relative humidity, thermoneutral ambient temperature, and some air movement) optimize evaporative cooling. Certain non-human primates (e.g., *Erythrocebus patas*) exhibit robust sweating when heat stressed, but the primates commonly used in laboratory studies (i.e., *Macaca mulatta* and *Saimiri sciureus*) have more limited sweating capabilities and can become hyperthermic when exposed to RF fields [18, 19, 28]. Under favorable conditions, however, the hyperthermia is modest and regulated at a level commensurate with the rate of energy absorption (SAR). The same cannot be said for rodents. As has been emphasized elsewhere [11], small mammals such as mice, rats, and hamsters are at great disadvantage when heat stressed because they lack efficient heat dissipation mechanisms.

Careful studies have demonstrated that partial-body exposure of human volunteers, at supra-resonant RF frequencies in a warm environment, stimulates robust sweating that helps to maintain the deep body temperature at the normal level [29]. An example of this result is shown in Figure 3, which shows group mean body temperatures (upper panel) and sweating rates (lower panel) for 7 subjects undergoing a standardized test protocol. The subjects were first equilibrated for 30 min to a 31 °C environment and then exposed for 45 min to a 2.45-GHz RF field at a power density of 35 mW/cm². At this power density, the peak local surface SAR was 7.7 W/kg, determined on the antenna boresight in the center of the subjects' back. In the upper panel, the top tracing shows that the deep body temperature, measured in the esophagus at the level of the heart, remained virtually constant ($\Delta T_{es} \leq 0.2$ °C) throughout the test. This was a consistent finding during a variety of exposure conditions at this frequency, confirming similar results reported earlier for a frequency of 450 MHz [30]. Figure 3 also indicates that the temperatures of the upper and lower back skin increased nearly 2.0 °C above the equilibrated level during the RF exposure, reflecting the superficial nature of RF energy deposition at 2450 MHz. Skin blood flow on the back (not shown here) also increased dramatically during the RF exposure period in all subjects tested. The lower panel of Figure 3 shows the associated group mean local sweating rates from back and chest that occurred during the RF exposure. The tracings show the pulsatile nature of the sweating response, the synchronous nature of sweat production at different skin sites, and the greater response on the skin that was irradiated directly. This vigorous heat loss response was largely responsible for the stability of the deep body temperature during RF exposure.

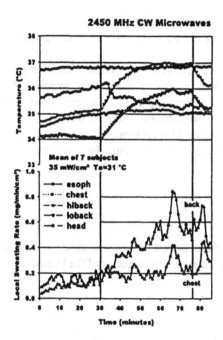

Figure 3. Mean data for a group of 7 subjects exposed to 2.45-GHz CW RF energy at a power density of 35 mW/cm^2 in a warm environment of 31 °C. Upper panel shows esophageal temperature (top tracing) and four skin temperatures; lower panel shows local sweating rate from back and chest.

7. Importance of Changes in Regional Blood Flow

Of all the thermoregulatory responses discussed here, it appears that the most important response for many endothermic species is a change in regional blood flow, especially when the endotherm undergoes exposure to RF energy. If the exposure is at a high frequency, (e.g., 3 GHz and above) the energy will be deposited primarily in the skin and the surface thermoreceptors that lie ~ 0.3 mm below the surface will be stimulated. Vasodilation of the peripheral vasculature will occur, with concomitant increases in skin blood flow. This autonomic response will increase the convective heat loss from the skin to the environment and facilitate the mobilization of sweating [31]. On the other hand, if the RF exposure is at a low frequency, especially close to resonance, the energy will be deposited at some depth below the skin surface and will cause heating in deep tissues that may harbor thermoreceptors. Neural input from these thermoreceptors, whether in the brain, the spinal cord, or other deep sites, will generate an integrated thermoregulatory signal for the mobilization of heat loss responses. The most important of these responses, as demonstrated in computerized models of the thermoregulatory system [32, 33], is an increase in convective heat transfer via blood

354

flow. This increase in blood flow produces a mixing and averaging of tissue temperatures over most of the body, thereby eliminating thermal hot spots that may accompany RF exposure.

It is important to note, however, that considerable time may be required for this thermal averaging to take place. Exposure to intense RF fields, such as are characteristic of high-power microwave systems, may result in regions of high local SAR in which the rate of tissue heating exceeds the maximal rate of convective heat transfer. The resultant localized hyperthermia [21, 22, 23] may be excessive and a true hazard against which human beings should be protected by appropriate exposure guidelines.

8. Disclaimer

The views expressed in this paper are those of the author and are not to be construed as official policy of the United States Air Force or of the United States Department of Defense.

9. References

1. Riu, P. J., Foster, K. R., Blick, D. W., and Adair, E. R. (1997) A thermal model for human thresholds of microwave-evoked warmth sensations, *Bioelectromagnetics* **18**, 578-583.
2. Benzinger, T. H. (1969) Heat regulation: Homeostasis of central temperature in man, *Physiol. Reviews* **49**, 671-759.
3. Stolwijk, J. A. J. and Hardy, J. D. (1966) Temperature regulation in man – A theoretical study, *Pflugers Archiv.* **291**, 129-162.
4. Jacobson, F. H. and Squires, R. D. (1970) Thermoregulatory responses of the cat to preoptic and environmental temperature, in J. D. Hardy, A. P. Gagge, and J. A. J. Stolwijk (eds.), *Physiological and Behavioral Temperature Regulation*, Charles C. Thomas, Springfield, IL, pp. 581-596.
5. Cabanac, M. (1970) Interaction of cold and warm temperature signals in the brain stem, in J. D. Hardy, A. P. Gagge, and J. A. J. Stolwijk (eds.) *Physiological and Behavioral Temperature Regulation*, Charles C. Thomas, Springfield, IL, pp. 549-561.
6. Stitt, J. T. and Hardy, J. D. (1971) Thermoregulation in the squirrel monkey (Saimiri sciureus), *J. Applied Physiology* **31**, 48-54.
7. Johnson, G. S. and Elizondo, R. S. (1979) Thermoregulation in Macaca mulatta: A thermal balance study, *J. Applied Physiology: Respiratory, Environmental, Exercise Physiology* **46**, 268-277.
8. Kuhnen, G. and Jessen, C. (1988) The metabolic response to skin temperature, *Pflugers Archiv.* **412**, 402-406.
9. Jessen, C. (1990) Thermal afferents in the control of body temperature, in E. Schonbaum and P. Lomax (eds.), *Thermoregulation: Physiology and Biochemistry*, Pergamon Press, New York, pp. 153-183.
10. Bligh, J and Johnson, K.G. (1973) Glossary of terms for thermal physiology, *J. Applied Physiology* **35**, 941-961.
11. Adair, E. R. (1995) Thermal physiology of radiofrequency radiation (RFR) interactions in animals and humans, in Klauenberg, B.J., Grandolfo, M. and Erwin, D.N. (eds.), *Radiofrequency Radiation Standards*, Plenum Press, New York, pp. 245-269.
12. Adair, E. R. (1974) Hypothalamic control of thermoregulatory behavior, in K. Lederis and K. E. Cooper (eds.), *Recent Studies of Hypothalamic Function*, S. Karger, Basel, pp. 341-358.
13. Hardy, J. D. (1971) Thermal comfort and health, *ASHRAE J.* **77**, 43-51.

14. Chatonnet, J. and Cabanac, M. (1965) The perception of thermal comfort, *Internat. J. Biometeorology* **9**:183-193.

15. Werner, J. (1986) Do black-box models of thermoregulation still have any research value? Contribution of system-theoretical models to the analysis of thermoregulation, *Yale J. Biology & Medicine* **59**, 335-348.

16. Rawson, R. O. and Quick, K. P. (1970) Evidence for deep-body thermoreceptor response to intra-abdominal heating of the ewe, *J. Applied Physiology* **28**, 813-820.

17. Gordon, C. J. (1983) Behavioral and autonomic thermoregulation in mice exposed to microwave radiation, *J. Applied Physiology* **55**,1242-1248.

18. Lotz, W. G. and Saxton, J. L. (1987) Metabolic and vasomotor responses of rhesus monkeys exposed to 225 MHz radiofrequency energy, *Bioelectromagnetics* **8**, 73-89.

19. Adair, E. R., Adams, B. W. and Hartman, S. K. (1992) Physiological interaction processes and radio-frequency energy absorption, *Bioelectromagnetics* **13**, 497-512.

20. Durney, C. H., Massoudi, H. and Iskander, M. F. (1986) *Radiofrequency Radiation Dosimetry Handbook, Fourth Edition"*, report **USAFSAM-TR-85-73**, USAF School of Aerospace Medicine, Brooks AFB, TX.

21. Chou, C. K., Guy, A. W., McDougall, J. A. and Lai, H. (1985) Specific absorption rate in rats exposed to 2,450-MHz microwaves under seven exposure conditions, *Bioelectromagnetics* **6**, 73-88.

22. Walters, T. J., Ryan, K. L., Tehrany, M. R., Jones, M. B., Paulus, L. A. and Mason, P. A. (1998) HSP70 expression in the CNS in response to exercise and heat stress in rats, *J. Applied Physiology* **84**, 1269-1277.

23. Walters, T. J., Ryan, K. L., Belcher, J. C., Doyle, J. M., Tehrany, M. R. and Mason, P. A. (1998) Regional brain heating during microwave exposure (2.06 GHz), warm-water immersion, environmental heating and exercise, *Bioelectromagnetics* **19**, 341-353.

24. Justesen, D. R. (1988) Microwave and infrared radiations as sensory, motivational, and reinforcing stimuli, in M.E. O'Connor and R.H. Lovely (eds.), *Electromagnetic Fields and Neurobehavioral Function*, Alan R. Liss, New York, pp. 235-264.

25. D'Andrea, J. A., DeWitt, J. R., Portuguez, L. M. and Gandhi, O. P. (1988) Reduced exposure to microwave radiation by rats: Frequency specific effects, in M. E. O'Connor and R. H. Lovely (eds.), *Electromagnetic Fields and Neurobehavioral Function*. Alan R. Liss, New York, pp. 289-308.

26. Adair, E. R. (1988) Microwave challenges to the thermoregulatory system, in M.E. O'Connor and R.H. Lovely (eds.), *Electromagnetic Fields and Neurobehavioral Function*, Alan R. Liss, New York, pp. 179-201.

27. Wenger, C. B. (1983) Circulatory and sweating responses during exercise and heat stress, in E. R. Adair (ed.), *Microwaves and Thermoregulation*, Academic Press, New York, pp. 251-273.

28. Lotz, W. G. and Saxton, J. L. (1988) Thermoregulatory responses in the rhesus monkey during exposure at a frequency (225 MHz) near whole-body resonance, in M. E. O'Connor and R. H. Lovely (eds.), *Electromagnetic Fields and Neurobehavioral Function*, Alan R. Liss, New York, pp. 203-218.

29. Adair, E.R., Cobb, B.L., Mylacraine, K.S. and Kelleher, S.A. (1998) Human exposure at two radio frequencies (450 and 2450 MHz): Similarities and differences in physiological response, *Bioelectromagnetics* (in press).

30. Adair, E.R., Kelleher, S.A., Mack, G.W. and Morocco, T.S. (1998) Thermophysiological responses of human volunteers during controlled whole-body radio frequency exposure at 450 MHz, *Bioelectromagnetics* **19**:232-245.

31. Nadel, E.R., Bullard, R.W. and Stolwijk, J.A.J. (1971) Importance of skin temperature in the regulation of sweating, *J. Applied Physiology* **31**:80-87.

32. Stolwijk, J.A.J. (1983) Thermoregulatory response to microwave power deposition, in E.R. Adair (ed.), *Microwaves and Thermoregulation*, Academic Press, New York, pp. 297-305.

33. Adair, E.R. and Berglund, L.G (1989) Thermoregulatory consequences of cardiovascular impairment during NMR imaging in warm/humid environments, *Magnetic Resonance Imaging* **7**:25-37.

THERMAL MODELS FOR MICROWAVE HEATING OF TISSUE

K. R. FOSTER
Department of Bioengineering
University of Pennsylvania
220 S. 33rd. St.
Philadelphia PA 19104

1. Introduction

Much work has been devoted to measuring the rate of microwave energy absorption in tissue (i.e. the SAR), and the SAR has become the principal dosimetric quantity by which exposures Much less work has been done on determining the resulting temperature rise. This is paradoxical because the as-yet established hazards of microwave energy are thermal in nature. The absorbed power does not represent a hazard in most cases of RF burns or thermal pain, but rather the resulting temperature rise in tissue.

Apart from hazard considerations, thermal modeling can complement and guide electromagnetic dosimetry in several respects:

First, calculating the absorption of electromagnetic energy in tissue is potentially an endless topic. Electromagnetic models can be developed at many levels of detail, ranging from uniform spheroidal models to block models of man to (most recently) detailed computer models of the human body with millimeter resolution. Some investigators have even developed "microdosimetric" models on the molecular level. Because the body is electrically heterogeneous and complex in geometry, the SAR will also be complex. However, the temperature rise in exposed tissue will be smoothed out by the effects of thermal diffusion. To the extent that the goal is to study thermal hazards, appropriate thermal models can be used to help determine the level of detail that is needed in electromagnetic models.

Second, thermal models can be very useful in setting standards for protection against thermal hazards. Current regulatory practice is shifting away from the use of fixed "safety factors" in setting standards, to the use of uncertainty factors that are based on the quality of the data. Thermal models can be useful in estimating thresholds for thermal hazards, or interpolating limited biological data to other exposure conditions.

In particular, the bioelectromagnetics literature has paid inadequate attention to thermal models for studies of RF hazards, particularly for partial body exposures or exposure at millimeter wave frequencies. By contrast, the hyperthermia literature pays

357

B.J. Klauenberg and D. Miklavcic (eds.), Radio Frequency Radiation Dosimetry, 357-366.
© *2000 Kluwer Academic Publishers. Printed in the Netherlands.*

much attention to thermal modeling, and includes some excellent didactic papers on the subject [1].

This article will review one standard formulation of heat transport in tissue, and discuss some of its implications with respect to microwave exposure of tissue. It is based on longer developments to be published elsewhere [2, 3].

2. Thermal Model

I consider the local temperature rise in tissue subject to partial body exposure, corresponding to exposure situations where the local temperature rise is the limiting thermal effect. This temperature rise is limited by heat conduction, heat convection by blood flow, and perhaps other heat transport mechanisms as well. For RF burns and thermal pain, the limiting effect is localized temperature rises, even though the total thermal burden to the body is modest.

A thermal model that is often adequate for such purposes is the Pennes' bioheat equation [4]:

$$k\nabla^2 T^* - \rho_b \rho_t C_b m_b T^* + \rho_t \; SAR^* = C_t \rho_t \frac{dT^*}{dt^*} \tag{1}$$

where variables marked with * have dimensions, and

$T*$ is the temperature of the tissue (°C) above mean arterial temperature
k is the thermal conductivity of tissue (0.6 W/m °C)
$SAR*$ is the microwave power deposition rate (W/kg)
C_b is the heat capacity of blood (4000 W sec/kg°C)
C_t is the heat capacity of tissue (4000 W sec/kg°C)
ρ_b is the density of blood (1000 kg/m^3)
ρ_t is the density of tissue (1000 kg/m^3)

and m_b is the volumetric perfusion rate of blood (arbitrarily assumed to be 40 mL/100 g of tissue per min, a typical value for soft tissue). The following discussion will assume for simplicity $\rho_b = \rho_t = \rho$ and $C_b = C_t = C$. I comment about the validity of Pennes' model elsewhere [2, 3].

It is useful to reformulate the problem in dimensionless form, which will remove the variability due to size and focus on the physics of the process. Equation 1 can be rewritten in terms of dimensionless quantities (x, t, T):

$$T = \frac{T^* k}{\rho \; SAR_o^* \; L^2} = \frac{T^*}{\tau_2} \frac{C}{SAR_o^*}$$

$$x = \frac{x^*}{L}$$

$$t = \frac{t^* k}{\rho C L^2} = t^* / \tau_2 \tag{2}$$

$$SAR = \frac{SAR^*}{SAR_o^*}$$

$$\nabla^2 T - \frac{\tau_2}{\tau_1} T + SAR(r, t) = \frac{\partial T}{\partial t} \qquad (3)$$

where L is a characteristic distance scale of the heating and SAR_o* characterizes the exposure (normally, it would be the maximum SAR in the heated volume) and τ_1 and τ_2 are time constants defined below. The resulting dimensionless equation becomes

$$\tau_2 = \rho C L^2 / k \qquad (4)$$

where
is a time constant for thermal conduction, and

$$\tau_1 = 1 / m_b \rho \qquad (5)$$

is a time constant for heat convection by blood flow. The source term in Equation 3 has been normalized by SAR_o.

Thus the thermal response (as modeled by the bioheat equation) is characterized by two time constants, τ_1 and τ_2, representing, respectively, heat convection by blood flow and heat conduction. These processes act in parallel, and that with the shortest time constant will dominate the thermal response in a region of tissue. It is evident from Equations 4 and 5 that the relative sizes of these two time constants will depend on the dimensions of the heated region. These are summarized in Table 1 for materials whose electrical and thermal properties are characteristic of those of tissue.

TABLE 1. Thermal and electrical properties of a typical soft tissue

Frequency, GHz	Energy penetration depth L, cm (for plane waves)	Energy transmission coefficient ς (for plane waves)	Convection time constant τ_1, sec	Conduction time constant τ_2, sec
1.0	1.74	0.41	150	2000
2.45	1.1	0.42	150	750
10	.15	0.43	150	15
35	.03	0.50	150	0.6
100	.015	0.64	150	0.14
300	0.01	0.8	150	0.07

* assuming thermal parameters given the body of the text, and using the Debye relaxation equation to approximate the electrical properties of tissue, using static permittivity 50, low-frequency conductivity 1 S/m, and relaxation frequency 20 GHz. There is considerable variability in tissue properties and these values are only approximate

The distance scale is appropriate for plane-waves incident on tissue, i.e. L is chosen as the 1/e energy penetration depth. The table was calculated using literature values for the dielectric properties of muscle at the respective frequency.

Steady-state temperature profiles. These can be obtained by solving Equation 3 with the right hand side set equal to zero. For short heating distances (appropriate for exposure at millimeter wave frequencies or for very localized heating), heat conduction will be the limiting effect. For regional heating (over distance scales of cm or larger) heat convection will be the limiting cooling mechanism. In the latter case, the steady-state temperature rise becomes simply

$$T(r,\infty) \approx \frac{\tau_1}{\tau_2} SAR(r) \tag{6}$$

or, in dimensional units,

$$T^*(r^*,\infty) \approx \frac{SAR^*}{C}\tau_1 = \frac{SAR^*}{\rho C_{mb}} \tag{7}$$

Transient regime. In the early transient limit, before heat conduction and convection effects become significant, the solution to Equation 3 is simply (in normalized units)

$$T = SAR\, t \quad \text{for} \quad t \ll 1$$

or, in dimensional units

$$T^*(r^*,t) \approx \frac{SAR^*}{C}t^* \tag{8}$$
$$t^* \gg \tau_1, \tau_2$$

This early transient period ranges from picoseconds (for heating over distance scales of macromolecules) to seconds or more (for heating over millimeter to centimeter distance scales). This is the basis for the thermal method of measuring SAR.

For longer times, the temperature rise will be smoothed out over distances comparable to the thermal diffusion distance. A simple correction for effects of heat conduction, assuming that the SAR is time-independent, is (again using nondimensional variables)

$$T \approx SAR\,t + \int \nabla^2 T dt$$
$$\approx SAR\,t + \frac{t^2}{2}\nabla^2 SAR \tag{9}$$

Thus, heat conduction effects will be more pronounced if the SAR pattern is inhomogeneous (more precisely, if its second derivative with respect to space is large). The effects will increase quadratically with time. Thus the time period in which the temperature increase accurately reflects the SAR will be shorter for nonuniform heating. At progressively longer times, because of thermal diffusion, the local temperature rise will reflect an average of the SAR and thermal properties of the tissue over progressively larger distances.

3. Applications

3.1. "POINT HEATING"

Over the years investigators have speculated that brief heating of a tissue will lead to nonuniform temperature rises, because of the nonuniform conductivity of the tissue on the distance scales of microns or less. This "point heating" has been conjectured as a mechanism for "microthermal" effects even though the spatially-averaged temperature rise might be very small.

This hypothesis can be examined by considering a simple model, consisting of a uniformly heated sphere of tissue surrounding by a large region of unheated tissue of otherwise similar thermal properties. The steady-state temperature rise within the heated region is limited by heat conduction to surrounding tissue. The thermal relaxation time (in this case determined by heat conduction) defines the time scale over which a structure approaches equilibrium with its surroundings if a quantity of heat is suddenly added to it. This and several other related cases (heated plane, heated cylinder) have been analyzed elsewhere [5].

In this case, a thermal relaxation time τ that gives a better indication of the time required to approach steady-state is

$$\tau = \frac{\rho C R^2}{4k} \tag{10}$$

where R is the radius of the heated region. Thermal relaxation times and steady state temperatures for tissue spheres of different dimensions are shown in Table 2.

Table 2 shows that local temperature fluctuations due to nonuniform heating over small distance scales are very small. It is evident that significant "microthermal" temperature fluctuations over subcellular distances are unlikely to be produced in the absence of significant heating of surrounding tissue, unless the local SAR is very high.

TABLE 2. Thermal relaxation times and steady state temperature rise for localized heating at SAR = 1 W/kg.

Radius of heated region, cm, surrounded by nonheated tissue region	Steady state temperature increase, K (conduction limited, i.e. no blood flow)	Thermal relaxation time, sec
0.01 cm	0.000008	0.02
0.1 cm	0.0008	2
1 cm	0.08	200
10 cm	8	20000

3.2. NONUNIFORM TEMPERATURE RISE DUE TO NONUNIFORM SAR IN THE BRAIN

An interesting example of localized temperature fluctuations due to high local SARs was reported recently by Walters et al. [6]. These investigators measured transient temperature rises in the rat brain subject to brief exposures to 2.06 GHz microwaves at high power levels. In this experiment, the SAR varied by a factor of more than two between the cortex and hypothalamus (1224 W/kg in the hypothalamus, 493 W/kg in the cortex). For short irradiation times (<25 sec) the temperature rise in the hypothalamus was about twice that in the cortex. At longer irradiation times (< 240 sec at one-tenth the exposure level given above) the temperature increases in the two brain regions were much closer. These results are clearly the result of heat conduction that becomes significant over a period of several tens of seconds.

A precise interpretation of this study would require a detailed thermal model of the rat brain based on anatomical data. However, simple considerations give insight into the results. The irradiation frequency was far below the resonant frequency for the head; thus the exposure corresponded to the low-frequency limit. Calculations based on a simple model for low-frequency microwave heating of ellipsoids similar in size to the rat brain [7] shows that twofold variations in SAR can occur over millimeter distances in the brain, because of the different contribution of magnetic- and electric-field components to the heating. Such distances correspond to conduction time constants τ_2 of a couple tens of seconds. Thus, the heat conduction effects, which are so clearly shown in Walters' study, can be understood in terms of a simple thermal model based on heat conduction.

The results of this study are clearly significant for animal exposures involving very high power levels for relatively short times, and will help to clarify previous results published by Walters et al. The extent of nonuniformity in SAR depends, obviously, on the size of the head and irradiation frequency, and Walters' results cannot be generalized to other animals or other irradiation frequencies. The thermal model can, provided that SAR patterns can be estimated. Thus the thermal model can be useful in helping to extend Walters' results to other exposure conditions.

3.3 SURFACE HEATING BY MILLIMETER OR INFRARED RADIATION

This is model is appropriate when the skin is exposed to microwaves above a few GHz, and leads to a simple one-dimensional formulation.

$$SAR^* = \frac{I_0\varsigma}{\rho L}\exp^{-x^*/L} = SAR_0\exp^{-x^*/L} \tag{11}$$

In terms of the intensity I_0 of the incident field, the SAR is
where ς is the fraction of incident energy that is absorbed in the tissue, L is the energy penetration depth and SAR_0 is the SAR at the surface of the tissue (which is defined by x = 0). Table 1 summarizes typical values for the parameters for typical tissues. The normalized bioheat equation becomes

$$\frac{\partial^2 T}{\partial x^2} - \frac{\tau_2}{\tau_1}T + e^{-x} = \frac{\partial T}{\partial t} \tag{12}$$

The full solution to the one-dimensional bioheat equation, including the case of heat loss from the surface, is given elsewhere [8].

Assuming an insulated boundary (no heat loss from the surface into space) the steady-state temperature T* at distance x* beneath the surface is:

$$T^*(x,\infty) = \frac{SAR_0\,\tau_2}{C}\left[\frac{\tau_1}{\tau_2-\tau_1}\right]\left[e^{-\frac{x}{L}} - \sqrt{\frac{\tau_1}{\tau_2}}e^{-\sqrt{\frac{\tau_2}{\tau_1}}\frac{x^*}{L}}\right] \tag{13}$$

At the surface (x* = 0) the temperature rise in dimensional units can be written:

$$T^*(0,\infty) = \frac{SAR_0}{C}\tau_{eff} \tag{14}$$

where

$$\tau_{eff} = \frac{\tau_2 - \sqrt{\tau_1\tau_2}}{\tau_2/\tau_1 - 1} \tag{15}$$

364

This takes on the limiting values of τ_1 and $(\tau_1 \tau_2)^2$ for the convection-limited case ($\tau_2 \gg \tau_1$) and conduction-limited cases, respectively. In dimensional units, the steady-state temperature increase at the surface is

$$T^*(0,\infty) = \frac{I_o S}{\rho \sqrt{m_b kC}} \qquad \tau_1 \ll \tau_2 \quad \text{(conduction limited)}$$

$$= \frac{I_o S}{LC_{m_b}\rho^2} \qquad \tau_2 \ll \tau_1 \quad \text{(convection limited)} \tag{16}$$

in the conduction-limited and convection limited cases, respectively.

In the absence of blood flow, this one-dimensional model has no steady-state solution. In reality, the steady-state temperature rise would be determined in that case by other heat transport mechanisms (e.g. evaporative cooling of the surface) which are not considered here.

The full closed-form solution for this problem is algebraically cumbersome. In the absence of blood flow, with insulated boundary conditions, the time-dependent surface temperature rise (in normalized units) reduces to

$$T(0,t) = 2\sqrt{\frac{t}{\pi}} + e^t erfc(\sqrt{t}) - 1 \tag{17}$$

The first term is the solution for surface heating (e.g. infrared heating of the skin). The second and third terms arises from the finite energy penetration depth of the radiation, and take into account the diffusion of heat over distances corresponding to the energy penetration depth; this contribution rapidly approaches zero when $t > 1$.

4. Discussion

It is interesting to consider the maximum temperature rise allowed by the major microwave standards, in particular ANSI/IEEE C95.1. These are summarized, for several conditions, in Table 3.

This standard is complex, since it is intended to cover several exposure situations. As applied to millimeter wave radiation, the standard has two provisions, for whole-body and partial-body exposure. The latter allows a maximum steady-state temperature rise in excess of 1 C at the surface of the skin (calculated on the basis of literature values for the dielectric properties of tissue). This is a factor of 8-10 below anticipated thresholds for thermal pain or thermal injury. Other provisions for partial-body exposure (e.g. from transmitters located near the body) limit SAR to 1.8 or 8 W/kg. This limit corresponds (with normal blood flow values) to steady-state temperature increases far below 1 C. Thus ANSI/IEEE C95.1 is clearly inconsistent in the safety factors that are built into it. I argue elsewhere (2) that a different provision in the standard, the low-power exclusion, corresponds to a safety factor close to 10 (which is more consistent with the rest of the standards).

The question, ultimately, is what hazard the standard is designed to protect against. If the limiting hazard is thermal damage from excessive local temperature rise in steady-state heating, the exact limit in the standard is unlikely to be very important. Because of the thermal inertia of the body, a substantial time is needed to approach the steady state (seconds to minutes). If the incident power density is sufficient to cause thermal damage to tissue from steady-state heating, the subject would perceive pain long before tissue damage occurred, and withdraw from exposure. The few reported injuries from microwave energy are generally associated with very high levels of exposure, in which tissue damage occurs before the victim can respond. Such exposure levels would exceed any reasonable safety limit. This is quite different from, for example, electric shock, where the threshold for hazard is rather sharp, and exposures slightly above the threshold are very quickly hazardous.

TABLE 3. Maximum temperature rise allowed by ANSI/IEEE C95.1 (1992) microwave standard.

Thermal Model	ANSI/IEEE C95.1 (controlled /uncontrolled environment)	Maximum Temperature Rise in the Steady State, Assuming Normal Blood Flow, °C
(millimeter waves) MPE for 15 - 300 GHz MPE for > 96 GHz, controlled environments, > 30 GHz for uncontrolled environments	Whole body exposure: 100 W/m^2 / 100 W/m^2 Partial Body Exposure 400 W/m^2 / 200 W/m^2	0.4 (whole body, controlled or uncontrolled) 0.7 (uncontrolled) 1.5 (controlled)
Local exposure	(Based on a limit of 1.6 or 8 W/kg for localized exposure)	0.06 (uncontrolled) 0.3 (controlled)

In short, thermal models can help define the thresholds for thermal injury, and refine exposure limits. If the limiting hazard is not a thermal effect, then the entire rationale for the exposure standard would need to be reevaluated in any event.

5. References

1. Roemer, R.B. (1990) Thermal dosimetry, in M. Gautherie (ed.) *Thermal Dosimetry and Treatment Planning, Clinical Thermology Subseries Thermography*, Springer-Verlag, pp. 119-314.
2. Foster, K.R. and Erdreich, L.S. (In press) Thermal models for microwave hazards, *Bioelectromagnetics*.
3. Foster, K.R., Lozano-Nieto, A., and Riu, P.J. (In press) Heating of tissue by microwaves: A model analysis, *Bioelectromagnetics*.
4. Pennes, H.H. (1948) Analysis of tissue and arterial blood temperature in resting forearm, *J. Appl. Physiol.* 1, 93-122.
5. Foster, K.R., Ayyaswamy, P.S., Sundararajan, T., and Ramakrishna, K. (1982) Heat transfer in surface-cooled objects subject to microwave heating. *IEEE Trans. Microwave Theory and Techniques*, MTT-30, 1158-1166.
6. Walters, T., Ryan, K.L., Beicher, J.C., Doyle, J.M., Tehrany, M.R., and Mason, P.A. (1998) Regional brain heating during microwave exposure (2.06 GHz), warm-water immersion, environmental heating and exercise, *Bioelectromagnetics* 19, 341-353.
7. Massoudi, H., Durney, C.H., and Johnson, C.C. (1977) Long-wavelength analysis of plane wave irradiation of an ellipsoidal model of man, *IEEE Trans. Microw. Theory Tech.* MTT-25, 41-46.
8. Foster, K.R., Kritikos, H.N., and Schwan, H.P. (1978) Effect of surface cooling and blood flow on the microwave heating of tissue, *IEEE Transactions on Biomedical Engineering* BME-25, 313-316.

SAR MEASUREMENTS IN THE RHESUS MONKEY ANKLE: IMPLICATIONS FOR HUMANS

R. G. OLSEN
Naval Health Research Center Detachment
8301 Navy Road
Brooks Air Force Base, Texas 78235-5365

1. Introduction

The introduction of limits on radiofrequency (RF) body currents in the most recent revision of the IEEE/ANSI exposure standard [1] has caused much discussion during recent years. Moreover, the revised standard covered a wider spectrum (down to 3 kHz), and for the first time brought the Navy's worldwide VLF submarine communication facilities under the same level of radiation hazard (RADHAZ) compliance as shipboard irradiation sources and commercial broadcasting.

Soon after the standard was published, a number of problems regarding the details of the allowed RF body currents were pointed out, mainly by individuals who assisted in drafting and in voting for the revised document. First, as printed, the standard imposed a uniform, 1-second averaging time for body currents over all covered frequencies, 3 kHz through 100 MHz. It had been assumed by many that the averaging time would be six minutes for RF body currents above 100 kHz by virtue of the predominant tissue-warming effect at higher frequencies. Second, the long-standing problem of point-touch versus grasping contact for body currents remained unresolved such that the prevailing opinion is that the IEEE/ANSI standard assumes a grasping contact. Third, the upper frequency limit of 100 MHz for RF body currents cuts through the FM broadcast band, 88 to 108 MHz, causing (technically) different compliance requirements for stations above or below 100 MHz. A fourth problem has been the least discussed; it concerns the 100-mA limit itself. In adopting that limit, localized "peak" SAR in the extremities was allowed to rise to 20 W/kg, and (according to accepted theoretical predictions) the limiting SAR would occur at approximately 100 mA for adult humans. The problem was that a limb current of 100 mA cannot typically be noticed as causing warmth for standing or grasping contact, a fact usually known only to those who have experience in personally evaluating RADHAZ situations. Furthermore, the problem of compliance with the 100-mA limit typically exists for a relatively small but important segment of workers, namely, those on broadcasting towers and those topside personnel aboard U. S. Navy ships [2].

B.J. Klauenberg and D. Miklavcic (eds.), Radio Frequency Radiation Dosimetry, 367-373.
© 2000 *Kluwer Academic Publishers. Printed in the Netherlands.*

368

Fortunately, most of the above-mentioned problems are well on the way toward being resolved. For example, changes in the language concerning induced current limits and averaging times have already been approved by the cognizant IEEE standards subcommittee. The problem of the current limit itself being overly conservative will require a comprehensive approach that uses several independent investigations and sites. The study described here was initiated to provide empirical evidence of RF-induced tissue heating in an animal ankle that can be easily compared to the human ankle.

2. Methods

The study used four adult male monkeys, *Macaca mulatta*, weighing 8.3 to 10.7 kg. The detailed procedures have been previously described [3]. Briefly, anesthetized subjects were seated in a plastic chair inside a partially RF-anechoic chamber with bare feet grounded to a metal-clad floor panel. Near-field irradiation (vertically polarized) at 100 MHz, CW was used at a power of 100 watts; the size of the seated subjects caused them to be close to the "grounded resonance" condition. Therefore, RF ankle currents in excess of 100 mA could be easily achieved when the subjects were located a distance of 1.5 m from the dipole/reflector antenna assembly. Assessment of localized ankle SAR was thermometrically determined through the use of "Vitek-type" thermal probes [4] that were inserted through implanted 16-gauge plastic catheters. Insertion and placement of the catheters involved a needle-puncture above the ankle joint to allow a nearly-full insertion of the barrel downward to the joint at various tissue depths. Taping the catheter body to the leg provided a rigid support for the flexible thermal probe tip. Measurement sites were typically anterior/posterior and medial/lateral for each ankle. To record RF-induced thermal increments inside the ankle joint capsule, a horizontal catheter was inserted from the front.

A typical procedure used a baseline thermal recording followed by a 120-s irradiation that was followed by an additional 2 to 5 min of thermal data collection. Temperatures were monitored continuously and recorded at 30-s intervals. Irradiations were repeated at least once, and the RF-induced temperature increments for a given session were averaged. At the end of each session, calipers were used to measure the size of each ankle from which the overall area was calculated.

Preliminary experimentation used a thermographic imaging system to monitor surface temperatures after brief, high-intensity exposures. Previously published predictions of relatively high ankle SAR [5,6] caused us to be cautious with respect to the possible injury of living subjects. However, after observing the first few thermographic scans of a shaved, post-irradiation rhesus ankle, we were convinced that no extremely high tissue temperatures had been produced, and we observed no evidence of any ankle injury to the subjects, The bony ankle region, indeed, was where the highest RF-induced heating was seen, but the thermal excursions, as observed thermographically, hardly exceeded 2° C for 60-s irradiations with ankle current between 100 and 200 mA.   For these conditions, SARs in excess of 1,000 W/kg were theoretically predicted.

3. Results

Figures 1–3 show thermal records from three regions of the ankle that are representative of the highest and lowest heating rates. The highest average heating occurred in the deep tissues of the posterior ankle region (Figure 1).

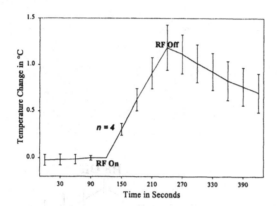

Figure 1. RF-induced heating (± 1 SD) for the posterior right ankle, intramuscular. Average RF current was 09 mA at 100 MHz; calculated wet-tissue SAR was 43.3 W/kg.

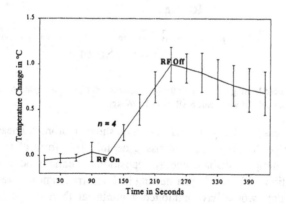

Figure 2. RF-induced heating (± 1 SD) for the anterior right ankle, near-bone depth. Average RF current was108 mA at 100 MHz; calculated wet-tissue SAR was 34.9 W/kg.

370

The initial heating rate of 0.62° C/min. predicted a localized SAR of 43.3 W/kg. Somewhat lower heating was measured at the ankle front surface (Figure 2) with a SAR of 34.9 W/kg. The lowest average heating was observed within the ankle joint capsule (Figure 3) where only two subjects were studied. Within the joint, the initial temperature slope was approximately one-third of that seen in Figure 2 from which an average SAR of 11.6 W/kg resulted. The area of the joint accounted for a sizable fraction of the ankle cross-section; therefore, the data show that even though RF-induced warming was relatively low inside the joint, much of the total energy deposition occurred there.

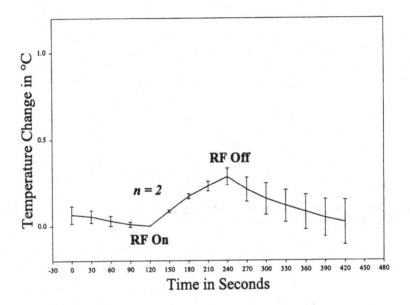

Figure 3. RF-induced heating (± 1 SD) for the intracapsular ankle, left and right. Average RF current 106 mA at 100 MHz; calculated wet-tissue SAR was 11.6 W/kg.

All of the measured locations near the ankle showed prompt, linear heating followed by immediate cooling regardless of tissue depth. This form of the heating-cooling curves implies simple, adiabatic energy deposition (essentially no heat loss or heat gain during irradiation.) If much higher heating had occurred in nearby regions of the ankle, the recorded data would have exhibited a nonlinear thermal increase instead of the observed nearly linear heating rate.

4. Discussion

These results suggest that RF body-to-ground current flows in a more-or-less uniform manner through all tissues of the ankle with no extreme "hot spots." The average overall area of our subjects' ankles was 9.6 cm², and our results show that if we were to construct an average "effective" cross section of high water-content tissue, about 60% of the total area would be appropriate. Localized SAR based on such an effective cross section is a strong function of the assumed diameter, being proportional to the inverse area squared. The implications for human exposure are significant, and extrapolation to wrist SAR is possible. For a given limb current, a SAR based on a 60% versus 15% cross section is reduced by a factor of 16. This is illustrated for the animal and human ankles in Figure 4 . Assumptions used in Figure 4 include the previously-accepted

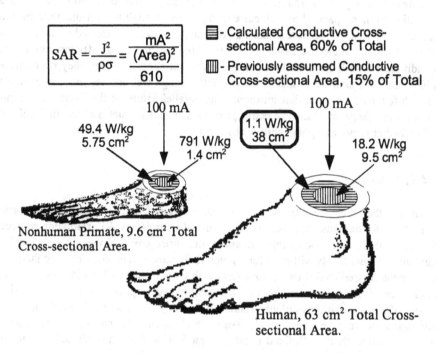

$$SAR = \frac{J^2}{\rho\sigma} = \frac{\frac{mA^2}{(Area)^2}}{610}$$

▤ - Calculated Conductive Cross-sectional Area, 60% of Total

▥ - Previously assumed Conductive Cross-sectional Area, 15% of Total

100 mA

49.4 W/kg
5.75 cm²

791 W/kg
1.4 cm²

Nonhuman Primate, 9.6 cm² Total Cross-sectional Area.

100 mA

1.1 W/kg
38 cm²

18.2 W/kg
9.5 cm²

Human, 63 cm² Total Cross-sectional Area.

Figure 4. Comparison of SAR at 100 mA RF current for two "effective" cross sections between the primate and human ankle.

equivalent human ankle cross section of 9.5 cm² that accounts for 15% of the total, a limb current of 100 mA, and the average overall rhesus ankle area of 9.6 cm². Figure 4 shows that when the larger cross sectional areas are used, lower SARs result with animal SARs being close to the present results. I, therefore, speculate that the lower human SAR (1.1 W/kg) is more appropriate for a limb current of 100 mA.

Although easy to apply, the use of an equivalent cross section to quantify RF-induced ankle and wrist warming is fraught with the difficulties of oversimplification regarding highly complex living structures. From our results, variations in localized SARs exceed a factor of three, which cannot be fitted into the concept of a single "effective" RF current cross section. Obviously, different amounts of various tissue regions account for significant variations in localized SAR. Extreme variations, however, were not observed in this study. The IEEE/ANSI C95.1-1991 standard is based on a localized SAR as averaged over a 10-g tissue mass, and our evidence does not show the predicted (extremely high) localized SARs. Recent evidence from RF-energized wrists from primate cadaver wrists have corroborated this conclusion (Dr. John D'Andrea, Private Communication). Also corroborative are our own preliminary results with *in vivo* primate wrist SARs. Although it would be difficult to demonstrate, I propose that there could be a net phase shift in the total RF current vector (with respect to the applied voltage) as it propagates through physically complex physical structures such as ankles and wrists. In this scenario, the total limb current is a summation of many parallel branch currents that flow through the various fluid-bathed tissues. Each branch current maintains a phase angle versus the applied voltage that is consistent with the dielectric tissue properties at that location. If a sufficient number of the individual branch currents flow through low-water-content tissues, they would have a phase angle different from that of the currents through wet tissues, and a measurable phase shift should occur at that location. The parallel nature of the summed currents and the wide range of tissue dielectric properties would predict a range of localized SARs whether or not any phase shift is observed.

5. Conclusions

I conclude that these results, when verified, replicated, and accepted, have major implications for humans, especially in occupational settings near RF irradiation sources. For adults, there is a wide range of ankle and wrist sizes; obviously, in smaller individuals, higher SARs will exist for a given limb current. Nevertheless, the factor of 16 reduction in predicted SAR, as consistent with these results, will be important for all adults. The disparity between these results and theoretical predictions probably exists because of the nature of physiological fluids in the body. No sharply defined regions of very high SAR were seen because all tissues are electrically "connected" to each other via a conductive fluid. Refined dosimetry models of the future should attempt to incorporate this important feature.

6. Acknowledgments

The author wishes to express sincere gratitude to Mr. Barry Van Matre, Senior Electronics Technician, who, in very little time, produced the highly informative and graphically pleasing SAR comparison illustration (Figure 4). The experiments reported herein were conducted according to the principles set forth in the "Guide for the Care and Use of Laboratory Animals," Institute of Laboratory Animal Resources, National Research Council, National Academy Press, 1996. This research was sponsored by the Naval Medical Research and Development Command under work unit 63706N-M00096.004-1515. The views expressed in this article are those of the author and do not reflect the official policy or position of the U. S. Department of the Navy, U. S. Department of Defense, or the U. S. Government.

7. References

1. IEEE Standards Coordinating Committee 28. (1992) *IEEE C95.1-1991 Standard for Safety Levels withRespect to Human Exposure to Radio Frequency Electromagnetic Fields, 3 kHz to 300 GHz,* Institute of Electrical and Electronics Engineers, New York.
2. Olsen, R. G., Griner, T. A., Van Matre, B. J. and King, J. J. (1993) *Measurements of Radiofrequency (RF) Body Potential and Body Currents in Personnel Aboard Two Classes of Navy Ships,* NAMRL-1384, Naval Aerospace Medical Research Laboratory, Pensacola. FL.
3. Olsen, R. G., Van Matre, B. J., and Lords, J. L. (In press) RF-induced ankle heating in a rhesus monkey at 100 MHz, in F. Bersani (ed.), *Electricity and Magnetism in Biology and Medicine* Plenum Press, New York.
4. Olsen, R. G. and Bowman, R. R. (1989) A simple nonperturbing temperature probe for microwave/RF dosimetry, *Bioelectromagnetics* 10, 209-213.
5. Gandhi, O. P., Chatterjee, I., Wu, D., and Yu, Y. G. (1985) Likelihood of high rates of energy deposition in the human leg at the ANSI-recommended 3-30 MHz safety levels, *Proc. IEEE* 73, 1145-1147.
6. Chen, J. Y. and Gandhi, O. P. (1988) Thermal implications of high SAR's in the body extremities at the ANSI-recommended MF-VHF safety levels, *IEEE Trans. Biomed. Eng.* 33, 435-441.

MICROWAVE DOSIMETRY AND LETHAL EFFECTS IN LABORATORY ANIMALS

V. G. PETIN, G. P. ZHURAKOVSKAYA and A. V. KALUGINA
Medical Radiological Research Center
Obninsk, Kaluga Region, Russia

1. Introduction

The traditional way used in Russia for the description of microwave effects consists in the description of these effects as they depend on the power flux density. As has become obvious, such a representation is not only insufficient, but also can be the reason for faulty conclusions and inadequate extrapolation [1, 2, 3]. The most acceptable parameter for a quantitative estimation of microwave effects is the specific absorption rate (SAR). This parameter, however, depends in a complex manner on the frequency of irradiation, the size, form, and weight of the object, and its orientation in a field [4]. The theoretical calculations of the average SAR are based on simplified models and thus require confirmation by the direct dosimetric measurements for real biological objects. The methods of experimental and theoretical numerical determination of the average SAR of electromagnetic radiations are widely used in various countries [5-8]. However, similar works in our country are absent. In this paper we describe the original calorimetric devices and the results of dosimetric measurements of the average SAR for various laboratory animals, frequencies, and polarizations. In addition, the comparison between our measured results and those calculated by others [4] will be presented.

One of the main biological effects of microwave radiation on living matter is an increase in the temperature of the irradiated object, which is directly proportional to the amount of energy absorbed [1, 9]. This is caused by the conversion of electromagnetic energy to thermal energy. Thermal responses of an animal to microwave radiation depend on environmental temperature, humidity and air flow [10]. The ultimate effect of the increased temperature can be animal death. Lethal effects are well reproduced and may be quantified. With sufficient energy absorption and duration of exposure, the animal's body temperature will rapidly rise to a lethal level and the animal will expire, death being due to hyperthermia. The total thermal loading of a subject exhibits two basic rules: it increases with exposure duration and a higher temperature corresponds to an increased thermal loading [3, 11]. Hence, the intensity of an incident field and exposure duration are important parameters that determine the quantity of energy absorbed by tissue.

375

B.J. Klauenberg and D. Miklavcic (eds.), Radio Frequency Radiation Dosimetry, 375-382.
© 2000 *Kluwer Academic Publishers. Printed in the Netherlands.*

The relative interspecies sensitivity to microwave exposure is also important in order to extrapolate the experimental biological results from animals to humans. The effect of microwave exposure rates in relation to survival during irradiation has been studied only for separate kinds of animals and within a comparatively narrow range of power flux density [3, 13-15]. In this paper, the survival during microwave exposure of various laboratory animals was obtained as a function of the power flux density and the average SAR, using a wide range of these values.

2. Dosimetric Data

To calculate SAR, the amount of energy absorbed was determined by measuring the amount of heat generated in a fresh animal carcass irradiated with microwaves by use of a differential or a quasiadiabatic calorimeter. The heating pattern should be measured before it is altered by blood flow and heat diffusion. Therefore dead animals have been used as a phantom material, eliminating the problems of physiological heat production and loss. Calorimetric techniques are usually used to measure the mean whole-body SAR. We have used a twin-well calorimeter similar to that used by McRee [15] but with some modifications. Twin-well calorimeters are instruments which measure mean whole-body SAR very accurately. The accuracy of these techniques has been estimated to be about 5%.

The principle of action of a differential calorimeter is based on measuring the energy difference required to support equal temperatures on the surface of two identical cells, in one of which an animal phantom with lower temperature was inserted. The condition of action of a quasiadiabatic calorimeter is based on the creation of an essential distinction between the thermodiffusion coefficient β_1 for the object in the calorimetric cell and the thermodiffusion coefficient β_2 for the calorimeter as a whole: $\beta_1/\beta_2 > 10$. To determine the absolute value of microwave energy absorbed by the whole body of an animal, the measurements were performed with animal cadavers. Two animal cadavers of identical mass were maintained in thermostatic conditions up to establishment of thermal balance at a given temperature. After this, one cadaver was placed in a measured cell and the amount of energy (Q_1) absorbed in a differential calorimeter or dissipated in an adiabatic calorimeter by the object was measured. Another phantom was placed in a radiotransparent and thermally-insulated container which was exposed to a microwave field during a certain fixed time t. After exposure, the energy Q_2 absorbed or diffused by the irradiated object in the corresponding calorimeter was measured. The difference between these energies, i.e. $\Delta Q = Q_1 - Q_2$, was the integral of absorbed microwave energy determined without taking into account the energy dissipated by the object during irradiation. The account of this part of energy resulted in the following formula for the total energy Q absorbed by the irradiated object:

$$Q = \beta t[\Delta Q/(1 - e^{-\beta t}) + \Delta T_0 mc] \, , \tag{1}$$

where ΔQ - the absorbed energy measured calorimetrically;

ΔT_0 - the difference between temperatures of the cadaver and surroundings just before the irradiation;

m - mass of the cadaver;

c - specific heat of the cadaver;

t - the duration of microwave exposure;

β - the thermodiffusion coefficient.

Then the average SAR was determined by

$$SAR = Q/tm. \tag{2}$$

Because one of the purposes of this work was to compare the dependence of animal lethality on the average SAR and on the power flux density, the last value was obtained by P3-9 apparatus (produced in Russia), the accuracy of measurement of the power flux density being only $\pm 30\%$.

Dosimetric investigations have been carried out using 90 mice, 105 rats, 18 rabbits and 18 dogs. The measured values of the average SAR (W/kg per mW/cm^2) for various animal mass, frequencies and polarizations are summarized in Table 1. Of particular interest is the comparison between measured and calculated values of the average SAR. Figure 1 and Figure 2 show our measured values compared with average SAR values calculated by others [4] for similar exposure conditions.

The data presented in Table 1, Figure 1 and Figure 2 lead to the following observations:

1. The values of the average SAR for various laboratory animals exposed to 7 GHz fields are almost identical both for E and H polarizations; the values for K polarization are about a factor of 1.4 lower.

2. In most cases, the values of average SAR for laboratory animals exposed to 2.4 GHz electromagnetic field are very close to those obtained at 7 GHz with the exception of the average SAR = 0.65 W/kg per mW/cm^2 obtained for mouse (E polarization) which is about a factor of 1.5 higher than the value for 7 GHz field.

3. For 460 MHz field, the highest microwave absorption is observed for E polarization; for H and K polarizations the average SAR in small animals is about a factor of 3-4 higher, while for large animals it is almost independent on orientation of the E and H-field vectors with respect to the absorbing object.

4. The values of the average SAR in dog exposed at 460 MHz field are about a factor of 2 higher than at 7 GHz while for mouse this dependence was quite opposite, i.e., the values of SAR at 460 MHz are about a factor of 2-5 lower than the corresponding values observed at 7 GHz.

5. The values of average SAR both at 7 and 2.4 GHz are altered in inverse proportion to the body mass of laboratory animals; this relationship was not observed at 460 MHz.
6. For all electromagnetic fields used in this investigation, for E and H polarizations, calculated planewave average SAR's are very close to the measured values, with the exception of some experimental measurements obtained for rabbits.
7. For K polarization, some inconsistencies at higher frequencies between measured and calculated values of average SAR are observed.

TABLE 1. Measured values of the average SAR (W/kg per mW/cm^2) for models of laboratory animals

Animal	Mass, g	Frequency, GHz	Polarization		
			E	H	K
Mouse	30 ± 5	7	0.41 ± 0.05	0.42 ± 0.03	0.30 ± 0.04
		2.4	0.65 ± 0.07	0.40 ± 0.06	0.30 ± 0.05
		0.46	0.23 ± 0.03	0.08 ± 0.04	0.08 ± 0.04
Rat	150 ± 20	7	0.24 ± 0.03	0.24 ± 0.02	0.17 ± 0.02
		2.4	0.26 ± 0.03	0.23 ± 0.03	0.19 ± 0.03
	250 ± 20	7	0.21 ± 0.02	0.22 ± 0.02	0.14 ± 0.02
		2.4	0.21 ± 0.02	0.21 ± 0.03	0.15 ±0.02
		0.46	0.41 ± 0.04	0.09 ±0.02	0.13 ± 0.03
	350 ±20	7	0.16 ± 0.02	0.16 ± 0.02	0.12 ± 0.02
		2.4	0.17 ± 0.02	0.18 ± 0.02	0.12 ± 0.02
Rabbit	3000 ± 500	7	0.07 ± 0.01	0.07 ± 0.01	0.045 ± .005
		0.46	0.13 ± 0.03	0.085 ± 0.02	0.12 ± 0.02
Dog	7500 ± 500	7	0.05 ± 0.005	0.05 ± 0.005	0.035 ± 0.005
		0.46	0.09 ± 0.02	0.08 ± 0.02	0.07 ± 0.02

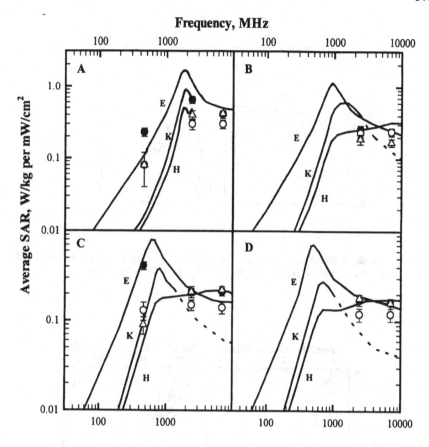

Figure 1. Calculated (lines) and measured (closed circles - E, open triangles - H, open circles - K polarizations) values of the average SAR for mouse (30 g, **A**) and rat (150 g, **B**; 250 g, **C**; 350 g, **D**).

3. Lethal Effects in Laboratory Animals

The experiments were performed on male mice of strain $C_{57}BL_6$, Wistar rats and Chinchilla rabbits. The animals were not anesthetized during exposure. The average body masses of the animals were 20 g for mice, 200 g for rats and 2.5 kg for rabbits. A total of 350 mice, 300 rats and 36 rabbits were used in this investigation. For irradiation, an exposure system in an anechoic chamber was used. Each animal was exposed individually to continuous-wave microwaves at a frequency of 7 GHz, mainly for H-orientation of body in the following ranges of power density: 0.08-25 W/cm^2 (mouse and rat) and 0.13-0.37 W/cm^2 (rabbit)

380

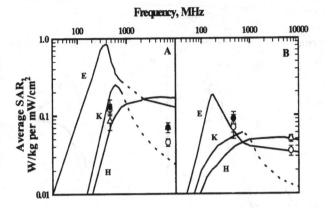

Figure 2. Calculated (lines) and measured (closed circles - E, open triangles - H, open circles - K polarizations) values of the average SAR for rabbit (3000 g, **A**) and dog (7500 g, **B**).

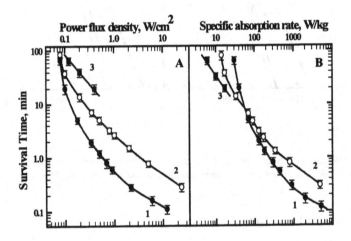

Figure 3. The dependence of the survival time of laboratory animals during microwave irradiation (7 GHz, curves 1 - mice, curves 2 - rats, curves 3 - rabbits) on microwave power density (**A**) and on the average SAR (**B**).

under following conditions. Room temperature was 22 ± 0.5^0C and humidity was $55\pm5\%$. Animal death during irradiation was registered by means of a television system after the end of tremor. Error bars represent the standard error of the mean. Each of the experimental points presented in Figure 3 is the average value for at least six animals.

The dependence of the survival of laboratory animals during irradiation (7 GHz) on power density is presented at Figure 3 (A). The data were obtained for mice (curve 1), rats (curve 2), and rabbits (curve 3). The survival during irradiation appears to increase with body mass for all power densities studied. From this result, animals in correspondence with their microwave sensitivity, may be arranged in the following range: mice, rats, rabbits. It can be also seen that the initial slopes of curves 1-3 differ from each other. Smaller animal sizes correspond to steeper slopes. One can expect that these curves will be crossed at low levels of power density where their relative interspecies microwave sensitivity may be different. This fact is more distinctly observed if we analyze the dependence of the survival on average SAR (Figure 3, B). Evidently, there are qualitatively different regions of SAR: at low SAR's the survival of rats was less than that of mice and *vice versa* at an SAR exceeding 50 W/kg.

The results presented illustrate a well known fact that to make a correct conclusion about the relative interspecies animal sensitivity, the average SAR must be used instead of power flux density. Indeed, the conclusion about the relative animal sensitivity to low level microwave radiation may be quite opposite using power flux density and SAR. The basic thermoregulatory profile of each species is also very important [16]. It is also apparent that biological responses are different before and after SAR = 50 W/kg at any given frequency: for lower magnitudes of SAR small animals are more resistant to microwave radiation and for higher values of SAR quite the reverse - small animals are more sensitive to microwave irradiation. It means that there is no simple correlation between animal sensitivity to microwave irradiation and their dimensions or body mass. In other words, the relative interspecies animal sensitivity to microwave irradiation is dependent on SAR. Thus, these results may be useful both for interpretation of microwave induced lethal animal effects and extrapolation of these data from animals to humans.

Taking into account the conclusion that whole-body SAR at frequencies between 3 MHz and 20 GHz may be used in developing new frequency-dependent radiofrequency radiation safety standards [4], one can hope that the results of this paper may be useful for further developing the dose-power approach in Russian hygienic standardization for microwave exposure [17] and diminishing the difference between various approaches in public health policy [18].

4. References

1. Johnson, C.C. and Guy, A.W. (1972) Non-ionizing electromagnetic wave effects in biological materials and systems, *Proc. IEEE* **60**, 692-718.
2. Massoudi, H., Durney, C.H., Barber, P.W., and Iskander, M.F. (1982) Postresonance electromagnetic absorption by man and animals, *Bioelectromagnetics* **3**, 333-339.

382

3. Michaelson, S.M. (1983) Thermoregulation in intense microwave fields, in E.R. Adair (ed.), *Microwaves and Thermoregulation*, Academic Press, New York, pp. 283-295.

4. Durney, C.H., Massoudi, H., and Iskander, M.F. (1986) *Radiofrequency Radiation Dosimetry Handbook*, USAFSAM-TR-85-73, Brooks Air Force Base, San Antonio, TX.

5. Ho, H.S. and Youmans, H.D. (1975) Development of dosimetry for RF and microwave radiation, *Health Physics* **20**, 325-329.

6. Durney, C.H. (1980) Electromagnetic dosimetry for models of humans and animals: A review of theoretical numerical techniques, *Proc. IEEE* **68**, 33-40.

7. Guy, A.W. (1987) Dosimetry associated with exposure to non-ionizing radiation: Very low frequency to microwave, *Health Physics* **53**, 569-584.

8. Gandhi, O.P. (1990) Electromagnetic energy absorption in humans and animals, in O.P. Gandhi (ed.), *Biological Effects and Medical Applications of Electromagnetic Energy*, Englewood Cliffs, New Jersey, Prentice Hall, pp. 174-195.

9. Chiabrera, A., Nicolini, C., and Schwan, H.P. Eds. (1985) *Interaction between Electromagnetic Fields and Cells*, Plenum Press, New York.

10. Adair, E.R. Ed. (1983) *Microwaves and Thermoregulation*, Academic Press, New York.

11. Sapareto, S.A. (1987) Thermal isoeffect dose: addressing the problem of thermotolerance, *Int. J. Hyperthermia* **3**, 297-305.

12. Rugh, 5.R., Ho, H., and McManaway M. (1976) The relation of dose rate of microwave radiation to the time of death and total absorbed dose in the mouse, J. *Microwave Power* **11**, 279-281.

13. Elder, J.A., Czerski, P.A., Stuchly, M.A., Mild, K.H., and Sheppard, A.R. (1989) Radiofrequency, in *Nonionizing Radiation Protection*, WHO Regional Publications, Europe, Ser. No. 25, Copenhagen, pp. 117-173.

14. Jauchem, J.R. and Frei, M.R. (1994) Cardiorespiratory changes during microwave-induced lethal heat stress and β-adrenergic blockade, *J. Appl. Physiol.* **77**, 434-440.

15. McRee, D.I. (1974) Determination of the absorption of microwave radiation by biological specimens in a 2450 MHz microwave field, *Health Physics* **26**, 385-390.

16. Adair, E.R. (1995) Thermal physiology of radiofrequency radiation (RFR) interactions in animals and humans, in Radiofrequency *Radiation Standards*, Plenum Press, New York, pp. 245-269.

17. Zhavoronkov, L.P. and Petin, V.G. (1998) The dose-power approach in hygienic standardization of electromagnetic fields. The concept, biophysical modeling, experimental estimation, in *Electromagnetic Fields. Biological Effects and Hygienic Standards, Moscow*, pp. 65.

18. Klauenberg, B.J., Grandolfo, M., and Erwin, D.N. (Eds.) (1995) *Radiofrequency Radiation Standards*. Plenum Press, New York.

TERATOLOGIC EFFECTS OF EXPOSURE TO RADIO FREQUENCY RADIATION

J. H. MERRITT AND L. N. HEYNICK
Radio Frequency Radiation Branch
Air Force Research Laboratory
Brooks Air Force Base, Texas 78235

1. Introduction

Teratology is the study of fetal malformations. Tissue in the embryo is rapidly proliferating and differentiating and because of this rapid growth, fetal tissue is considered to be extremely sensitive to environmental conditions. The type and degree of fetal malformations depend largely on the stage of development at which the detrimental environmental condition is applied. For instance, exposure to a harmful influence in the early stage of fetal development may result in fetal death and resorption, while exposure during organogenesis will result in malformations. And exposure to harmful environments can also lead to changes in postnatal development. It has been suggested that the fetus might be sensitive to exposure to radio frequency radiation (RFR). There have been a significant number of teratologic studies on the effect of such exposure on the fetuses of various species.

2. Non-Mammalian Species

Teratogenesis is more properly applied to mammalian species, but some of the early studies of the teratogenic effects of exposure examined avian and insect species. For instance, a study by Hills et al. [1] on the hatchability of chicken and turkey eggs showed deleterious effects of 2450 MHz exposure on viability. These studies were carried out at "high" power, but no explanation of dosimetry was provided. A series of studies on quail eggs has spanned almost two decades. In 1975, McRee, et al. [2] exposed quail eggs to 2450 MHz radiation such that the internal egg temperature was maintained at about 37° C. No differences were noted with respect to hatchability or body weights of the neonates. No deformities were noted. The only significant difference between the exposed and sham groups was a lower blood hemoglobin content in the exposed chicks. This research group subsequently carried out a number of studies on quail eggs. In 1977 McRee and Hamrick [3] exposed eggs to 2450 MHz radiation for 24 hours per day for the first 12 days of incubation. Exposures were at

383

B.J. Klauenberg and D. Miklavcic (eds.), Radio Frequency Radiation Dosimetry, 383-391.

two ambient temperatures - 35.5 and 37° C. Though the hatchability of the exposed eggs at the lower incubation temperature was higher than the sham-exposed control, the authors ascribed this difference to the higher internal egg temperature in the exposed group. At the higher incubation temperature, hatchability in both groups was nearly zero. Another report from this group [4] indicated an effect of fertility on quails who had been exposed to 2450 MHz radiation *in ovo*. Fertility of eggs from females mated with exposed males was lower than sham exposed eggs. Motility of sperm from exposed males was also lower. An extension of this study by Gildersleeve et al. [5] failed to confirm these findings with respect to reduced fertility.

Inouye et al. [6] studied the effect of continuous exposure from incubation day 1 to day 12 to 2450 MHz radiation. The specific absorption rate (SAR) in the exposed eggs was approximately 4 W/kg. Groups of eggs were removed from the exposure chamber on days 12, 13, and 14. The embryos were removed from the eggs and body and brain weight determined. Some brains were sectioned, stained, and histological examinations were made. The brains of the exposed embryos were significantly smaller than the sham exposed group at days 12 and 14, but not at day 13. There were no differences in the brain/body ratios. In addition, the brains of the day 12 embryos were judged to be developmentally retarded. No effects of prenatal exposure on body or brain weight or brain development were seen in 8-day-old chicks, however. Another study [7] examined effects of exposure for 8 hours per day for incubation days 1-15 at two different power densities and at four ambient incubation temperatures. No effects on viability or retarded or abnormal embryonic development were noted. On the other hand, Saito, et al [8] exposed chick embryos to 428 MHz RFR at 5.5 mW/cm^2 for more than 20 days and reported lethal and teratogenic effects. These investigators indicated that these effects were seen in the absence of frank heating of the embryos.

Other early studies examined effects of RFR exposure on the darkling beetle (*Tenebrio molitor*). Carpenter and Livstone [9] studied effects of 10 GHz radiation on *Tenebrio* pupae, the so-called mealworm. Exposures were for two hours at rather high SAR; up to 40 W/kg. Significant teratogenic effects were noted such as 4% death and 76% with gross abnormalities. When the pupae were exposed to conventional heating to obtain the same temperature as those exposed to RFR, 75% developed normally. These investigators concluded that the abnormalities observed in the RFR-exposed pupae were induced by a non-thermal mechanism. This study led to a number of other studies attempting to replicate the original experiments and to extend them. Studies by Pickard and Olsen [10] and Olsen [11] revealed that culture methods influenced results and that there was a threshold for deleterious effects of about 40° C. Clearly then, teratogenic effects in *Tenebrio* were thermally induced and not related to non-thermal effects of RFR exposure.

3. Mammalian Species

Many studies have examined RFR effects on the embryos of mice, hamsters, and rats. These studies have used various exposure regimens with respect to frequency, orientation in the field, SAR, days of gestation, and other parameters. It is, therefore, difficult to reconcile differences in experimental results reported in the various studies. Among the earliest studies of mammals were those of Rugh et al. [12, 13] who used mice. Exposures were at 138 mW/cm^2 (123 W/kg) for 2 to 5 min on day 8.5. At 18 days of gestation, the dams were killed and the fetuses examined for abnormalities. The data in these studies are rather confusing, but the authors claimed that they could not find a threshold for abnormalities. However, in an analysis of the data, Heynick and Polson [14] determined that there was a threshold total energy of about 12.6 J/g for exencephaly. Subsequently, a number of investigators [15, 16, 17, 18, 19] studied the effects of RFR on pregnant mice. These studies examined *inter alia* gross abnormalities, fetal weight, resorptions, and skeletal maturity. Though there were some differences in some of the examined endpoints when RFR-exposed were compared to sham-exposed, most of these differences could be ascribed to thermal insult to the dams induced by the exposure. For instance, Berman, et al [17] found significantly lower brain weights in the neonates from the exposed dams. The authors indicated that these effects were due to the SAR used (16.5 W/kg). Other studies show no significant differences in endpoints that could be ascribed to exposure [19].

A report of a large number of studies with mice by Polish workers has appeared [20]. The investigations consisted of exposure of dams during the entire period of gestation (days 1 thorough 19) or during the first and second halves of gestation (days 1-8 and days 9-19). In addition, some mice were killed at 4 days of gestation and the fetuses examined. Microwave exposures (2450 MHz) were given for 2 hrs daily at power densities from 1 to 40 mW/cm^2. On the 19th day of gestation, the dams were killed and the fetuses removed. Living and dead fetuses were counted, weighed, and examined for macroscopic malformations such as cleft palate and limb formation. After fixation and staining, the skeleton was examined for anomalies. Exposure at 16.7 W/kg (40 mW/cm^2) resulted in a lower per cent of full-term fetuses; per cent of full-term fetuses was unaffected by exposures at 2.5 W/kg. Four-day fetuses showed significant inhibition of development after exposure of the dams to 16.7 W/kg but not at 2.5 W/kg. The results from dams killed at 19 days of gestation are shown in Table 1. Exposure at the highest SAR resulted in significantly lower number of implantations

Table 1. Implantations, living fetuses, and mean fetal body mass in litters exposed to 2450 MHz radiation. Data are from Ref. 20.

SAR	Exposure period (days)	Number of implantations mean ± S.D.	Number of living fetuses mean ± S.D.	Body mass g ± S.D.
Sham	--	10.2 ± 2.1	10.0 ± 1.9	1.271 ± 0.132
5 W/kg	1 - 7	10.2 ± 1.9	10.1 ± 1.9	1.170 ± 0.153
	8 - 19	9.9 ± 1.8	9.8 ± 2.0	1.115 ± 0.079*
	1 - 19	9.2 ± 1.6	9.1 ± 1.6	1.134 ± 0.153*
16.7 W/kg	1 - 7	8.4 ± 2.4*	8.1 ± 1.9*	1.249 ± 0.1142
	8 - 19	9.5 ± 1.9	8.0 ± 2.0*	1.124 ± 0.188*
	1 - 19	7.7 ± 2.3*	6.1 ± 2.6*	1.090 ± 0.104*

*P <0.05

and number of living fetuses, while exposure at both SARs lowered the body mass of the fetuses. There were no significant differences with respect to gross malformations between the experimental groups. The authors of this report indicated that most of the effects seen in their studies were induced thermally. They considered the lower body mass of the exposed fetuses to be trivial, since weight gain postnatally was the same for both groups. It should be noted that the fetal body mass was based upon the mean of the individual animals, rather than the litter mean, which is the appropriate metric. A study by Fukui, et al. [21] showed changes in brain development to pregnant mice exposed to 2450 MHz radiation on gestation day 13 such that core temperature of fetuses increased to 42° C. As a thermal control, some of the dams were immersed in hot water to increase core temperature; changes in brain development were also seen in the thermal control experiments.

Kubinyi, et al. [22] studied the effect of 2450 MHz radiation (continuous wave and 50 Hz amplitude modulated) on the liver and brain aminoacyl-transfer synthetase in mouse fetuses exposed *in utero*. Dams were exposed throughout gestation at 4.23 W/kg. No effects were observed on body or organ weight in the pups after delivery. Brain enzyme activity decreased after continuous wave exposure but not after amplitude modulated radiation.

In a study with Syrian hamsters, Berman et al [23] showed significantly higher resorptions, lower fetal body mass, and delayed skeletal maturity when pregnant dams were exposed at an SAR that caused a 1.6° C increase in body temperature (9 W/kg). At 6 W/kg (0.4° C increase) no significant abnormalities were noted.

A very large number of studies of the effects of RFR exposure on the rat fetus have been made. Among the first of these studies was that reported by Dietzel [24]. This investigator exposed rats with a diathermy device (27.12 MHz) between day 1 and day 16 of gestation such that maternal rectal temperatures reached from 39 to 42° C. On day 20 of gestation, the fetuses were removed and examined for external malformations. Abnormalities in the cranium and limbs were noted predominantly for exposures on days 13 and 14. These abnormalities were correlated with elevated maternal core temperature. Brown-Woodman et al. [25] also exposed pregnant rats on day 9 of gestation to radiation from a diathermy device at 27.12 MHz. The exposures were such that core temperatures of the dams were raised 5, 4.5, 4, 3.5, 3 and 2.5° C for varying lengths of time. Significant teratologic effects (decreased fetal weight, increased abnormality rate) were noted that were correlated with core temperature and exposure duration.

Other early studies include those of Chernovetz, et al [26] and Shore et al. [27], in which rats were exposed to 2450 MHz radiation. Fetal mass, live or dead fetuses, brain weight, brain norepinephrine and dopamine, and evidence of morphological abnormalities were examined. In one study, infra red heating was used as a control [26]. The authors ascribed the effects noted (lower fetal mass, lower brain norepinephrine) to microwave-induced increase in maternal core temperature.

Jensh et al. [28, 29] and Jensh [30] carried out a series of studies to examine the teratologic potential of microwave exposure. In the first series of the studies, these investigators exposed dams to 915 MHz radiation for 6 hrs per day at 10 mW/cm^2. The mean SAR was estimated to be about 3.7 W/kg. The dams were killed and fetuses examined; there were no significant differences in any of the endpoints studied (litter size, fetal mass, resorptions, or gross abnormalities). In another study, this research group exposed pregnant dams to 2450 MHz radiation at 20 mW/cm^2. Mean SARs changed as the dams grew; on days 0-1 of gestation, the SAR was 5.2 W/kg; on days 7-8, 4.8 W/kg; on day 20, 3.6 W/kg. No significant differences were noted between the RFR-exposed and the sham-exposed groups with respect to fetal mass, resorptions, or malformation rates.

Merritt et al. [31] exposed pregnant rats in a circular wave guide system to 2450-MHz radiation. Exposure was continuous from day 2 of gestation to day 18. The power density was varied so as to maintain an SAR of 0.4 W/kg during the entire exposure period. A similar group of sham-exposed dams was maintained in unpowered wave guides. After 18 days of exposure, the dams were killed and the fetuses removed, weighed, and the brains removed and weighed. Brain protein, RNA, and DNA were determined. There were no significant differences between the exposed and sham-exposed groups in any of the endpoints. The effect of exposure on brain development was determined by regressing mean litter brain weight on mean litter body weight of the sham exposed group. Then these values for the RFR exposed group were plotted on the same graph. The RFR values were scattered about the regression line for the sham, an indication that RFR exposure did not result in microencephaly (Figure 1).

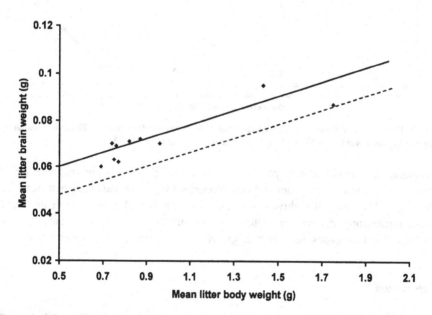

Figure 1. Effects of 2450 MHz exposure on fetal brain weight. Brain weight is regressed on body weight. The solid line is the sham-exposed condition. the dotted line is based on 2 standard deviations of the estimate of the solid line. The dots are the values for microwave exposed fetuses. A criterion of microencephaly is taken as values falling below the 2 S.D. line. Data are from Ref. 31.

388

In another study conducted in circular wave guides by Schmidt, et al. [32], pregnant rats were exposed to 0.4 W/kg from day 2 to day 18 of gestation. On day 18, the dams were killed and the fetuses were removed, fixed, stained, and examined by light microscope. No significant abnormalities were seen with respect to mean fetal weight, length, resorption rate, or abnormality rate.

Lary et al [33] studied the incidence of birth defects in rats exposed to 27.12 MHz as a function of maternal core temperature. The dams were exposed on gestation day 9 at a whole-body SAR of 10.8 W/kg until core temperatures reached values between 41 and 43° C. On day 20, the rats were killed and the fetuses examined. Fetal abnormalities and fetal mortality were plotted against maternal core temperature (Figure 2). The authors indicated a threshold temperature for teratogenic effects was 41.5° C.

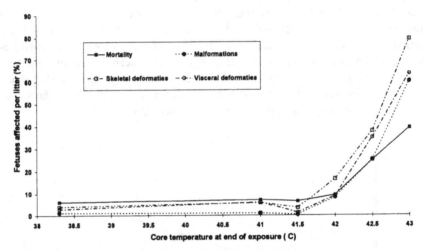

Figure 2. Threshold temperatures for fetal mortality and deformities in rats induced by 27.12 MHz radiation exposure. Exposure was for 10-40 min on gestation day 9. Data are from Ref. 33.

Cultured 9.5 day old rat embryos were irradiated to RFR electric and magnetic fields at various modulation frequencies and field strengths [34]. SAR values ranged from 0.2 to 5 W/kg. The cultured embryos were examined on day 11.5 using both light and electron microscopy and protein content was determined. This *in vitro* study did not reveal any effects of exposure on normal growth and differentiation of the embryos.

4. Discussion

Though some studies reviewed here appear to show teratogenic effects of low-level RFR exposures, most investigators indicate that such effects as are found on the fetus are induced by maternal hyperthermia. It is well documented that hyperthermia induces congenital defects. Many studies have shown that rapidly proliferating tissue, such as in the early embryo, are particularly susceptible to high temperatures. Edwards [35] has pointed out the toxic effect of hyperthermia on the developing fetus. This concept has been well established by numerous studies with methods of heating other than exposure to RFR. Studies with RFR show that some species are more susceptible than others to elevated temperatures, and this is borne out in studies using other methods of inducing hyperthermia.

Many of the early studies in particular, were deficient with regard to dosimetry, and this may help explain some of the differences in results from the studies reviewed here. Assessment of SAR was often made by estimation only. As pregnant animals grow during gestation, SAR will change with size. Merritt et al. [31] have shown that mean power absorption in a circular wave guide system at constant power density will increase as a function of animal weight and orientation (Figure 3). In these studies power density was changed to maintain a constant SAR of 0.4 W/kg.

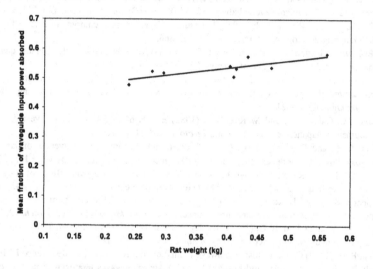

Figure 3. Measurement of differential power as a function of animal weight. Each point was the mean of 14 different orientations of animal carcasses in the wave guide. Data are from Ref 31.

Jensh et al. (29) estimated the mean SAR on gestation days 0-1, 7-8, and 20 to be 5.2, 4.8, and 3.6 W/kg respectively. In most studies, SAR was either estimated or measured by calorimetry, but usually on the basis of mean weight of the animals during the entire exposure period. Thus, significant errors in actual SAR in many of the studies are likely. In addition, the issue of nonuniform heating during RFR exposure adds to the uncertainties on energy absorption in experimental animals, with the possibility that the temperature of the embryo may exceed that of the dam. This has been pointed out in terms of the human condition by Fleming and Joyner (36). These authors indicated that the human fetus could be overexposed with respect to current safety standards at frequencies in the 80-100 MHz and 300-1500 MHz regimens due to a partial body resonance effect.

390

From the studies reviewed here, we conclude that the teratogenic potential of RFR exposure lies strictly in the ability of this type of energy to increase the maternal temperature of test animals and not in some intrinsic quality of the radiation per se. Taken together, these studies indicate that teratogenic effects occur only from RFR exposures at levels that induce significant body temperature increases. And, in mammals at least, specific temperature increases are necessary for inducing such teratogenic effects.

5. References

1. Hills, G. A., Kondra, P. A., and Hamid, M. A. (1974) Effects of microwave radiations on hatchability and growth in chickens and turkeys, *Can. J. Animal Sci.* **54**, 573-578.
2. McRee, D. I., Hamrick, P. E., and Zinkl, J. (1975) Some effects of exposure of the Japanese quail embryo to 2.45-GHz microwave radiation, in P. W. Tyler (ed.), *Ann. N. Y. Acad. Sci.* **247**, 377-390.
3. McRee, D. I. and Hamrick, P. E. (1977) Exposure of Japanese quail embryos to 2.45-GHz microwave radiation during development, *Radiat. Res.* **71**, 355-366.
4. McRee, D. I., Thaxton, J. P., and Parkhurst, C. R. (1983) Reproduction in male Japanese quail exposed to microwave radiation during embryogeny, *Radiat. Res.* **96**, 51-58.
5. Gildersleeve, R. P., Galvin, M. J., McRee, D. I., Thaxton, J. P., and Parkhurst, C. R. (1987) Reproduction of Japanese quail after microwave irradiation (2.45 GHz CW) during embryogeny, *Bioelectromagnetics* **8**, 9-21.
6. Inouye, M., Galvin, M., and McRee, D. I. (1982) Effects of 2.45 GHz microwave radiation on the development of Japanese quail cerebellum, *Teratology* **25**, 115-121.
7. Spiers, D. E. and Baummer, S. C. (1991) Thermal and metabolic responsiveness of Japanese quail embryos following periodic exposure to 2,450-MHz microwaves, *Bioelectromagnetics* **12**, 225-239.
8. Saito, K., Suzuki, K., and Motoyoshi, S. (1991) Lethal and teratogenic effects of long-term low-intensity radio frequency radiation at 428 MHz on developing chick embryo, *Teratology* **43**, 609-614.
9. Carpenter, R. L. and Livstone, E. M. (1971) Evidence for nonthermal effects of microwave radiation: Abnormal development of irradiated insect pupae, *IEEE Trans. Microwave Theory Tech.* **19**, 173-178.
10. Pickard, W. F. and Olsen, R. G. (1979) Developmental effects of microwaves on Tenebrio: Influences of culturing protocol and of carrier frequency, *Radio Science* **14**, 181-185.
11. Olsen, R. G. (1982) Constant-dose microwave irradiation of insect pupae, *Radio Science* **17**:145-148.
12. Rugh, R., Ginns, E., Ho, H., and Leach, W. (1974) Are microwaves teratogenic? in P. Czerski, et al. (eds.) *Biological Effects and Health Hazards of Microwave Radiation*, Polish Medical Publishers, Warsaw, pp. 98-107.
13. Rugh, R., Ginns, E., Ho, H., and Leach, W. (1975) Responses of the mouse to microwave radiation during estrous cycle and pregnancy. *Radiat. Res.* **62**, 225-241.
14. Heynick, L. N. and Polson, P. (1996) *Radiofrequency Radiation and Teratogenesis: A Comprehensive Review of the Literature Pertinent to Air Force Operations*, Armstrong Laboratory Technical Report AL/OE-TR-1996-0036, Brooks Air Force Base, Texas.
15. Berman, E., Kinn, J. B., and Carter, H. B. (1978) Observations of mouse fetuses after irradiation with 2.45 GHz microwaves, *Health Physics* **35**, 791-801.
16. Berman, E., Carter, H. B., and House, D. (1982) Reduced weight in mice offspring after in utero exposure to 2450-MHz (CW) microwaves, *Bioelectromagnetics* **3**, 285-291.
17. Berman, E., Carter, H. B., and House, D. (1984) Growth and development of mice offspring after irradiation in utero with 2,450-MHz microwaves, *Teratology* **30**, 393-402.
18. Nawrot, P. S., McRee, D. I., and Staples, R. E. (1981) Effects of 2.45 GHz CW microwave radiation on embryo fetal development in mice, *Teratology* **24**, 303-314.

19. Inouye, M., Matsumoto, Galvin, M. J., and McRee, D. I. (1982) Lack of effect of 2.45-GHz microwave radiation on the development of preimplantation embryos of mice, *Bioelectromagnetics* **3**, 275-283.
20. Development and Teratogenic Effects of 2450 MHz Microwaves in Mice. M. Troszynski and S. Szmigielski, (eds.) Report FDA/CDRH-87/44, U. S. Food and Drug Administration, Center for Devices and Radiological Health, Rockville, MD, 1987.
21. Fukui, Y., Hoshino, K., Inouye, M., and Kameyama, Y. (1992) Effects of hyperthermia induced by microwaves on brain development in mice, *J. Radiat. Res. (Toyko)* **33**, 1-10.
22. Kubinyi, G., Thuróczy, Bakos, J., Bölöni, Sinay, H., and Szabó, L. (1996) Effect of continuous-wave and amplitude-modulated 2.45 GHz microwave radiation on the liver and brain aminoacyl-transfer RNA synthetases of in utero exposed mice, *Bioelectromagnetics* **17**, 497-503.
23. Berman, E., Carter, H. B., and House, D. (1982) Observations of Syrian hamster fetuses after irradiation with 2450-MHz microwaves, *J. Microwave Power* **17**, 107-112.
24. Dietzel, F. (1975) Effects of electromagnetic radiation on implantation and intrauterine development of the rat. In P. W. Tyler (ed.), *Ann. N. Y. Acad. Sci.* **247**, 367-376.
25. Brown-Woodman, , P. D., Hadley, J. A., Waterhouse, J., and Webster, W. W. (1988) Teratogenic effects of exposure to radiofrequency radiation (27.12 MHz) from a shortwave diathermy unit, *Indust. Health* **26**, 1-10.
26. Chernovitz, M. E., Justesen, D. R., and Oke, A. F. (1977) A teratological study of the rat: Microwave and infrared radiations compared, *Radio Sci.* **12**, 191-197.
27. Shore, M. L., Felten, R. P., and Lamanna, A. (1977) The effect of repetitive prenatal low-level microwave exposure on development in the rat, in D. G. Hazzards (ed.) *Symposium on Biological Effects and Measurement of Radio Frequency/Microwaves.* U. S. Department of Health, Education, and Welfare, HEW Publication (FDA) 77-8026, Washington, D. C., pp. 280-289.
28. Jensh, R. P., Weinberg, I., and Brent, R. L. (1982) Teratologic studies of prenatal exposure of rats to 915-MHz microwave radiation, *Radiat. Res.* **92**, 160-171.
29. Jensh, R. P., Weinberg, I., and Brent, R. L. (1983) An evaluation of the teratogenic potential of protracted exposure of pregnant rats to 2450-MHz microwave radiation. I. Morphologic analysis, *J. Toxicol. Environ. Health* **11**:23-35.
30. Jensh, R. P. (1984) Studies of the teratogenic potential of exposure of rats to 6000-MHz microwave radiation. I. Morphologic analysis at term, *Radiat. Res.* **97**:272-281.
31. Merritt, J. H., Hardy, K. A., and Chamness, A. F. (1984) In utero exposure to microwave radiation and rat brain development, *Bioelectromagnetics* **5**, 315-322.
32. Schmidt, R. E., Merritt, J. H., and Hardy, K. A. In utero exposure to low-level microwaves does not affect fetal development, *Int. J. Radiat. Biol.* **46**, 383-386.
33. Lary, J. M., Conover, D. L., Johnson, P. H., and Hornung, R. W. (1986) Dose-response relationship between body temperature and birth defects in radiofrequency-irradiated rats, *Bioelectromagnetics* **7**, 141-149.
34. Klug, S, Hetscher, M., Giles, S., Kohlsmann, S., and Kramer, K. (1997) The lack of effects of nonthermal RF electromagnetic fields on the development of rat embryos grown in culture, *Life Science* **61**, 1789-1802.
35. Edwards, M. J. (1978) Congenital defects due to hyperthermia, *Adv. Vet. Sci. Comp. Med.* **22**, 29-52.
36. Fleming, A. H. J. and Joyner, K. H. (1992) Estimates of absorption of radiofrequency radiation by the embryo and fetus during pregnancy, *Health Phys.* **63**, 149-159.

SUMMARY OF SESSION G: RESPONSES OF MAN AND ANIMALS II

J. A. D'ANDREA
Naval Health Research Center Detachment
Brooks AFB, TX 78235 U.S.A.

Five papers were presented in this session with each investigating unique interactions of electromagnetic fields and biological function. To even begin to understand these interactions requires interdisciplinary teamwork. Each of the papers presented has its own unique dosimetry challenge.

J. D'Andrea began the session with a review of the effects of microwaves and millimeter waves on vision. He pointed out that microwaves in the range of 1 to 10 GHz could produce cataracts if sufficient exposure intensity and exposure duration is used. He reviewed some of the studies that have reported damage to the retina at 1.25 GHz with an SAR of 4 W/kg. Other studies have failed to find similar results. Millimeter waves may be particularly injurious to the cornea because energy is deposited within the first millimeter of tissue.

J. deLorge pointed out that behavioral research on the effects of radiofrequency radiation has decreased substantially over the last ten years, which he found surprising since most safety standards in the west are based on behavioral effects. Behavior is a final common expression of virtually all physiological processes with, perhaps, the exception of disease, and even that endpoint is often initially expressed as a behavioral response. If low level RF radiation produces adverse health effects, behavioral changes will continue to be observed first.

A. Lerchl presented investigations that were planned to answer the following questions: does long-term exposure of Djungarian hamsters with a typical mobile communication signal influence the melatonin synthesis and the testicular cell composition as estimated after exposure? Also, what are the effects with respect to body weight development during exposure? To achieve high statistical power, a large number of animals per group were used. To handle large groups of animals, a sophisticated exposure system was developed which allowed the simultaneous exposure of 120 hamsters with well-defined exposure parameters.

A. Chiabrera discussed the influence of electromagnetic field exposure on ligand binding to receptor proteins as part of an interaction mechanism leading to biological effects. He placed specific emphasis on ion binding as a first step of this interaction. The biochemical output is the change of the ligand binding probability due to low intensity electromagnetic exposure at radio frequencies.

B.J. Klauenberg and D. Miklavcic (eds.), Radio Frequency Radiation Dosimetry, 393-394.
© 2000 *Kluwer Academic Publishers. Printed in the Netherlands.*

K.S. Nikita presented a rigorous analysis for investigating the focusing properties of short baseband pulses emitted by a TEM concentric array of applicators in a layered cylindrical biological tissue model. The results provide an enhanced understanding of pulse propagation inside layered lossy media and can be used to achieve improved focusing inside biological tissues.

EFFECTS OF MICROWAVE AND MILLIMETER WAVE RADIATION ON THE EYE

J. A. D'ANDREA
Naval Health Research Center Detachment
Brooks AFB, TX 78235

S. CHALFIN
Ophthalmology Department
University of Texas Health Science Center at San Antonio
San Antonio, TX 78284

1. Introduction

Exposure of the eye to microwave radiation can lead to intraocular temperature increase sufficient to damage tissues. The eye of mammalian species does not efficiently remove heat. Within the anterior segment of the eye, active thermal transport is not known to occur. Conduction through the sclera and convection from the surface of the cornea is the primary avenue for heat dissipation which is poor compared to many other tissues in the body. The anterior segment can be closer to ambient temperature than the posterior segment. On the other hand, a primary avenue of heat dissipation in the posterior segment of the eye is bloodflow through the choroidal vascular system which lies just inside the sclera. It is thought to act as a heat sink and maintain a stable thermal environment for rod and cone cell metabolism in the retina [1, 2, 3]. Bloodflow throughout the retina may also serve to maintain a stable thermal environment for the photoreceptors. Many investigators have postulated that the poor heat dissipation capability from within the anterior segment of the eye of humans and other animals may lead to heat buildup and subsequent thermal damage [4]. For many, this has implied that the eye is sensitive to heating and especially sensitive to heat deposited within the anterior segment of the eye by deeper penetrating microwaves. Early investigations of microwave produced eye damage focused on the production of cataracts. In the past several years, new studies have investigated microwave effects on the retina, iris vasculature, and corneal endothelium.

At the much shorter wavelengths of millimeter waves, damage to the eye may occur near the radiated surface such as the cornea [5, 6]. Absorption of energy from this portion of the spectrum occurs within less than a millimeter resulting in very high specific absorption rates and rapid temperature increases. Few studies have been conducted at millimeter wavelengths to evaluate either acute exposure effects, or to

B.J. Klauenberg and D. Miklavcic (eds.), Radio Frequency Radiation Dosimetry, 395-402.
© 2000 *Kluwer Academic Publishers. Printed in the Netherlands.*

determine if changes occur with prolonged exposures. This paper presents a short summary of many of the reported ocular effects of microwave and millimeter wave radiation.

2. Cataract Formation

Much of the early laboratory research on microwave damage to the eye was conducted at 6 GHz and below [7, 8]. The research focused mainly on acute exposure damage to the lens. The lens lies just behind the iris and pupil of the eye in the anterior segment. The lens can become cloudy or opaque due to injury, aging, or disease. Opacification resulting from microwave irradiation was usually found localized in the posterior portion of the lens capsule, but at higher frequencies (above 6 GHz) they were localized in the anterior portion of the lens. Most of the early research was carried out in the lower portion of the microwave spectrum (at 2.45 GHz) and demonstrated a high dose response relationship between microwave exposure and cataract induction [7]. For example, Carpenter and Van Ummersen [9] irradiated anesthetized rabbits at 2.45 GHz and showed a decreasing threshold for cataractogenesis from 4 minute exposure at 400 mW/cm² to 40 minutes at 80 mW/cm². Guy et al. [10] repeated some of the earlier research and found essentially the same threshold for cataract production in rabbits exposed with a near field applicator at 2.45 GHz. At minimum, they determined that 150 mW/cm^2 was required for 100 min. to produce a cataract. The maximum specific absorption rate (SAR) associated with cataract production was 138 W/kg. Kramar et al., [11] found nearly the same threshold (180 mW/cm^2 for 140 min.) for rabbits exposed to 2.45 GHz radiation. Interestingly, they found that exposures even higher in power density (500 mW/cm^2 for 60 min.) could not produce a similar cataract in rhesus monkeys. This was attributed to poor absorption by the monkey skull due to shape and the more recessed eye sockets that found on the rabbit. It is important to note that at the power densities and exposure durations used in these studies, far field exposure would be lethal for the exposed animal long before a cataract could be formed [7].

3. Retinal and Corneal Damage

Research and clinical findings over the past 10 years have indicated that absorption of microwave radiation at power densities much lower than induce cataracts can cause damage to the retina. A recent report of an accidental overexposure described retinal damage [12]. The authors presented a case report of retinal damage resulting from microwave exposure of a 44-yr-old male who received two 15-min accidental exposures to 30-W, 6-GHz CW microwave radiation while inspecting a 3.2-m satellite transmitter dish. He developed facial erythema, bilateral foreign body sensation, and blurred vision. When given an ophthalmologic examination 5 days later, the subject had corrected visual acuity of 20/20 in the right eye and 20/25 in the left. A slit lamp examination showed superficial eyelid erythema, bilateral superficial punctate keratopathy, endothelial pigment dusting, and aqueous cells. The keratitis resolved after 2 days and the eyelid burns after 8 days. Pelli-Robson contrast sensitivity demonstrated a 2-line difference (left worse than right) between the left and right eyes [12].

Over the last several years a series of studies conducted at the Johns Hopkins University Applied Physics Laboratory and the Wilmer Ophthalmological Institute [13-15] found several effects of 1.25 and 2.45 GHz microwave exposure in animal eyes. These effects include electroretinogram changes, corneal endothelial lesions, macula detachment and pigment epithelial pyknosis. Corneal endothelial abnormalities were produced in anesthetized cynomolgus monkeys exposed for four hours to 2.45 GHz CW or pulsed microwaves (10 mW/cm^2 or 20-30 mW/cm^2, respectively)[13]. In another study retinal damage was reported following exposure to an average power density of near 10 mW/cm^2 that was pulsed with a peak power of near 1 MW at 1 or 16 pps with pulse duration's of 0.5 μs or 10 μs [14, 15]. The findings were significant in that they were observed at an SAR in the retina of 4 W/kg which has been thought to produce no lasting effects. Three recent studies have sought to evaluate some of the effects reported by Kues and coworkers. First, Kamimura et al. [16] attempted to repeat the findings of corneal endothelial changes in monkeys exposed to 2.45 GHz radiation reported by Kues et al. [13]. They exposed alert nonanesthetized cynomolgus monkeys for four hours to 2.45 GHz CW microwaves in the near field of an antenna at 15.9 to 43 mW/cm^2. Forty-eight hours after exposure the monkeys were examined by slit-lamp, specular microscope, and fundus camera. The only finding was minor conjunctivitis. Changes in corneal endothelium, as reported earlier by Kues et al. [13], were not observed.

Second, a study by Lu et al. [17] reported failure to find retinal damage to rhesus monkeys exposed to 4 W/kg and up to 5 times the SAR reported by Kues et al. [14, 15]. This was a research project designed to evaluate retinal effects of 1.25 GHz high peak power microwave exposure in monkeys. Fundus photography, retinal angiograms and electroretinograms (ERG) were used prior to exposure to screen for normal ocular structure and function of the monkeys' eyes, and after exposure as endpoints of the experiment. Histopathology of the retina was also used as an endpoint. Nineteen monkeys were randomized to receive sham exposure or pulsed microwave exposures. Microwaves were delivered cranially at 4, 8, or 20 W/kg average retinal specific absorption rates (R-SAR). The pulse characteristics were 1.04 MW (1.30 MW/kg peak SAR), 5.59 μs pulse width at 0 Hz, 0.59 Hz, 1.18 Hz and 2.79 Hz pulse repetition rates. The exposure regimen was 4 hours per day and 3 days per weeks for 3 weeks, for a total of 9 exposures. No evidence of histopathological changes were seen in these animals. The response of cone photoreceptor to light flash was enhanced in monkeys exposed at 8 or 20 W/kg R-SAR but not in monkeys exposed at 4 W/kg R-SAR. The authors concluded that retinal injury was very unlikely at 4 W/kg, and functional changes that occur at higher R-SAR are probably reversible because of lack of histopathologic correlates.

Third, D'Andrea et al. [18] exposed four monkeys to 1.25 GHz microwave pulses and measured visual performance on a contrast sensitivity task. Contrast sensitivity functions are widely used as a measure of basic visual performance. Four male rhesus monkeys (Macaca mulatta) were trained on an operant task to establish spatial contrast sensitivity functions. The operant task required monkeys to press one plastic lever reinforced on a variable interval schedule (VI-20 s) to produce a visual stimulus (vertical sinusoidal grating) and then respond on a second lever to obtain food. Gratings of five different spatial frequencies were tested; 1.5, 3, 6, 12, and 18 cycles per degree. Using a titration procedure, various contrasts of the five gratings were presented to determine thresholds. The monkeys were exposed to microwave pulses (16 pps, 6 μs pulse duration) produced by a military radar (FPS-7B). Microwave exposures were delivered over nine sessions, 4 h per session, at an average whole-body SAR of 1 W/kg and local SAR at the retina of 4 W/kg. Peak waveguide output power was approximately 1 MW. Pulsed microwaves used in this study did not significantly alter contrast sensitivity thresholds compared to preexposure thresholds.

The research findings presented by Kues et al. [13-15] have not been repeated by other investigators. Differences in exposure systems, pulse characteristics, monkey handling and exposure procedures are likely causes. Nevertheless, it appears unlikely that an SAR of 4 W/kg can cause retinal pathology or alteration of function.

4. Corneal Damage

Because of the short wavelength, millimeter waves only superficially penetrate biological tissue. The energy that penetrates the body, for example, is almost completely absorbed in the first few tenths of a millimeter of tissue. The depth of penetration varies inversely with frequency resulting in rapid temperature increase in a relatively small volume of tissue. At lower microwave frequencies the greater penetration depth allows absorption in a greater volume of tissue with rates of heating per unit of tissue that are much lower than at millimeter wavelengths. In addition to high temperatures, the pattern of millimeter wave energy that is absorbed in tissue can be nonuniform with localized areas exceeding average values by over 10 times. These excesses can be exceedingly high in a small volume of tissue. Khizhnyak and Ziskin [18], using infrared (IR) thermography, showed that SAR in hotspots of the nonuniform heating pattern can exceed 1 kW/kg at incident powers as low as 10 mW/cm^2. Using saline models they evaluated millimeter wave heating patterns in saline "thin film" models and observed frequency-specific effects and multiple hotspot patterns. In the 32 to 59 GHz frequency range they observed average spatial SARs of 0.4 kW/kg with hotspot SARs as high as 5kW/kg. Interestingly the pattern of heating varied in a cyclic manner at 5.5 GHz intervals changing from a single hotspot to two hotspots as frequency varied. In the range of 57-78 GHz multiple hotspot peaks were observed. As many as 6 hotspots were observed during irradiation and these depended on the size and shape of the horn irradiator used as well as frequency. This experiment emphasizes the fact that heating patterns at millimeter wavelengths need to be empirically characterized as these change with size and type of irradiator and frequency, especially under near field exposure conditions. Most important is the observation that at relatively low incident powers peak SAR values can be high enough to significantly alter biological structures.

The results of the experiments described above were previously predicted. Gandhi and Riazi [5] calculated power reflection, absorption coefficients, depth of penetration and SAR for skin in the 30-300 GHz range. They observed that for 5 mW/cm^2 incident power, SAR varied from 65.5 W/kg at 30 GHz to 357.1 W/kg at 300 GHz which is considerably in excess of current safety guidelines. In addition, they cite research on the perception of IR radiation on the back of the hand and predict that a sensation of very warm to hot would occur at millimeter wavelengths with an incident power of 8.7 mW/cm^2. This predicted field power density is very close to the current recommend maximum permissible exposure (MPE). Recently, however, Blick et al. [19] measured cutaneous threshold of warmth on the back of subjects exposed to 94 GHz and IR and found a threshold for sensation of 4.5 mW/cm^2 and 5.34 mW/cm^2, respectively. The threshold for very warm to hot at 94 GHz must be much higher than Gandhi and Riazi [5] predicted.

Skin is highly vascularized and can likely remove much of the deposited energy. The cornea and lens of the eye, on the other hand, do not remove heat well and may be especially vulnerable to millimeter wave injury. The cornea is sensitive to small thermal changes and absorption of 1 kW/cm^2 at IR wavelengths can produce a temperature increase of 10° C, which elicits a pain response in humans in a fraction of a second [21]. Eye blink and head aversion from an IR or millimeter wave source is reflexive and will occur for intense sources of heat [22]. However, thermal damage may still occur at lower intensities. The extent of lasting damage will depend on the involvement of corneal layers. At IR wavelengths damage limited to the corneal epithelium will undergo a normal repair process. Damage to the underlying stromal layer, however, results in the development of corneal opacities [22]. Since the depth of penetration of millimeter waves is deeper than IR the likelihood of greater damage to the stromal layer at millimeter wave frequencies is plausible. One research study has shown that stromal layer damage occurs at millimeter wave frequencies and depends on the wavelength. Rosenthal et al. irradiated anesthetized rabbits and found that at 40 mW/cm^2 extensive epithelial damage was observed at 35 GHz and 107 GHz [23]. Stromal injury was only observed for the deeper penetrating energy of 35 GHz irradiation. They reported longer lasting damage to rabbit eyes for 35 GHz, but a lower threshold for damage at 107 GHz, which usually disappeared the next day. It may be important to note that the threshold levels reported by Rosenthal et al. [23] were determined under anesthesia. The corneas were deprived of their normal convective cooling as no blinking occurred during irradiation. As the authors reported, their data pertained only to acute exposures. Data on the chronic exposure effects of millimeter waves to the eye is almost nonexistent.

Recently, Kues et al. [24] reported chronic exposures of rabbits and rhesus monkeys to 60 GHz millimeter waves at an incident power density of 10 mW/cm^2. A variety of diagnostic tests including slit lamp examination, specular microscopy, and iris angiography were used before and after exposure to millimeter waves to evaluate damage in each subject. After single 8 hour or five separate 4 hour exposures, damage to the exposed eye of rabbits and monkeys was not observed. Light and transmission electron microscopy of eye tissues failed to reveal any damage from the chronic exposures to millimeter waves.

Also, Chalfin et al. [25] evaluated exposures to millimeter wave irradiation at 94 GHz. They used a four element rating scale to establish thresholds for corneal damage at 94 GHz in rhesus monkeys. They also used much higher power densities (2 W/cm^2) and shorter exposure durations (1 to 4 seconds) than the Kues' rabbit study [24]. Five juvenile rhesus monkeys were given baseline eye exams using slit lamp examination, corneal topography, specular microscopy, and pachymetry. Anesthetized monkeys were then exposed in one eye to pulsed 94 GHz microwaves at different intensities and exposure duration with the other eye serving as a control. The fluence necessary to produce a threshold superficial corneal lesion (faint epithelial edema and fluorescein staining) at 94 GHz was 6 J/cm^2 (3 second exposure to 2 W/cm^2).

Some studies have reported different thresholds of eye damage for pulsed (PW) and continuous wave (CW) radiation. Trevithick et al., [26] and Creighton et al., [27] found lower millimeter wave threshold to eye damage for pulsed (22.6 W/kg) than for CW exposures (550 W/kg). Differences were also observed between PW and CW by Kues et al. [13]. Corneal endothelial abnormalities were discovered in cynomolgus monkeys. They concluded that pulsed microwave radiation (2.45 GHz, 10 ms pulse width, 100 pulses per second) were twice as effective in producing effects as continuous wave (CW) microwaves of the same average power density.

5. Conclusions

Microwaves in the range of 1 to 10 GHz can produce cataracts if sufficient exposure intensity and exposure duration are used. These seem to be limited to near field exposures with a threshold near 150 mW/cm^2 at 100 min. exposure duration. Experimentally induced cataracts have not been produced by far field exposures or in non-human primates. Exposures at these power densities encompassing the whole-body would be lethal for the laboratory animal. Some evidence exists suggesting that PW is more damaging than CW microwaves. Damage to the retina has been reported for exposures at 1.25 GHz at a SAR of 4 W/kg. Other studies have failed to find similar results. A study to evaluate visual performance in rhesus monkeys after exposure [18] did not find effects of 1.25 GHz pulses that would corroborate the findings of Kues et al. [14, 15]. Damage to the corneal endothelium reported at 2.45 GHz could not be repeated. Millimeter waves may be particularly injurious to the cornea because energy is deposited within the first millimeter of tissue. Rapid increase and high temperatures can be achieved in tissues because of the limited volume of tissue that millimeter wave energy is deposited in. One study has documented the threshold for corneal damage by exposure to 94 GHz millimeters waves at near 6 J/cm^2 [25]. Continued research to establish thresholds for biological effects will be required as new uses are discovered for energy in the microwave and millimeter wave portions of the electromagnetic spectrum.

6. Disclaimer

The views expressed in this article are those of the author and do not reflect the official policy or position of the Department of the Navy, Department of Defense, nor the U.S. Government. Trade names of materials and/or products of commercial or nongovernment organizations are cited as needed for precision. These citations do not constitute official endorsement or approval of the use of such commercial materials and/or products. The animals used in this work were handled in accordance with the principles outlined in the *Guide for the Care and Use of Laboratory Animals*, prepared by the Committee on Care and Use of Laboratory Animals of the Institute of Laboratory Animal Resources, National Research Council, DHHS, NIH Publication No. 85-23, 1985; and The Animal Welfare Act of 1966, as amended 1970 and 1976. This review was sponsored by the Naval Medical Research and Development Command under work unit 61153NMRO4101.001.1709 DN240853.

7. References

1. Cameron, J.D. and Ryan, E.H. (1997) Retinal vascular occlusive disease: An often-unsuspected thief of visual activity, *The Medical Journal of Allina* 6, Winter. (www.allina.com/Allina_Journal/Winter1997/cameron.html).
2. Parver, L.M., (1991) Temperature modulating action of choroidal blood flow, *Eye* 5, 181-5.
3. Stiehl, W.L., Gonzalez-Lima, F., Carrera, A., Cuebas, L.M., and Diaz, R.E. (1986) Active defence of retinal temperature during hypothermia of the eye in cats, *J Physiol (Paris)* 81, 26-33.
4. Al-Badwaihy, KA. and Youssef, A.B. (1976) Biological thermal effect of microwave radiation on human eyes, in C. C. Johnson and M. L. Shore (eds.) *Biological Effects of Electromagnetic Waves*. Vol. I, Selected papers of the USNC/URSI Annual Meeting, Washington DC: U.S. Government Printing Office, HEW publication (FDA) 77-8011, 61-78.
5. Gandhi, O.P. and Riazi, A. (1986) Absorption of millimeter waves by human beings and its biological implications, *IEEE Trans. Microwave Theory Tech.* 34, 228-235.
6. Motzkin, S.M. (1990) Biological effects of millimeter waves, in O. P. Gandhi (ed.) *Biological Effects and Medical Applications of Electromagnetic Energy*, Englewood Cliffs, New Jersey: Prentice Hall pp. 373-413.
7. Elder, J. A. (1984) Special Senses: Cataractogenic Effects, in J. A. Elder and D. F. Cahill (eds.) *Biological Effects of Radiofrequency Radiation*, Environmental Protection Agency Report, EPA-600/8-83-026F, pp5-64 to 5-68.
8. Tengroth, B.M. (1983) Cataractogenesis induced by RF and MW energy, in M. Grandolfo, S. M. Michaelson, and A. Rindi (eds.) *Biological Effects and Dosimetry of Nonionizing Radiation. Radiofrequency and Microwave Energies*, New York: Plenum Press, pp. 485-500.
9. Carpenter, R.L. and Van Ummersen, C.A. (1968) The action of microwave radiation on the eye, *J. Microwave Power* 3, 3.
10. Guy, A.W., Lin, J.C., Kramar, P.O., and Emery, A.F. Effect 2450 MHz radiation on the rabbit eye, *IEEE Trans. Microwave Theory Techniques* MTT-23, 492-498.
11. Kramar, P.O., Harris, P.C., Emery, A.F., and Guy, A.W. (1978) Acute microwave irradiation and cataract formation in rabbits and monkeys, *Journal of Microwave Power* 13, 239-249.
12. Lim, J. I., Fine, S. L., Kues, H. A., and Johnson, M. A. (1993) Visual abnormalities associated with high-energy microwave exposure, *Retina* 13, 230-233.

13. Kues, H.A., Hirst, L.W., Lutty, G.A., D'Anna, S.A., and Dunkelberger, G.R. (1985) Effects of 2.45 GHz microwaves on primates corneal endothelium, *Bioelectromagnetics* **6**, 177-188.

14. Kues, H.A. and McLeod, D.S. (1988) Ocular changes following exposure to high peak pulsed 1.25 GHz microwaves, presented at the Tenth Annual Meeting of the Bioelectromagnetics Society, Stamford, CN, June 19-23.

15. Kues, H.A. and McLeod, D.S. (1989) Pulsed microwave-induced ocular pathology in non-human primates, presented at the Eleventh Annual Meeting of the Bioelectromagnetics Society, Tucson, AZ, June 18-22.

16. Kamimura, Y., Saito, K., Saiga, T., and Amemiya, Y. (1994) Effect of 2.45 GHz microwave irradiation on monkey eyes, *IEICE Trans. Commun.* **E77B**, 762-765.

17. Lu, S.T., Mathur, S.P., Stuck, B., Zwick, H., D'Andrea, J.A., Ziriax, J.M., Merritt, J.H. Lutty, G., McLeod, G., and Johnson, M. (1998) Retinal effects of high peak power L-band microwaves in monkeys, presented at the Twentieth Annual Meeting of the Bioelectromagnetics Society, St. Petersburg, FL, June 1998.

18. D'Andrea, J. A., Thomas, A., Hatcher, D. J., and DeVietti, T. L. (1993) Rhesus monkey contrast sensitivity during extended exposure to high peak power 1.3 GHz microwave pulses, presented at the Fifteenth Annual Meeting of the Bioelectromagnetics Society, Los Angeles, CA, June 13-17.

19. Khizhnyak, E.P. and Ziskin, M.C. (1994) Heating patterns in biological tissue phantoms caused by millimeter wave electromagnetic irradiation, *IEEE Trans on Biomedical Engineering* **41**.

20. Blick, D. W., Adair, E. R., Hurt, W. D., Sherry, C. J., Walters, T. J., and Merritt, J. H. (1997) Thresholds of microwave-evoked warmth sensations in human skin, *Bioelectromagnetics* **18**, 403-409.

21. Moss, C. E., Ellis, R. J., Murray, W. E., and Parr, W. H. (1982) Infrared radiation, in *Nonionizing Radiation Protection* WHO Regional Publications European Series No. 10.

22. Lipshy, K. A., Wheeler, W. E., and Denning, D. E. (1966) Opthalmic thermal injuries, *Am. Surg.* **62**, 481-483.

23. Rosenthal, S. W., Birenbaum, L., Kaplan, I. T., Metlay, W., Snyder, W.Z., and Zaret, M. M. (1977) Effects of 35 and 107 GHz CW microwaves on the rabbit eye, in C. C. Johnson and M. L. Shore (eds.) *Biological Effects of Electromagnetic Waves.* Vol. I, Selected papers of the USNC/URSI Annual Meeting, October 20-23, 1975, Boulder, Colorado. HEW publication (FDA) 77-8011, U.S. Government Printing Office, Washington DC, pp. 110-128.

24. Kues, H. A., D'Anna, S. A., Osiander, R., Green, W. R., and Monahan, J. (1998) Absence of ocular effects in the rabbit and non-human primate from single or repeated 60 GHz CW exposure at 10 mW cm^{-2}, presented at the Twentieth Annual Meeting of the Bioelectromagnetics Society, St. Petersburg, FL, June 1998.

25. Chalfin, S., D'Andrea, J. A., Comeau, P. D., and Belt, M. E. (1998) Millimeter wave absorption in the nonhuman primate eye, presented at the Twentieth Annual Meeting of the Bioelectromagnetics Society, St. Petersburg, FL, June 1998.

26. Trevithick, J. R., Creighton, M. O., Sanwal, M., Brown, D.O., and Bassen, H.I. (1987) Histopathological studies of rabbit cornea exposed to millimeter waves, Conference Proceedings, *IEEE Engineering in Medicine and Biology*.

27. Creighton, M.O., Trevithick, J. R., Dzialoszynski, T., Sanwal, M., Brown, D.O., and Bassen, H. I. (1988) Comparison of histological effects of pulsed and continuous millimeter waves on rabbit corneas, presented at the Tenth Annual Meeting of the Bioelectromagnetics Society, Stamford, CT, June 19-23.

CONTEMPORARY RESEARCH ON THE BEHAVIORAL EFFECTS OF RADIO FREQUENCY RADIATION

J. O. DE LORGE
McKessonHBOC BioServices
8308 Hawks Rd, Bldg. 1184
Brooks AFB, TX 78235

1. Introduction

The most recent and widely followed standards for safe levels of human exposure to radio frequency (RF) electromagnetic fields, the IEEE C95.1-1991 guide [1] relied on behavioral research as the most sensitive biological measure of potentially harmful effects. Research on the behavioral effects of RF exposure has continued to be important following the publication of that guide although there has been a substantial decrease in the number of facilities and published articles involved in RF biology.

The present chapter will examine much of the behavioral work published since 1990 and attempt to draw some conclusions as to its contribution to an understanding of the biological effects of exposure to electromagnetic fields in the RF range of the spectrum. Many of these recent studies can be categorized into two groups. The first group consists of research on the effects of pulses of high-powered energy. The second contains research with more conventional sources of radiation or at least of sources used in many studies conducted in the previous decade.

2. High Power Pulses

One of the first high-peak-power experiments of this decade is a study by Akyel *et al.* [2]. Pulses at a peak power of 1 MW and a transmitter frequency of 1.25 GHz were used to illuminate constrained rats for 10 min. Specific absorption rates (SARs) of 0.84, 2.5, 7.6, and 23 W/kg were produced. The rats had been trained to lever press and following exposure they were placed back into a performance chamber for behavioral testing. Disruption of performance was found at the highest SAR (23 W/kg) but not at the lower levels. The authors concluded that the observed effects were based on a thermal effect.

A second study at a similar frequency, 1.3 GHz, by D'Andrea and his colleagues [3] was published the following year in a technical report. Rhesus monkeys were the subjects and performance was measured during irradiation at three SAR levels, 16, 26,

403

B.J. Klauenberg and D. Miklavcic (eds.), Radio Frequency Radiation Dosimetry, 403-407.

and 35 W/kg. Animals were illuminated for 25 minutes by an open-end waveguide focused on the rear of the monkey's head. Behavioral response rates on a vigilance task were reduced by the 26 and 35 W/kg SARs but not the 16 W/kg level. The authors concluded, amongst other things, that the current safety standard for localized maximum SAR of 8 W/kg in any one gram of tissue fails to provide for a safety factor of 10 and should be re-examined.

A series of studies with rhesus monkeys was conducted by John D'Andrea and colleagues [4, 5]. Two different frequencies, 5.62 GHz [4] and 2.37 GHz [5] were used. The animals were trained on operantly conditioned tasks and exposed to microwaves while performing the tasks. Even though the peak powers were quite high, the pulses were of short duration, e.g., 50 ns [4] and 93 ns [5]. The peak power densities ranged from 518 W/cm^2 at 5.62 GHz to 11.3 kW/cm^2 at 2.37 GHz. The exposures at the lower frequency resulted in relatively small SARs of 0.075 W/kg or less and no behavioral effects were observed. The exposures at 5.62 GHz at various energy levels produced SARs of 2, 4, and 6 W/kg. Only the 4 and 6 W/kg doses produced behavioral changes, a decrease in response rate and an increase in reaction time. These findings agreed with much of the behavioral work cited in the IEEE guide [1]. Two types of pulses, an enhanced pulse (9 times the level of the radar pulse) and the standard radar pulse, were examined in the higher frequency study [4]. No different behavioral effects occurred as a consequence of these two different pulse strengths.

A unique behavioral response of pulsed high power microwaves was observed by Brown *et al.* [6]. A mouse was confined to a plastic tube and its head and neck were exposed to single pulses of 1.25 GHz radiation at various power levels and number of pulses. It was observed that each animal would emit a twitch or movement of the body when it was exposed at levels ranging from 0.0072 to 200 kW. A dose dependency was found which leveled off at the higher doses (1 kW/kg). Again, a thermal mechanism seemed to be responsible with a threshold near an increment of 1.2 °C from baseline.

Several behavioral investigations have been conducted with a TEMPO vircator located at the microwave facilities of the Walter Reed Army Institute of Research. The device generates an 80 ns, 3000 MHz pulse with 700 MW peak transmitted power. In one study treadmill performance of rats was studied [7]. The subjects were trained to steady rates of daily running and then exposed in 25 min sessions to 1 pulse every 8 seconds. Following an exposure session the animals were placed on a treadmill and their performance assessed. Exposures produced SARs of 0.072 W/kg; nevertheless, a reduction in running time occurred. Similarly, another experiment with rats using the TEMPO [8] reported that decision-making was disrupted after a 25 min exposure session (200 pulses). Both of these experiments observed behavioral effects at power levels substantially below those indicated as a safe level in the IEEE guide [1]. Such effects could be specific to unique properties of the TEMPO, thus they are difficult to replicate with other sources.

Another group of devices producing high-power pulses covering the RF area involves ultra-wide band (UWB) and electromagnetic pulse (EMP) generators. UWB generators typically produce 60 Hz, 1-10 ns pulses at a bandwidth of 0.25-2.50 GHz, with a peak E field strength of 250 kV/m. The EMP generators contain most of their energy in the 20-100 MHz region at E-field strengths of 100 kV/m. Exposures in both instances are generally of short duration, but the EMP pulse is typically much longer, e.g., 900 ns.

Results of behavioral experiments with EMP exposures vary. One study disclosed that exposure to 200 EMP pulses for about 30 minutes resulted in a group of rats that preferred the exposure and another group whose performance was actually enhanced [9]. However, the enhancement was not a statistically significant change .

Orientation behavior in birds 3 to 6 hours after undergoing EMP exposure was examined by Moore and Simons [10]. Some disruption in selection of migratory direction was found, but the investigators concluded that free-flying migratory birds should not be strongly influenced by EMP pulses.

Research dealing with animals exposed to UWB pulses has been conducted on rats [11] and monkeys [12]. A variety of both cognitive and motivational behaviors was explored in rats after exposure to UWB radiation between 0.25 and 2.50 GHz for 2 minutes. No statistically significant differences in behavior were found between sham and exposed animals [11]. The monkey experiments [12] were designed with a narrower bandwidth (100 MHz to 1.5 GHz) and continued the 2 min exposures prior to testing. Still, no differences were found in a very complicated cognitive task involving vestibular canal functions.

3. Conventional Pulses

In the current decade there has been a decrease in the number of experiments with animals exposed to energy from microwave sources. Furthermore, the results of these few experiments have not been consistent and seem to add little information for the documentation of potential health effects of RF exposure. Each study is unique in both the frequency explored and the behavior examined. For example, Lai et al. [13] exposed rats for 45 minutes to pulsed 2450 MHz microwaves at 1 mW/cm^2 (SAR = 0.6 W/kg) in cylindrical waveguides. Following exposure the subjects were placed in a radial arm maze and tested for learning and memory. Several independent variables were employed, and in the main treatments of sham versus microwave exposure, a deficit in learning was found in the exposed rats.

In another study with a 2450 MHz source rats were also exposed at a power density of 1 mW/cm^2. When they were tested in a shuttle-box following exposure they were similarly found to have a performance decrement [14].

A recent report deals with the operantly conditioned behavior of the offspring of RF exposed rats [15]. Dams were exposed continuously from post-conception days 1-20 to 900 MHz radiation modulated at 217 Hz at 0.1 mW/cm^2. When the offspring became adults they were trained on an operant task under various contingencies. No differences in the performance of the offspring of exposed versus the offspring of the sham-exposed mothers were observed.

A rather unique study is that of Gapeev et al. [16]. They examined the effects of a range of continuous wave radiation centered on 42.25 GHz on the movement of protozoa. Cells of *Paramecium caudatum* were exposed in a 2 mm saline layer located in a cuvette. Cells were irradiated for 12 minutes at power densities from 0.1 to 20 mW/cm^2. A motility index of the animal's activity was calculated. The carrier frequency of the radiation was modulated from 0.05 to 16 Hz. Locomotor activity increased and reached a maximum at 0.0956 Hz modulation and 0.1 mW/cm^2. Different power densities produced the same effect. The effect decreased when the carrier frequency was either increased or decreased on either side of 42.25 GHz. This experiment illustrates a relatively simple preparation for investigating RF behavioral effects and should be attempted by others even though isolating the effects on the organism as opposed to the effects on the organism in saline will present some problems.

4. Conclusions

Behavioral research on the effects of radiofrequency radiation has decreased substantially over the last ten years. This is a surprising phenomenon in light of the fact that exposure standards in place in most countries are based on the results of behavioral experiments. Perusal of the program for the recent annual meeting of the Bioelectromagnetics Society held in St. Petersburg, Florida, reveals that only three RF behavioral studies were presented. Only two of these concentrated on UWB effects. It may be that behavior is no longer perceived as a sensitive indicator of RF health effects. Certainly, most of the research concerning cellular telephones is on non-behavioral biological effects.

Behavior is a final common expression of virtually all physiological processes with, perhaps, the exception of disease, and even that endpoint is often initially expressed as a behavioral response. If low level RF radiation produces adverse health effects, behavioral changes will continue to be the first to be observed.

5. References

1. IEEE C95.1-1991 (1992) *Safety Levels With Respect To Human Exposure to Radio Frequency Electromagnetic Fields, 3 kHz to 300 GHz*, IEEE, Piscataway, NJ.
2. Akyel, Y., Hunt, E.L., Gambrill, C., and Vargas, C. Jr., (1991) Immediate post-exposure effects of high-peak-power microwave pulses on operant behavior of Wistar rats, *Bioelectromagnetics* 12, 183-195.
3. D'Andrea J.A., Cobb B.L., and Knepton J. (1992) *Behavioral effects of high peak power microwave pulses: Head exposure at 1.3 GHz.* Report NAMRL-1372; Pensacola, FL: Naval Aerospace Medical Research Laboratory, NAMRL - 1372.
4. D'Andrea, J.A., Thomas, A., and Hatcher, D.J. (1994) Rhesus monkey behavior during exposure to high-peak-power 5.62-GHz microwave pulses, *Bioelectromagnetics* 15, 163-176.
5. D'Andrea, J.A. (1995) Effects of microwave radiation exposure on behavioral performance in nonhuman primates, in B.J. Klauenberg, M. Grandolfo, and D.N. Erwin (eds.), *Radiofrequency Radiation Standards. Biological Effects, Dosimetry, Epidemiology, and Public Health Policy*, Plenum Press, New York, pp. 271- 277.

6. Brown, D.O., Lu, S-T., and Elson, E.C. (1994) Characteristics of microwave evoked body movements in mice, *Bioelectromagnetics* **15**, 143-161.

7. Akyel, Y., Belt, M., Raslear, T.G., and Hammer, R.M. (1993) The effects of high-peak power pulsed microwaves on treadmill performance in the rat, in M. Blank (ed.), *Electricity and Magnetism in Biology and Medicine*, San Francisco Press, San Francisco, pp. 668-670.

8. Raslear, T.G., Akyel, Y., Bates, F., Belt, M., and Lu, S-T. (1993) Temporal bisection in rats: The effects of high-peak-power pulsed microwave radiation, *Bioelectromagnetics* **14**, 459-478.

9. Akyel, Y. and Raslear, T.G. (1993) The effects of acute EMP fields on the behavior of rats, in J. de Lorge and W. Mick (eds.), *EMP Human Health Effects Science Review Panel Proceedings, 16-18 Mar,* Theater   Nuclear Warfare Program, Crystal Plaza 5, Rm. 866, Washington, DC 20362-5101, pp. 45-57.

10. Moore, F.R. and Simons, T. (1993) Orientation behavior of migratory birds in response to Empress II-simulated EMP, in J. de Lorge and W. Mick (eds.), *EMP Human Health Effects Science Review Panel Proceedings, 16-18 Mar,* Theater Nuclear Warfare Program, Crystal Plaza 5, Rm. 866, Washington, DC 20362-5101, pp. 97-112.

11. Walters, T.J., Mason, P.A., Sherry, C.J., Steffen, C., and Merritt, J.H. (1995) No detectable bioeffects following acute exposure to high peak power ultra-wide band electromagnetic radiation in rats, *Aviat. Space and Environ. Med.* **65**, 562-567.

12. Sherry, C.J., Blick, D.W., Walters, T.J., Brown, G.G., and Murphy, M.R. (1995) Lack of behavioral effects in non-human primates after exposure to ultrawide electromagnetic radiation in the microwave frequency range, *Radiat. Res.* **143**, 93-97.

13. Lai, H., Horita, A., and Guy, A.W. (1994) Microwave irradiation affects radial-arm maze performance in the rat, *Bioelectromagnetics* **15**, 95-104.

14. Varetski, V.V., Rudnev, M.I., Degtyar, V.N., and Redshod'ko, T.L. (1991) Combined influence of gamma- rays and superhigh-frequency radiation on conditioned reflex behavior in the rat, *Radiobiologiia* **31**, 246-251.

15. Bornhausen, M., Kinkel, D., Wu, X.N., and Scheingraber, H. (1997) Operant behavior tests in rats after prenatal exposure to high frequency electromagnetic fields, presented at the Second World Congress for Electricity and Magnetism in Biology and Medicine, 8-13 June, Bologna, Italy, pp. 322-323.

16. Gapeev, A.B., Chemeris, N.K., Fesenko, E.E., and Khramov, R.N. (1994) Resonance effects of a low intensity modulated EHF field. Alteration of the locomotor activity of the protozoa paramecium caudatum, *Biofizika,* **39**, 74-82.

FOCUSING PROPERTIES OF PULSED SIGNALS INSIDE BIOLOGICAL TISSUE MEDIA

K. S. NIKITA
Department of Electrical and Computer Engineering
National Technical University of Athens
9, Iroon Polytechniou Str, Zografos 15773, Athens, Greece

1. Introduction

A highly interesting topic in using RF/microwave signals in biomedical applications is to develop techniques and systems achieving "focusing" inside tissues. To this end, positive interference arising from various sources has extensively been employed during the last decades, based on the use of continuous wave signals. In this context, phased array principles and optimization techniques have been applied to develop hyperthermia systems for the treatment of malignant tumors and mainly the low microwave spectrum (100-1000 MHz) has been employed [1]-[4]. However, these techniques suffer from the excessive attenuation of each wave radiated from each individual source and from the side effects created by the coupling phenomena between array source elements [5].

The possibility of employing pulsed signals to improve focusing properties has been suggested by various researchers, using as an additional system parameter the time delay between array signals. Although limited information is presently available on the behavior of propagation of pulsed signals, the behavior arising from precursor phenomena is expected to be useful in focusing electromagnetic waves.

The propagation of a pulsed signal inside dispersive dielectric media has received widespread treatment by the use of asymptotic analysis [6], transform techniques [7], time-domain integral equation solvers [8], the finite-difference time-domain algorithm [9-11] and Fourier series based approaches [12]. All these works have been restricted to the propagation of a pulsed electromagnetic plane wave in homogeneous dispersive dielectric media and the main interest has been focused on the study of the associated precursor fields from the point of view of possible hazardous health effects.

In the present work, the use of pulsed signals (~1ns pulse width) in order to achieve focusing inside biological tissues is examined rigorously. Focusing properties of both short baseband pulses and pulse modulated microwave signals with a high frequency (9.5 GHz) carrier inside a layered lossy cylinder are investigated, by using sources fed from TEM waveguides or TE_{10} rectangular waveguides, respectively. The high frequency of the carrier in the latter case enables the use of a large number of applicators compared to low frequency systems. This fact together with the significantly different

409

B.J. Klauenberg and D. Miklavcic (eds.), Radio Frequency Radiation Dosimetry, 409-418.
© 2000 *Kluwer Academic Publishers. Printed in the Netherlands.*

behavior of pulsed signals have motivated the initiation of the present study. In order to treat both problems, detailed electromagnetic models are developed in the following sections, which take into account the modification of the field on each waveguide aperture resulted from the other radiating elements of the array as well as from the presence of the lossy, layered, dielectric body standing at the near field region. Time coincidence of the pulses originated from each individual applicator of the array and constructive phase interference of the microwave carrier in the case of pulse modulated signals are applied, in order to achieve focusing at a specific point of interest within biological tissue.

2. Mathematical Formulation and Analysis

A Fourier based methodology is adopted in order to investigate the focusing properties of both pulsed baseband signals and pulse modulated microwave signals inside a three-layer cylindrical lossy model of circular cross section. The three layers can be used to simulate different biological media with dispersive characteristics, such as skin, bone and brain tissues. Alternatively, the two internal layers may be used to simulate biological media (e.g. brain and bone tissues) with the external layer simulating a lossless dielectric medium, which is commonly used to prevent excessive heating of the tissue surface. The dielectric properties of the layers are denoted with the corresponding frequency-dependent relative complex permittivities $\varepsilon_1(\omega)$, $\varepsilon_2(\omega)$, $\varepsilon_3(\omega)$. The free-space wavenumber is $k_0 = \omega\sqrt{\varepsilon_0\mu_0}$, where ε_0 and μ_0 are the free-space permittivity and permeability, respectively.

In order to examine the focusing properties of pulsed baseband signals, sources fed from TEM waveguides are assumed. In this direction, the selection of the structure shown in Figure 1, is a natural selection based on the use of an arbitrary number (L) of identical parallel-plate waveguide applicators, with an aperture size of D circulating around the cylindrical body's surface and their infinite dimension being parallel to the cylindrical body's axis. In order to investigate the possibility of improving the focusing properties by using a large number of applicators, the use of pulse modulated microwave signals with a high frequency carrier (9.5 GHz) is also examined. To this end, a concentric array of L identical rectangular aperture waveguide

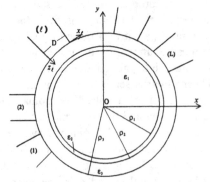

Figure 1. Cross section of a three-layer tissue model irradiated by a TEM concentric array of parallel-plate waveguide applicators.

applicators is considered, having an aperture size of a×b (b<a), placed at the periphery of the lossy model with the large dimension at the transverse direction circulating around the cylindrical body's surface and the small dimension parallel to the cylinder's axis (Figure 2). In both configurations, it is assumed that the apertures are not completely planar, conforming to the cylinder's surface, and are separated by perfectly conducting flanges.

The methodology of the adopted approach is first to predict the medium response to time harmonic excitation of the TEM and TE_{10} arrays, by adopting integral equation techniques in conjunction with Galerkin's procedures. Then, the medium response to pulsed excitation of the array elements is considered and the time dependence of the fields produced at any point within tissue is obtained as a Fourier series representation, if individual pulses are considered as members of a pulse train, or in the form of an inverse Fourier integral for the case of single, compact incident pulses.

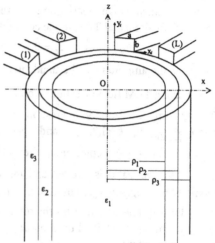

Figure 2. Three-layer tissue model irradiated by a TE_{10} rectangular aperture waveguide array.

2.1. BASEBAND PULSES - TEM WAVEGUIDE ARRAY

In this section, the transmission of baseband pulses radiated from a concentric TEM waveguide array (Figure 1) is analyzed rigorously. A Gaussian pulse train with a pulse repetition period T_0, is considered to be driven to the applicators. The pulses in the incident pulse train are assumed to be spaced widely enough apart so that there is no mutual interference of the pulses within the target structure. One period of the time variation of the incident signal of the ℓ th applicator is written as

$$g_\ell(t) = \exp(-\frac{\pi(t-t_\ell)^2}{\tau^2}), \quad \ell = 1,2,...,L \tag{1}$$

that is centered around the time $t_\ell > 0$, with τ a constant related to the pulse width in time. The associated incident pulse train $g_{P,\ell}(t)$ has a Fourier series representation,

$$g_{P,\ell}(t) = \sum_{n=-\infty}^{+\infty} c_\ell(n)\exp(jn\omega_0 t) \tag{2}$$

with $\omega_0 = 2\pi/T_0$ the pulse train fundamental angular frequency. The nth member of the discrete spectrum of the signal is given by

$$c_\ell(n) = \frac{\tau}{T_0} \exp(-\frac{n^2 \omega_0^2 \tau^2}{4\pi}) \exp(-jn\omega_0 t_\ell) \qquad (3)$$

and defines the complex amplitude of a time harmonic wave incident to the applicator.

In order to achieve in practice a stable operation and a good match to the power generator, it is desirable to have only a single propagating mode inside the waveguides. Thus, the incident wave distribution on the waveguide apertures is obtained as a spectrum of fundamental TEM mode components,

$$\underline{E}^w_{\ell,t(incid)}(x_\ell, z_\ell = 0; n) = p_\ell c_\ell(n) \underline{e}^{EM}_\ell(x_\ell) \qquad (4)$$

where p_ℓ is the amplitude of the incident TEM mode driven to the ℓ th applicator and $\underline{e}^{EM}_\ell(x_\ell)$ is the TEM field distribution on the aperture.

The quantity of primary interest in the present analysis is the complex transfer function $\underline{F}_\ell(\underline{r}; \omega)$, $\ell = 1, 2, ..., L$ and specifically the discrete spectrum components $\underline{F}_\ell(\underline{r}; n\omega_0) = \underline{F}_\ell(\underline{r}; n)$, representing the field produced at point $\underline{r}$ inside tissue, when only the ℓ th applicator is excited and the field on its aperture is a continuous time harmonic field ($\exp(+jn\omega_0 t)$) of unit amplitude. If this response can be predicted, then the time-domain representation for the total electric field due to the pulsed excitation of the array elements is obtained by summing the contributions made by each transmitted frequency component

$$\underline{E}(\underline{r}; t) = \text{Re}\left\{ \sum_{n=1}^{N} \left(\sum_{\ell=1}^{L} p_\ell c_\ell(n) \underline{F}_\ell(\underline{r}; n) \right) \exp(jn\omega_0 t) \right\} \qquad (5)$$

where Re(.) denotes the real part of the complex exponential form of the time-domain representation of each transmitted frequency component and N is the mode number of the highest frequency component retained for the Fourier series.

Thus, the strategy of the adopted approach is first to predict the medium response to the excitation of the array consisting of a single time harmonic component $\exp(+j\omega t)$ at a fixed angular frequency $\omega = n\omega_0$. The associated boundary value problem is solved by applying an integral equation technique.

The solution of the wave equation in cylindrical polar coordinates, inside the tissue layers ($i = 1,2,3$) is expressed in terms of cylindrical vector wave functions [13],

$$\underline{E}_i(\underline{r}) = \sum_{m=-\infty}^{m=+\infty} (a_{im} \underline{M}^{(1)}_m(\underline{r}, k_i) + b_{im} \underline{N}^{(1)}_m(\underline{r}, k_i) + a'_{im} \underline{M}^{(1)}_m(\underline{r}, k_i) + b'_{im} \underline{N}^{(1)}_m(\underline{r}, k_i)) \qquad (6)$$

where $k_i = k_0 \sqrt{\varepsilon_i(\omega)}$ and $a_{im}, b_{im}, a'_{im}, b'_{im}$ are to be determined.

Next, the fields inside the waveguide applicators are described, by using the waveguide normal modes. Thus, the transverse electric field in the ℓ th applicator ($\ell = 1, 2, ..., L$) can be written as the superposition of the incident TEM mode and an infinite number of all the reflected TEM and TM modes, since TE modes are not excited due to the geometry of the structure,

$$\underline{E}_{t,t}^{w}(x_{\ell},z_{\ell}) = A_{\ell}\underline{e}_{-t}^{EM}(x_{\ell})e^{-jkz_{\ell}} + A'_{\ell}\underline{e}_{-t}^{EM}(x_{\ell})e^{jkz_{\ell}} + \sum_{m=1}^{\infty}[\frac{j\lambda_{m}}{v_{m}}B'_{\ell,m}\underline{e}_{-m,t}^{M}(x_{\ell})e^{j\lambda_{m}z_{\ell}}] \tag{7}$$

where the subscript t is used to denote the transverse field components, A_{ℓ} is the complex amplitude of the excited TEM mode in the ℓ th waveguide and A'_{ℓ}, $B'_{\ell,m}$ are the complex amplitudes of the reflected TEM, mth order TM modes, respectively, in the ℓ th waveguide and k, λ_{m} are the corresponding propagation constants [13].

By satisfying the continuity of the tangential electric and magnetic field components on the $\rho = \rho_{1}$ and $\rho = \rho_{2}$ interfaces and on the $\rho = \rho_{3}$ contact surface between the cylindrical lossy model and the radiating apertures, the following system of L coupled integral equations is obtained, in terms of an unknown transverse electric field $\underline{E}_{a}$ on the waveguide apertures,

$$\sum_{q=1}^{L}\int_{\Gamma_{q}}\underline{\overline{K}}_{\ell q}(x/x')\underline{E}_{\alpha}(x')dx' = 2A_{\ell}(\frac{k}{\omega\mu_{0}})\underline{h}_{t}^{EM}(x), \quad \ell = 1,2,...,L \tag{8}$$

where $\underline{h}_{t}^{EM}$ is the incident TEM mode transverse magnetic field on the aperture of the ℓ th waveguide, and the kernel matrices $\underline{\overline{K}}_{\ell q}(x/x')$, $\ell = 1,...,L/q = 1,...,L$ indicate the effect of coupling from the qth aperture $(x') \in \Gamma_{q}$ to the ℓ th aperture $(x) \in \Gamma_{\ell}$. In order to determine the electric field on the waveguide apertures, the system of integral equations (8) is solved. To this end, a Galerkin's technique is adopted by expanding the unknown transverse electric field on each aperture $\underline{E}_{q,a}$ into waveguide normal modes. Therefore, with respect to the qth $(q=1,2,...,L)$ aperture's local Cartesian coordinate system, the electric field on the same aperture is expressed in the following form,

$$\underline{E}_{q,a} = g_{q}\underline{e}_{t}^{EM} + \sum_{m=1}^{\infty}f_{q,m}\underline{e}_{-m,t}^{M}, \quad q = 1,2,...,L \tag{9}$$

By substituting Equation 9 into the system of coupled integral equations (8), and making use of the waveguide modes orthogonality [13], the system of integral equations (8) is converted into an infinite system of linear equations. Assuming the aperture fields are determined, then the coefficients $a_{im}, b_{im}, a'_{im}, b'_{im}$ $(i=1,2,3)$ are determined easily. Substituting the values of these coefficients into Equation 6, the electric field (complex transfer function) at any point inside tissue can be easily computed and then the temporal dependence of the field is obtained by computing the Fourier series of Equation 5.

2.2. PULSED MICROWAVE SIGNALS - TE$_{10}$ WAVEGUIDE ARRAY

The radiating system examined in this section is shown in Figure 2. An input pulse modulated harmonic signal of fixed carrier frequency ω_{0} is considered to be driven to the applicators. Thus, the signal driven to the ℓ th applicator may be represented as

$$g_{\ell}(t) = u_{\ell}(t)\cos(\omega_{0}t + \psi_{\ell}), \quad \ell = 1,2,...,L \tag{10}$$

414

where $u_\ell(t) = \exp(-\pi(t-t_\ell)^2/\tau^2)$ is the Gaussian envelope function and ψ_ℓ is a phase term of the carrier.

The spectrum of frequencies contained in the pulse modulated microwave signal is obtained from its Fourier transform,

$$G_\ell(\omega) = \frac{1}{2}e^{j\psi_\ell}e^{-j(\omega-\omega_0)t_\ell}U(\omega-\omega_0) + \frac{1}{2}e^{-j\psi_\ell}e^{-j(\omega+\omega_0)t_\ell}U(\omega+\omega_0) \qquad (11)$$

where

$$U(\omega) = \frac{\tau}{2\pi}\exp(-\frac{\tau^2\omega^2}{4\pi})\exp(j\omega t_\ell) \qquad (12)$$

is the Fourier transform of the Gaussian envelope.

Since, it is desirable to have only the fundamental TE_{10} mode propagating inside the waveguides, the instantaneous distribution of the excited field on the aperture of the ℓ th applicator is given by the equation

$$\underline{e}_\ell(x_\ell,y_\ell,z_\ell=0;t) = p_\ell g_\ell(t)\underline{e}_{1,\ell}^{TE}(x_\ell,y_\ell) \qquad (13)$$

where p_ℓ is the real amplitude of the incident fundamental TE_{10} mode driven to the ℓ th applicator and $\underline{e}_{1,\ell}^{TE}(x_\ell,y_\ell)$ is the TE_{10} field distribution on the aperture.

The strategy of the adopted approach is first to analyze the propagation of each frequency component individually into the structure of interest, by computing the associated complex transfer function $\underline{F}_\ell(\underline{r};\omega)$, $\ell=1,2,...,L$, representing the field produced when only the ℓ th applicator aperture is excited by a continuous time harmonic field, of unit amplitude and zero phase. Then, the instantaneous field at a point of interest inside tissue due to the pulse modulated excitation of the array, is obtained in the form of a Fourier inversion integral [14],

$$\underline{E}(\underline{r};t) = \frac{1}{2}\text{Re}\left\{\exp(j\omega_0 t)\sum_{\ell=1}^{L}p_\ell\exp(j\psi_\ell)\int_{-\Delta\omega/2}^{\Delta\omega/2}d\omega\underline{F}_\ell(\underline{r};\omega_0+\omega)U(\omega)\exp(j\omega(t-t_\ell))\right\} \qquad (14)$$

where $\Delta\omega$ is the frequency bandwidth of the incident Gaussian pulses.

The medium response to a continuous wave excitation of the array is predicted, by adopting an integral equation technique [5]. To this end, in a way similar with that explained in section 2.1, the fields inside the tissue layers are described in terms of cylindrical wave functions [13],

$$\underline{E}_i(\underline{r}) = \int_{-\infty}^{+\infty}dk\sum_{m=-\infty}^{m=+\infty}\left(a_{im}\underline{M}_{m,k}^{(1)}(\underline{r},k_i)+b_{im}\underline{N}_{m,k}^{(1)}(\underline{r},k_i)+a'_{im}\underline{M}_{m,k}^{(2)}(\underline{r},k_i)+b'_{im}\underline{N}_{m,k}^{(2)}(\underline{r},k_i)\right) \qquad (15)$$

where in this case, a $\int_{-\infty}^{+\infty}dk$ integral is encountered due to the z-dependent geometry.

The fields inside each waveguide are described as the superposition of the incident TE_{10} mode and an infinite number of all the reflected TE and TM modes. Following the notation of [13], the transverse electric field inside the ℓ th waveguide applicator

($\ell = 1,2,...,L$) can be written, with respect to the local cartesian coordinates system x_ℓ, y_ℓ, z_ℓ, attached to the aperture's corner (see Figure 2), as follows:

$$\underline{E}^w_{\ell,t}(x_\ell, y_\ell, z_\ell) = p_\ell e^{j\psi_\ell}\, \underline{e}^{TE}_{1,t}(x_\ell, y_\ell)\frac{j\omega\mu_0}{u_1} e^{-j\gamma_1 z_\ell}$$

$$+ \sum_{m=1}^{\infty}\left(A'_{\ell,m}\underline{e}^{TE}_{m,t}(x_\ell, y_\ell)\frac{j\omega\mu_0}{u_m} e^{j\gamma_m z_\ell} + B'_{\ell,m}\underline{e}^{TM}_{m,t}(x_\ell, y_\ell)\left(-\frac{j\lambda_m}{v_m}\right)e^{j\lambda_m z_\ell}\right) \quad (16)$$

By imposing the boundary conditions for the tangential electric and magnetic field components, the following system of L coupled integral equations is obtained,

$$\sum_{q=1}^{L} \iint_{\Gamma_q} dx' dy' \overline{K}_{\ell q}(x, y / x', y')\underline{E}_a(x', y') = 2p_\ell e^{j\psi_\ell}\, \underline{h}^{TE}_{1,t}\left(\frac{j\gamma_1}{u_1}\right), \quad \ell = 1,2,...,L \quad (17)$$

where $\underline{h}^{TE}_{1,t}$ is the incident TE_{10} mode transverse magnetic field on the aperture of the ℓ th waveguide, and the kernel matrices $\overline{K}_{\ell q}(x, y / x', y')$ indicate the effect of coupling via the radiating apertures and are given in detail in [5]. In order to solve this system, a Galerkin's procedure is adopted by expressing the unknown electric fields on the apertures in terms of the corresponding waveguide mode fields,

$$\underline{E}_{q,a} = \sum_{m=1}^{\infty}\left(g_{q,m}\underline{e}^{TE}_{m,t} + f_{q,m}\underline{e}^{TM}_{m,t}\right), \quad \ell = 1,2,...,L \quad (18)$$

Once the aperture fields are determined, the electric field distribution within tissue can be easily computed and then the dynamic field evolution can be obtained by computing the Fourier inversion integral of Equation 14.

3. Numerical Results and Discussion

The analysis presented in Section 2, has been applied to investigate the focusing ability of both a 8 - element TEM waveguide array and a 30 - element TE_{10} waveguide array at a point of interest inside a cylindrical tissue model, 16 cm in diameter. The body model consists of two layers, simulating bone and brain tissues and it is surrounded by a 2 cm thick lossless dielectric layer. The thicknesses of the bone and the external dielectric layers are assumed to be $\rho_2 - \rho_1 = 0.5cm$ and $\rho_3 - \rho_2 = 2cm$, respectively ($2\rho_3 = 20cm$). The dielectric constant of the external layer used in the calculations is taken to be $\varepsilon_3 = 2.1$. The numerical values of tissue complex permittivities used in the calculations are defined at the frequency range of interest by using the data compiled from the relevant literature [15]. In order to provide focusing at a point of interest, within the brain tissue, located at 2 cm depth from the tissue surface, time coincidence and constructive phase interference principles are used and numerical results are presented in the following sections.

3.1. BASEBAND PULSES

A concentric array consisting of eight (8) TEM waveguides with an aperture size of $D = 7.5cm$ placed symmetrically at the periphery of the external dielectric layer is considered. The input signal driven to each applicator is assumed to be a Gaussian pulse train, with an individual pulse duration of 1 ns and a pulse repetition interval of 10 ns.

In an attempt to focus the electromagnetic radiation at the point of interest within the brain tissue, located on the axis of an applicator, time coincidence of the fields originated from the eight (8) waveguides of the array is used. To this end, the discrete spectrum components of the transfer function of each individual applicator are computed and the temporal evolution of the main vector component E_φ of the field originated from each individual applicator at the point of interest is obtained. The latter is used to determine the appropriate time delays to be introduced to the individual pulses, in order to achieve time coincidence of the fields originated from the individual applicators.

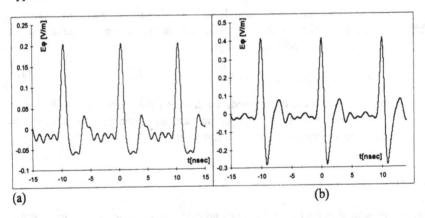

(a)   (b)

Figure 3. Temporal evolution of the main field component Eφ of a 8-element TEM array, at a point of interest, located on the axis of an applicator, at 4 cm depth from its aperture. (a) Uniform array excitation. (b) Array excitation adjusted to provide focusing at this point.

The time dependence of the field produced at the point of interest is shown in Figure 3a, for uniform array excitation ($p_1=...=p_8=1$, and $t_1=...=t_8=0$ and in Figure 3b for time excitation, adjusted to achieve time coincidence of the signals at the point of interest. By comparing Figure 3a with Figure 3b, a 100% increase of the main peak amplitude of the pulse is achieved by adjusting the excitation of the pulsed signals driven to the individual applicators. Moreover, by integrating the squared magnitude of the electric field for a period of $T_0 = 10 \, ns$, it can easily be observed that a 340% increase of the deposited power at the point of interest is achieved, by adjusting the array temporal excitation.

418

4. Conclusions

A rigorous analysis has been presented for investigating the focusing properties of short baseband pulses emitted by a TEM concentric array of applicators in a layered cylindrical biological tissue model. Furthermore, the possibility of improving focusing properties by using a large number of array elements has been examined, by considering a TE_{10} rectangular aperture waveguide array, excited by short pulses with a high frequency microwave carrier. Numerical results have been computed and presented for a bone-brain tissue model irradiated by a 8 - element TEM array and a 30 – element TE_{10} array. By adjusting the time delay of the signals injected to the individual applicators and the carrier phase in the case of modulated microwave signals, focusing at a target point within brain tissue has been achieved. The obtained results provide an enhanced physical insight of pulse propagation inside layered lossy media and can be used in order to achieve focusing inside biological tissues.

5. References

1. Chen, J. and Gandhi, O.P. (1992) Numerical simulation of annular phased arrays of dipoles for hyperthermia of deep seated tumors, *IEEE Trans. Biomed. Eng.* **39**, 209-216.
2. Nikita, K.S., Maratos, N.G., and Uzunoglu, N.K. (1993) Optimal steady-state temperature distribution for a phased array hyperthermia system, *IEEE Trans. Biomed. Eng.* **40**, 1299-1306.
3. Boag, A., Leviatan, Y., and Boag, A. (1993) Analysis and optimization of waveguide multiapplicator hyperthermia systems, *IEEE Trans. Biomed. Eng.* **40**, 946-952.
4. Nikita, K.S., Maratos, N.G., and Uzunoglu, N.K. (1998) Optimization of the deposited power distribution inside a layered lossy medium irradiated by a coupled system of concentrically placed waveguide applicators, *IEEE Trans. Biomed. Eng.* **45**, 909-920.
5. Nikita, K.S. and Uzunoglu, N.K. (1996) Coupling phenomena in concentric multi-applicator phased array hyperthermia systems, *IEEE Trans. Microwave Theory Tech.* **44**, 65-74.
6. Oughstun, K.E. and Laurens, J.E.K. (1991) Asymptotic description of electromagnetic pulse propagation in a linear causally dispersive medium, *Radio Science* **26**, 245-258.
7. Wyns, P., Fotty, D.P., and Oughstun, K.E. (1989) Numerical analysis of the precursor fields in linear dispersive pulse propagation, *J. Opt. Soc. Am. A* **6**, 1421-1429.
8. Bolomey, J., Duri, C., and Lesselier, D. (1978) Time domain integral equation approach for inhomogeneous and dispersive slab problems, *IEEE Trans. Antennas Propag.* **26**, 658-667.
9. Joseph, R., Hagness, S., and Taflove, A. (1991) Direct time integration of Maxwell's equations in linear dispersive media with absorption for scattering and propagation of femtosecond electromagnetic pulses, *Opt. Lett.* **16**, 1412-1414.
10. Luebbers, R.J. and Hunsberger, F. (1992) FD-TD for n-th order dispersive media, *IEEE Trans. Antennas Propag.* **40**, 1297-1301.
11. Petropoulos, P.G. (1995) The wave hierarchy for propagation in relaxing dielectrics, *Wave Motion* **21**, 253-262.
12. Blashank, J.G. and Frazen, J. (1995) Precursor propagation in dispersive media from short-rise-time pulses at oblique incidence, *J. Opt. Soc. Am. A* **12**, 1501-1512.
13. Jones, D.S. (1964) *Theory of Electromagnetism*, Pergamon Press, Oxford.
14. Champeney, D.C. (1973) *Fourier Transforms and their Physical Applications*, Academic Press, New York.
15. Gabriel, C., Gabriel, S., and Corthout, E. (1996) The dielectric properties of biological tissues, *Med. Phys.* **41**, 2231-2293.

900 MHZ ELECTROMAGNETIC FIELDS: EXPOSURE PARAMETERS AND EFFECTS ON DJUNGARIAN HAMSTERS

A. LERCHL, H. BRENDEL, M. NIEHAUS, H. KRISHNAMURTHY
Institute of Reproductive Medicine
University of Münster
Domagkstr. 11, D-48129 Münster, Germany

V. HANSEN, J. STRECKERT, A. BITZ
Department of Electromagnetic Theory
University of Wuppertal
Gaussstr. 20, D-42097 Wuppertal, Germany

1. Introduction

It is generally accepted that radio frequency (RF) electromagnetic fields (EMF) can result in adverse effects if the intensities of the fields are high enough to increase the temperature of the biological target above the level of thermal noise. The mechanisms by which non-thermic EMF may have biological effects, however, are not yet understood. In the view of increasing public concerns about electromagnetic radiation („electro-smog") it is important to address this point, especially because of the steadily increasing use of mobile telephones and other electronic equipment working in the MHz to GHz range and operating with signal modulations in the frequency range of biological activities.

Among the possible links between EMF and adverse biological effects is the so-called „melatonin hypothesis" [1, 2]. This theory is based on epidemiological studies and experimental work in the low-frequency (*i.e.*, 50 Hz or 60 Hz), non-thermic EMF range [3, 4, 5, 6, 7]. According to this theory, the synthesis of the pineal hormone melatonin is diminished as a consequence of EMF exposure. Since melatonin is known to suppress estrogen production which is, in turn, responsible for the growth of certain breast cancers, it was argued that the tumor-promoting effect of EMF is an indirect consequence of melatonin suppression [1]. This model has changed over time: first, it was shown that melatonin has direct effects on the proliferation of cancer cell lines [8, 9]. Second, melatonin was identified as a potent endogenous scavenger for oxygen radicals, especially acting against the hydroxyl radical, a substance capable of damaging DNA [10-12]. It appears therefore possible that any suppression of melatonin production has direct consequences with respect to the promotion of malignant diseases.

419

B.J. Klauenberg and D. Miklavcic (eds.), Radio Frequency Radiation Dosimetry, 419-428.
© *2000 Kluwer Academic Publishers. Printed in the Netherlands.*

Another effect of EMF on testicular cell composition was recently described [13]. Here it was shown that exposure of male Djungarian hamsters (*Phodopus sungorus*) to 50 Hz magnetic fields lead to a significant increase of testicular cells. Since the testis is an organ with a high proliferation rate of cells, the effects of magnetic fields are an interesting counterpart to the observed promoting effects of 50 Hz magnetic fields on chemically induced cancer [14-16]. Hence, it may be that the testis is a well suited organ for studies addressing the possible effects of EMF on proliferating cells.

2. Aims

The present investigation was planned to answer the following questions: does long-term exposure of Djungarian hamsters with a typical mobile communication signal influence the melatonin synthesis and the testicular cell composition as estimated after exposure? Additionally, what effects are observed with respect to body weight development during exposure? Since we intended to have a statistically high power, a large number of animals per group was considered an important point. Consequently, a sophisticated exposure system was developed allowing the simultaneous exposure of 120 hamsters with well-defined exposure parameters. An additional exposure chamber without an EMF source connected ensured true sham exposure conditions.

3. Material and methods

3.1. EXPOSURE SYSTEM

3.1.1. *Requirements*
The essential prerequisite is that the electromagnetic field distribution can be reproduced. Therefore, it is necessary to select a uniquely defined and stable type of field with a known field distribution inside the test objects [17].

The experiment is arranged as a double blind test with one exposure and one sham-exposure setup each containing 120 animals in 40 cages and three hamsters per cage (Figure 1). Both exposure devices should be shielded against each other by a factor less than −60 dB. The frequency is 900 MHz with a modulation according to a typical GSM signal including the 2 Hz, 8 Hz, 217 Hz, and 1733 Hz frequency components. The RF power has to be adjusted in such a way that an averaged SAR per animal of approximately 80 mW/kg is achieved. The illumination and ventilation of the cages must be provided. Furthermore, removing the hamsters from the cages should be as simple as possible.

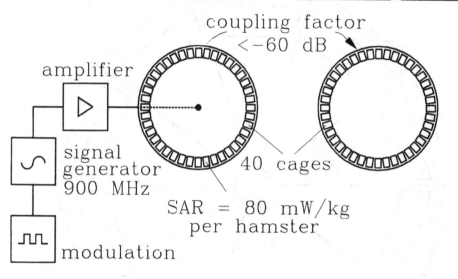

Figure 1. Requirements of the exposure system

3.1.2. *Exposure setup*

A circular parallel plate waveguide was chosen as exposure device (Figure 2). Diameter and height of this waveguide are 4 m and 0.14 m, respectively. By the choice of the height is guaranteed that no higher order waveguide modes with respect to the z-direction can occur at 900 MHz. Modes with an azimuthal dependency are avoided by the symmetric excitation through the feed at the center of the waveguide. It was designed in such a way that in a distance ρ greater than a wavelength the field distribution of the empty waveguide is given by the fundamental TEM-mode. The radial boundary of the waveguide is terminated by flat metal-backed foam absorbers. The cages are inserted through openings in the upper plate which can be closed by wire mesh lids in order to guarantee aeration and light. The design of the lids was chosen with respect to a minimum disturbance of the field inside the waveguide and a sufficient degree of decoupling from the outer space.

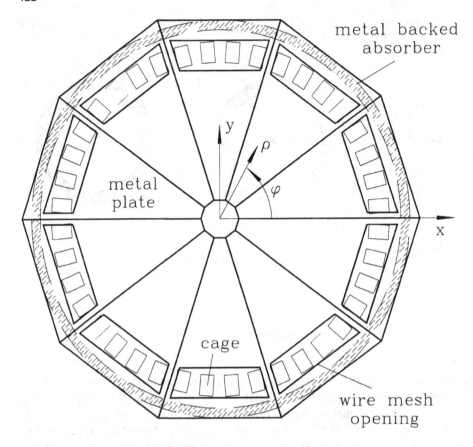

Figure 2. Top view of the circular parallel plate waveguide

3.1.3. *Performance*

The main advantages of this setup are the circular uniformity of the exposure, the efficient exploitation of the RF power and the fact that it is fully-shielded, thus no RF shielded room is required that surrounds the setup.

The measured coupling factor between the setups for exposure and for sham-exposure is less than −75 dB.

Figure 3 shows the calculated electric field distribution in a quarter of the waveguide. The radial decay of the field strength following a $1/\sqrt{\rho}$ – dependency is superimposed by a standing wave ratio of 1.22 due to the reflectivity of the absorber. Furthermore, the azimuthal field distribution is very uniform and no higher order waveguide modes are excited.

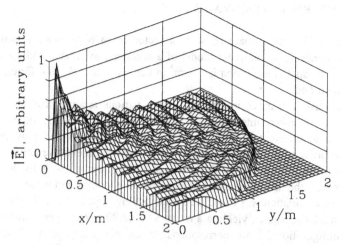

Figure 3. Calculated magnitude of the electric field strength in a quarter of the empty waveguide in arbitrary units. An absorber-model with a reflectivity of –20 dB is located at the radial boundary.

Figure 4 gives the field distribution with 30 computer-models of the hamsters placed into a 90°-sector of the waveguide. Obviously, the insertion of the hamsters leads only to localized field perturbations in the closer environment of each hamster. The reflections at the hamsters slightly increase the standing wave ratio in front of them. Furthermore, small variations with respect to φ are observed, but the azimuthal field distribution is still almost uniform.

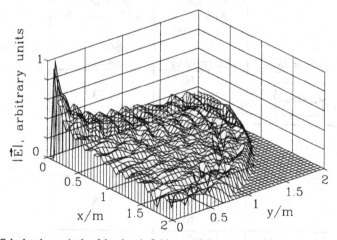

Figure 4. Calculated magnitude of the electric field strength in a quarter of the waveguide with inserted hamster models. Cut at the center of the models.

424

3.2. DOSIMETRY AND SAR-VARIATION

One adult male Djungarian hamster was subjected to an NMR scan. Based on these data, a three-dimensional computer model of a hamster was created and used for SAR calculations. The spatial resolution is 1 mm^3 in the entire model. 18 different materials can be distinguished.

As the animals can move inside their cages, various relative positions of the test object have to be taken into account. Due to mutual shadowing different animal positions lead to different field distributions inside the cages and thus to different specific absorption rates inside the hamsters. For some representative configurations of a triple of hamsters, Table 1 gives the averaged SAR-values of the individual hamster, the variation of the averaged SAR-values of each configuration and the overall variation. This calculation was based on a total RF input power of 3.5 W in a burst of the GSM modulation signal, yielding a mean SAR of 80 mW/kg per hamster averaged over all configurations. This corresponds to an electric peak field strength of approximately 43 V/m and a power density of 2.5 W/m^2 in front of the cages.

Additionally it was checked whether a certain threshold value of the local SAR within the animals was exceeded or not. According to international recommendations for humans, a value of 2 W/kg was taken as a reference. The obtained local SAR remained below 0.6 W/kg. This ensures that the experiments are performed within the non-thermic EMF exposure range.

TABLE 1. Averaged SAR-values and SAR-variation of four different configurations of hamster triples.

Configuration	Averaged SAR – value per hamster SAR [mW/kg]	averaged SAR-variation $\frac{SAR_{max}}{SAR_{min}}$	overall variation $\frac{SAR_{max}}{SAR_{min}}$
a:	1: 110 2: 80 3: 87	1.33	
b:	1: 80 2: 80 3: 75	1.07	1.93
c:	1: 65 2: 57 3: 65	1.38	
d:	1: 104 2: 78 3: 79	1.14	

3.3. EXPERIMENTAL PROCEDURE

Male Djungarian hamsters were born and raised in our colony [18]. A total of 240 adult animals was randomly assigned to two groups of 120 animals each. Groups of 3 litter mates were kept in macrolon cages. Food was supplied *ad libitum*. Water was supplied as apples since conventional water bottles contain metal parts which were not allowed inside the chambers. Before and during exposure, animals were weighed every two to three days.

After exposure, animals were sacrificed during night-time hours. To prevent any light exposure which would have influenced melatonin concentrations night-vision goggles were used. Trunk blood was collected and kept at 4°C before centrifugation. Serum was separated and kept at -20°C until analysis. Pineal glands were removed from the carcasses and snap frozen on solid CO_2. They were also kept at -20°C until analysis. Melatonin was measured in serum and pineal glands as described previously [19].

One testis per animal was removed in the course of the experiment from 20 individuals of each group every 10 days. Small tissue pieces were minced in buffer and treated with trypsin to yield individual cells. A DNA-specific fluorescent dye was added and the cells were individually analyzed by a flow cytometer. According to the DNA content and the respective fluorescence intensities, cells were discriminated to be haploid (1C), diploid (2C), or tetraploid (4C). Additionally, condensed haploid cells (1CC) were identified which represent elongated spermatids. Details of the method are given elsewhere [19].

3.4. STATISTICAL ANALYSIS

The body weights were analyzed both as absolute data and as percentages, setting the individuals' body weights as 100%. Statistical comparisons were performed with two-way analysis of variance (factors time and exposure). In the case of body weights, no transformations were necessary since these data were normally distributed. During the experiment, none of the persons directly involved in the measurements were aware which group was exposed and which was sham-exposed. Only after complete analysis, the code for the respective parameter was broken.

4. Results

At the time of manuscript preparation, the experiment has been finished, and the data on body weight are available. It is clear from both the absolute body weights and the relative changes from the starting values, respectively, that the exposure leads to an increase in body weight (Figure 5). The remaining parameters are currently being analyzed, and the results will be presented at the conference.

Absolute body weight

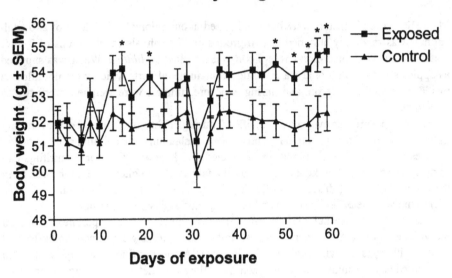

Relative body weight

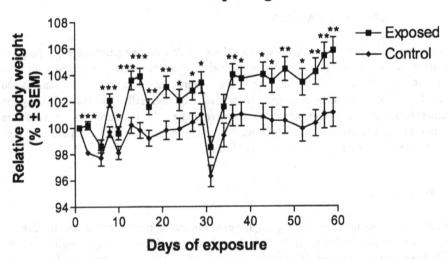

Figure 5. Effects of exposure to 900 MHz EMF on absolute and relative body weight in Djungarian hamsters (Phodopus sungorus). Relative data are related to day 1 of exposure (= 100 %). Asterisks represent levels of significance: *, $p < 0.05$; **, $p < 0.01$; ***, $p < 0.001$, ANOVA.

5. Discussion

This experiment shows that it is possible to expose 120 Djungarian hamsters simultaneously to a 900 MHz pulsed electromagnetic field while another group of 120 animals is sham-exposed at the same time. Due to the design of the chambers, the supply of light is ensured while almost no radio-frequency EMF is escaping. This leads to a difference in field intensities of exposed to sham-exposed animals of more than 75 dB.

The results obtained so far show that the exposure of Djungarian hamsters to 900 MHz pulsed EMF over 60 days affects body weight significantly although the average SAR of 80 $mW \cdot kg^{-1}$ belongs to the category of non-thermal exposure. It is not yet clear which biological mechanism(s) may be responsible for such effects.

6. Acknowledgements

This work was supported by a grant from the Forschungsgemeinschaft Funk (Bonn, Germany). Elke Kößer, MTA, is acknowledged for expert technical assistance.

7. References

1. Stevens, R. G. (1987) Electric power use and breast cancer: A hypothesis, *Am. J. Epidemiol.* **125**: 556-561.
2. Stevens, R. G., and Davis, S. (1996) The melatonin hypothesis: Electric power and breast cancer, *Environ. Health Perspect.* **104** Suppl. 1, 135-140.
3. Hatch, M. (1992) The epidemiology of electric and magnetic field exposures in the power frequency range and reproductive outcomes, *Paediatr. Perinat. Epidemiol.* **6** 198-214.
4. Loomis, D. P., Savitz, D. A., and Ananth, C. V. (1994) Breast cancer mortality among female electrical workers in the United States, *J. Natl. Cancer Inst.* **86**, 921-925.
5. Stevens, R. G. (1993) Biologically based epidemiological studies of electric power and cancer, *Environ. Health Perspect.* **101** Suppl. 4, 93-100.
6. Tynes, T., Hannevik, M., Andersen, A., Vistnes, A. I., and Haldorsen, T. (1996) Incidence of breast cancer in Norwegian female radio and telegraph operators, *Cancer Causes Control* **7**: 197-204.
7. Wood, A. W. (1993) Possible health effects of 50/60Hz electric and magnetic fields: Review of proposed mechanisms, *Australas. Phys. Eng. Sci. Med.* **16**: 1-21.
8. Harland, J. D., and Liburdy, R. P. (1997) Environmental magnetic fields inhibit the antiproliferative action of tamoxifen and melatonin in a human breast cancer cell line, *Bioelectromagnetics* **18**: 555-562.
9. Liburdy, R. P., Sloma, T. R., Sokolic, R., and Yaswen, P. (1993) ELF magnetic fields, breast cancer, and melatonin: 60 Hz fields block melatonin's oncostatic action on ER+ breast cancer cell proliferation, *J. Pineal Res.* **14**: 89-97.
10. Poeggeler, B., Reiter, R. J., Tan, D. X., Chen, L. D., and Manchester, L. C. (1993) Melatonin, hydroxyl radical-mediated oxidative damage, and aging: A hypothesis, *J. Pineal Res.* **14**, 151-168.
11. Reiter, R. J. (1993) Interactions of the pineal hormone melatonin with oxygen-centered free radicals: A brief review, *Braz. J. Med. Biol. Res.* **26**, 1141-1155.
12. Reiter, R. J., Melchiorri, D., Sewerynek, E., Poeggeler, B., Barlow Walden, L., Chuang, J. Ortiz, G. G., and Acuna Castroviejo, D. (1995) A review of the evidence supporting melatonin's role as an antioxidant, *J. Pineal Res.* **18**, 1-11.

428

13. Niehaus, M., Brüggemeyer, H., Behre, H. M., and Lerchl, A. (1997) Growth retardation, testicular stimulation, and increased melatonin synthesis by weak magnetic fields (50 Hz) in Djungarian hamsters, *Phodopus sungorus, Biochem. Biophys. Res. Commun.* **234**, 707-711.
14. Löscher, W., Wahnschaffe, U., Mevissen, M., Lerchl, A., and Stamm, A. (1994) Effects of weak alternating magnetic fields on nocturnal melatonin production and mammary carcinogenesis in rats, *Oncology* **51**: 288-295.
15. Mevissen, M., Haussler, M., Lerchl, A., and Löscher, W. (1998) Acceleration of mammary tumorigenesis by exposure of 7,12- dimethylbenz[a]anthracene-treated female rats in a 50-Hz, 100-microT magnetic field: Replication study, *J. Toxicol. Environ. Health* **53**, 401-418.
16. Mevissen, M., Lerchl, A., Szamel, M., and Löscher, W. (1996) Exposure of DMBA-treated female rats in a 50-Hz, 50 microTesla magnetic field: Effects on mammary tumor growth, melatonin levels, and T lymphocyte activation, *Carcinogenesis* **17**, 903-910.
17. Hansen, V. (1997) *Guidelines for Experiments to Investigate the Effect of Radio-Frequency Electromagnetic Fields on Biological Systems – Radio-Frequency Aspects*, Edition Wissenschaft, Forschungsgemeinschaft Funk, No. 11/E.
18. Lerchl, A. (1995) Breeding of Djungarian hamsters (*Phodopus sungorus*): Influence of parity and litter size on weaning success and offspring sex ratio, *Lab. Anim.* **29**, 172-176.
19. Lerchl, A., and Schlatt, S. (1992) Serotonin content and melatonin production in the pineal gland of the male Djungarian hamster (*Phodopus sungorus*), *J. Pineal Res.* **12**, 128-134.
20. Krishnamurthy, H., Weinbauer, G. F., Aslam, H., Yeung, C. H., and Nieschlag, E. (1998) Quantification of apoptotic testicular germ cells in normal and methoxyacetic acid treated mice as determined by flow cytometry, *J. Androl.* (in press).

LIGAND BINDING UNDER RF EM EXPOSURE

A. CHIABRERA[1], B. BIANCO[1], S. GIORDANO[1], S. BRUNA[1], E. MOGGIA[1] and J. J. KAUFMAN[2]

[1]ICEmB at the Department of Biophysical and Electronic Engineering, University of Genoa, Via Opera Pia 11a, 16145 Genoa, Italy.
[2]Bioelectrochemistry Laboratory, Department of Orthopaedics, Mount Sinai School of Medicine, 1 Levy Place, New York, N. Y. 10029, USA and CyberLogic, Inc., New York.

1. Abstract

The influence of electromagnetic exposure on ligand binding to receptor proteins is a putative early event of the interaction mechanism leading to biological effects. The most recent development of the quantum Zeeman-Stark model is reviewed, addressing the following points: losses due to the collisions of the ligand ion inside the hydrophobic binding crevice and thermal noise; evaluation of the attracting endogenous force of the binding site from the protein data base; out of equilibrium state of the ligand-receptor system due to the basal cell metabolism.

The biochemical output is the change of the ligand binding probability due to low intensity electromagnetic exposure at radio frequencies.

2. Introduction

The scientific interest in the biological effects induced by the exposure of living systems to an electromagnetic field (e.m.f.) is related to biomedical applications and to a new database for safety standards of non-ionizing e.m.f., going beyond the current mechanistic assumption, based on the electromagnetic (e.m.) power deposition in biological tissues (Specific Absorption Rate, S.A.R. [W kg^{-1}]), in order to incorporate the experimental evidence of the biological effects of the e.m. exposure (see, for example [4, 6, 9, 12, 21, 22, 25, 31, 32, 63-65, 67-69, 73, 82, 84, 86, 90, 91, 98, 101-106, 108, 110, 112, 116, 120]). Therefore there is the need of clarifying the underlying interaction mechanisms [2, 3, 5, 24, 25, 36, 52, 70, 71, 78, 121, 122] and the of improving the reproducibility and the quality of the experiments [49]. Toward this goal most researchers have concentrated their experimental and theoretical efforts on the early steps of the e.m. interaction, at the molecular level [9, 33, 61-63, 66-70, 73, 83, 84, 86-91, 93, 107-110, 116].

B.J. Klauenberg and D. Miklavcic (eds.), Radio Frequency Radiation Dosimetry, 429-447.

In this respect, one of the most widely studied biochemical processes is the binding of light ligands (e.g. metal ions, like Ca^{++}) to receptor proteins. Two general theoretical models of ion binding are available in the literature: the classical Langevin-Lorentz (L-L) model [14, 15, 17, 35-37, 42, 43, 48, 53, 56, 57, 66, 87, 88, 96, 97] and the quantum Zeeman-Stark (Z-S) model [15-17, 20, 23, 41, 44, 48, 52, 59, 83, 94]. They are simplified in such a way as to retain the essential features of the e.m. interaction with the binding process and to neglect all the details of the complete molecular dynamics simulation of the ion-protein system [54, 76].

The purpose of this paper is to review the state of the science concerning the aforesaid quantum Z-S model and to offer a predictive example of its application to radio frequency (r.f.) sinusoidal e.m. exposures [7, 20, 48, 50, 124].

The e.m.f. intensities considered in this paper are low, i.e., intensities below the current safety standard based on thermal effects [28, 29].

3. Ligand Binding to a Receptor Protein

Before analysing in detail the Z-S model, it is worthwhile to review the simplest possible description of the binding process, that could be linked to experimental data [10, 11, 72, 81, 99, 111, 113-115, 118, 126, 128].

The example we discuss concerns an idealized protein of the cell membrane, with a single type of binding site attracting a ligand ion. The cell is considered as a sphere of radius R_0 [m] and therefore area $4\pi R_0^2$ $[m^2]$.

The number of receptor proteins embedded in the cell membrane is S. Their binding sites are located, for example, on the extracellular side of the cell membrane. Each receptor site can be occupied by one ligand only, or it is empty. Letting P_B be the probability for a receptor to be occupied and L $[m^{-3}]$ the concentration of the ligands near the cell surface, the simplest first order mass-action law which gives the time course of P_B is:

$$dP_B/dt \approx K^+L(1-P_B) - K^-P_B \tag{1}$$

where K^- $[s^{-1}]$ and K^+ $[s^{-1}\ m^{-3}]$ are, respectively, the so-called dissociation and association rate "constants" in SI units.

In biochemistry L is measured in $[M^{-1}]$ and K^+ in measured in $[min^{-1}\ M^{-1}]$, where 1 $M = N_A\ /\ (1\ dm^3) = 6\ 10^{26}\ m^{-3}$, being N_A the Avogadro's constant.

In general, K^- and K^+ depend not only on the endogenous attractive force exerted by the binding site on the ligand, but they may depend also on the exogenous e.m. exposure, i.e., on the electric field vector $\vec{E}$ $[Vm^{-1}]$ and the magnetic induction vector $\vec{B}$ [T]. Therefore, strictly speaking, K^- and K^+ may depend on time, via the time dependence of $\vec{E}$ and $\vec{B}$.

The purpose of this paper is to outline a procedure, based on the aforesaid Z-S model, for evaluating the changes of K^- and K^+ due to the e.m.f.. Hence, the theory can be linked by means of equation (1) to binding experiments, i.e., to the measurement of the total number of bound ligands (i.e., SP_B), once L and S are known.

The model assumes, as a further simplification, that the binding crevice of the protein is isotropic, and that the ligand ion is a point charge Q [C] and mass M [kg], without any magnetic property. The site is occupied if the ion is inside a sphere of radius R_C [m], whose centre coincides with the binding crevice centre, chosen as origin of the coordinates.

It is clear, from equation (1), that in order to fully predict the influence of the e.m.f. on P_B during any binding experiment, one should be able to evaluate K^- and K^+ from first principles.

This can be obtained by means of a "gedanken" experiment performed by choosing a special value for L, say L_P, such that the corresponding value P, assumed by P_B when $L = L_P$, can be theoretically computed.

The peculiar concentration value L_P of L is chosen in such a way that there is always just one ligand interacting with one site, which is occupied with probability $P_B = P$ or is empty with probability $(1-P)$. In order to clarify the issue, one could consider the S receptors as uniformly distributed on the surface $4\pi R_0^2$ of the spherical cell membrane.

The next step is the practical evaluation of L_P which can be obtained from the computation of the average R_P [m] of the ion displacement given by:

$$\lim_{t \to \infty} < \vec{r}(t) \cdot \vec{r}(t) > = (R_P)^2 \qquad (2)$$

where <...> means expectation value of the "observable" argument, i.e., the observable ensemble average.

Typically, such a limit always exists, because of the attracting endogenous force of the site. In order to be consistent with the conceptual framework developed in this section, one can conclude, by assuming a conservative radius $2R_P$ that inside the volume $(4/3)\pi(2R_P)^3$ centred around each site there should always be just one ligand ion (bound or unbound), so that one can directly assume that

$$L_P \approx 3(1-P) / [4\pi(2R_P)^3] \qquad (3)$$

The modelling approach discussed in the next section allows the theoretical evaluation of $P(t)$, i.e., the value of P_B in the case $L = L_P$ under e.m. exposure. Consequently, for the introductory purposes of this section, one can assume that P and R_P (i.e., L_P), can be theoretically evaluated.

A way of evaluating K^- is to perform another "gedanken" experiment, i.e., releasing at time t=0 the ion at the crevice centre with some initial velocity $\vec{v}_{bm}$ [ms^{-1}] and computing both its displacement $\vec{r}(t)$ [m] and the time t^- needed to reach the binding distance R_C in the mean square sense :

$$< \vec{r}(t^-) \cdot \vec{r}(t^-) >=_{(R_C)^2} < (R_P)^2 \qquad (4)$$

Once t^- is computed from equation (4) it offers an estimate of the value of K^- according to the following relationship

$$K^- \approx 1/t^- \qquad (5)$$

In conclusion, knowing P , L_P and R_P, one gets

$$K^+ = (K^-P + dP/dt) / [L_P(1 - P)] \approx (K^-P + dP/dt)(4/3)\pi(2R_P)^3 / (1 - P)^2 \qquad (6)$$

so that the value of P_B corresponding to general value L can be obtained by substituting equation (6) in equation (1):

$$dP_B/dt \approx [(K^-P + dP/dt) / (1 - P)](L/L_P)(1 - P_B) - K^-P_B \qquad (7)$$

If the microscopic process is slow enough to average the time variations of P due to the e.m. exposure, then $dP/dt \approx 0$ and the corresponding term can be dropped out from equation (6, 7).

In general, once the values of P, K^-, L_P, i.e., R_P are theoretically evaluated with and without exogenous exposure, equation (7) can be applied to the analysis of a real binding experiment.

If it is $dP_B/dt \approx 0$, e.g. in a steady state experiment, so that both time derivatives can be neglected in equation (7), we obtain

$$P_B = P\{L(4\pi/3)(2R_P)^3/[1+PL(4\pi/3)(2R_P)^3-P^2]\} \qquad (8)$$

In practice, the changes of P due to the e.m. exposure can be already considered, per se, a reasonable assessment of the potential biological effectiveness of the e.m.f.. These changes are sufficient to offer the experimentalist the possibility of an educated guess about the susceptibility of the ligand-receptor under consideration of the various parameters which characterize the e.m.f..

4. The State of the Science for the Zeeman-Stark Quantum Model

The most general approach to the study of ligand binding to a receptor under e.m. exposure is based on quantum modelling of the process (Z-S model). Adopting a scheme similar to the classical one, the problem is to find the so called reduced density operator ρ [1, 20, 51, 80, 117, 119] which describes the ion motion in the attracting (isotropic) potential energy well $U_{end}(r)$ [J], in presence of exogenous e.m. potentials, i.e., a scalar potential ϕ [V] and a vector potential $\vec{A}$ [T m] such that $\vec{E} = -\nabla\phi - \partial\vec{A}/\partial t$ and $\vec{B} = \nabla \wedge \vec{A}$.

A typical first order approximation for isotropic $U_{end}(r)$ can be obtained by fitting the parameters U_0 [J], ω_{end} [Hz], R_B [m], ξ_B [Jm] of the relationship

$$U_{end}(r) \approx -\xi_B/r + \left\{ \xi_B/R_B - U_0 + \xi_B/r + \left(\xi_B/2R_B^2 - U_0/R_B\right)r + \right.$$
$$\left. + \left(M\omega_{end}^2/2 - U_0/2R_B^2 + \xi_B/6R_B^3\right)r^2\right\}\exp(-r/R_B) \tag{9}$$

to the available data of the protein of interest, as obtained from the Brookhaven Protein Data Bank.

The energy ($-U_0$) is the depth of the potential energy well at the centre of the binding crevice ($\vec{r} = 0$), whereas $R_B > R_C$ is related to the protein size. For small values of $\vec{r}$, the above expression gives:

$$-\nabla U_{end} = Q\vec{E}_{end} \approx -M\omega_{end}^2\vec{r} \qquad (r \ll R_B) \tag{10}$$

which is coincident with the typical "linear" endogenous attractive force (spring like) used by most authors [15, 43, 47, 54, 56, 66]. The nabla operator is ∇. Therefore ($M\omega_{end}^2$) plays the role of the spring constant.

For large value of r, the above expression gives:

$$-\nabla U_{end} = Q\vec{E}_{end} \approx -\xi_B\vec{r}/r^3 \qquad (r \gg R_B) \tag{11}$$

which is the typical "coulombic" endogenous attractive force originally used in the Z-S model.

The time evolution of ρ must obey the following relationship:

$$\partial\rho/\partial t = \left(-j/\hbar\right)\left[H_{end} + H_{bm} + H_1, \rho\right] - \left(j\beta/2\hbar\right)\sum_{i=1}^{3}\left[r_i, \Theta_i\rho + \rho\Theta_i\right] +$$
$$- \left(\beta K_B TM/\hbar^2\right)\sum_{i=1}^{3}\left[r_i, \left[r_i, \rho\right]\right] \tag{12}$$

where $r_1 = x$, $r_2 = y$ and $r_3 = z$.

The Hamiltonian $H_{end} = -\left(\hbar^2/2M\right)\nabla^2 + U_{end}$ refers to the ion motion in the potential energy U_{end}. The Hamiltonian H_{bm}, takes into account the contributions of the endogenous basal force $\vec{F}_{bm} = -\nabla H_{bm}$, which emulates the effects of the basal metabolism of the living cell on the ion receptor system [19, 20, 45, 48, 50]. The need of such a force is consistent with the exponential macroscopic evidence that across the membrane of any living cell it exists an excess voltage drop sustained by the biochemically driven ion pumps. The related excess electric field is $\vec{E}_{bm} = \vec{F}_{bm}/Q$.

We assume for simplicity sake that the spatial force is spatially uniform and constant in time.

The Hamiltonian H_1 takes into account the contribution of ϕ and $\vec{A}$. We adopt the gauge condition $\nabla \cdot \vec{A} = 0$ so that $H_1 \approx j\hbar\gamma\vec{A} \cdot \nabla$, where $\gamma = Q/M$. A typical assumption is that $\vec{A}$ is small enough so that the term proportional to $\vec{A} \cdot \vec{A}$ in H_1 can be neglected. The commutator [S,R] means, by definition, SR-RS.

Care must be paid in fitting the above parameters to the protein data. A common practice is to evaluate, from the protein data bank, the endogenous electric potential $\phi_{end} = U_{end}/Q$ [V] generated by the surrounding atoms (the contribution of the protein embedding medium should be included if necessary) inside the binding crevice, in a static conformation [75].

In reality, when the ligand ion is approaching the binding site, the electric field due to its charge displaces the protein atoms in a very fast time scale, so that the actual ϕ_{end} to be used takes into account the "instantaneous" rearrangement of the protein atoms corresponding to the actual ion position [26, 27, 95]. The reaction field resulting from such displacement lowers the actual value of the endogenous force which attracts the ion toward the crevice centre, so that ω_{end} can assume values which could be orders of magnitude lower than those computed by assuming the protein atom in static position.

A procedure for obtaining these more realistic values of ω_{end}, without performing a detailed molecular dynamics simulation of the protein, is outlined in [26, 27, 95].

A rather effective and simpler approach is to obtain, from the protein data bank the value of ϕ_{end} in presence and absence of the ligand.

The schematic diagrams of figures. 1 and 2 offer a clear example of the different conformations assumed by a binding site of calmodulin in presence and in absence of Ca^{++}.

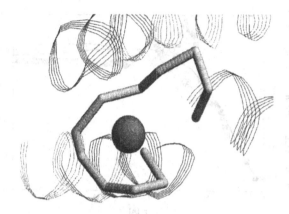

Figure 1. Backbone of one binding site of calmodulin, with bound ligand (Ca^{++}) (Brookhaven Protein Data Bank).

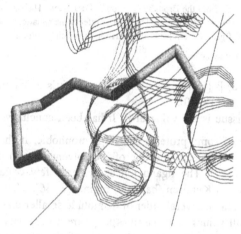

Figure 2. Backbone of one binding site of calmodulin, without bound ligand (Ca^{++}) (Brookhaven Protein Data Bank).

From the first set of values we can obtain U_0 and ω_{end} in the limit $r \ll R_B$. From the second set of values we obtain ξ_B in the limit $r \gg R_B$. From both sets we obtain an estimate of R_B. A typical result is shown in Figure 3 for the same site sketched in Figures. 1 and 2.

436

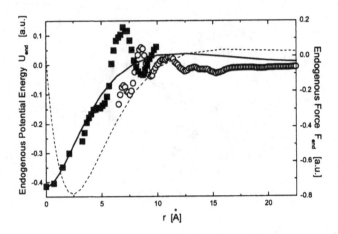

Figure 3. Example of the endogenous potential energy for Ca^{++} in one of the four binding sites of calmodulin, as obtained from the Brookhaven Protein Data Bank. The continuous curve has been obtained by fitting equation(9) to the protein data. The squares refer to the situation in which the ion is at the centre of the receptor site (see fig.1), the circles to the situation in which the ion is outside it (see fig.2). The dashed curve is the attractive endogenous force $-dU_{end}/dr$

The parameter β [Hz] is the classical Langevin's collision frequency of the ion in the binding crevice [18, 30].

A practical issue is the value of β. It has been conclusively demonstrated that the binding crevice of some proteins can be hydrophobic, if the modulus of $-\nabla U_{end}$ is large and negative dielectrophoresis of the solvent (water) dipolar molecules occurs [38-40, 60, 74, 127]. The ligand ion experiences few collisions inside the crevice, where it moves in a Knudsen (ballistic) regime [8, 46]. Therefore β can assume local values which could be several order of magnitude smaller than in bulk water ($\beta_{water}\approx0.5$ 10^{14} [Hz]). Small values of β, i.e. $\beta<<\beta_{water}$, are a necessary prerequisite for possible bioeffects of low intensity e.m.f.. The value of the initial velocity mentioned in the previous section can be approximated by $\vec{v}_{bm} = \vec{F}_{bm}/\beta M$.

The operators Θ_i play the role of appropriate quantum analogues of the classical drag terms in the Fokker-Plank equation which gives the time evolution of the classical probability density of the ligand. Their physical meaning become apparent when the system relaxes to thermal equilibrium, in the limit of $(1/T) \to 0$ and $\vec{F}_{bm} = 0$, when Θ_i becomes coincident with the i-th momentum component of the ligand. The last term in equation (12), which is proportional to the product of the Boltzmann's constant K_B with T, is the quantum counterpart of the thermal (white) noise effects in the classical Fokker-Plank equation.

The novel result is that equation (12) takes into account all the various aspects of the interaction of the quantum system with the thermal bath, as a function of T and of one fitting parameter only, i.e. β, which has a classical physical meaning.

It is beyond the scope of this paper to further discuss this point. It is enough to clarify that the operators Θ_i are chosen in such a way that the steady state value ρ_0 assumed by ρ when $H_1=0$ is the same as given in [20]. Furthermore, equation (12) is consistent with the so-called Generalised Master Equation [51]. In this case, by using the secular approximation, one can retrieve the link among the drag operators Θ_i and the lifetimes introduced in [20, 44].

Once a complete set of suitable orthonormal basis functions $\psi_m(x,y,z)$ has been chosen, the integration of equation (12) leads to the evaluation of the reduced density matrix entries $\rho_{mn}(t)$ of ρ so that the observable expectation value R of any quantum operator R can be computed from the trace expression

$$R = Tr\,(R\rho) \tag{13}$$

Note that we neglect from now on the notation <...> which is implicit in the trace expression above.

We evaluate, as a representative output of the ion protein system, the binding probability $P(t) = Tr\,(P\rho)$, with $H_{bm}\neq0$, as discussed in the previous section. The value of the quantum operator P, actually a function, is 1 inside the binding sphere, and 0 outside, so that the entries of its matrix representation are

$$P_{mn} = \int \psi_m^* \psi_n \, dx\,dy\,dz \tag{14}$$

where the integration domain is a binding sphere of radius $R_C \leq R_B$.

In practice, the solution of equation (12) with the boundary condition $\rho(0)=\rho_0$ gives the system transient behaviour ρ(t), in terms of the matrix entries $\rho_{mn}(t)$, corresponding to the onset of the e.m. exogenous exposure H_1 at t = 0 [20, 48, 50]. Then, the time evolution of ρ(t), can be obtained and P(t) can be computed from equation (13) and finally introduced in equation (6, 7). Sometimes it is more interesting to compute the time average

$$P_{av} = (1/t) \int P(t) \, dt \tag{15}$$

where the integration domain is [0,t], and to compare its asymptotic value

$$P_{av,\infty} = \lim_{t\to\infty} P_{av} \tag{16}$$

with its value P(0) in the absence of any exposure, being

$$P(0) = \text{Tr}(P\rho_0) \tag{17}$$

The value of $(R_P)^2$ of equation (2) can be computed as $(R_P)^2 = \lim_{t \to \infty} \text{Tr}[\bar{r}(t) \cdot \bar{r}(t)\rho]$.

The value of $t^- \approx 1/K^-$ of equation (4) can be computed from $\text{Tr}[\bar{r}(t^-) \cdot \bar{r}(t^-)] = R_c^2$.

We pointed out previously [43, 47, 77, 96] that any bioelectromagnetic model must include thermal noise as input. Then, the first task to be accomplished is the evaluation of the output $P(t)$ when the exogenous e.m. exposure is absent in equation (12), i.e., $H_1 = 0$, so that noise is the only input acting on the system. The second task is the evaluation of the output $P(t)$ when the exogenous e.m. exposure is active, and noise is still present. The third task is to compare, in relative terms. the outputs obtained in the two situations. Any conclusion about the effectiveness of the e.m. exposure on the ion-protein system must be drawn only as consequence of such a comparison.

In the literature, some theoretical papers do not consider noise at all, and their authors perform the second task only in the absence of noise. These studies provide some information about the output dependence on the, e.m. parameters (e.g., frequency, amplitude, etc.) but nothing can be inferred concerning the effectiveness of the e.m. exposure [14, 23, 35, 56-58, 61, 62, 66, 83, 87-89, 92].

A further aspect is the possibility of stochastic resonance [13, 55, 79, 100, 109, 125], which was briefly reviewed in [47]. The ion-protein system retains all the necessary features for stochastic resonance so that one could expect that an optimal range of characteristic parameter values exist where the signal-to-noise ratio of the output is enhanced. The evaluation of the system state equation (12) does naturally include stochastic resonance, whose study does not require any inherently different model.

5. Bioeffect of RF Exposure

The improved Z-S model outlined in the previous section can be applied to analyse the bioeffects of the e.m.f. produced by mobile telecommunications equipment [7, 85] adopting the same approach outlined in [20]. In this case, the exogenous e.m. input to the ion-protein system is described by $\vec{A}(x, y, z, t)$ and $\phi(x, y, z, t)$ and is classically known. It is adequate to consider a linearly polarized TEM wave [16, 20, 48, 50] which can be described in terms of $\vec{A}$ only, letting $\phi = 0$. A reasonable approximation is to consider the r.f. carrier alone, at $f_c = \omega_c/2\pi$ [Hz], propagating in a biological medium, whose average conductivity is σ [S m^{-1}], whose electric permittivity is $\varepsilon_0\varepsilon_r$ [F m^{-1}] and whose magnetic permeability is μ_0 [H m^{-1}]. The vector potential is given by

$$A_{rf} \approx \sqrt{2\rho_t S/\sigma} \left(\exp(-\alpha_c y)/\omega_c \right) \cos[\omega_c(t - y/v_c)] \tag{18}$$

where S [W kg^{-1}] is the local S.A.R. and ρ_t [kg m^{-3}] is the local tissue density.

The attenuation coefficient is

$$\alpha_c = \left(\sigma/\sqrt{2}\right)\left[\left(\varepsilon_0/\mu_0\right)\left(\varepsilon_r + \sqrt{\varepsilon_r^2 + \sigma^2/\omega_c^2\varepsilon_0^2}\right)\right]^{-1/2}$$

(19)

and the phase velocity is

$$v_c = \left[\left(\varepsilon_0\mu_0/2\right)\left(\varepsilon_r + \sqrt{\varepsilon_r^2 + \sigma^2/\omega_c^2\varepsilon_0^2}\right)\right]^{-1/2}$$

(20)

The TEM wave is incident from the half space y < 0 (air) into the lossy semi-infinite medium (σ = 1 S m^{-1}, ε_r = 80 F m^{-1} and ρ_t = 10^3 kg m^{-3}), which fills the half space y≥0.

The carrier frequency is f_c = 915 MHz (i.e., in the range of interest for cellular telephones [7]). The putative process under consideration is the binding of Ca^{++} ion to a receptor protein located at x=z=0 and y=0$^+$. The e.m. sinusoidal exposure is switched on at t=0$^+$. Five Coulombic eingenfunctions have been used in the computer simulations.

We choose an ideal putative protein characterized by $\omega_0 = 3\xi_B^2 M/8\hbar^3 \approx 2\pi f_c$, so that the e.m. photon matches, in energy, the depth of the ligand potential energy well.

In these conditions, after the initial transient, the binding probability $P(t)$ reaches an asymptotic behaviour $P_{as}(t)$ which is almost constant and differs from $P(0)$. Therefore it is convenient to consider the time average of P_{as}, i.e. $P_{av,\infty}$, which is constant, and to plot $[P(0) - P_{av,\infty}] / P(0)$ versus the incident power density I.P.D. [W m^{-2}] as a measure of the biological effectiveness of the r.f. exposure. A typical result is shown in fig. 4. It is apparent that if $\vec{F}_{bm} = -\nabla H_{bm}$ goes to zero (so that $\rho_0 = \rho_{th}$) there is no effect, irrespective of the level of the incident e.m. power. If $\vec{F}_{bm}$ is increased, the effect on the binding probability of the TEM exposure becomes significant, at power (or S.A.R.) values which are below the current safety standards. This result proves that low-intensity r.f. exposure can affect an elementary biological process in a living cell.

440

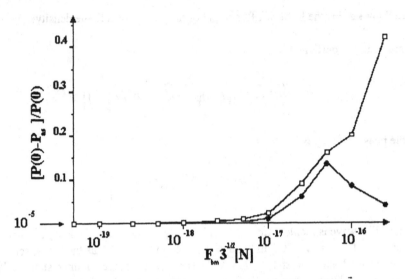

Figure 4. Relative excess change of the binding probability versus the modulus of $\vec{F}_{bm}$, assuming that $F_{bm,x}=F_{bm,y}=F_{bm,z}$. The exposure intensities are, respectively, 1 mW cm^{-2} (S.A.R. = 0.148 W kg^{-1}) (full circles) and 10 mW cm^{-2} (S.A.R. = 1.486 W kg^{-1}) (open squares). Some representative values of P(0) are 0.33 at $F_{bm,x,y,z}=10^{-17}$ N, 0.43 at $F_{bm,x,y,z}=5\ 10^{-17}$ N and 0.53 at $F_{bm,x,y,z}=10^{-16}$ N.

6. Conclusions

We have laid down a biophysical basis for assessing the effects of low-intensity e.m. r.f fields on ligand binding to receptors, with specific emphasis on ion binding to a receptor protein as a first step of interaction.

Several topics have been analyzed by means of the quantum Z–S model:

1) The endogenous field inside a molecular structure has been characterized according to the protein database. The related endogenous force provides a strong nonlinearity in the state equations for the ion-protein system.

2) Any protein with a hydrophobic crevice is a putative candidate for hosting an effective interaction between low-intensity exposure and a binding ion, by providing low values of the classical collision frequency β, i.e., long quantum lifetimes.

3) Basal metabolism maintains the cell out of thermodynamic equilibrium [1, 2, 3]. At the molecular level, the metabolic activity maintains the ion-protein system itself out of thermodynamic equilibrium sustaining an excess ion velocity inside the binding crevice, and it supplies power to the system. This power can be converted, via the nonlinearity provided by the endogenous force, into signalling power "controlled" by the low-intensity e.m. exogenous field.

4) The contribution of thermal noise to the ion-protein binding probability has been taken into account in the presence and in the absence of e.m. exposure, whose effectiveness has been judged from the comparison of the two situations.

These results seem in contrast with those reported in [2, 3, 5] irrespective of the similar physical approach adopted. The differences can be better understood by resuming the electronic jargon.

The metabolic activity can *bias* the ion-protein system far enough from thermodynamic equilibrium, at an *operating point* of the nonlinear binding *characteristic* where the system may be potentially able to detect small e.m. signals. The system takes advantage of the *power supply* provided by the basal metabolism of the cell, much like transistor uses its power supply to amplify the time-varying signal applied to its input gate.

Therefore, in this paper we deal with a *transistor* analogy of processes in *living* cells ($\vec{E}_{bm} \neq 0$), whereas the approach developed in [5] deals with a *diode* analogy of processes in *dead* cells ($\vec{E}_{bm} = 0$). In fact, if the exogenous e.m. exposure is switched off, the systems considered in [3, 5] return to thermodynamic equilibrium, so that "it is difficult to make consistent biological effects with low fields strengths" in this case [5].

In conclusion, we have offered a plausible biophysical basis for potential effects of low-intensity e.m. fields.

7. References

1. Abragam, A. (1961) *The principles of nuclear magnetism*, Oxford , Clarendon Press, 264-353.
2. Adair, R.K. (1991) Constraints on biological effects of weak extremely low-frequency electromagnetic fields, *Phys. Rev.* A **43**, 1039-1048.
3. Adair, R.K. (1992) Criticism of Lednev's mechanism for the influence of magnetic fields on biological systems, *Bioelectromagnetics* **13**, 231-235.
4. Adey, W.R. (1980) Frequency and power windowing in tissue interactions with weak electromagnetic fields, *Proc. IEEE* **68**, 119-125.
5. Astumian, R.D., Weaver, J.C., and Adair, R.K. (1995) Rectification and signal averaging of weak electric fields by biological cells, *Proc. Natl. Acad. Sci. USA* **92**, 3740-3743.
6. Azanza, M.J. and Del Moral, A. (1994) Cell membrane biochemistry and neurological approach to biomagnetism, *Progress in Neurobiology* **44**, 517-601.
7. Bach Andersen, J., Johansen, C., Frolund Pedersen, G., and Raskmark, P. (1995) On the possible health effects related to GSM and DECT transmissions, A tutorial study for the European Commission, Aalborg Univ., Denmark.
8. Balian, R. (1992) *From Microphisics to Macrophisics*, Vols. I and II, Springer Verlag, Berlin, 331.
9. Bawin, S.M. and Adey, W.R. (1976) Sensitivity of calcium binding in central tissue to weak environmental electric fields oscillating at low frequency, *Proc. Natl. Acad. USA* **73**, 1999-2003.
10. Bell, G.I. (1978) Model for the specific adhesion of cells to cells, *Science* **200**, 618-627.
11. Berg, H.C. and Purcell, E.M. (1977) Physics of chemoreception, *Biophys. Journal* **20**, 193-239.
12. Berman, E., Chacon, L., House, D., Koch, B.A., Koch, W.E., Leal, J., Lovtrup, S., Mantiply, E., Martin, A.H., Martucci, G.I., Mild, K.H., Monahan, J.C., Sandstrom, M., Shamsaifar, K., Tell, R., Trillo, M.A., Ubeda, A., and Wagner, P. (1990) Development of chicken embryos in a pulsed magnetic field, *Bioelectromagnetics* **11**, 169.
13. Bezrukov, S.M. and Vodyanoy, I. (1995) Noise-induced enhancement of signal transduction across voltage dependent ion channel, *Nature* **378**, 362-364.
14. Bianco, B., Chiabrera, A., Morro, A., and Parodi, M. (1988) Effects of magnetic exposure on ions in electric fields, *Ferroelectrics* **83**, 355-365.

15. Bianco, B. and Chiabrera, A. (1992) From the Langevin-Lorentz to the Zeeman model of electromagnetic effects on ligand-receptor binding, *Bioelectrochem. Bioenerg.* **28**, 355-365.

16. Bianco, B., Chiabrera, A., Moggia, E., and Tommasi, T. (1993) Interaction mechanisms between electromagnetic fields and biological samples under a TEM exposure system, 2nd Int. IEEE-URSI Scient. Meet. Microwave in Medicine, Rome, Italy, Oct. 11-14.

17. Bianco, B., Chiabrera, A., D'Inzeo, G., Galli, A., and Palombo, A. (1993) Comparison between classical and quantum modelling of bioelectromagnetic interaction mechanisms, in *Electricity and Magnetism in Biology and Medicine*, M. Blank Eds., San Francisco Press, San Francisco, 537-539.

18. Bianco, B. (1994), Internal Report, ICEMmB at DIBE. University of Genoa.

19. Bianco, B., Chiabrera, A., and Kaufman, J.J. (1995) A new paradigm for studying the interaction of electromagnetic fields with living systems: an out-of-equilibrium characterization, BEMS 7th Annual Meet., Boston, USA, June 18-22.

20. Bianco, B., Chiabrera, A., Moggia, E., and Tommasi, T. (1997) Enhancement of the interaction between low-intensity R.F. e.m. fields and ligand binding due to cell basal metabolism, *Wireless Networks* **3**, 477-487.

21. Blackman, C.F., Benane, S.G., Robinovltz, J.R., House, D.E., and Joines, W.T. (1985) A role for the magnetic field in the radiation-induced efflux of calcium ions from brain tissue in vitro, *Bioelectromagnetics* **6**, 327-337.

22. Blackman., C.F., Benane, S.G., and House, D.E. (1991) The influence of temperature during electric and magnetic field induced alteration of calcium-ion release from in vitro brain tissue, *Bioelectromagnetics* **12**, 173-182.

23. Blackman, C.F., Blanchard, J.P., Benane, S.G., and House, D.E. (1995) The ion parametric resonance model predicts magnetic field parameters that affect nerve cells, *FASEB J.* **9**, 547-551.

24. Blanchard, J.P. and Blackman, C.F. (1994) Clarification and application of an ion parametric resonance model for magnetic field interactions with biological systems, *Bioelectromagnetics* **15**, 217-238.

25. Cancer risk and electromagnetic fields (1995) Scientific correspondence, *Nature* **375**, 22-23.

26. Cavanna, M. (1996) Master Thesis (in Italian), DIBE, Univ. of Genoa.

27. Cavanna, M., Moggia, E., and Chiabrera, A. (1996) Reaction of a receptor protein to ligand binding under e.m. exposure, 18th Annual Int. Conf. IEEE Engineering in Medicine and Biology Soc., Amsterdam, The Netheriands, Oct.31-Nov.3.

28. CENELEC (1995) ENV-50166-1: Human exposure to electromagnetic fields - low frequency, European prestandard.

29. CENELEC (1995), ENV-50166-2: Human exposure to electromagnetic fields - high frequency, European prestandard.

30. Chandrasekhar, S. (1943) Stochastic problems it physics and astronomy, *Rev. Mod. Phys.* **15**, 1-89.

31. Chiabrera, A., Hinsenkamp, M., Pilla, A.A., Ryaby, J., Ponta, D., Belmont, A., Beltrame, F., Grattarola, M., and Nicolini, C. (1979) Cytofluorometry of electromagnetically controlled cell dedifferentiation, *J. of Histochemistry and Cytochemistry* **27**, 375.

32. Chiabrera, A., Viviani, R., Parodi, G., Vemazza, G., Hinsenkamp, M., Pilla, A.A., Ryaby, J., Beltrame, F., Grattarola, M., and Nicolini, C. (1980) Automated absorption image analysis of electromagnetically exposed frog erythrocytes, *Cytometry* **1**, 42.

33. Chiabrera, A., Grattarola, M., and Viviani, R. (1984) Interaction between electromagnetic fields and cells microelectrophoretic effect on ligands and surface receptors, *Bioelectromagetics* **5**, 173-191.

34. Chiabrera, A. and Rodan, G.A. (1984) The effect of electromagnetic fields on receptor-ligand interaction: A theoretical analysis, *Journ. of Bioelectricity* **3**, 509-521.

35. Chiabrera, A., Bianco, B., Caratozzolo, F., Giannetti, G., Grattarola, M., and Viviani, R. (1985) Electric and magnetic field effects on ligand binding to cell membrane, in A. Chiabrera, C. Nicolini, and H.P. Schwan (eds)., *Interaction between Electromagnetic Fields and Cells*, Plenum, New York and London, 253-280.

36. Chiabrera, A. (1987) Comments on the dynamic characteristics of membrane ions in multifield cofigurations of low-frequency electromagnetic radiation, BEMS Annual Meet., June 21-25, Portland. USA.

37. Chiabrera, A. and Bianco, B. (1987) The role of the magnetic field in the e.m. interaction with ligand binding, in M. Blank and E. Findi (eds.) *Mechanistic Approaches to Interactions of Electric and Magnetic Fields with Living Systems*, Plenum Publishing Corporation, New York and London, 79-95.

38. Chiabrera, A., Morro, A., and Parodi, M. (1989) Water concentration and dielectric permittivity in molecular crevices, *Il Nuovo Cimento sect. IID* 7, 981-992.

39. Chiabrera, A., Bianco, B., Liebman, M.N., Kaufman, J.J., and Pilla, A.A. (1990) Movement of ions near macromolecules in the presence of electromagnetic exposure, BRAGS 10th Annual Meet., Philadelphia, USA, Oct. 14-17.

40. Chiabrera, A., Bianco, B., Parodi, M., Morro, A., and Liebman, M.N. (1991) Hydrophobicity of ion binding sites in proteins, BEMS 13th Annual Meet., Salt Lake City, USA, June 23 -27.

41. Chiabrera, A., Bianco, B., Kaufman, J.J., and Pilla, A.A. (1991) Quantum dynamics of ion in molecular crevices under electromagnetic exposure, in C.T. Brighton and S.R. Pollak, (eds.), *Electromagnetics in Biology and Medicine*, San Francisco Press, San Francisco, 21-26.

42. Chiabrera, A., Bianco, B., Tommasi, T., and Moggia, E. (1992) Langevin-Lorentz and Zeeman-Stark models of bioelectromagnetic effects, *Acta Pharm.* 42, 315-322.

43. Chiabrera, A., Bianco, B., Kaufman, J.J., and Pilla, A.A. (1992) Bioelectromagnetic resonance interactions: endogenous field and noise, in B. Norden and C. Ramel (eds.), *Interaction Mechanisms of Low-Level Electromagnetic Fields in Living Systems*, Oxford Science Publications, Oxford, 164-179.

44. Chiabrera, A., Bianco, B., and Moggia, E. (1993) Effects of lifetimes on ligand binding modelled by the density operator, *Bioelectrochem. Bioenerg.* 30, 35-42.

45. Chiabrera, A., Bianco, B., Moggia, E., and Tommasi, T. (1994) The out-of-equilibrium steady state of a cell as reference for evaluating bioelectromagnetic effects, BEIMS 16th Annual Meet., Copenhagen, Derunark, June 12-17.

46. Chiabrera, A., Bianco, B., Moggia, E., and Tommasi, T. (1994) The interaction mechanism between e.m. fields and ion adsorption: Endogenous forces and collision frequency, *Bioelectrochem. Bioenerg.* 35, 33-37.

47. Chiabrera, A., Bianco, B., and Kaufman, J.J. (1995) Biological effectiveness of low intensity electromagnetic exposure: Non-linearitv, out-of-equilibrium state and noise, Electromagnetic Compatibility EMC 95, Invited paper, URSI Open Meet., Commission K, Zurich, Switzerland, March 7-9.

48. Chiabrera, A., Bianco, B., Moggia, E., Tommasi, T., and Kaufman, J.J. (1995) Recent advances in biophysical modelling of radio frequency electromagnetic field interactions with living systems, Invited Paper, Proceedings of the State of the Science Colloquium, WTR and ICWCHR, Rome, Nov. 13-15.

49. Chiabrera, A., Hamnerius, Y., Bianco, B., Berquist, B., and Kenny, T. (1996) Design guidelines for "in vitro" and "in vivo" exposure conditions at sub-ELF/LF and their quality control, Position Paper, COST 244 European Commission DGXII and18th Annual Meeting of BEMS, Victoria, Canada, June 9-14.

50. Chiabrera, A., Bianco, B., Moggia, E., and Tommasi, T. (1996) Down-conversion of mobile telecommunications frequencies at ligand-receptors binding site, Symposium K1: Biological effects and mechanism of interaction, Invited paper, URSI XXV General Assembly, Lille, France, August 28-September 5.

51. Cohen-Tannoudji, C., Diu, B., and Laloe, F. (1977), *Quantum Mechanics*, Vols. I and 11, J. Wiley & Sons New York, 305-307.

52. Comments on clarification and application of an ion parametric resonance model for magnetic interactions with biological systems, *Bioelectromagnetics* 16, 268-275.

53. D'Inzeo, G., Galli, A., and Palonbo, A. (1993) Further investigations on non-thermal effects referring to the interaction between ELF fields and transmembrane ionic fluexes, *Bioelectochem. Bioenerg.* 30, 93-102.

54. D'Inzeo, G., Palombo, A., Tarrico, L., and Zago, M. (1995) Molecular simulation studies to understand non-thermal bioelectromagnetic interaction, BEMS 17th Annual Meet., Boston, Massachusetts, June 18-22.

55. Duglass, J.K., Wilkwns, L., Pantazaleou, E., and Moss, F. (1993) Noise enhancement of information transfer in crayfish mechanoreceptors by stochastic resonance, *Nature* **385**, 337-340.

56. Durney, C.H., Rushforth, C.K., and Anderson, A.A. (1988) Resonant dc-ac magnetic fields: Calculated response, *Bioelectromagnetics* **9**, 315-330.

57. Edmonds, D.T. (1993) Larmor precession as mechanism for the detection of static and alternating magnetic fields, *Bioelechem. Bioenerg.* **30**, 3-12.

58. Eichwald, C. and Kaiser, F. (1995) Model of external influences on cellular signals transduction pathways including cytosolic calcium oscillations, *Bioelectromagnetics* **16**, 75-85.

59. Engstrom, S. (1996) Dynamic properties of Lednev's parametric resonance mechanism, *Bioelectromagnetics* **16**, 58-70.

60. Ernst, J.A., Clubb, R.T., Zhou, H.X., Gronenborn, A.M., and Clore, G.M. (1995) Demonstration of positionally disordered water within a protein hydrophobic cavity by N.M.R., *Science* **267**, 1813-1817.

61. Eichwald, C. and Kaiser, F. (1993) Model for receptor-controlled cytosolic calcium oscillations and for external influences on the signal pathways, *Biophysical J.* **65**, 2047-2058.

62. Eichwald, C. and Kaiser, F. (1995) Model for external influences on cellular signal transduction pathways including cytosolic calcium oscillations, *Bioelectromagnetics* **16**, 75-85.

63. Falugi, C., Grattarola, M., and Prestipino, G. (1987) Effects of low-intensity pulsed electromagnetic fields on the early development of sea urchin, *Biophysical J.* **51**, 999-1003.

64. Fitzsimmons, R.J., Ryaby, J.T., Magee, F.P., and Baylink, D.J. (1995) IGF-II Receptor number is increased inTE-85 osteosarcoma cells by combined magnetic fields, *J. Of bone and Mineral Research* **10**, 812-817.

65. Fitzsimmons, R.J., Ryaby, J.T., Mohan, S., Magee, F.P., and Baylink, D.J. (1995) Combined magnetic fields increase Insulin-like Growth Factor-II in TE-85 human osteosarcoma bone cell cultures, *Endocrinology* **136**, 3100-3107.

66. Galt, S., Sanblom, J. and Hamnerius, Y. (1993), Theoretical study of the resonance behaviour of an ion confined to a potential well in a combination of ac and dc magnetic fields, *Bioelectromagnetics* **14**, 299-314.

67. Grattarola, M., Viviani, R., and Chiabrera, A. (1982) Modelling of the perturbation induced by low frequency electromagnetic fields on the membrane receptors of stimulated human lymphocyte, I: Influence of the fields on the system's free energy, *Studia Biophysica* **91**, 117-124.

68. Grattarola, M., Viviani, R., and Chiabrera, A. (1982) Modelling of the perturbation induced by low frequency electromagnetic fields on the membrane receptors of stimulated human lymphocyte, II: Influence of the fields on the mean lifetimes of the aggregation process, *Studia Biophysica* **91**, 125-131.

69. Grattarola, M., Chiabrera, A., Bonanno, G., Viviani, R., and Raveane, A. (1985) Electromagnetic field effects on phytohemagglutinin (PHA) induced lymphocyte reactivation, in A. Chiabrera, C. Nicolini, and H.P. Schwan (eds.), *Interactions between Electromagnetic Fields and Cells*, Plenum Press, New York, 401-421.

70. Grundler, W., F. Kaiser, Keilman, F., and Walleczek, J. (1994) Mechanism of electromagnetic interaction with cellular systems, *Naturwissenschaften* **79**, 551-559.

71. Halle, B. (1988) On the cyclotron resonance for magnetic field effects on trans-membrane ion conductivity, *Bioelectromagnetics* **9**, 381-385.

72. Hill, T. H. (1975) Effect of rotation on the diffusion controlled rate of ligand-protein association, *Proc. Natl. Acad. Sci. USA* **72**, 4918-4922.

73. Hinsenkamp, M., Chiabrera, A., and Bassett, C.A.L. (1978) Cell behaviour and DNA modification in pulsing electromagnetic fields, *Acta Orthop. Belgica* **44**, 636.

74. Hollfelder, F., Kirby, A.J., and Tawfik, D.S. (1996) Off-the shelf proteins that rival tailor-made antibodies as catalysts, *Nature* **383**, 60-63.

75. Honig, B. and Nicholls, A. (1995) Classical electrostatics in biology and chemistry, *Science* **268**, 1144-1149.

76. Karplus, M. (1984) Dynamic Aspects of Protein Structure, *Ann NY Aca.of Sci*, 107-123.

77. Kaufman, J.J., Chiabrera, A., Hatem, M., Bianco, B., and Pilla, A.A. (1990) Numerical stochastic analysis of Lorentz force ion binding kinetics in electromagnetic bioeffects, BRAGS 10th Annual Meet., Philadelphia USA, Oct. 14-17.

78. Kinouchi, Y., Tanimoto, S., Ushita, T., Sato, K., Yamaguchi, H., and Miyamoto, H. (1988) Effects of static magnetic fields on diffusion in solution, *Bioelectomagnetics* **9**, 159-166.

79. Kruglkov, I.L. and Dertinger, H. (1994) Stochastic resonance as a possible mechanism of amplification of weak electric signals in living cells, *Bioelectromagnetics* **15**, 539-547.

80. Landau, L. and Lifschitz, E. (1966) *Quantum Mechanics*, Moscow MIR.

81. Lauffenburger, D.A. and Linderman, J.J. (1993) *Receptors*, Oxford University Press, Oxford.

82. Leal, J., Trillo, M.A., Ubeda, A., Abraira, B., Shamsaifar, K., and Chacon, L. (1986), Magnetic environment and embryonic development: a role of the earth's field, *IRCS Med. Sci.* **14**, 1145.

83. Lednev, V.V. (1991) Possible mechanism for the influence of weak magnetic fields on biological systems, *Bioelectromagnetics* **12**, 71-75.

84. Lednev V.V. (1994) Interference with the vibrational energy sublevels of ions bound in calcium-binding proteins as the basis for the interaction of weak magnetic fields with biological systems, in A. H. Frey (ed.), *On the Nature of Electromagnetic Field Interactions with Biological Systems*, RG Landes Company, Medical Intelligens Unit, Boca Ranton, FL, 59-72.

85. Li, V.O.K. and Qiu, X. (1995), Personal communication system (PCS), *Proc. IEEE*, **83**, 1210-1243.

86. Liboff, A.R.,. Williams , T, Strog, D.M, and Wistar, R. (1984) Time varying magnetic fields: Effect on DNA synthesis, *Science* **223**, 818-820.

87. Liboff, A.R. (1985) Cyclotron resonance in membrane transport, in Interaction between electromagnetic field and cells, A. Chiabrera, C. Nicolini, and H.P. Schawn (eds.), Plenum Press, New York, 281.

88. Liboff, A.R. and McLeod, B.R. (1988) Kinetics of channelized membrane ions in magnetic fields, *Bioelectrotnagnetics* **9**, 39.

89. Liboff A.R. (1995) Geomagnetic cyclotron resonance in living cells, *J. Biol Phys.* **12**, 99-102.

90. Luben, R. A., Cain, C. D., Chi-Yun Chen, M., Rosen, D.M., and Adey, W.R. (1982) Effects of electromagnetic stimuli on bone and bone cells in vitro: inhibition of responses to parathyroid hormone by low-energy low-frequency fields, *Proc. Natl. Acad. Sci. USA* **79**, 4180-4184.

91. Markov, M.S., Wang, S., and Pilla, A.A. (1993) Effects of weak low frequency sinusoidal and DC magnetic fields on myosin phosphorylation in a cell-free preparation, *Bioelectrochem. Bioenerg.* **30**, 119-125.

92. McLeod, B.R. and Llboff, A.R. (1986) Dynamics characteristics of membrane ions in multifield configurations of low frequency electromagnetic radiation, *Bioelectromagnetics* **7**, 117.

93. Moggia, E. (1993) Dynamic properties of ions in solutions in the presence of magnetic fields, Internal Report and (1996) Ph.D. Thesis (in Italian), ICEmB at DIBE, University of Genoa.

94. Moggia, E., Tommasi, T., Bianco, B., and Chiabrera, A. (1993) Comparison of 5-state vs. 3-state coulombian Zeeman model of e.m.f. effects on ligand binding, in M. Blanc (ed.), *Electricity and Magnetism in Biology and Medicine*, San Francisco Press, San Francisco, 556-558.

95. Moggia, E., Cavanna, M., and Chiabrera, A. (1996) The reaction component of the endogenous field in receptor proteins, EBEA '96, COST 244 Congress Nancy, France, Feb. 28-March 2.

96. Moggia, E., Chiabrera, A., and Bianco, B. (1997) Fokker-Plank analysis of the Langevin-Lorentz equation : Application to ligand receptor binding under electromagnetic exposure, *J. Appl. Phys.* **82**, 4669-4677.

97. Muehsan, D.J. and Pilla, A.A. (1996) Lorentz approach to static magnetic field effects on bound-ion dynamic and binding kinetics: Thermal noise considerations, *Bioelectromagnetics* **17**, 89-99.

98. Noda, M., Johnson, D., Chiabrera, A., and Rodan, G.A. (1997) Effect of electric currents on DNA synthesis in raosteosarcoma cells: Dependence on conditions that influence cell growth, *J. of Orthopaedic Research* **5**, 253-260.

99. Northrup, S.H (1988) Diffusion-controlled ligand binding to multiple competing cell-bound receptors, *J. Phys. Chem.* **92**, 5847-5850.

100. Ott, E., Spano, M. (1995) Controlling chaos, *Physic Today*, 34-40, May.

101. Papers in *Biological Effects and Electric and Magnetic Fields*, D.O. Carpenter and S. Ayrapetyan (eds.) Vols. I and II, Academy Press, San Diego (1994).

102. Papers of the Proceedings of the Second EBEA Congress in *Advances in Bioelectromagnetics*, D. Miklavčič, R Karba, L. Vodovnic, and A. Chiabrera (eds.), Special issue of Bioelectrochem. Bioenerg. **35** (1994).

103. Papers of the Proceedings of the COST 244 Workshop on Mobile Communications and Extremely Low Frequency Fields, D. Simunic (ed.), European Commission, DGXIII, Bled, Slovenia, Dec. 10-12 (1993) and papers of the Proceedings of the COST 244 Workshop on Biomedical Effects Relevant to Amplitude Modulated RF Fields, D. Simunic (ed.), European Commission, DGXIII, Kuopio, Finland, Sept. 3-4 (1995).

104. Papers of the Radiofrequency Radiation Standards: Biological Effects, Dosimetry, Epidemiology, and Public Health Policy, Edited by B.J. Klauenberg, M. Grandolfo, and D.N. Erwin, NATO ASI Series A274, Plenum Press (1995)

105. Papers of the Proceedings of the State of the Science Colloquium, WTR and ICWCHR, Nov. 13-15 (1995).

106. Pilla, A.A. (1974), Electrochemical information transfer at living cell membrane, *Ann NY Acad. Sci.* **238**, 149-170.

107. Pilla, A.A. (1974), Mechanism of electrochemical phenomena in tissue growth and repair, *Bioelectrochem. Bioenerg.* **1**, 227-243.

108. Pilla, A.A., Nasser, P.R., and Kaufman, J.J. (1993) On the sensitivity of cell and tissues to therapeutic and environmental electromagnetic fields, *Bioelectrochem. Bioenerg.* **30**, 161-169.

109. Poponin, V. (1994) Non-linear stochastic resonance in weak e.m.f. interactions with diamagnetic ions bound within proteins, in M.J. Allen, S.F. Clearly, and A.F. Sowers (eds.), *Charge and Field Effects in Biosystems-4*, World Scientific, Singapore, 306-319.

110. Relter, R.J (1994) The pineal gland and melatonin synthesis: Their responses to manipulations of static magnetic fields, in D.O. Carpenter and S. Ayrapetyan (eds.), *Biological Effects of Electric and Magnetic Fields*, Academy Press, San Diego, 261-285.

111. Rodan, S.B., and Rodan, G.A. (1981) Parathyroid hormone and isoproterenol stimulation of adenylate cyclase in rat osteosarcoma clonal cells, *Biochem. Biophys. Acta* **46**, 673.

112. Rodan, G.A., Bourret, L.A., and Norton, L.A. (1987) DNA synthesis in cartilage cells is stimulated by oscillating electric fields, *Science*, **199**.

113. Rodan, S.B. and Rodan, G.A. (1984) Hormone-adenylate cyclase coupling in osteosarcoma clonal cell lines, in P. Greengard and G.A. Robison (eds.), *Advances in Cyclic Nucleotides and Protein Phosphorylation Research*, Raven Press, New York, 127-134.

114. Rodbell, M. (1980) The role of hormone receptors and GTP-regulatory proteins in membrane transduction, *Nature* **284**, 17.

115. Rubinow, S.I. (1975) *Introduction to Mathematical Biology*, J. Wiley and Sons Eds., New York.

116. Ryaby, J.T., Fitzsimmons, R.J., Ni Aye Khin, Culley, P.I., Magee, F.P., Weinstein, A.M., and Baylink, D.J. (1994) The role of insulin-like growth factor II in magnetic regulation of bone formation, *Bioelectrochem. Bioenerg.* **35**, 87-91.

117. Sargent III, M., Scully, M.O., and Lamb, W.E. (1974) *Laser physics*, Addison-Wesley Publ., Co. Reading, 79-95.

118. Shoup, D. and Szabo, A. (1982) Role of diffusion in ligand-binding to macromolecules and cell-bound receptors, *Biophys. J.* **40**, 33-39.

119. Ter Haar, D. (1961) Theory and applications of the density matrix, *Rept. Prog. Phys.* **24**, 304-362.

120. Trillo, M.A., Ubeda, A., House, D.E., and Blackman, C.F. (1996) Magnetic fields at resonant conditions for the hydrogen ion affect neurite outgrowth in PC-12 cells: A test of the ion parametric resonance model, *Bioelectromagnetism* **17**, 10-20.

121. Weaver, J.C. and Astumian, R.D. (1990) The response of living cells to very weak electric fields: the thermal noise limit, *Science* **247**, 459-462.

122. Weaver, J.C. and Astumian, R.D. (1994) *The thermal noise limit for threshold effects of electric and magnetic fields in biological systems*, D.O. Carpenter and S. Ayrapetyan (eds.), Academic Press, San Diego, 83-104.

123. Weinans, H. and Prendergast, P.J. (1996) Tissue adaptation as a dynamical process far from equilibrium, *Bone* **19**, 143-149.

124. Wickelgren, I.J. (1996) Local-area networks go wireless, *IEEE Spectrum* **33**, 34-40.

125. Wiesenfeld, K. and Moss, F. (1995) Stochastic resonance and the benefits of noise from ice ages to cryfish and SQUIDS, *Nature* **373**, 33-36.

126. Wyman, J., Gill, S.J. (1990), Binding and Linkage, Univ. Science Books, Mill Valley, CA.

127. Yamashita, M.M., Wesson, L., Eisenman, G., and Eisenberg, D. (1990) Where metal ions bind in proteins, *Natl. Acad. Sci. USA* **87**, 5648-5652.

128. Zwanzig, R. (1990) Diffusion-controlled ligand binding to spheres partially covered by receptors: an effective medium treatment, *Proc. Natl. Acad. Sci. USA* **87**, 5856-5857.

SUMMARY OF SESSION H: APPLICATIONS OF DOSIMETRY IN BIOLOGY & MEDICINE

D. MIKLAVČIČ
University of Ljubljana
Faculty of Electrical Engineering
Tržaška 25, SI-1000 Ljubljana
Slovenia

1. Dose Determination in Epidemiological Studies, Kjell Mild
2. Applications of Dosimetry in Military Epidemiological Studies, Stanislaw Szmigielski
3. Use of a Full-size Human Model for Evaluating Metal Implant Heating During Magnetic Resonance Imaging, Chung-Kwang Chou
4. Visualization of electromagnetic field exposure using radio-frequency current density imaging, K.Beravs, R. Frangež, F. Demsar
5. Possible Effects of Electromagnetic Fields on the Nervous - Endocrine - Immune Interactions, D. Miklavčič, T. Kotnik, L. Vodovnik

Public concern regarding exposure to non-ionising electromagnetic radiation initiated several studies. In evaluating possible health-risk, the initial knowledge is obtained from epidemiological studies. In the case of non-ionising electromagnetic radiation, low if any increased risk is associated, however, the number of exposed people encompasses basically the whole population. In setting the limits of exposure, two predominant mechanisms were taken into account: low-frequency range – the excitation of muscles and nerves, and in the high-frequency range - thermal effects due to energy absorption in the tissue. In the low-frequency range, however, more subtle effects than tissue excitation were reported and increased risks for some cancers were associated with different surrogates and historic measures of exposure to electromagnetic radiation. The lesson from low-level low-frequency electromagnetic story tells us that the concept of the dose is not an easy task to solve. Although in high-frequency range specific rate of absorption may not be the most appropriate dose, it is definitely the most widely accepted concept of the dose, it is energy based – already in the first years of our education we learned about the importance of energy, and until we find exact mechanisms of electromagnetic radiation to biological system interaction, it should be used. In addition, the exact description of the exposure in terms of power, amplitudes, pulse or waveform and geometry should be reported as precisely as possible.

In this chapter the first two of the papers are addressing the question of the dose of exposure in epidemiological studies. The difficulties in assessing the dose of non-ionising electromagnetic radiation are discussed as well as the sole concept and

B.J. Klauenberg and D. Miklavcic (eds.), Radio Frequency Radiation Dosimetry, 449-450.
© 2000 *Kluwer Academic Publishers. Printed in the Netherlands.*

understanding of the dose in investigating different health-risks. The conclusion from this two papers is that the question of the SAR as the dose concept may not be valid for different health risks. We also need to take into consideration whether the whole body exposure or only local exposure is at play.

In the paper by C.K. Chou a thorough investigation of extensive heating during MRI examination is reported due to the presence of metallic implants in the body. The study is very informative as it shows that the dysfunction of the implant itself can drastically change the heating pattern.

In the paper by K. Beravs et al. a relatively new MRI technique is described, the radio-frequency current density imaging. As the technique is new, the SAR in different body regions are not yet fully determined. Numerical SAR analysis is shown to be suitable for prediction of peak SAR values and temperature rise for various RF-CDI imaging parameter setting. In addition, this current density imaging allows in vivo current density imaging which corresponds to anatomical structures with different dielectric and conductive properties.

In the last paper, by Miklavčič et al., an attempt was made to offer a plausible explanation for the effects of low-level non-ionising electromagnetic radiation on human health. Numerous reports describe the use of ELF, RF and MW radiation in treatment of a variety of illnesses. At least some of all these different effects, beneficial and harmful could be explained through nervous-endocrine-immune interactions.

In summary, until the exact biological mechanisms of electromagnetic interaction with biological systems are fully elucidated, the concept and understanding and determination of the dose will be questioned as well as the effects which are (should) be caused to human health by low-level non-ionising electromagnetic radiation.

DOSE DETERMINATIONS IN EPIDEMIOLOGICAL STUDIES

K. HANSSON MILD
National Institute for Working Life
S-907 13 Umeå, Sweden

1. Introduction

Studies of health effects of radiofrequency (RF) electromagnetic fields have so far not been using the concepts of *exposure* and *dose* to any great extent. The problem of dose assessment in epidemiological studies have been discussed briefly by Repacholi [1] in connection with the WHO International EMF project, and he points out that satisfactory dosimetry has been a shortcoming in many studies. This is due to several factors, but mainly because the interaction mechanism(s) are today not well understood, especially concerning weak fields and non-thermal effects. See also Bergqvist [2] for a recent review of some RF epidemiological studies.

The major concern with RF exposure is tissue heating and the thermal effects. It has been a common practice to relate laboratory findings to measured or estimated SAR (Specific Absorption Rate) values. Most present Western standards and guidelines are based on levels of SAR less than 0.4 W/kg for whole body average. The SAR and its distribution in the body depends on several factors such as intensity of the electric and magnetic fields, frequency, polarization, and shape and orientation of the body, and possible contact with RF ground. The situation of estimating SAR in an occupational study is thus difficult since each situation needs individual attention. For instance, near RF sealers and glue dryers measurements of field strengths in air need to be combined with measurements of induced currents. In the use of mobile phones local SARs need to be addressed. For low-level RF field effects SAR might not be generally applicable but at present it is the best starting point for the *dose* concept.

In the following I will give some examples of studies were attempts have been made to measure or estimate the exposure for radiofrequency electromagnetic fields. However, direct estimate of SAR has never been made in any study, and usually the exposure has been proxied by other means. In the further research in this area it is necessary to work more on these questions.

2. Study of RF Sealer Workers

In an epidemiological study of RF sealer workers Kolmodin Hedman et al. [3] measured the E and H fields at the various workplaces. Since no near field dosimetry

B.J. Klauenberg and D. Miklavcic (eds.), Radio Frequency Radiation Dosimetry, 451-457.

was available at the time, and the electromagnetic fields varied widely both in strengths and in combination of E and H fields, it was impossible to combine the measured values into a dose-value for each individual. The best classification of exposure that could be obtained was a categorization of the workers according to the various RF sealers they worked with. The categories used were tarpaulin machines, machines for ready-made clothing industry, and automatic machines. In the first category the workers are usually getting whole body exposure with the highest field on the trunk of the body, and since the workers usually are operating the machine in a standing position with little or no separation from the factory floor, which can be considered to be RF ground, the induced body current can be rather substantial (see further Williams and Hansson Mild [4]). Working with ready-made machines, the highest exposure usually occurs on the hands but both the head and the trunk of the body can get substantial exposure. However, the operator is usually sitting down when operating the machine and the body current is therefore much lower than in the previous case. The third category, the automatic machines, often give rise to very low field strength at the operator's position, which is normally located 1-3 m away from the electrodes. Both sitting and standing positions can occur with these machines. A significantly impaired two-point discrimination on the fingertip of the index finger and in the palm of the hand was found in the exposed operators of ready-made machines as compared to a control group consisting of sewing machine operators. The total group of welders reported a higher frequency (19%) of neurasthenic symptoms in comparison with the control group (9%). The consumption of headache pills was also much higher in the exposed group, on the average 1 pill per week versus 0.2 in the control group.

In some workshops with several RF sealers and with reinforced steels in the concrete in the floor, one can see body currents occurring in the operators not only when they are working with their "own" machine but also when the other machines are activated. This needs to be studied further.

3. Radio Communication Systems

Tynes et al. [5] studied breast cancer among female telegraph operators onboard ships. They measured the RF fields in the radio room on some of the vessels. An excess risk was seen for breast cancer (SIR= 1.5) but not for all cancer (SIR= 1.2). However, they never used the measured values in the evaluation of the outcome; the measurements were merely used as a confirmation that there was RF exposure in the radio room, and the values found were not particularly high and within present standard.

Altpeter et al. [6] found a high frequency of disorders of neurovegetative nature among residents near a short-wave transmitter. Sleep interruptions were directly associated with the electromagnetic field strengths from the transmitter. Sleep quality apparently improved after interruption of broadcasting. No effect was seen on melatonin metabolism.

Bortkiewicz et al. [7, 8] studied workers occupationally exposed to RF in broadcast stations and in radioservices of mobile radiocommunication network. The exposure was assessed by measurements of the electric field and the dose was given as maximum E field encountered during the day or as a daily dose given by Vh/m. In the highest exposed group - AM station workers - the abnormalities in ECG recordings (resting and/or 24-hour) was significantly higher (83%) than among both Radioservices workers (55%) and the control group Radio Link Stations workers (40%). The authors conclude by saying that at the present stage it is impossible to determine which of the parameters (frequency, max E field, daily dose, etc) may be responsible for the observed cardiovascular impairment.

Hocking et al. [9] studied the cancer incidence and mortality and the proximity to TV towers. Based on calculations they estimated the power density in the exposed areas to 8.0 $\mu W/cm^2$ near towers, 0.2 $\mu W/cm^2$ at a radius of 4 km, and 0.02 $\mu W/cm^2$ at 12 km. Leukemia incidence and mortality were significantly increased in the inner area (near the tower). The authors point out that the mechanism whereby such low intensities could cause biological effects is a matter of intense research, and more detailed studies are required before any conclusions can be drawn. See also Goldsmith [10] for a discussion of the findings of TV towers and cancer.

4. Mobile Phones

In a recent epidemiological study of subjective symptoms among mobile phone (MP) users Hansson Mild et al. [11] used several factors to asses the exposure and to get an estimate of dose.

The study was set up to see if there were differences between users of the analogue (NMT) and the digital (GSM) systems. During 1995 many people with complaints of symptoms they experienced while using the MP contacted manufacturers, net operators and researchers working with electromagnetic fields. The complaints were about symptoms such as headaches, feeling of discomfort, warmth behind/around and on the ear, and difficulties to concentrate. The number of complaints from users of MPs was larger for GSM users; i.e. with pulse modulated fields. In the scientific literature of biological effects of weak microwaves there is a tendency for lower thresholds for reported biological effects from exposure to modulated fields. The hypothesis was therefore that GSM users experienced more symptoms than did NMT users, but this was falsified by the study. Actually, GSM users reported warmth sensation on the ear and behind or around the ear less frequently than did NMT-users. The study also found statistical associations between both *Calling time* and *Number of calls per* day and the occurrence of warmth sensation as well as headache and fatigue both among NMT users and GSM users. Those people using phones for 15 to 60 minutes per day were 1.6 times more likely to complain of fatigue and 2.7 times more likely to complain of headache than people who used their phones for two minutes or less per day. Users of phones who talked more than 60 minutes per day were 4.1 times more likely to complain of fatigue and 6.3 times more likely to have a headache than those who talked for less than two minutes per day.

Whether this association also demonstrates a causal relation between MP use and the genesis of the different symptoms could not be determined. The finding, however, gives rise to the hypothesis that the *Calling time* and *Number of calls* are associated with the sensation of warmth and some vegetative symptoms. Further studies are required to test this hypothesis and to explore the role of various physical factors in genesis of the observed symptoms.

The technical details of analogue and digital systems used for MPs have been described in detail (McKinlay et al. [12], Bach Andersen et al. [13]) and only a short background is given here. The NMT systems operates at 900 MHz with a continuous carrier wave. The maximum output from the handheld NMT900 phones is 1 W. The NMT phones have their output power regulated through the base station in two levels, 0.1 or 1 W; the closer to the station the lower is the output power.

In the digital GSM system the information is sent in pulses with a repetition rate of 217 Hz. The pulse length and repetition frequency give a duty cycle of 1/8. The maximum output power is 2 W which gives a mean value of 0.25 W maximum. The GSM system also provides a battery saving function, which practically turns down the output power to half of the maximum. The output power is also regulated from the base station, from a maximum of 2 W down to a minimum of 20 mW (with the phones sold today an even lower value of 5 mW is used). The mean output power is, thus, normally well below 0.1 W.

Different phones have different design for the antenna position and physical dimensions, for instance, a dipole antenna or a helical antenna. Recently Kuster [14] measured 16 different European digital phones, and he found a very wide variation in the SAR-values. The phone giving the lowest value, when averaged over 10 g tissue, had a SAR of 0.28 W/kg and the one with the highest value had 1.33 W/kg; all normalized to an antenna input power of 0.25 W, which is the maximal value for a GSM phone. If the averaging was done over 1 g tissue the span was from a low of 0.42 W/kg to a high of 2.0 W/kg.

Generally the GSM phones have a lower output power than the NMT900 phones. The antenna systems used are similar for the two systems, and the SAR values, thus, are slightly higher for the NMT900 users.

The currents from the battery also gives rise to magnetic fields near the phone. For GSM phones magnetic flux densities of a few μT near the phone have been measured (Bach Andersen et al. [11], Linde and Mild [15]). The fields are pulsed DC fields with a frequency of 217 Hz. For the NMT phones the magnetic field from the battery current are to be regarded as pure DC fields.

The SAR- measurements were done under normal user conditions. However, when the phone is slightly tilted towards the head of the user Kuster [14] exemplifies that the value can go from 0.2 to 3.5 W/kg. Thus, for different phones under maximal output, we have a factor of about 5 between the extremes, and to this the personal handling of the phone gives a factor of tenfold or more.

It should be noted that all given values are maximum SAR values found regardless of the anatomical localization. An equal weight is given the values regardless if it is obtained on the external ear, middle or inner ear, or behind the ear. In the future, it will be necessary to make the comparison at the same anatomical localization. Presumably, then the values as given by Kuster [14] might differ even more.

Taken together we have a rather large uncertainty in the actual SAR determination for a specific situation, with a factor of 100 from the nearness to the base station and at least a factor 10-50 depending on make and model and personal style of use.

It is difficult to see how the actual exposure (if measured as SAR) can be proxied by "billing record", a procedure that in Europe would only give the total time for outgoing calls, not showing incoming calls, nor the power settings of the phone. Instead of billing records or estimates of SARs the exposure was assessed by transmitter system and the estimated number of minutes on the phone per day and the number of calls per day.

For future studies of MP users it would be useful to perform a study of the base station regulation for a number of users. This would not be possible to do retrospectively, but it could be done prospectively for a selected number of users. For these it would be possible to obtained detailed records of their phone use (with their permission of course) both number of calls, length of each call, and the actual power settings of the phone as ordered by the base station. These records could then be compared with the subjective estimates of number and time of calls from the users themselves.

5. Discussion

Since the interaction mechanism(s) are not well understood it is not surprising that the concepts *exposure* and *dose* are not clearly defined. As I have shown with examples from the RF epidemiological studies none of these have related the effect to the SAR and its distribution in the body. However, this issue is of great importance for the continued research in bioelectromagnetics. One of the questions we need to address is for instance how *time* comes into the connection between *exposure* and *dose*, and here we need to distinguish between different aspects of time: very short times - order of minutes, daily averages, and total time in the actual occupation - number of years with exposure.

Let us assume that we only have to worry about thermal effects and that we know all we need about the SAR-distribution in an RF-sealer operator. In the present guidelines there is a 6 min time-averaging based on thermal factors and for preventing heating effects. How would we best construct a *dose* from this? Would the thermal effects have an accumulating effect and then *dose* would simple be an integration of the SAR over time, or are we here to deal with a threshold effect, i.e. if the 6 min averaging gives whole body values below 0.4 W/kg than it does not count into the total daily *dose*?

In a recent paper by Persson and coworkers [16] the effect of various exposure times and power densities were investigated with respect to blood-brain-barrier changes in rats. They obtained the Specific Absorption (SA) from the known SAR distribution and the exposure time. SA's are expressed with the unit J/kg, which in ionizing radiation is better known as Gray (I am thereby not implying that the effects are the same for the two different types of "radiation"). Persson et al. [16] does not give any details to what

456

combinations they used, so this does not help us in answering the questions brought up in connection with the use of mobile phones: Is there a difference in the effect of one ten min call and ten one min calls? I think it would be of value to look into how the dose concept has developed regarding exposure to ionizing radiation. We need to better understand things like dose-rate, fractioned dose, etc also for non-ionizing radiation.

Another aspect of time is in the long term exposure. In some of the ELF epidemiological studies number of years in each occupation has been used as part of the *exposure/dose* determination, but in the RF papers this has not been applied with any success. In the recent Scandinavian mobile phone study [11] the MP-users were asked how many months they had used a MP. When the answers were analyzed versus the experienced symptoms no correlation was found with this parameter, and this indicates that at least with regard to the subjective symptoms the long term exposure has no influence.

Presumably we need to look at different meanings to the dose concept with regard to different symptoms and diseases due to different organ/tissue sensitivity. Probably, for thermal effects we will need one way of expressing the *dose*, in cancer studies another meaning might be given to *dose*, and for nervous system effects still another. For mobile phone studies we need to be more detailed with the use of SAR´s and not just use the highest value found anywhere near the phone and paying no attention to the anatomical localization. We need more information from the medical/biological side as to what sites are of interest for which symptom and/or disease.

Presently, regarding RF epidemiological studies we have more questions about what constitutes *dose* than we have answers, and it is clearly time for us to start working with these questions. Dosimetry is only a part of the total picture here.

6. References

1. Repacholi, M.H. (1997) Low-level exposure to radiofrequency fields: Health effects and research needs, *Bioelectromagnetics* 19, 1-19.
2. Bergqvist, U. (1977) Review of epidemiological studies, in N. Kuster, Q. Balzano, and J.C. Lin (eds.), *Mobile Communications Safety*, Chapman & Hall, London, pp.149-170.
3. Kolmodin-Hedman, B., Hansson Mild, K., Hagberg, M., Jönsson, E., Andersson, M-C., and Eriksson, E. (1988) Health problems among operators of plastic welding machines and exposure to radiofrequency electromagnetic fields, *Int. Arch. Occup. Environ. Health* 60, 243-247.
4. Williams, P., and Hansson Mild, K. (1991), Guidelines for the measurement of RF welders, Undersökningsrapport, Arbetsmiljöinstitutet, 8.
5. Tynes, T., Hannevik, M., Andersen, A., Vistnes, A.I., and Haldorsen, T. (1996) Incidence of breast cancer in Norwegian female radio and telegraph operators, *Cancer Causes and Control* 7, 197-204.
6. Altpeter, E.S., Krebs, Th., Pfluger, D.H. et al. (1995) *Study on Health Effects of Shortwave Transmitter Station of Schwarzenburg, Berne, Switzerland* (Major Report). Universität Bern. BEW Publ. Series Study No 55.
7. Bortkiewicz, A., Zmyslony, M., Palczynski, C., Gadzicka, E., and Szmigielski, S. (1995) Dysregulation of autonomic control of cardiac function in workers at AM broadcast stations (0.738 - 1.503 MHz). *Electro- and Magnetobiology* 14, 177-191.

8. Bortkiewicz, A., Zmyslony, M., Gadzicka, E, Palczynski, C., and Szmigielski, S. (1997) Ambulatory ECG monotoring in workers exposed to electromagnetic fields, *Journal of Medical Engineering & Technology* **21**, 41-46.

9. Hocking, B., Gordon, I.R., Grain, H.L. and Hatfield, G.E. (1996) Cancer incidence and mortality and proximity to TV towers. *MJA* **165**, 601-5.

10. Goldsmith, J.R. (1997) TV broadcast towers and cancer: The end of innocence for radiofrequency exposure, *Amer. J. Indust. Med.* **32**, 89-92.

11. Hansson Mild, K., Oftedal, G., Sandström, M., and Wilén, J. (1998) Comparison of analogue and digital mobile phone users and symptoms. A Swedish-Norwegian epidemiological study. Nordic Radio Symposium, Umeå, May 14, 1998, extended abstract.

12. McKinlay, A.F., Andersen, J.B., Bernhardt, J.H., Grandolfo, M., Hossman, K-A., van Leeuwen, F.E., Hansson Mild, K., Swerdlow, A.J., Verschaeve, L., and Veyret, B. (1996) Possible health effects related to the use of radiotelephones. Proposal for a research programme by a European Commission Expert Group.

13. Bach Andersen, J., Johansen, C., Frölund Pedersen, G., and Raskmark, P. (1995) On the possible health effects related to GSM and DECT transmissions, Aalborg, Aalborg University, Report to the European Commission.

14. Kuster, N. (1997) Swiss tests show wide variation in radiation exposure from cell phones. *Microwave News* **1**, 10-11.

15. Linde, T. and Hansson Mild, K. (1997) Measurement of Low Frequency Magnetic Fields from Digital Cellular Telephones, *Bioelectromagnetics* **18**, 184-186.

16. Persson, B. Salford, L.G., and Brun, A. (1997) Blood-brain barrier permeability in rats exposed to electromagnetic fields used in wireless communication, *Wireless Networks* **3**, 455-61.

APPLICATION OF DOSIMETRY IN MILITARY EPIDEMIOLOGICAL STUDIES

S. SZMIGIELSKI, R. KUBACKI. and Z. CIOLEK
Department of Microwave Safety
Military Institute of Hygiene and Epidemiology
PL-01-163 Warsaw, Poland

1. Introduction

Epidemiological studies of military personnel which operate and/or repair RF/MW emitting devices need valid assessment of individual exposure levels during working shift, as well as calculation of daily and periodic doses of the radiation. However, there exist only limited possibilities for precise measurements of real exposure levels, as typical situations in the military environment comprise large quantities of electronic and/or high-power RF/MW equipment confined to a relatively small space, with the servicing personnel being exposed in the near field zone. Additionally, the exposure conditions are frequently altered by the presence of the operator. This results in inhomogeneous absorption of RF/MW energy in the operator's body and thus, impedes precise calculation and application of SAR levels for epidemiological and/or medical studies. Therefore, in practice the assessment of exposure in epidemiological studies of RF/MW-exposed personnel and residents living close to high-power RF/MW military sources (e.g. radar bases) has to be limited to measurements of field power density (FPD) and to valuation of radiation doses (RD).

In the present paper we describe our 15-year experience with assessment of RF/MW exposure of military personnel in Poland, based on improved measurements of FPD, mainly of pulse-modulated MW fields and calculations of individual daily RD. The results of the exposure assessment serve as the basis for respective epidemiological and/or medical studies of RF/MW-exposed personnel [1, 2].

During 1980-1985 in Poland about 3500 - 4000 career military servicemen were considered as occupationally exposed to RF/MW [3. This population can be divided into five subgroups, according to type and level of exposure (Figure 1). There are two main types of RF/MW-emitting devices which contribute to exposure of military personnel - radio transmitting centers (RTC) (1-100 MHz) and radars (1-20 GHz). In relation to specific service and levels of exposure the RTC can be divided into stationary (S-RTC operating at 1-30 MHz) and mobile (M-RTC using a wider range of frequencies 1 - 150 MHz), while radars operate as defence radars (DR) or fire control radars (FCR) devices, both using the MW spectrum of 1-20 GHz.

459

B.J. Klauenberg and D. Miklavcic (eds.), Radio Frequency Radiation Dosimetry, 459-471.
© 2000 *Kluwer Academic Publishers. Printed in the Netherlands.*

Assessment of RF/MW exposure of Polish military personnel revealed that about 50% of subjects serve in RTC and about 45% in DR and FCR. There exists also a small subpopulation (about 3% of all exposed personnel) of servicemen from RF/MW repair workshops, but the exposure levels of these subjects appear to be considerably higher than of those servicing RCS or radars (Figure 1).

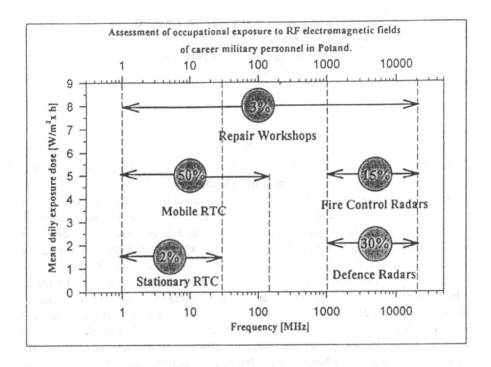

Figure 1. RF/MW exposure of military career personnel in Poland. Note two different levels of exposure for personnel servicing radio transmitting centres (RTC) and radars and high exposure levels for personnel of repair workshops.

Exposure doses presented in Figure 1 were valued from measurements of FPD, as described below (see equations 7 and 8). It should be stressed however that at least five subgroups of RF/MW-exposed military personnel can be differentiated; for two of them the daily radiation dose RD was valued for 1.5-2 W/m^2 x h, for other two for about 5 W/m^2 x h, while personnel of RF/MW repair workshops received a daily RD of about 8 W/m^2 x h (Figure 1).

2. Application of Dosimetry for Assessment of Exposure to Electromagnetic Fields

For evaluation of EMF exposure of workers and residents it is essential to establish parameters of the radiation which penetrates the investigated subject. Standards in most countries require measurement of electric field strength (E) for RF and power flux density (S) for MW fields. However, in non-uniform RF/MW fields (fields which change their intensity during exposure and/or when exposed subjects move in the space with changeable field distribution) measurements of basic field parameters (E or S) appear to be inadequate for assessment of exposure levels.

In case of laboratory animals reliable assessment of exposure is enabled with calculation of SAR (specific absorption rate), e.g. by the twin-well calorimetry approach [4, 5]. Although several experiments with a full-sized human models have been also performed [6, 7] the SAR calculation cannot be from obvious reasons applied in epidemiological/medical investigations of RF/MW-exposed personnel.

In practice, a parameter, which may be applied for more precise expression of exposure conditions, appears to be a properly calculated radiation dose (RD). A classical definition of RD (as applied for ionizing radiation) includes the amount of energy loss per unit of body mass. In case of non-ionizing radiation, safety standards in most countries require measurements of field parameters (E or S) in the place of residence (working post, dwelling, etc.) but in absence of human beings. Therefore, in this case a modification of definition of RD is required, because there is no possibility to access the energy-absorbing by humans. On the other side, it is well established that absorption of RF/MW energy by living organisms is dependent from one side upon the frequency and orientation of the incident field and, from the other side upon size of the body, its geometry and dielectric differentiation of its parts.

Calculation of RD, although in some cases based on evaluations, allows better assessment of RF/MW exposure, mainly in case of complex fields and fields changeable in time (e.g. from rotating radar antennas), than application of basic field parameters (E or S). Therefore, it is our strong feeling that RD is essential for valid correlation with biological effects and/or health risks of RF/MW radiation.

2.1. THE AVERAGED VALUE OF POYNTING'S VECTOR AND ELECTRIC FIELD STRENGTH

The equations for mean values of S and E, as well as relations to RD for EM fields changeable in time are summarized below. The Poynting's vector (S) appears to be in fact the mean value over ωt [8].

$$S = \frac{1}{2} Re \left(E_0 \times H_0^* \right) \tag{1}$$

where:

S - Poynting's vector, known also as power flux density
(value measured and related to safety standards in case of CW),

E_0 - magnitude of electric field strength,

H_0^* - magnitude of magnetic field strength (complex conjugate).

At the far-field the relation E to H remains constant and independent from distance to the antenna:

$$\frac{E_0}{H_0} = Z_0 = 376,7\,\Omega \tag{2}$$

In this case the equation 1 can be expressed as:

$$S = \frac{E^2_{rms}}{Z_0} \tag{3}$$

where:
  E_{rms} - the rms value of electric field strength (measured in V/m).

Measurements of power flux density are performed in agreement with (3) – exclusively by measurement of the rms value of electric field strength.

The value of Poynting's vector averaged over any period of time T (e.g., T may be equal to time of work shift T = 8 h) is expressed by definition as:

$$S_{av} = \frac{1}{T}\int_0^T S(t)\,dt \tag{4}$$

The equation 4 may be also simplified as:

$$S_{av} = \frac{1}{T}\sum S_i\,t_i \quad \left[\frac{W}{m^2}\right] \tag{5}$$

where:
  S_i - the values of samples Poynting's vector over time t_i,
  t_i - the sampled time; during t_i the value of S_i must be different from
         zero and remain constant for the whole period of t_i.

The value of E averaged over any period of time T can be expressed as:

$$E_{av} = \sqrt{\frac{1}{T}\sum E_i^2\,t_i} \quad \left[\frac{V}{m}\right] \tag{6}$$

where:
  E_i - values of sampled rms electric field strength during time t_i,
  t_i - sampled time; during t_i the value of E_i has to be different from
         zero and remain constant for the whole period of t_i .

According to the above equations, it may be concluded that EM fields changeable in time may be presented as fields at equivalent averaged value S_{av} and E_{av}.

2.2. RADIATION DOSE

The radiation dose (RD) may be expressed with the following equation, based on values of sampled Poynting's vector S_i:

$$D = \sum S_i t_i \quad \left[\frac{W}{m^2}h\right] \tag{7}$$

RD may be also calculated on base of E_i:

$$D = \frac{1}{Z_0} \sum E_i^2 t_i \quad \left[\frac{W}{m^2}h\right] \tag{8}$$

and on base of average values of S or E (S_{av} and E_{av}):

$$D = S_{av} T = \frac{1}{Z_0} E^2_{av} T \tag{9}$$

The above equations for calculation of RD are practically useful for assessment of exposure in variable EM fields.

2.3. RADIATION DOSE INDEX

For assessment of RF/MW exposure in epidemiological studies it seems useful to introduce an additional value of radiation dose index (RDI) which can be defined as relation of individual RD to maximal acceptable RD, calculated from standards being in force for the investigated subjects:

$$RDI = \frac{D}{D_n} \tag{10}$$

where:
   D - real RD, calculated from equation 7 or 8,
   D_n - maximal permissible RD, calculated from operative EM safety standards with the following equations:

$$D_n = S_n T \quad \text{or} \quad D_n = \frac{1}{Z_0} E_n^2 T \tag{11}$$

where:
S_n and E_n - respective values of S or E established in safety standards for EM frequencies for which the RDI is calculated.

Final equations for calculation of RDI have the following form:

$$RDI = \frac{\sum S_i t_i}{S_n T} \tag{12}$$

$$RDI = \frac{\sum E_i^2 t_i}{E_n^2 T} \tag{13}$$

RDI calculated from equations 12 and 13 appears to be useful in assessment of exposure levels in comparison to permissible levels and to levels acceptable in other countries. It remains obvious that RDI < 1.0 indicates exposure levels lower than permissible for the investigated population, while all RDI values exceeding 1.0 occur when workers were exposed to unacceptably high field levels.

3. Assessment of EM Exposure of Military Personnel

Table 1 summarizes data of RF/MW exposure assessment, calculated daily RD and RDI for career military personnel in Poland. As it was already mentioned, the population of RF/MW-exposed servicemen can be divided into five subgroups, according to type and level of exposure.

TABLE 1. RF/MW exposure levels of military personnel in Poland (1985-1990).

Type of equipment	Percent of exposed population	Frequency of EM radiation	Highest field levels during shift	Permissible levels for workers (3)		RD [W/m^2 x h]	Dn [W/m^2 x h]		RDI	
	(1)		(2)	PL	EC		PL	EC	PL	EC
RTC-stationary	2 %	1 - 20 MHz	30 V/m	70 V/m	77 V/m	1.5	104	120	0.015	0.012
RTC-mobile	50 %	1 - 100 MHz	100 V/m	70 V/m	77 V/m	5	104	120	0.05	0.04
Defence Radars	30 %	1 - 20 GHz	50 W/m^2	10 W/m^2	100 W/m^2	2	80	800	0.025	0.002
Fire Control Radars	15%	1 - 20 GHz	50 W/m^2	10 W/m^2	100 W/m^2	5	80	800	0.06	0.006
EM Repair Workshops	3 %	1 MHz 20 GHz	50-100 W/m^2	10 W/m^2	100 W/m^2	> 8	80	800	0.08 - 0.1	0.01

Explanations for TABLE 1.

RD	- radiation dose (daily), calculated for 8 h shift from equations (7) and (8);
Dn	- dose permissible (daily), based on safety levels for occupational exposure, calculated for 8 h shift from equation (11),
RDI	- radiation dose index (daily), calculated for 8-h shift from equation (12) or (13),
PL	- data and calculations based on EM safety levels operative in Poland,
EC	- data and calculations based on EM safety levels recommended by IRPA (1988),
(1)	- distribution of the RF/MW-exposed population of military personnel in Poland; the whole population of exposed personnel is considered as 100%,
(2)	- highest levels of E or S at typical places where personnel is exposed during shift the exposure at these intensities lasts typically for few-several minutes during work and may be repeated two-four times daily,
(3)	- field levels, RD, D_n and RDI were given for 8 MHz (RF) and for 1500 MHz (MW) for Polish workers (PL) the permissible safety levels are specified for 8-h work (intermediate/hazardous zone, without limitations of time), according to Polish acts.

Table 1. gives only the general expression of exposure levels and radiation doses that are experienced by military personnel servicing RTC and radars. In fact, individual exposure conditions of particular subjects may differ considerably. This occurs because the subjects work in the vicinity of RF/MW-emitting devices and move during the working shift quite frequently between locations with high and low intensities of EM radiation [2]. Therefore, each case should be considered individually. For epidemiological studies of relatively small groups (e.g. specialized medical examinations of 50 - 100 selected RF/MW-exposed subjects) the best advice is to register the exposure of each subject during the whole shift (including breaks) and to calculate the cumulated daily RDs.

For military personnel, as for all other RF/MW-exposed workers, a considerable help in assessment of the exposure may come from good knowledge of the peculiar character of the post and schedule of service/work.

3.1. RTC - STATIONARY

Radio Transmitting Centers (RTC) are composed of several to tens of broadcasting stations (1-30 MHz) which are located inside of special buildings and of the antenna field of constant configuration. For proper work of RTC is responsible a team of technicians, who work in the 12/24 h shift system. Two groups of personnel of RTC are being exposed to RF - tuning control technicians and guardsmen of the antenna field.

466

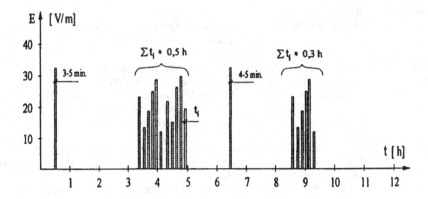

Figure 2. Exposure of technicians at the stationary RTC during a 12 h shift. Note relatively high peak exposure levels (up to 30 V/m) but lasting only 3-5 minutes each. Total time of exposure during shift remains below 60 min.

During the last decade (1987 - 1997) a considerable lowering of exposure levels of personnel of stationary RTC in Poland has been noted. This results, among other, from introduction of modern communication systems, including the communication-satellite systems and from lowering of output power of emitting devices.

A typical RF exposure of a technician working at the stationary RTC during 12-h shift is presented in Figure 2. Although the maximal exposure levels during this shift exceeded 30 V/m, a total period of abiding in the range of measurable RF fields was limited to 50 minutes, and still more, the exposure occurred in form of short (lasting 3-5 minutes) peaks. Therefore, the daily RD for this worker was calculated for about 1 W/m^2 x h, much below RDI permissible in Poland (compare data in Table 1).

3.2. RTC MOBILE

In the vicinity of mobile RTC, located at the tactical level of command, intensities of RF fields (up to 150 MHz) of 80-120 V/m can be noted quite frequently. The highest levels were recorded close to coaxial junctions, antenna junctions and/or elements of antennas mounted to masts (Figure 3).

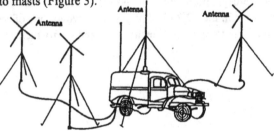

Figure 3. General view of mobile RTC.

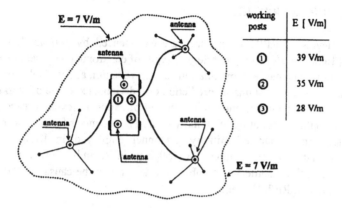

Figure 4. Distribution of electric field at the height of 1.8 m around operating mobile RTC; E values on working posts inside the vehicle are also given.

These elements are located directly on vehicles or close to their structures; therefore mobile RTC may be a source of considerable RF exposure of personnel e.g., on service posts inside vehicles E values close to 40 V/m were recorded (Figure 4).

3.3. RADARS

Due to variety of types and considerable differences in specific of radiation, only the most representative groups of radar devices can be mentioned here. Two basic groups can be differentiated:

a) defensive radars (DR) -e.g. search, tactical, surveillance radars, which are permanently located at selected places. DR emits high power MW radiation and may cause considerable levels of exposure for personnel and for the environment. Still more, this exposure tends to be regular, as the emissions of DR may be maintained around the clock;

b) fire control radars (FCR) - e.g. battlefield, air defence, artillery locating, missile control radars, which have their antennas located on arms (gibbets) on mobile vehicles. FCR have generally lower power outputs than DR and are mainly used periodically, during trainings and manoeuvres. Therefore, the exposure of personnel from FCR tends to be periodic and, although the short-lasting maximal exposure levels may be relatively high, the cumulative doses of radiation remain quite low.

3.4. REPAIRING WORKSHOPS

The highest levels of RF/MW exposure are experienced by personnel of repairing workshops (Table 1.). However, a variety of types of electronic equipment used at these posts, different frequencies and modulations of EM radiation and variable configuration of tested/repaired equipment unable prediction of individual exposure of workers and/or calculation of RD without continuous monitoring of exposure levels in each case. The best solution would be mounting of an individual MW dosimeter for each worker.

Fortunately, the group of military personnel employed in RF/MW repairing workshops is relatively small (about 100-150 men in Poland). However only in this group different health disorders were observed, which may be causally related to high levels of exposure to RF/MW fields.

4. Basic Problems in Measurement of RF/MW Radiation Emitted from Military Equipment

Calculation of RD and valid assessment of exposure of personnel for correlation with results of medical/epidemiological investigations requires valid measurements of RF/MW radiation at working posts and/or in the environment, close to the radiating sources. In case of RF/MW-emitting equipment used by military units the measurements may became quite complicated and may require a sophisticated metrologic equipment.

RF radiation emitted from RTC is relatively easy for measurement; any sensitive meter that is standardized for the 1-150 MHz frequency may be used with equal satisfaction.

Much more complicated appears to be valid measurement of radiation emitted by military radars [9]. Characteristics of radar MW radiation, space distribution of the field around antenna and rotation of antennas in certain types of radars may cause unacceptable errors in measurements, especially if meters applied for these measurements are not sensitized for recording of pulsed radiation.

4.1. MEASUREMENTS OF POYNTING'S VECTOR IN FRESNEL REGION OR IN NEAR FIELD

Most of safety standards require measurements of Poynting's vector (S) around radars (E x H, see equation 1). In the far field the relation E/H (2) remains constant and independent from distance from the antenna. Therefore, the equation for S can be expresses, as given in equation 3; in this case measurement of E is sufficient for determining of S. In fact, these relations are applied in all commercially available MW meters and their probes appear to be electric field probes.

However, it should be strongly stressed that all the above relations are valid only for the far field, where the distance from radiating antenna is larger than:

$$r_D = \frac{2 D^2}{\lambda} \qquad (14)$$

where:

  r_D - far field criterion,
  D - largest diameter of antenna,
  λ - length of the EM wave.

For large radar antennas the far field criterion is fulfilled at distances larger than r_D. In practice, this distance ranges from 1.0 to 1.5 km and for most military DR the far field region may be identified at distances r > 1.5 km from the antenna. This indicates that most of measurements is performed at the Fresnel region or even at the near field region. In this regions the relation (2) is no more valid and measurements have to be individually analyzed in terms of error of the method. Fortunately, from analysis of field distribution around typical DR it became evident that in this case the error of measurement in the Fresnel region is relatively small and in most cases does not exceed 1 dB.

4.2. MEASUREMENTS OF PULSE-MODULATED MW RADIATION

MW radiation from radars is pulse-modulated with the energy being emitted in form of short pulses lasting few thousand times less than the period of their repetition. For safety standards this form of MW radiation has in most cases determined the value of averaged power flux density (S) or electric field strength (E), according to equations 5 and 6, where t_i should be considered as pulse width, while T as time of pulse repetition.

For measurement of pulse-modulated MW radiation special meters, which allow for averaging over any time of pulse repetition, have to be applied. For this aim special electric field probes (with thermistor mount or with thermocouple array) were designed [10]. These probes allow the averaging automatically due to their relatively large time constant (τ). However, the results of measurements are correct only when pulse-modulated MW radiation can pass the probe for a sufficiently long period of time. This condition can be provided only for radars with fixed antennas, while for those with rotating antennas (which are quite frequent among military DR) additional problems with valid measurements arrive.

4.3. MEASUREMENTS OF PULSE-MODULATED MW RADIATION FROM ROTATING ANTENNAS

In case of radars with rotating antennas each object is irradiated with only trains of pulses (several to few tens of microsecond pulses of very high energy) when the main lobe of a beam passes the object. At the remaining time in practice there is no irradiation, as the side lobes have very low energy (less than 20 dB in comparison to the main lobe). The above very rapid variation in the exposure levels cause that the meters designed for measurement of pulse-modulated MW cannot provide valid readings, because of the relatively large time constant (τ) of their probes. Therefore, none of the commercially available MW meter provides valid measurements of pulse-modulated MW radiation from radars with rotating antennas.

A reasonable solution of this problem may be immobilization of antenna during measurements and directing the main lobe of the beam toward probe of the meter. In this case, all meters which provide averaging of MW energy over any time of pulse repetition (see above) can be used for measurements. It has be stressed however, that the value measured after immobilization of radar antenna represents the highest mean MW energy, which occurs in the beam. A real exposure should be lower and would depend on beamwidth and frequency of antenna rotation.

In Poland the safety standards which are in force require immobilization of antenna during measurements and directing the main lobe of the beam toward probe of the meter. Additionally, for radar antennas there are existing separate safety levels for workers and for residents; these levels serve for direct comparison of exposure data which were recorded after immobilization of antenna. This allows avoiding further calculations of exposure data and fastens the procedure.

However, it should be stressed that measurement of radar fields with immobilized antennas provides only partial assessment of the exposure levels and has numerous disadvantages. Not to mention the fact, that each measurement needs a proper cooperation with personnel servicing radar stations, the situation becomes totally unpredictable in case of complex MW fields, when several radars with rotating antennas are located on small space (e.g. in bases). Therefore, a real progress in assessment of exposure from radars with rotating antennas would be construction of a meter which allow for averaging of EM energy over any time of pulse repetition but is equipped with a probe with very short time constant (τ). Construction of such meter is now at its final phase in the present authors' Institute in Warsaw, Poland.

4.4. ASSESSMENT OF MW EXPOSURE OF PERSONNEL FROM MILITARY RADARS

Assessment of MW exposure of personnel from military radars, including calculation of RD, appears to be a difficult task, because of the above-mentioned lack of proper meters, need for immobilization of antennas during measurements and generally large mobility of workers during shift. In general, the exposure strongly depends on number and output power of devices generating MWs, their location and concentration in certain places, as well as on location of buildings and military posts in bases equipped with strong MW stations.

Measurements of field distribution from radar antennas are regularly performed in all Polish military units, both inside and outside the units for bordering the "hazardous" (10 - 100 W/m^2), "intermediate" (2 - 10 W/m^2) and "safe" (below 2 W/m^2) zones for personnel, according to the rules established in Poland in 1972.

In the 70's and 80's only about 3.5 - 4% of the career military personnel in Poland was considered as permanently exposed to MW fields stronger than 1 W/m^2 during at least 2 h daily. Of this group about 80 - 85% experienced maximal exposures not exceeding 1 - 5 W/m^2, while only 10 - 15% were exposed to stronger fields. Only about 5% were exposed to fields exceeding 10 W/m^2. However, the detailed analysis of exposure levels, including calculation of RD, needs continuous measurements during the whole working shift and thus, it can be performed only in selected subjects.

In most cases, for epidemiological reasons, a group of RF/MW-exposed personnel can be divided only arbitrary into three classes of exposure - low, medium and high - basing on service records, knowledge of typical exposures on service posts and the above mentioned distribution of exposure (see also *Figure 1.*). This analysis of exposure data is used by the present authors for epidemiological studies of Polish military personnel exposed to RF/MW radiation for several years.

5. Acknowledgements

This work was supported by the Research Grant 4P-05D-01111/96 from the National Committee of Scientific Research (KBN) in Poland and by the Joint Research Project PL 973038 from European Commission INCO-COPERNICUS Program in Brussels, Belgium.

6. References

1. Bassen, H.J. (1983) Electric field probes - A review, *IEEE Trans. Antenna Propag.* **AP-31**, 710-718.
2. Chou, C-K. and Guy, A.W. (1984) SAR in rats exposed in 2 450 MHz circularly polarized waveguides, *Bioelectromagnetics* **5**, 389-398.
3. Chou, C-K., Guy, A.W., McDougall, J.A. and Lai, H. (1985) Specific absorption rate in rats exposed to 2450 MHz microwave under seven exposure conditions, *Bioelectromagnetics* **6**, 73-88.
4. Christman, C.L. and Ho, H.S. (1974) A microwave dosimetry system for measured sampled integral-dose rate, *IEEE Trans. Microwave Theory Tech.* **MTT-22**, 1267-1272.
5. Hill, D.A. (1985) Further studies of human whole-body radiofrequency absorption rates, *Bioelectromagnetics* **6**, 33-40.
6. Jordan, E.C. and Balmain, K.G. (1968) *Electromagnetic Waves and Radiating System*, Prentice Hall, New Jersey.
7. Kubacki, R. and Szmigielski, S. (1995) Methodologic problems with RF pulse-power measurements, in D. Simunic (ed.), *Methods of Exposure Assessment Related to Standards, Proc. COST-244 Meeting, Athens, European Union (DG XIII), Brussels*, 42-47.
8. Olsen, R.G. and Griner, T.A. (1989) Outdoor measurements of SAR in a full-sized human model exposed to 29.9 MHz in the near field, *Bioelectromagnetics* **10**, 161-171.
9. Szmigielski, S. (1996) Cancer morbidity in subjects occupationally exposed to high frequency (radio frequency and microwave) electromagnetic radiation, *Science Total Environm. (STOTEN)* **180**, 9-17.
10. Szmigielski, S. (1997) Analysis of cancer morbidity in Polish career military personnel exposed occupationally to radiofrequency and microwave radiation, *2nd World Congress, Bologna, Abstr.*, **D-4**, 101.
11. Szmigielski, S. and Kubacki, R. (1996) Bioeffects and health risks of occupational exposure to microwave radiation on base of experience of the Polish military health services, *Int. Rev. Armed Forces Med. Services* **3**, 40-43.

USE OF A FULL-SIZE HUMAN MODEL FOR EVALUATING METAL IMPLANT HEATING DURING MAGNETIC RESONANCE IMAGING

C. K. CHOU
Motorola Florida Research Laboratories
8000 West Sunrise Blvd.
Fort Lauderdale, Florida 33322
U.S.A.

1. Introduction

Magnetic resonance imaging (MRI) and computer tomography (CT) have been used routinely in medical diagnosis. MRI is often preferred to CT because of its better soft-tissue resolution and lack of risk from ionizing radiation. However, the use of MRI involves magnetic and radio frequency (RF) exposures. In a strong static magnetic field and a pulsed RF field, five potential adverse effects are anticipated with metallic implants: 1) force on the implant by the strong static magnetic field, 2) current induced in the implants by the RF field, 3) damage of implant electronic circuitry by RF exposure, 4) MRI image distortion caused by the implant, 5) implant and adjacent tissue heating due to absorption of RF energy. Only the last of these, RF tissue heating, will be addressed in this paper. The present Food and Drug Administration (FDA) recommended limits of RF heating during MRI scanning are 1°C in the brain, 2°C in the torso and 3°C in the extremities [1].

Since RF energy absorption is a function of body size, the implant heating can only be tested in a full-size phantom human model. FDA does not approve tests in small and simple geometry phantoms. Although the RF field is strongest inside the MRI scanner, there is no guaranty that body parts outside the scanner do not affect the energy absorption in the exposed region. Therefore, a whole body model is necessary. The determination of implant heating in an RF field is similar to the measurement of specific absorption rate (SAR). The difference is the low RF power and long duration used during an MRI procedure. For SAR determination, high power and short duration are necessary to minimize thermal diffusion [2]. The tissue temperature during long RF exposures finally reaches a steady state when heat loss (due to thermal diffusion) equals power dissipation.

While at the City of Hope National Medical Center in Duarte, California, the author worked on three projects evaluating the safety of three medical implants during MRI. Two of the studies on auditory implants and spinal fusion stimulators have been published [3, 4]. A third study on the tissue heating due to cervical fixation devices is

473

B.J. Klauenberg and D. Miklavcic (eds.), Radio Frequency Radiation Dosimetry, 473–482.
© 2000 *Kluwer Academic Publishers. Printed in the Netherlands.*

not yet published [5] and therefore will not be included in this paper. In these studies, temperature rises adjacent to the implants during MRI exposures were measured. Because blood flow was not simulated in the static phantom, the temperature rise in the phantom around the implants would be a worst case scenario.

2. Materials and Methods

2.1. AUDITORY IMPLANTS

The cochlear implant (CI) has been widely accepted as a method for providing "sound" to the profoundly hearing impaired [6]. More than 10,000 patients have benefited from this implant. One example, the Nucleus Mini System 22 CI (Cochlear Corporation, Englewood, Colorado) with 22 electrodes is implanted into the cochlea to stimulate spiral ganglion cells at various locations in the cochlea to produce sound sensations of different pitches. The conventional CI has an implanted magnet for anchoring an external antenna behind the ear. The magnet can never be exposed to a strong magnetic field (1.5 T) due to the intense force generated. Therefore, for the MRI test, this magnet was replaced with a Silastic plug. If all tests show the modified CI is safe during MRI procedure, then the magnet can be removed before the MRI scans. In a separate study, possible heating of a newer Nucleus 24 implant and a modified Nucleus 22 implant (magnetic plug is replaced with a titanium plug) was tested.

The auditory brainstem implant (ABI) is a new device (also from Cochlear Corporation) that enables completely deaf patients to regain some hearing after the removal of neurofibromatosis type II (NF2) tumors [7,8]. The implant is surgically placed in the lateral recess of the fourth ventricle, adjacent to the cochlear nucleus. The device consists of a receiver/stimulator, lead wires, and eight electrode contacts attached to a Dacron mesh carrier. NF2 tumor patients often develop spinal tumors and therefore require MRI. Readers are referred to Figure 2 of Chou et al. [3] for the pictures of the Nucleus 22 CI and ABI.

2.2. SPINAL FUSION STIMULATOR

The spinal fusion stimulator (SpF®, Electro-Biology, Inc., Parsippany, New Jersey) has been used to increase bone fusion in more than 50,000 patients. The stimulator consists of a small direct current generator in a titanium shell that acts as an anode. Two insulated nonmagnetic silver/stainless steel leads provide a connection to two titanium electrodes which serve as the cathodes. A continuous 20-μA current is produced. The cathodes are embedded in pieces of bone grafted on the lateral aspects of the fusion sites, usually at the lumbar section. A schematic of the SpF® is shown in Fig. 1 of Chou et al. [4].

2.3.   PHANTOM MODEL

A fiberglass-shell human model 1.6 m tall was constructed. The model was bisected frontally. Flanges at the cut edge were held by nylon screws. The model had phantom brain, muscle, lungs and bone. The phantom skull was filled with phantom brain and the chest with phantom lungs; the remaining space between the skeleton and fiberglass shell was filled with phantom muscle. The phantom head was removable at the neck. The two arms were positioned close to the body to fit the MRI tunnel.

The phantom bone was made according to the formula of Hartsgrove et al. [9] and was a mixture of 36% two ton epoxy, 36% hardener (Devcon) and 28% 2 M KCl. The materials, enough to make an entire skeleton, were sent to Medical Plastic Laboratories in Gatesville, Texas, who specializes in making plastic skeletons for teaching. The skeleton was kept in a moisture-tight plastic bag and periodically soaked in 2 M KCl solution. The dielectric constant of the bone material is 14, and its conductivity is 0.8 S/m at 100 MHz. The dielectric properties at the MRI RF frequency of 63.8 MHz should be close to that at 100 MHz. The disks between the vertebrae were made of PVC with a dielectric constant of 2.8. New mixes were developed to simulate brain and muscle at 64 MHz [3]. The brain was mixed with 93% H_2O, and 7% TX-151 (a gelling agent), the muscle with 91.48% H_2O, 8.4% TX-151 and 0.12% NaCl. At 20 °C room temperature, the phase shift and attenuation of the mixtures inside a coaxial slotted line were measured with a network analyzer, as described by Chou et al. [10]. The dielectric constant and conductivity of the phantom brain and muscle at 64 MHz were calculated from the measurements to be 78.12 and 0.52 S/m, and 79.8 and 0.8 S/m, respectively. These dielectric properties are well within the range of published data [11]. For the lung, a mixture of 47% phantom muscle and 53% polyethylene powder, following Hartsgrove et al. [9], was used.

To make the human head model, starting from the skull, regions of the head and neck were built up with paraffin to compensate for missing tissues. Anatomic details were scaled from references [12-14]. From the paraffin head, two sagittally bisected rubber female molds were formed. Then two epoxy male molds were made from the rubber molds. Using the epoxy molds, heated 1 mm thick PET-G plastic sheets were vacuum suctioned to form sagittally bisected shells of heads. The two half shells were glued to reconstruct a whole head shell. The skulls were filled with phantom brain and the space between the skull and plastic shell was filled with phantom muscle. Figure 1 shows the back part of the fiberglass shell with the skull, skeleton and lungs.

2.3.1.   *For Testing Auditory Implants*

Since the temperature measurement was in the head region, the phantom from the neck down was a full size fiberglass shell model filled with phantom muscle tissue. The head to body phantom connection at the neck region was through silk screen and tight contact was provided with filament strapping tapes to assure electrical continuity. ABI or modified CI was placed in the phantom similar to the actual implant. The receiver/ stimulator coil was fixed behind the ear on the skull. The electrodes and lead wire were pulled into the skull through a drilled hole. The ABI electrodes were positioned at the

476

brainstem. The CI electrodes were stretched across the cranium. Teflon tubes (AWG 19&22) were placed next to the implants for positioning the fiberoptic temperature sensors at the end of the catheters. AWG19 was for the linear array (4 sensors at 1 cm spacing) and AWG22 was for single sensors. The slack ABI electrode lead was coiled inside the cranium. Phantom brain material was first poured into the cranium to immerse the electrodes, then the phantom muscle was poured to fill the rest of the head.

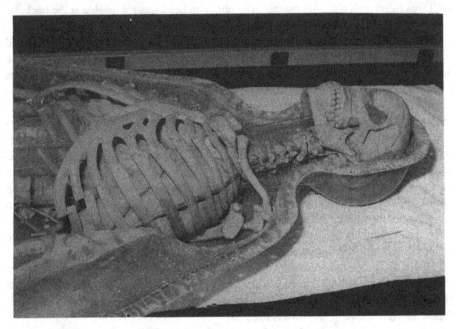

Figure 1. Inside view of the phantom model with skeleton and lungs, before filling of the muscle.

2.3.2. For testing Spinal Fusion Stimulators

The temperature measurement site was at the lumbar section, a full skeleton was used. An oval window (14 x 21 cm) on the back of the fiberglass model was opened for implant modification. Catheter tubings made of thin wall AWG 22 Teflon were positioned at 9 locations on the electrodes and generator shell. The window was covered with a fiberglass plate tightened with nylon screws.

Four conditions were simulated: 1) Intact implant: The SpF® (model 2T) was implanted as in a patient. 2) One wire broken. 3) The generator shell and leads were removed, and the two electrodes remained. Most implant patients stay in this condition. 4) No implant. This was the control case.

2.4. MRI EXPOSURES

According to the manufacturer (General Electric) a maximum RF heating condition on a GE Signa 4X, 1.5 tesla MRI unit could be obtained by setting the following parameters: Scan plane, axial; Pulse sequence, spin-echo; Echo time, 25 ms; Repetition time, 134 ms; Field of view, 48 cm; slice thickness, 20 mm; Matrix, 128; Direction of slice acquisition, Anterior to Posterior; Number of excitation, 94; Number of slices, 1; Number of echoes, 4; Phasing, Anterior to Posterior; Transmitter gain, 200; and Scan time, 26 minutes. Maximum switched magnetic gradient was 16.7 T/m/sec. The console-predicted, whole-body, averaged SAR for these settings and a body weight of 68 kg was 1.095 W/kg, using the body coil.

For the auditory implant test, the head was positioned in the center of bore without using the head coil. The line across the center of eyes was taken as the center of the RF field. Two heatings for each implant were performed to assure repeatability. For the spinal fusion stimulator study, eight worst case MRI RF heating tests were conducted for the 4 implant conditions. For every condition, two consecutive scans were done to assure repeatability. For condition 2, two additional regular MRI setting scans, as used on patients, were taken to compare with the worst case RF heating. The settings were as follows for basic spin echo: scan plane, axial; pulse sequence, spin-echo; Echo time, 17 ms; Repetition time, 650 ms; Field of view, 14 cm; Slice thickness, 5 mm; Matrix, 192; Direction of slice acquisition, Anterior-Posterior; Number of excitation, 2; Number of slices, 1; Number of echos, 1; Phasing, Anterior-Posterior; Transmitter gain, 200; and Scan time, 4 minutes and 18 seconds. The whole-body averaged SAR for these settings was 0.914 W/kg.

2.5. TEMPERATURE MEASUREMENTS

Luxtron (Santa Clara, CA) Model 3000 fiber optic sensors were used to measure temperatures. Twelve sensors were inserted into pre-implanted thin wall AWG22 Teflon catheters (for single sensors) or AWG19 tubes (for array sensors) to record the temperatures with a notebook computer. The sensors were calibrated with a gallium cell temperature standard (YSI Models 17401, 17402). To stabilize the phantom tissue temperature, the night before the test, the model and all associated equipment were set up in the MRI and the control rooms. Detailed sensor locations are shown in the original papers [3,4]. Figure 2 shows the phantom model entering the MRI scanner.

3. Results

3.1 AUDITORY IMPLANTS

Figure 3 shows an example of the temperature recording on a cochlear implant. The temperatures were recorded every 2 seconds and the raw data were smoothed to eliminate the high frequency noise ($\pm$ 0.1 °C). The durations of the 26 minute MRI scans are indicated. While the temperatures were changing over the 80 minute recording, the MRI scans did not change the slopes of these curves during or after the scan. The upward drift of temperature during the tests was shown to be caused by room

478

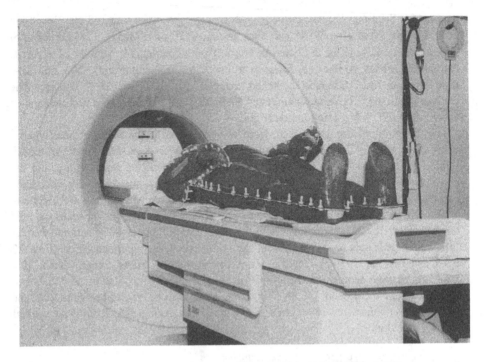

Figure 2. Full-size phantom human model was positioned into an MRI scanner.

warming. Sensor 4 was used to monitor the room air temperature near the nose of the phantom. The saw tooth temperature recordings were a result of the air conditioner cycling. During the day, the temperature increased. This room warming explains the upward drift of temperature. In both runs, there was no heating observed during the MRI. This result was true for all auditory implants.

3.2 SPINAL FUSION STIMULATOR

Fig. 4 shows the temperature results in the model with an intact implant. The highest temperature rise was at the center of the titanium generator surface (from 22.5 °C to 24.3 °C), and the second highest was at the edge of the generator (from 22.5 °C to 24 °C). Note, the room temperature rise was 0.5 °C (trace #4). Therefore, with implanted SpF® the maximum temperature rise was less than 2 °C measured at the center of the generator surface. The temperature rises at the electrodes were less than 1 °C. The temperature changes at the tip of the electrode (Sensor #8) were smaller than other locations.

When one of the generator leads was broken at the middle section, the temperature rise was as high as 11 °C (from 23 °C to 34 °C), as shown in Figure 5. This occurred at Sensor #1 which was connected to the end of the broken lead connected to the generator. The other end of the broken lead (Sensor #3) had only 3 °C rise (23 °C to 26 °C). The initial peaks shown in Figure 5 were due to RF heating during MRI tuning. Although the temperature of Sensor #2 (generator center) remained high, Sensor #11 (at the center of the electrode of the uncut side) was the third highest in temperature rise. In a repeated run (data not shown), the highest temperature rise was about 14 °C. When a regular spinal examination was scanned for 4 minutes, the maximum increase was also about 12 °C. Measurements made after the generator and the associated leads were removed from the phantom showed a maximum temperature rise of about 1.5 °C at one of the electrode pins. All other sensors showed insignificant temperature changes.

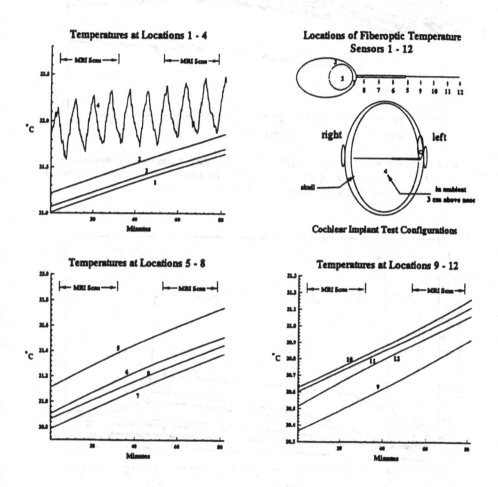

Figure 3. Temperatures during MRI RF heating test on Cochlear Implant [from Chou et al., 1995].

480

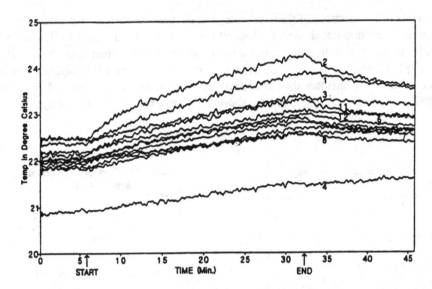

Figure 4. Temperature data of an intact SpF® implant in a phantom model during a worst- case MRI scan. The highest temperature was recorded at the center of the generator. The second highest was at the edge of the generator [from Chou et al., 1997].

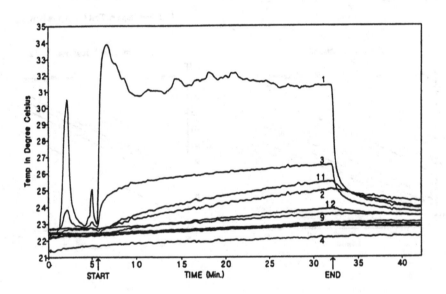

Figure 5. Temperature data of SpF® implant in a phantom model with a broken lead during a worst-case MRI scan. A rise of 11 °C was observed at the tip of the broken lead connected to the generator. The initial peaks were due to RF heating during MRI tuning [from Chou et al., 1997].

4. Discussion

A full-size phantom model was used to study the possible tissue heating due to the induced RF current near metallic implants during magnetic resonance imaging. Dielectric properties and geometry of the human tissues as well as the implants were modeled as close to real conditions as possible. A thermographic method was initially used to scan internal surfaces of the phantom head with auditory implants to locate any possible hot spots, so as to properly place the fiberoptic sensors. No heating was observed with that method [3]. In later studies, the temperature sensor method was used directly.

With auditory implants, there was no heating detected with fiberoptic temperature sensors. For the spinal fusion stimulator, when properly implanted, the maximum temperature rise in a static phantom model was less than 2 °C, which was within the FDA guidelines for the torso. However, if an electrode lead was broken, the temperature rise could be very high (14 °C). This exceeded the FDA regulation. Careful radiological examinations must be done to ensure that there are no broken leads in the patient before an MRI scan, although company records show that lead breakage is rare (10 broken leads in 50,000 patients). After the generator was removed, electrode heating was within the FDA guidelines. These studies were done on a static phantom and, therefore, no blood flow was simulated. In a patient, due to blood flow, maximum temperature rise should be lower than the phantom model measurements. Therefore, these results provide a worst-case estimation. These studies also show that RF tissue heating is difficult to predict. A case by case evaluation is necessary.

5. Acknowledgments

Projects were supported by Cochlear Corporation (Englewood, Colorado) and Electro-Biology Inc. (Parsippany, New Jersey). Mr. John McDougall and Kwok Chan were the co-authors of the original publications.

6. References

1. FDA (1989) Recommendations and report on petitions for magnetic resonance reclassification and codification. Final rule, 21 CFR Part 892, *Fed. Reg.* 54, 5077-5088.
2. Chou, C. K., Bassen, H., Osepchuk, J., Balzano, Q., Petersen, R., Meltz, M., Cleveland, R., Lin, J.C., and Heynick, L. (1996) Radio frequency electromagnetic exposure: A tutorial review on experimental dosimetry, *Bioelectromagnetics* 17, 195-208.
3. Chou, C. K., McDougall, J. A., and Chan, K. W. (1995) Absence of heating from auditory implants during magnetic resonance imaging, *Bioelectromagnetics* 16, 307-316.
4. Chou, C. K., McDougall, J. A., and Chan, K. W. (1997) RF heating of implanted spinal fusion stimulator during magnetic resonance imaging, *IEEE Transaction of Biomedical Engineering* 44, 367-373.
5. Chou, C. K., McDougall, J. A., and Ren, R.L. (1997) *RF Tissue Heating Due To Cervical Fixation Devices During Magnetic Resonance Imaging*, Final Report.

6. Brimacombe, J. A., and Beiter, A. L. (1994) The Application of Digital Technology to Cochlear Implants, in R. E. Sandlin (ed.), *Understanding Digitally Programmable Hearing Aids*, Allyn and Bacon Publishers, Needham Heights, MA, pp. 151-170.

7. Brackmann, D. E., Hitselberger, W. E., Nelson, R. A., Moore, J., Waring, M. D., Portillo, F., Shannon, R. V., and Telischi, F. F. (1993) Auditory brainstem implant: I. Issues in surgical implantation, *Otolaryngol. Head Neck Surg.* **108**, 624-633.

8. Shannon, R. V., Fayad, J., Moore, J., Lo, W. W. M., Otto, S., Nelson, R. A., and O'Leary, M. (1993) Auditory brainstem implant: II Postsurgical issues and performance, *Otolaryngol. Head Neck Surg.* **108**, 634-642.

9. Hartsgrove, G., Kraszewski, A., and Surowiec, A. (1987) Simulated biological materials for electromagnetic radiation absorption studies, *Bioelectromagnetics* **8**, 29-36.

10. Chou, C. K., Chen, G. W., Guy, A. W., and Luk, K. H. (1984) Formulas for preparing phantom muscle tissue at various radiofrequencies, *Bioelectromagnetics* **5**, 435-441.

11. Durney, C. H., Massoudi, H., and Iskander, M. F. (1986) *Radiofrequency Radiation Dosimetry Handbook*, (4th Edition), **USAFSAM-TR-85-73**, Brooks AFB, TX.

12. Eycleshymer, A. C., and Shoemaker, D. M. (1911) *Across-Section Anatomy*. D. Appleton & Company, New York.

13. Pernkoph, E. (1980) *Atlas of Topographical & Applied Human Anatomy*. Urban & Schwarzenberg, Baltimore-Munich.

14. Schnitzlein, H. N., and Murtagh, F. R. (1985) *Imaging Anatomy of the Head and Spine*. Urban and Schwarzenberg, Munich.

VISUALIZATION OF ELECTRO MAGNETIC FIELD EXPOSURE USING RADIO-FREQUENCY CURRENT DENSITY IMAGING

Specific Absorption Rate Study

K. BERAVS[1], R. FRANGEŽ[2] and F. DEMSAR[1]
[1]Institute "Jožef Stefan", Ljubljana, Slovenia, [2]Veterinary Faculty, University of Ljubljana, Ljubljana, Slovenia
Institute "Jožef Stefan", Jamova 39, 1000 Ljubljana, Slovenia

1. Introduction

Exposure to electro magnetic (EM) fields is monitored through specific absorption rate (SAR) measured in watts of energy per kilogram of body weight (W/kg). The biological effects of power deposition are usually tissue heating and effects on the nervous system. Pulsed and modulated EM radiation has been shown to synchronize the firing of neurons in the brain [1, 2]. The neurological effects are highly dependent on the frequency of excitation: direct current stimulus would cause cell depolarization, while RF stimulus would mainly tend to capacitatively charge the cellular membrane. The exposure to EM field can be visualized using a magnetic resonance (MR) imaging technique called Radio-Frequency Current Density Imaging (RF-CDI) [3, 4, 5].

1.1. RADIO-FREQUENCY CURRENT DENSITY IMAGING

RF-CDI is a magnetic resonance technique that can image the Larmor frequency current parallel to static magnetic field B_0 in one sample orientation [3, 4, 5]. An externally applied RF current at the Larmor frequency flowing in a sample can cause the net nuclear magnetization M of the sample to precess. The RF current creates a magnetic field whose left circularly polarized component transverse to the static field $\tilde{H} = \tilde{H}_x \tilde{a}_x + \tilde{H}_y \tilde{a}_y$ will cause precession of M about $\tilde{H}$ at a rate proportional to $|\tilde{H}|$. The strength of the components $\tilde{H}_x$ and $\tilde{H}_y$ are thus encoded in the phase of the magnetic resonance image and are used in a post processing curl operation to compute the current density magnitude and phase in a pixel region. The current density can be computed by the approximation [4]:

$$J_z \approx 2\left[\frac{\partial \tilde{H}_y}{\partial x} - \frac{\partial \tilde{H}_x}{\partial y}\right] + 2j\left[\frac{\partial \tilde{H}_x}{\partial x} + \frac{\partial \tilde{H}_y}{\partial y}\right] \tag{1}$$

483

B.J. Klauenberg and D. Miklavcic (eds.), Radio Frequency Radiation Dosimetry, 483-492.
© 2000 *Kluwer Academic Publishers. Printed in the Netherlands.*

Equation (1) assumes that the RF current flows predominantly parallel to the B_0.

Although other methods are possible, RF-CDI is usually implemented using the modified spin-echo pulse sequence shown in Figure 1 [3]. The RF current is applied for a time T_C. Simultaneously, a rotary echo, linearly polarized, magnetic field pulse B_1 that is synchronous with the RF current pulse and is aligned along one or other transverse rotating frame axis ($\tilde{a}_x$ or $\tilde{a}_y$) is applied. This rotary echo field must be large so that it determines the axis of precession. The component of the left rotating field which is parallel to the rotary echo field will thus determine the net precession angle. After excitation by the $\tilde{H}$ field, the net magnetization is projected into the transverse plane by a hard 90° pulse and measured (modulo 2π) as the phase in a conventional magnetic resonance image. Thus, for the rotary echo RFCDI method the RF current must flow predominantly parallel to the static field B_0 and the B_1 pulse must dominate over the field $\tilde{H}$ due to this current and any static field inhomogeneities ΔB_0.

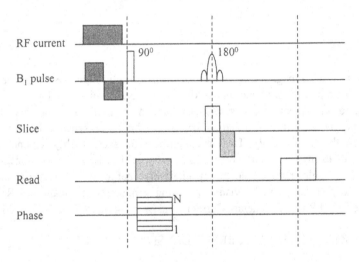

Figure 1. The RF current density imaging pulse sequence. A B_1 pulse that is synchronous with the RF current is added to a spin echo pulse sequence as shown.

1.2. POWER CONSTRAINTS

Compared to conventional MR imaging, RF-CDI uses two additional sources of RF power: the rotary echo B_1 and RF current pulses. Rotary echo RF-CDI depends greatly on the strength of the B_1 pulse, which is the dominant source of extra absorbed RF power. Its power deposition must be carefully considered. The SAR due to induced electric field from the B_1 pulse can be estimated as:

$$SAR = \frac{T_C}{T_R} \frac{\sigma}{2\rho} \frac{16\pi^4}{\gamma^2} R^2 f_0^2 f_1^2 \tag{2}$$

where f_0 is the Larmor frequency in Hz, f_1 is the rotary echo strength ($\gamma B_1/2\pi$), R is the loop radius of the cross section, σ is the tissue conductivity and ρ the tissue density [6]. The SAR contributed by externally applied RF current is:

$$SAR = \frac{T_C |J|^2}{T_R \, 2\sigma\rho} \tag{3}$$

where J is the current density [3]. The power deposition by the technique is the sum of the SAR induced by the B_1 pulse and the RF current.

To predict the SAR due to induced electric field from the B_1 pulse and externally applied RF current before *in vivo* imaging, a finite element model of the rat brain was built to simulate the distribution of absorbed RF power and tissue heating as a function of RF current pulse magnitude and length T_C, repetition time T_R, tissue conductivity σ and density ρ, tissue specific heat C_P, Larmor frequency f_0, magnitude of the B_1 field and loop radius of the cross section R. Current density distribution in the rat brain was then evaluated *in vivo* by RF-CDI.

2. Power Deposition and *in vivo* RF-CDI Imaging

This section outlines the methods and results from the RF-CDI experiments to image *in vivo* rat brain. To complement the RF-CDI brain images, a finite element simulation was performed to obtain power deposition estimates due to the application of the B_1 and RF current pulses and to estimate the RF current flow pattern and pathways.

2.1. FINITE ELEMENT SIMULATION

A quasi-static first order finite element model (FEM) was used to verify that RF current will flow parallel to B_0 during imaging. The model was used to obtain approximations of power deposited due to the B_1 pulse and application of RF current. It was built using Borland®C++ (Version 4.5) and Matlab® (Version 4.2b) software packages. The geometry was based on the MRI cross section scans of the rat brain and incorporated the complex conductivity of skin, connective tissues, bone, cerebrospinal fluid, white matter and gray matter regions with electrodes to couple the RF current. The characteristics of tissues, such as tissue thickness, length and complex permittivity used to generate the model were collected from the literature [7]. Resulting model was made of 1833 first order triangular elements that were defined by 1934 grid points (Figure 2).

The FEM model was used to calculate and/or plot conductivity, current density: electric field multiplied by conductivity, SAR/temperature rise: a function of duty cycle (T_C versus repetition time T_R), tissue conductivity σ and density ρ, current density J, effects of the B_1 field, and specific heat of tissues C_P. Temperature was estimated by:

$$\Delta T = \frac{SAR \cdot t}{C_P} \qquad (4)$$

where t is the exposure time.

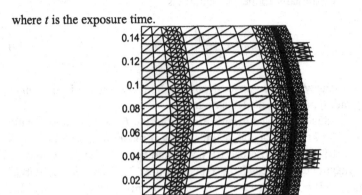

Figure 2. Mesh of first order triangular elements with electrodes to couple RF current (arbitrary units). Incorporated tissues are: skin, connective tissues, bone, cerebrospinal fluid, white matter and gray matter regions.

SAR was monitored through T_C, T_R and B_1 field. Figure 3 shows SAR as a function of T_C and B_1 field.

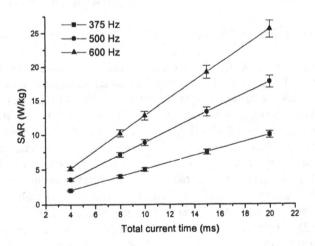

Figure 3. SAR as a function of T_C and B_1 field.

The simulation results were plotted for an 8 ms 375 Hz B_1 pulse and an 8 ms RF current pulse. Other parameters were: $T_R = 1000$ ms, $\rho \approx 1000$ kg/m³, $C_P = 3.47$ kJ/kg°C [6], and $f_0 = 100$ MHz. Figure 4a shows FEM mesh of first order triangular elements with the corresponding regions of varying conductivity. Figure 4b shows the magnitude of RF current density with the vectors denoting the flow direction. Figure 4c shows the distribution of power deposition. Simulation results predict a peak SAR due to RF current and B_1 pulse of 2.25 W/kg (40 m°C/min, not accounting for bioheat dissipation) occurring at the electrode edge-tissue boundary.

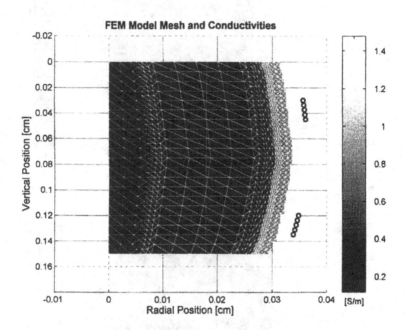

a.

488

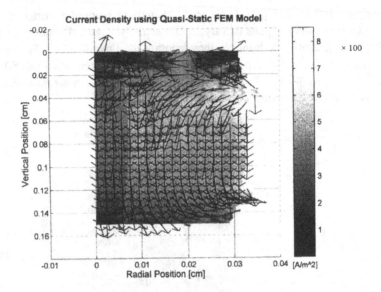

b.

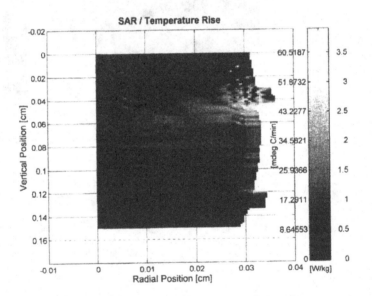

c.

Figure 4. a. FEM mesh of first order triangular elements with the corresponding regions of varying conductivity. b. The magnitude of RF current density with the vectors denoting the flow direction. c. The distribution of power deposition. Grid units are arbitrary.

2.2. RF-CDI IMAGES OF THE RAT BRAIN

In vivo RF-CDI experiment was performed with the same parameters as those to plot the SAR simulation.

Four male Wistar rats were anesthetized with chloralhydrate (300 mg/kg i.p.). After a midline incision, two screw electrodes were placed onto the skull over the frontal and parietal cortex (Fig. 5).

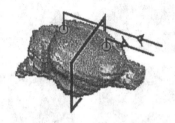

Figure 5. Electrode position and imaging slice orientation.

The electrodes were connected with a $\lambda/4$ coaxial cable to a matching network that assured the system resistivity of 50 Ω. Distance between the electrodes was 1.5 cm. The electric current generated by a RF amplifier was of *mA* order. Conductive gel was used to assure good contact between tissues and electrodes. RF-CDI was performed on a 100 MHz Bruker Biospec system. The imaging parameters were: repetition time T_R = 1000 ms, echo time T_E = 25 ms, field of view *FOV* = 6 cm, slice thickness 2 mm, matrix 256×256, 8 ms 375 Hz rotary echo B_1 pulse and duration of RF current pulse T_C = 8 ms.

Two millimeter thick coronal-cross sections through the rat brain (starting from the neck toward the eyes) were imaged by conventional MR. Contrast within the brain is clearly visible (Fig. 6 - left column). With the predicted peak SAR of ~ 2.5 W/kg, RF-CDI images show differences in current density through all the cross-sections between white matter (J was 160 ± 23 A/m^2 30 pixel sample), gray matter regions (J was 199 ± 14 A/m^2 23 pixel sample) and lateral ventricles (J was 350 ± 25 A/m^2 10 pixel sample) (Fig. 6 - right column).

490

MR images   **RF-CDI maps**

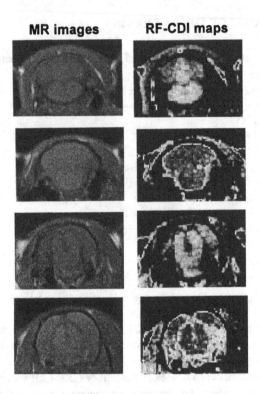

Figure 6. Two millimeter coronal-cross sections of the rat brain as seen by MRI and RF-CDI. The contrast of RF-CDI maps is proportional to the current density scale in figure 4b.

The differences are due to evoked excitation in the brain. The total current I through the images was approximately 60 mA over 72.5 cm^2. It was determined by integrating the current density J over the sample area without including edge pixels:

$$I = \int_S |\vec{J}| \cdot ds \qquad (5)$$

3. Discussion

RF-CDI is a relatively new MR technique that images tissue conductivity contrast. Compared to conventional MR, RF-CDI uses two additional sources of RF power to be absorbed and therefore must be evaluated in terms of proper parameter optimization to prevent excessive tissue heating and effects on the nervous system.

The SAR of a tissue is determined by the operating frequency, RF pulse amplitude and duration, conductivity and dielectric constant of the tissue and proximity and geometry of the transmitter. While MR is considered a relatively safe imaging technique, the manufacturers are required by the Food and Drug Administration (FDA) to monitor the RF power absorbed by the patient so that excessive patient heating does not occur both over the excited tissue volume and over the entire patient. MRI systems are designed to operate at or below the SAR recommended values, which are set to limit the patient heating to approximately 1°C or less [8]. In the United States and Canada, the suggested EM field exposure limits for occupational exposure is restricted to 0.4 W/kg for whole body average SAR and a spatial average of 8 W/kg when averaged over any gram of tissue for any region except at the body surface or limbs where the limit is 25 W/kg. For the eye, the SAR is not to exceed 0.4 W/kg.

Presented SAR analysis (Fig. 3 and Figs. 4a,b,c) is to provide an approximation of the power deposition due to the RF-CDI and to address additional issues concerning the clinical feasibility of RF-CDI. With the proper parameter optimization and the development of low SAR pulse sequences the SAR of RF-CDI can reach comparable values to those in conventional MR imaging and obtain good signal to noise ratio. Also, substantial sensitivity of the technique and image quality is achieved by using parameters that produce low SAR as is shown in Figure 6.

Our RF-CDI images clearly show anatomical structures in the brain. For example gray and white matter and ventricles are clearly visible in Figure 6. In reference [5], bone muscle and fat have been distinguished in the human leg. Similar results in rabbit brain are reported in [9] but at very low frequencies. Conductivity contrast is, therefore, observable at RF frequencies as well.

4. Conclusions

Two main conclusions can be drawn from this study. Numerical SAR analysis can be used to predict peak SAR values and temperature rise for various RF-CDI imaging parameter settings. These predictions can then be verified and refined when the actual current densities have been measured by RF-CDI. Furthermore, the RF-CDI technique is capable of producing *in vivo* images with current density contrast that corresponds to anatomical structures.

It is widely assumed that the main biological effect of the application of a combination B_1 pulse and RF current to biological tissues is heating. Although this chapter only addresses power deposition considerations, it is evident that future applications of RF-CDI will require careful monitoring for any additional, non-thermal biological effects.

5. Acknowledgment

The authors thank Dr. M.L.G. Joy from the University of Toronto, Toronto, Ontario, Canada for stimulating discussions and for the critical reading of the manuscript.

492

6. References

1. Gandhi, O. (1990) *Biological Effects and Medical Applications of Electromagnetic Energy*. Prentice Hall, New Jersey.
2. Schiff, S., Jerger, K., Duong, D., Chang, T., Spano, M., and Ditto, W. (1994) Controlling chaos in the brain, *Nature* **370**, 615-620.
3. Scott, G.C., Joy, M.L.G., and Henkelman,R.M. (1995) Rotating frame current density imaging, *Magn. Reson. Med.* **33**, 355-369.
4. Scott, G.C., Joy, M.L.G., Armstrong, R.L., and Henkelman, R.M. (1995) Electromagnetic considerations for RF current density imaging, *IEEE Trans. Med. Imag.* **14**, 515-524.
5. Gerkis, A.N., Yoon, R., Joy, M.L.G., and Kwan, H. (1996) Conductivity contrast in biological media using RF current density imaging, *Proc., ISMRM, 4th Annual Meeting, New York*.
6. Persson, B.R. and Stahlberg, F. (1989) *Health and Safety of Clinical NMR Examinations*, Boca Raton, CRC Press Inc., Florida.
7. Bronzino, J.D. (1995) *The Biomedical Engineering*, CRC Press, Florida.
8. Brown, M.A. and Semelka, R.C. (1995) *MRI Basic Principles and Applications*, John Wiley & Sons, New York.
9. Joy, M.L.G, Lebedev, V.P., and Gatti, J. (1994) Current density in sections through rabbit brain. *Proc., SMRM, 2nd Annual Meeting, San Francisco*.

POSSIBLE EFFECTS OF ELECTROMAGNETIC FIELDS ON THE NERVOUS - ENDOCRINE - IMMUNE INTERACTIONS

D. MIKLAVČIČ, L. VODOVNIK, and T. KOTNIK
University of Ljubljana
Faculty of Electrical Engineering
Tržaška 25, SI-1000 Ljubljana
Slovenia

1. Introduction

In recent years, complex interactions between the nervous, the endocrine and the immune system have been thoroughly investigated resulting in a large database of knowledge. On the other hand, the effectiveness of electromagnetic stimulation of the neurons has also been widely demonstrated, and the techniques of such stimulation are well established. A combination of the two knowledges offers a possibility to artificially influence the nervous - endocrine - immune (NEI) interactions, in cases when the immunity of a person is impaired, or its glandular functions are disordered. There is a wealth of reports on the use of electromagnetic fields to treat various pathological conditions. The number of diagnostic procedures that rely on the use of electromagnetic fields is also increasing. Therapeutic and diagnostic approaches include: surgical knife, cancer treatment (both hyperthermia and sub-hyperthermal EMF treatment, alone or in combination with other antitumor treatments), chronic and acute wound healing, bone healing, magnetic resonance imaging and spectroscopy, catheter ablation for cardiac arrhythmia, pain treatment, thermotherapy of musculo-skeletal tissue in physical medicine and rehabilitation, edema treatment, etc [1-5]. A number of these therapeutic procedures rely on excessive local heating of the tissue by means of radio frequency electromagnetic field. Other therapeutic procedures however do not have fully established mechanisms of EMFs. It is also very interesting that various signals in terms of amplitudes, pulse shape and frequencies are successfully used. At the moment a large database of EMF interactions with biological systems on various levels are available. Yet, due to very complex interactions between different levels, it is difficult if not impossible to explain reported therapeutic effects. Additional problem is that dosimetric determinations are only available in electrosurgery, catheter ablation, cancer hyperthermia and MR imaging and spectroscopy.

This paper suggests a different approach in explanation of such therapies and offers at the same time the possibility of finding new effective therapies. After a general review of the known NEI interactions, we synthesize the presented knowledge into a model. Based

B.J. Klauenberg and D. Miklavcic (eds.), Radio Frequency Radiation Dosimetry, 493-499.
© 2000 *Kluwer Academic Publishers. Printed in the Netherlands.*

on this model, we then present a concept of modification of nervous - endocrine - immune interactions by means of electromagnetic stimulation of the autonomic nervous system. Finally, using this model, we attempt to explain the global effect of electromagnetic stimulation on wound healing, which has been observed on several occasions, and possible use of electrical stimulation for cancer treatment.

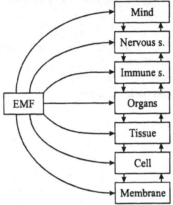

Figure 1. Levels of electromagnetic fields interaction with human body.

2. Nervous-Endocrine-Immune Interactions

Neuroendocrinoimmunology (NEI) is a fast-growing field of research that is gaining the interest of many scientists in the fields of biochemistry, physiology, pharmaceutics, and medicine. It deals with the complex interactions between the nervous, immune and endocrine systems. During the last decade, significant progress has been achieved in this field. Some of the advances have consolidated the established concepts in NEI, and some have added new ones.

The established concepts, often also referred to as "classical", are based on the following series of interactions:

1. the nervous system stimulates or suppresses glandular activity by means of neural signaling, which acts through the innervation of the glands;
2. the glands of the endocrine system respond with increased or decreased production of the hormones;
3. modified hormone production then affects the activity of other glands, or the immune system, which responds by modifying the production of its own regulatory molecules - the cytokines;
4. modified cytokine production affects the activity of the immune system by regulating proliferation and activation of lymphocytes, inflammatory response, phagocytosis, as well as production of erythrocytes.

The described concept is depicted in Figure 2.

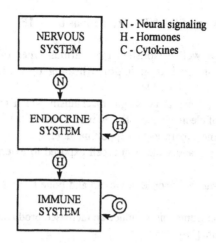

Figure 2. The "classical" pathways of nervous-endocrine-immune interactions.

A substantial evidence exists, however, that some of the roles presumed in the "classical" concept can also be reversed, making the scheme more complex. Namely, it has been known for a long time that some of the neurons of the hypothalamus function as endocrine glands, secreting hormones into the blood [6]. In addition, a series of recent studies suggests that the immune system is also able under some circumstances to produce hormones [7], and endocrine system to produce cytokines [8]. Also, it has been demonstrated that some cytokines can inhibit the activity of the hypothalamus [9]. These pathways are named "alternative" or "auxiliary".

Addition of these pathways yields a modified model, shown in Figure 3. The solid arrow-lines represent the "classical" pathways of the interactions, while the dashed arrow-lines designate the "alternative" pathways.

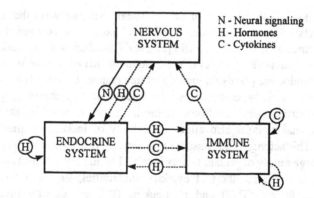

Figure 3. "Classical" (solid) and "alternative" (dashed) pathways of nervous-endocrine-immune interactions.

3. Modulation of the NEI Interactions by EMF

It is by now well established that various types of electric currents using different techniques of application improve healing of chronic wounds and reduce or even eradicate tumors [1,10,11].

However, there is no single mechanism which could explain and justify the large assortment of electrotherapeutic regimes. Some of the hypotheses that have been advanced and have some experimental support include:
- Transmembrane voltage in a cell exposed to external electric fields is modified [12-14].
- Host immune response in wound and bone healing and tumor treatment is stimulated [1,11,15].
- Electric currents might induce an increased production of peptide signaling molecules - growth factors [16].
- Electric currents might also enhance the processes of activation of immune cells [17].

All these phenomena are based on local application of electromagnetic fields. However, there are documented reports that wound healing occurs even when electrical stimuli are applied quite distantly from the wound site. Illis and co-workers observed that after spinal cord stimulation that was originally intended for pain suppression and improvement of locomotion, indolent ischaemic wounds on the leg began to heal [18]. Kaada applied electroacupuncture to a point on the hand and observed wound healing on the legs [19]. Regarding cancer, a large body of evidence is connecting the disease to neural and psychological factors [20]. These observations suggest an involvement of a global mechanism. In the cases of electroacupuncture and spinal cord stimulation, neural pathways have to be involved in healing of distant wounds. As these pathways alone do not suffice to produce such an effect, it is probably the whole NEI system that achieves the healing.

Figure 4 shows in detail some of the "classical" NEI pathways that could lead to systemic effects of neurostimulation. The model takes into account the previously published schemes by Husband [21], Old [22], and Chambers with co-workers [23].

Within the "classical" pathways, the autonomic nervous system influences the activity of the endocrine glands through direct innervation. Coordination and regulation of the glandular activity is based on the release of hypothalamic, pituitary and peripheral endocrine hormones. Some of the hormones released by thymus, adrenal cortex and adrenal medulla also affect the activity of leukocytes, mainly T and B lymphocytes. This activity is demonstrated by increase or decrease in the production of cytokines, a large family of protein factors involved in numerous aspects of the immune response. Some of the members of cytokine subfamilies, such as interferons (IFN), tumor necrosis factors (TNF) and interleukins (IL), are directly involved in the processes of tumor growth inhibition, including the destruction of tumor cells. Some of the cytokines reputedly also regulate the synthesis of growth factors in different cell types, such as fibroblasts and endothelial cells [22]. Growth factors then accelerate the pace of growth and division of tissue cells, thus promoting the repair of damaged tissues, including wound healing. It has also been reported that some cytokines, especially interleukin-1 (IL-1), influence the activity of the hypothalamus [4].

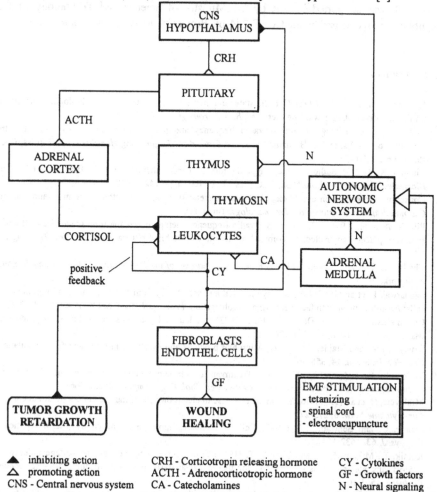

Figure 4. Some of the possible ways to affect wound healing and tumor growth via the immune system.

The model presented in Figure 4 shows possible pathways leading from electromagnetic stimulation to wound healing and tumor reduction. Neural activity, which is influenced by means of electromagnetic stimulation, then affects the immune system, which can finally induce accelerated healing of wounds, as well as decrease of tumor size, or at least the retardation of tumor growth.

Based on this example we can therefore conclude that electromagnetic stimulation of the nervous system could be an additional modality for systemic treatment of various pathologies.

498

4. Acknowledgement

This work was supported in part by the Ministry of Science and Technology of the Republic of Slovenia and by the COST-244-bis Project of the European Community.

5. References

1. Pakhomov, A.G. et al. (1998) Current state and implications of research on biological effects of millimeter waves: A review of the literature, *Bioelectromagnetics* **19**, 393-413.
2. Elson, E. (1995) Biologic effects of radio frequency and microwave fields: In vivo and in vitro experimental results, in J.D. Bronzino (ed.), *The Biomedical Engineering Handbook*, CRC Press, Boca Raton, pp.1417-1423.
3. Chou, C.K. (1995) Radiofrequency hyperthermia in cancer therapy, in J.D. Bronzino (ed.), *The Biomedical Engineering Handbook*, CRC Press, Boca Raton, pp.1424-1430.
4. Consensus conference on use of electrical and magnetic stimulation in orthopedics and traumatology, *Giornale Italiano di Ortopedia e Traumatologia* (in press).
5. Pilla, A.A. et al. (1996) Effect of pulsed radio frequency therapy on edema from grades I and II ankle sprains: a placebo controlled, randomized, multi-site double blind clinical study, *J. Athl. Training* **31**, S53.
6. Chadwick, D. J. and Marsh, J. (1992) *Functional Anatomy of the Hypothalamus*, John Wiley & Sons, New York.
7. Sabharwal, P. et al. (1992) Prolactin synthesized and secreted by human peripheral blood mononuclear cells: An autocrine growth factor for lymphoproliferation, *Proc. Natl. Acad. Sci. USA* **89**, 7713-7716.
8. Cunningham, E.T.Jr. and De Souza, E.B. (1993) Interleukin 1 receptors in the brain and endocrine tissues, *Immunol. Today* **14**, 171-176.
9. Krueger, J.M. and Majde, J.A. (1994) Microbial products and cytokines in sleep and fever regulation, *Crit. Rev. Immunol.* **14**, 355-362.
10. Vodovnik, L. and Karba, R. (1992) Treatment of chronic wounds by means of electric and electromagnetic fields. Part 1: Literature review, *Med. Biol. Eng. Comput.* **30**, 257-266.
11. Miklavčič, D. et al. (1997) Host's immune response in electrotherapy of murine tumors by direct current, *Eur. Cytokine Netw.* **8**, 275-279.
12. Grosse, C. and Schwan, H.P. (1992) Cellular membrane potentials induced by alternating fields, *Biophys. J.* **63**, 1632-1642.
13. Kotnik, T., Miklavčič, D., and Slivnik, T. (1998) Time course of transmembrane voltage induced by time-varying electric fields – a method for theoretical analysis and its application. *Bioelectrochem. Bioenerg.* **45**, 3-16.
14. Fear, E.C. and Stuchly, M.A. (1998) Biological cells with gap junctions in low frequency electric fields, *IEEE Trans. Biomed. Eng.* **45**, 856-866.
15. Spadaro, J.A. (1997) Mechanical and electrical interactions in bone remodelling, *Bioelectromagnetics* **18**, 193-202.
16. Sorrentino, V. (1989) Growth factors, growth inhibitors and cell cycle control - A review, *Anticancer Res.* **9**, 1925-1936.
17. Chandy, K.G. et al. (1985) Electroimmunology: The physiologic role of ion channels in the immune system, *J. Immunol.* **135**, 787-791.
18. Illis, L.S., Sedgwick, E.M., and Tallis, R.C. (1980) Spinal cord stimulation in multiple sclerosis: Clinical results, *J. Neurol. Neurosurg. Psych.* **43**, 1-14.
19. Kaada, B. (1983) Promoted healing of chronic ulceration by transcutaneous nerve stimulation (TNS), *VASA J. Vascular. Dis.* **12**, 262-269.
20. Sabbioni, M.E.E. (1993) Psychoneuroimmunological issues in psycho-oncology, *Cancer Invest.* **11**, 440-450.

21. Husband, A.J. (1993) Role of central nervous system and behaviour in the immune response, *Vaccine* **11**, 805-814.
22. Old, L.J. (1988) Tumor necrosis factor, *Sci. Am.* **258(3)**, 41-49.
23. Chambers, D.A., Cohen, R.L., and Perlman, R.L. (1993) Neuroimmune modulation: Signal transduction and catecholamines, *Neurochem. Int.* **22**, 95-100.

SUMMARY FOR SESSION I: STANDARDS AND APPLICATIONS

J. M. OSEPCHUK
Full Spectrum Consulting
248 Decon Haynes Road
Concord, Massachusetts 01742

In this session speakers from four different countries surveyed work on RF safety standards and their applications, one speaker reviewed the recent Guidelines of an international organization, ICNIRP, and one speaker reviewed current work on RF protective clothing. These presentations in sum illustrate the considerable differences in standards around the world. The interchange and communication among these authors, however, is certain to promote future developments towards international harmonization. We can point out a few bases for such advances which were aired at this session.

Osepchuk reviewed the standards work in the United States. This is dominated by the large consensus group under the IEEE SCC-28* which is very active in all areas with considerable emphasis on due process and thorough preparation for new standards within not only the SCC-28 community but most other groups in the world through the process of coordination. The IEEE is the largest technical professional society in the world and its supervision and support of SCC-28 is broad and continuing.

Bernhardt reviewed the 1998 Guidelines issued by ICNIRP**. This organization with WHO support draws on the expertise of experts from around the world, albeit in small elite committees. The Guidelines are built around "basic restrictions' on current density, SAR and power density as appropriate for the part of the spectrum under study. "Reference levels" then are developed in terms of more easily measured quantities such as fields and power density in the environment.

Hofmann reviewed work in Germany with reference to a practical example of a high-power radar station. In addition to governmental regulation of environmental fields, Hofmann makes reference to DIN VDE documents as well the pre-standard ENV 50166 developed by CENELEC. Although standards in the microwave range do not differ very much for applications described by Hofmann, he looks forward to more international guidelines, such as the ICNIRP guidelines. Though these guidelines will

* IEEE: Institute of Electrical and Electronics Engineers
SCC: Standards Coordination Committee
** ICNIRP: International Commission on Non-Ionizing Radiation Protection.

501

B.J. Klauenberg and D. Miklavcic (eds.), Radio Frequency Radiation Dosimetry, 501-502.
© 2000 *Kluwer Academic Publishers. Printed in the Netherlands.*

be the basis for a recommendation by the parliament of the European Union (EU), Hofmann suggests that documents such as the NATO STANAG 2345 may offer some advantages as a basis for more "obligatory" regulation. This may be connected to the mandatory nature of a "standards' under the IEEE rules which were a large input into the NATO STANAG development.

Both Thuroczy of Hungary, and Musil of the Czech Republic, make reference to the pre-standard ENV 50166 in addition to their own regulations. Thuroczy emphasizes the current focus on safety of wireless base stations and the use of hand-held phones. He reports on considerable survey of environmental fields around such base stations. Musil stresses the principle long held in Eastern Europe that exposure, especially for long duration should be based on a limit to "irradiation", or energy density, in analogy with ionizing radiation. All other guidelines in the world are based on limits for fields or power density along with the concept of an averaging time of some minutes. Musil does allow, however, that for short periods of time--e.g. less than an hour, limits on power density around the world are roughly in agreement. Since most occupational exposures are less than one hour per day, these observations perhaps are a harbinger of future agreement on occupational exposure limits. Agreement on environmental limits involves more complicated issues and is more distant.

Musil, Thuroczy and Hofmann all mentioned concern about hazardous RFI, e.g. to medical devices, and the need to consider special exposure limits for protection against this hazard. They do not, however, emphasize the primary need, however, to impose susceptibility limits on the device suffering the interference.

Again as in many of these international meetings, presentations from various countries and interactions among the representatives of these countries, contributes to interchange of ideas, an airing of possible areas of agreement and another step towards international harmonization of radio frequency safety standards.

RADIO FREQUENCY EXPOSURE STANDARDS IN THE UNITED STATES

J. M. OSEPCHUK
Full Spectrum Consulting
248 Deacon Haynes Road
Concord, Massachusetts 01742

The past, present and future of radio frequency (RF) exposure standards in the United States is surveyed. The central role of the broad consensus process through the Standards Coordinating Committee 28 of the IEEE (Institute of Electrical and Electronics Engineers) is stressed. This process also will play an important role in the trend toward international harmonization of such standards.

1. Introduction

This subject is only one part of the grand fabric of rules for the safe use of electromagnetic energy or non-ionizing radiation. It is useful to review some definitions that will help relate the present subject to an integrated overview of the larger scope of such rules.

1.1. FREQUENCY SPECTRUM

The frequency spectrum associated with non-ionizing radiation ranges from 0 (static fields) through the ultraviolet. Radio frequency is sometimes used to characterize the whole spectrum from 0 to 3000 GHz. Here, however, we will restrict the RF range to 3 kHz through 300 GHz.

1.2. CLASSIFICATION OF SAFETY RULES

Safety rules are characterized by varying degrees of authority or importance. We recognize the classification scheme used by the IEEE, viz.: (1) Standards—mandatory rules, using words like *shall*; (2) Recommended Practice—suggested preferences using words like *should*; and (3) Guides—listing acceptable alternatives using words like *may*.

The scope of safety is broad including not only the need to avoid hazardous exposure of people to RF energy but also the need to avoid hazardous exposure of flammable materials, electro-explosive devices and electronic systems—particularly when they are part of critical systems such as medical devices either in hospitals or in

503

B.J. Klauenberg and D. Miklavcic (eds.), Radio Frequency Radiation Dosimetry, 503-511.
© 2000 *Kluwer Academic Publishers. Printed in the Netherlands.*

the field and possibly implanted into patients. The scheme of safety rules accordingly also obeys the following classification:

(1) Exposure standards (or other-level rule): rules for people to follow in order to avoid potentially hazardous exposures. These rules involve exposure duration, averaging time and maximum permissible exposure (MPE) levels (power or fields.)

(2) Product performance standards: performance criteria which will insure that the users of products will experience exposures well below acceptable limits set forth in exposure standards.

(3) Environmental standards: limits on environmental fields in the vicinity of transmitting sites. Exposure time is not a factor here except for acknowledgment of relaxation of limits for transient exposure situations. Relevant phenomena include interference to all electronics systems, hazards to fuels and EEDs, effects on all biota (flora and fauna) as well as societal considerations including the accommodation of involuntary nature of environmental exposures.

(4) Susceptibility (or immunity) standards: criteria imposed on electronic systems that are potential "victims" of interference, whether hazardous or not. Clearly such rules must exist if there is to be a balance of responsibilities between manufacturers of radiating systems and electronics technology, in general, that lead to true *electromagnetic compatibility* in modern society which is increasingly dependent on electronics.

2. History

A capsule review of the history of EM safety rules in the United States is important if we are to understand the key roles of the military and the voluntary standards systems in the United States in radio frequency exposure standards.

In the United States the military has been a major and key supporter of research and development since the intimate relationship between the military and the scientific community was formed in World War II [1]. Since then all quarters of the science and technology communities have enjoyed military funding, even though since the end of the cold war the funding has been diminished somewhat. The military, in particular, would fund longer-range and other tasks which yield little profit potential for private industry. One such task is to investigate the potential hazards of electromagnetic energy and to develop standards for protection against such hazards.

There is a striking dichotomy, however, in the history of standards development depending on whether the EM energy is developed above 300 GHz or below 300 GHz In both cases the military has played a leading role. Above 300 GHz, the principal source of EM energy that caused concern was the laser. There was no Electrophobia, however. Thus a standards committee, accredited by ANSI (the American National Standards Institute) has since the 1960s been the world leader for the development of laser safety standards. The secretariat for this committee has always been the Laser Institute of America and although military personnel play leading roles in this committee thorough participation and support from Federal health and safety agencies has always been evident. Included have been the FDA (Food and Drug Administration), OSHA (Occupational Safety and Health Administration) and NIOSH (National Institute of Occupational Safety and Health) and private safety and professional societies. The standard [2] for safe use of lasers is quite detailed and complex but it enjoys worldwide acceptance.

Below 300 GHz, however, there is a long history of Electrophobia [3] that dates back to fears in World War II that the newly-developed ability to form highly collimated beams of microwave energy presaged the approach of the *death ray*. Although for many years RF standards were developed by an accredited standards committee, analogous to that for lasers, events peculiar to the lower frequencies and associated technologies led in the late 1980s to a conversion to a Standards Coordinating Committee under the IEEE. Litigation fears associated with Electrophobia ostensibly were one reason for the conversion in which all volunteers in IEEE committees are indemnified by the IEEE. Another reason was to enhance the credibility of the standards process by giving a professional society rather than military agencies the key role of Secretariat. Coupled with the stringent oversight by the IEEE Standard Board the development of standards for RF in the United States is on a high level of credibility as regards due process and input from all stakeholders. The extension of these standards down to zero frequency has been delayed by a the-year period of cancer scare associated with low frequencies below 3 kHz. This fear now seems to be dissipating and progress towards standards at low frequencies will now accelerate. The past and present, however, has been dominated by the C95 series of standards for RF safe exposure. Over the last 30 years there have been attempts by the EPA, NIOSH, OSHA to develop federal standards or guidance on safe RF exposure, but all have failed. Federal agencies, in the main have relied upon the C95 series of standards. An important factor in this practice has been and is the existence of a Federal policy, OMB A-119, mandating support of and participation by Federal agencies in the voluntary standards-setting process [4]. A notable reference on the major role of the military and the C95 standards for RF safety is a report by the Inspector General of the Department of Defense [5].

3. Status

In his recent review [6] of RF safety standards-setting in the U.S., Ronald Petersen described the dominance of the IEEE SCC28 community and its procedures that are a key to its being the platform for the broadest (by far) consensus on safe exposure standards in the use of electromagnetic energy at frequencies below 300 GHz. In our review here, we will restrict ourselves to the highlights of the existing standards and activities for revised and new standards (or lower-level rules).

3.1. IEEE

The current IEEE C95 standard [7] was reaffirmed by the IEEE and ANSI in 1997. Its highlights are depicted in the capsule guide shown in Figure 1. Key features are the minimum MPE in the resonance (human body resonance) frequency range, 30 to 300 MHz, the agreement with laser standard, ANSI Z136, at 300 GHz in both MPE and averaging time, the relaxation of MPEs for magnetic field at low frequency, and two tiers for controlled and uncontrolled environments in the resonance range. Other features depicted are limits on induced and contact (grasping) currents and average and spatial peak values of SAR (specific absorption rate.)

A Supplement to C95.1-1991 is in the late stages of approval. It will introduce 6 minute time averaging of induced and contact currents above 0.1 MHz, exclusion from required current measurements in sufficiently weak fields, change of minimum measurement distance to 5 cm and a few revised definitions to better express the intent of the drafters of the C95.1 standard.

Work towards the revision of C95.1 has included an extensive and computerized literature review. This process is important in view of the fact that much of the literature is unacceptable or useful for serious standards setting. A documented consensus judgment on such papers is necessary. Work is underway to tighten up (lower) averaging times at high frequencies based on thermal modeling and at low frequencies based on minimal averaging time consistent with the nature of electrostimulation which is the dominant physical process in the body upon exposure at low frequencies. Work is also underway to review existing limits on temporal peak MPEs and improve discussions on whole-body spatially-averaged MPEs and relaxed partial-body exposures. An existing caveat on inapplicability to exposure of eyes and testes will become unnecessary after averaging times are tightened up at high frequencies. Also in the coming revision there will be a review of the desirability of the two tiers. One possible outcome is to revert to one tier in an exposure standards for all people while shifting the question on what environmental levels should be to a new forum--i.e. a new Standards Coordinating Committee on environmental standards.

Also in late stages of approval is an expanded C95.2 Standard for warning symbols (icons). This will expand on the use of the well-known C95 symbol as well as introduce a symbol to discourage contacting metal surfaces that could result in undesirable contact currents. New Recommended Practice documents are in early stages of balloting for measurement techniques (for 3 kHz to 300 GHz) and safe distance from transmitters for safe use of EEDs. Finally a revitalized Subcommittee is beginning work on an IEEE standard for frequencies between 0 and 3 kHz.

IEEE SCC28 has been engaged in considerable expansion of its international liaison activities. Almost 10% of its membership is now from outside the U. S. and active communications now exist between SCC28 and many international bodies including the WHO, ICNIRP and the IEC. Note that IEEE standards are considered international. The IEEE has membership from all of the world and is the world's largest technical professional society with over 300,000 members.

3.2. NCRP

The organization and activities of the National Council on Radiation Protection and Measurements has been well reviewed recently [8]. Documents from the NCRP are produced by small (e.g. 8 - 10 people) expert committees in a process that is kept confidential until final publication--in contradistinction to the open process that characterizes the process in the IEEE. The documents must be approved by a 90-member council, most of which is not expert in non-ionizing radiation. The NCRP, however, does invite critical review of documents by outside collaborating organizations of which the IEEE is one.

The latest relevant document [9] presented "guidelines" for MPEs that generally agree with the latest IEEE C95 standard in SAR limits and MPEs in the resonance range. The NCRP document covers a smaller frequency range, 0.3 MHz to 100 GHz, and presents more restrictive MPEs for magnetic fields at low frequencies and power densities at the high end. Thus at 0.3 MHz, NCRP recommends an MPE of 1.63 A/m for both occupational and general population groups but at 100 GHz an MPE of 5 mW/cm^2 is presented for occupational exposure and 1 mW/cm^2 for the general population. Averaging times, however, are not reduced as in C05.1 but remain at 6 minutes for occupational exposure and 30 minutes for exposure of the general public. The NCRP guidelines also recommend reduction of occupational MPEs if amplitude modulation index is over 50%. NCRP is alone among organizations throughout the world in making this recommendation.

An NCRP committee, chaired by Prof. Om Gandhi, was created about four years ago to review the possibility and relevance of modulation effects to standards setting. This committee has not completed its work. Meanwhile, a committee of 8, chaired by Prof. James Lin, was created to update the 1986 literature review and then recommend possible changes to the 1986 criteria. The target date for completion of this work--by the committee is 1999.

As pointed out by Adair [10] existing NCRP publications on measurements and practical exposure assessment comprise a valuable asset to the practicing professional in this field

3.3. FEDERAL COMMUNICATIONS COMMISSION (FCC)

During 1996-1997, the FCC adopted and published its new set of Rules for certification by its licensees that their transmitters meet criteria for safe exposure of occupational and general population groups. Although the initial intent in 1993 was for the adoption of IEEE C95.1-1991, after considerable review by other agencies, particularly the EPA, the FCC adopted Rules that agree with IEEE on basic SAR limits and current limits but follow NCRP on MPEs and frequency range. The FCC rejected, however, the caveat on modulation which is in the NCRP guidelines. The FCC Rules thus form its "pick and choose" mixture of ideas and concepts from the two sources, NCRP and IEEE. There are a large number of documents related to the FCC Rules that are admirably catalogued and reproduced in part by the ARRL (the American Radio Relay League) [11].

Although the FCC Rules contain specific values taken from either NCRP or IEEE, they do not address the subtle question of definitions which had been treated only partially in either NCRP or IEEE and remain an area of continuing improvement within IEEE today. Thus questions of spatial averaging, partial-body exposure and peak limitations are not addressed adequately.

The FCC Rules, however, are a key reference in the U. S. for assessment of safety of the wireless technologies which enjoy explosive growth. As such it has been a lightning rod for attacks by activists who oppose siting of new transmitting towers for wireless services. A collection of lawsuits have been filed in Federal [12] in which the plaintiffs allege that the FCC Rules are inadequate for protection of the general public. A key part of their allegations is that the FCC neglected consideration of "non-thermal" effects of RF radiation, presumed by the plaintiffs to be harmful. This viewpoint is a challenge to the view by most authorities that scientific consensus bodies must judge scientific literature before it can be applied to judgments on safety as in standards.

3.4. AMERICAN CONFERENCE OF GOVERNMENTAL AND INDUSTRIAL HYGIENISTS

This organization, though headquartered in the U. S., has stated worldwide scope in its activities. Thus the 1997 Physical Agents TLV Committee included three non-U. S. members in its complement of 12. The most recent publication [13] of threshold limit values (TLVs) for electromagnetic energy generally is in agreement with the laser standards, ANSI Z136-1993 and the RF standard, IEEE-C95.1-1991. In particular the MPE for far infrared, above 300 GHz, is 100 mW/cm^2 for small areas (<100 cm^2) as averaged over 10 seconds and 10 mW/cm^2 for large areas (>1000 cm^2). This latter MPE is in agreement with IEEE C95.1-1991. The publications of ACGIH stress that they are designed for use by professional industrial hygienists.

3.5. OCCUPATIONAL SAFETY AND HEALTH ADMINISTRATION (OSHA)

Personnel from OSHA have a long history of participation in and contributions to the work of the C95 committees. Thus, in general, the positions of OSHA with respect to compliance questions on RF exposure have generally been in agreement with C95 standards. From a legal viewpoint, however, there always has been some ambiguity which has not bee resolved. Thus, even today there are on the books of OSHA regulations the C95 standards for safe exposure and warning symbols generated in the 1960s. Courts have ruled, however, that the 1966 exposure standards is unenforceable because its language is not mandatory (it used the word *should* and not *shall*). OSHA has never replaced these regulations with updated versions despite a few sporadic moves in this direction. It has agreed that use of updated C95 standards, including that for warning symbols is generally acceptable in a responsible RF safety program in the workplace.

The full position of OSHA, from a technical viewpoint, can be gained from an examination of its well-stocked web site: www.osha.gov. One can download five pages of extensive reference material on radiofrequency radiation. Reference is made to the guidelines from the IEEE, NCRP, the FCC, ACGIH and many publications from governmental and private sources in this subject area. Among RF Safety Programs, there are listed not only an OSHA document (developed in 1995 by Robert Curtis) but comparable information from the FCC, Motorola and NATO.

In terms of official OSHA standards there are listed, in addition to the unenforceable 10 mW/cm^2 standard, two regulations developed by OSHA which apply the 10 mW/cm^2 standard to telecommunication workers and to tower climbers.

3.6. FOOD AND DRUG ADMINISTRATION (FDA)

The FDA has had a key role in this field starting with the passage of the 1986 Radiation Control for Health and Safety Act of 1968 (P. L. 90-602). The performance standard for microwave ovens , which was developed by the FDA, has long since become universally adopted throughout the world--an emission limit of 5 mW/cm^2 at 5 cm in the field. In about 1980, FDA was about to promulgate an emission standard for diathermy equipment. This was withdrawn by the FDA and since then the FDA has chosen not to promulgate performance standards on products emitting RF energy. It does support, however, the work of the recently created committee , SCC34, on such standards. The FDA is also a significant contributor to the work of SCC28.

In recent years, the FDA, although involved in some research and oversight on questions of safety from RF exposure, it has emphasized the need for new measures to control hazardous radio-frequency interference (RFI), especially when medical devices are involved. [14].

3.7. DEPARTMENT OF DEFENSE (DOD)

The DoD in the U. S. has adopted the C95.1 standard in its regulations and directives as well as having worked to make it the model for the recently adopted STANAG 2345. [15]. As pointed out by the Inspector General [5], the U. S. military make major contributions to voluntary standards-setting, and in particular to IEEE SCC28, without dominating the committees. Notable contributions have included the realization of NATO Workshops, like the present one and specialized workshops like the one bringing the laser and microwave communities together in 1997 [16]. The U. S. military now probably carry out the major portion of research on RF bioeffects at the tri-service facilities located at Brooks Air Force base in San Antonio, Texas.

DoD is responsible for the valuable resource the *Dosimetry Handbook* which is one subject being reviewed at this meeting.

Leaders in the military, such as Dr. Michael R. Murphy, have played a leading role in spreading the outreach of the IEEE SCC28 community worldwide through various liaison arrangements and activities.

3.8. OTHER AGENCIES

Other federal agencies have shown little activity in the area of RF safety standards. NIOSH did conduct a workshop on workplace hazards involving non-ionizing radiation in March 1998. Recommendations emanating from this conference seem to focus on "needed research" and not standards. [17].

Several states and other jurisdictions have adopted environmental and occupational exposure limits generally based on C95.1.

4. Future Trends

It is clear that in the U. S., although not monolithic, there is one large connected community concerned with RF standards, dominated by the IEEE SCC28 community. Indeed, various experts participate in many of the organizations in parallel. The work on revising C95.1-1991 along with the substantial literature review denotes a coming milestone of accomplishment. Various improvements are expected in the application of averaging times and the formulas used to make transitions from the resonance-frequency range, dominated by SAR considerations to low frequencies where electrostimulation dominates and the laser range above 300 GHz where surface phenomena dominate and the potential hazard is associated with the possibility of well-focused beams of energy. Advances in how to relax limits in case of partial-body exposure are critically needed. Finally more precise definitions and more thorough rationales in the standards documents will insure easier application of standards in practice and less need for recurring requests for "interpretations" of standards because of ambiguous wording.

The DoD [5] and other scientists have shown dissatisfaction with the two-tier concept which now pervades standards worldwide. The history [18] of this concept makes it clear that it was introduced (first in Massachusetts) into an ostensible exposure regulation in order to serve as an environmental limit. One solution for this problem is to make a logical separation of standards for safe exposure and the environment. This will require a new committee (SCC) for development of environmental standards.

Thanks to research supported by the military and others the database, particularly on human exposure, is increasing dramatically [10]. The basis for "basic restrictions"—i.e. parameters associated with bioeffect or hazard thresholds is becoming more precise. The art of adding safety factors is not a scientific exercise but must incorporate the input of all affected (stakeholders) to insure that standards are rational as well as science-based. Society is not well served when excessive safety factors are arbitrarily introduced into standards.

In the U. S. there is a trend towards stronger and more consistent support of the voluntary standards activities (SCC28 and SCC34) through Congressional oversight of Federal agency policies and greater financial support from all sectors of society for the voluntary standards activities.

In the end, a vibrant and broad standards community in the U.S. is playing an increasing role towards international harmonization of RF safety standards, both through greater non-U.S. participation in U.S. based organizations but also greater liaison between U. S. based organizations and all the world's organizations which develop such standards and guidelines.

5. References

1. Bush, V. (1945) *Science the Endless Frontier: A Report to the President*, U. S. Govt. Printing Office, Washington, D. C.
2. ANSI (1993) *Z136.1-1993 American National Standard for the Safe Use of Lasers*, American National Standards Institute, New York.
3. Osepchuk, J. M. (1996) COMAR after 25 years: still a challenge, *IEEE Engineering in Medicine and biology, 15(3)*, 120.
4. OMB (1993) *Revision of OMB Circular No. A-119, Notice of Implementation, Federal Register, Vol. 58, No. 205, 58 FR 57643, Tuesday, October 26, 1993*, Executive Office of the President, Office Of Management and Budget, Washington, D. C.
5. DoD (1994) *Review of the Biological Effects of Radiofrequency Radiation Exposure*, Inspector General, Department of Defense, Washington , D. C.
6. Petersen, R. C. (1998) Radiofrequency safety standards-setting in the United States, in *Electricity and Magnetism in Biology and Medicine*, F. Bersani, ed., Plenum Publishing Co., Ltd., London.
7. IEEE (1991) *IEEE Standard for Safety Levels with Respect to Human Exposure to Radio frequency Electromagnetic Fields, 3 kHz to 300 GHz*, Institute of Electrical and Electronics Engineers, New York.
8. Adair, E. R. (1998A) The NCRP standards-setting process: advantages and disadvantages for international harmonization, *International Seminar on Human Health and Electromagnetic Fields (EMF)--Global Need for Harmonization of EMF Standards*, Ljubljana, Slovenia, October 9, 1998.
9. NCRP (1986) *Report No. 86, Biological Effects and Exposure Criteria for Radiofrequency Electromagnetic Fields*, National Council on Radiation Protection and Measurement Bethesda, Maryland.
10. Adair, E. R., Kelleher, S. A. , Berglund, L. G. , and Mack, G. W. (1998B) Physiological and perceptual responses of human volunteers , in *Electricity and Magnetism in Biology and Medicine*, F. Bersani, ed. , Plenum Publishing Co., Ltd., London.
11. Hare, E. (1998) *RF Exposure and You*, American Radio Relay League, Newington, Connecticut.
12. Cellular Phone Taskforce et al. (1997) *Case No. 97-4328*, Cellular Phone Taskforce et al. vs. Federal Communications Commission and the United States of America, U. S. Court of Appeals, Second Circuit, New York.
13. ACGIH (1998) *1998 TLVs and BEIS*, Threshold Limit Values for Chemical Substances and Physical Agents, Biological Exposure Indices, American Conference of Governmental and Industrial Hygienists, Cincinnati, Ohio.
14. Barron, J. (1996) Electromagnetic compatibility of radiation- emitting products with medical devices, *Proceedings of the 31st Microwave Power Symposium*, International Microwave Power Institute, Manassas, Virginia.
15. NATO STANAG 2345 (1997) Control and Evaluation of Personnel Exposure to Radio Frequency Fields - 3 kHz to 300 GHz; North Atlantic Treaty Organization; Brussels, Belgium.
16. USAF (1997) Digest (2 volumes), *Infrared Lasers & Millimeter Waves Workshop: The Links Between Microwaves and Laser Optics*, Brooks AFB, Texas, Jan. 21 -22, 1997.
17. Lotz (1998)
18. Osepchuk, J. M. (1995) Impact of public concerns about low-level electromagnetic fields on interpretation of electromagnetic fields/radiofrequency database, in B. J. Klauenberg (ed.) *Radiofrequency Radiation Standards*.
19. M. Grandolfo and D. Erwin, (ed.), Plenum Press, New York

THE NEW ICNIRP GUIDELINES: CRITERIA, RESTRICTIONS, AND DOSIMETRIC NEEDS

J. H. BERNHARDT
ICNIRP Chairman
Institute for Radiation Hygiene
D- 85764 Munich-Oberschleissheim, Germany

Abstract

The International Commission on Non-ionizing Radiation Protection (ICNIRP) is the independent, non-governmental, scientific organization, comprising all essential scientific disciplines, which is qualified to assess health effects of exposure to electromagnetic fields together with the WHO. ICNIRP uses the results of this assessment to draft health based exposure guidelines. This assessment is free from vested interest. The paper describes the criteria used for this assessment, the classes of guidance (basic restrictions and reference levels) and the applied dosimetric models. Since the reference levels are given for the condition of maximum coupling of the fields to the exposed individual, they provide maximum protection. For a less conservative assessment, basic restrictions on the induced current density, whole-body average and localized SAR should be used. The determination of these quantities is a complicated function of various exposure parameters, the biophysical properties of the exposed object, design of the specific equipment and the field coupling to the body. Mainly due to these reasons the ICNIRP guidelines do not address product performance standards or guidance concerning computational methods or measurements techniques. It is the opinion of the ICNIRP that technical standards bodies should perform the provision of technical advice concerning the practical implementation of standards. There is a clear need to analyze systematically the general classes of equipment in order to decide whether the application of the equipment meet the basic restrictions and to determine whether additional protective measurements are necessary.

1. Introduction

The development of international EMF standards requires a critical in-depth evaluation of the established scientific literature. Using established scientific literature allows exposure restrictions to be determined with a higher degree of confidence about their protective value. Literature for review should have been published in scientific peer

B.J. Klauenberg and D. Miklavcic (eds.), Radio Frequency Radiation Dosimetry, 513-521.
© *2000 Kluwer Academic Publishers. Printed in the Netherlands.*

review journals. While peer review adds confidence in the study results, for health risk assessment, additional review is necessary to evaluate study design, conduct and analysis of each report, and to compare them with the results of other studies. Detailed guidelines on the conduct of high quality laboratory research can be found in [5, 12, 14, 16]. Essential points for the conduct of high quality are:

- Experimental techniques, methods, and conditions should be as completely objective as possible and based on biological systems appropriate to the endpoints studied.
- Environmental conditions should be measured and recorded periodically (i.e., temperature, humidity, light, vibration, sound, background EMF's). EMF's should be fully characterized and remeasured periodically (i.e., time-varying and static field components, field directions and polarization, modulation, waveform, pulse shape and timing, frequency spectrum, harmonics and transients from switching on and off, etc.).
- All data analyses should be fully and completely objective, with no relevant data deleted from consideration, and with uniform use of analytical methods.
- Published descriptions of methods should be given in sufficient detail that a critical reader would be convinced that all reasonable precautions were taken and that other researchers can reproduce them.
- Results should demonstrate an effect of the relevant variable at a high level of statistical significance using appropriate tests.
- Results should be quantifiable and susceptible to confirmation by independent researchers. Preferably, the experiments should be repeated and the data confirmed independently.
- Results should be viewed with respect to previously accepted scientific principles before ascribing them new ones. Research findings pointing to previously unidentified relationships should be carefully evaluated and appropriate additional studies should be conducted before the findings are further considered.

Criteria that have been widely accepted when evaluating epidemiological studies have been recently summarized by Repacholi and Cardis [15]. Under these criteria, strength and consistency of the association between EMF exposure and biological effects, evidence of a dose-response relationship, evidence provided by laboratory studies, and plausibility that biological systems exposed to EMF fields manifest biological effects, are all examined.

For exposure to any physical agent, there exists a range of biological effects that needs to be assessed for determining health effects. From the assessment of health hazards from exposure to NIR, a dose response relationship exists. At very high doses clearly adverse physiological effects occur, and a threshold can generally be defined for producing these effects. A consensus of scientific opinion frequently exists which enables one to delineate the range of exposures where no adverse effects occur or appears possible. Controversy sometimes occurs however with regard to the grey area in between these two dose ranges. Once threshold exposures are established, safety factors may be incorporated into the limits depending on the degree of certainty about the threshold dose. Safety factors in health protection standards do not guarantee safety, but represent an attempt to compensate for unknown and uncertainties.

An important part of the rationale for any exposure standard is the definition of the population to be protected. Occupational health standards are aimed at protecting healthy adults exposed as a necessary part of their work, who are aware of the occupational risk and who are likely to be subject to medical surveillance.

General population standards must be based on broader considerations, including health status, special sensitivities, possible effects on the course of various diseases, as well as limitations in adaptation to environmental conditions and responses to any kind of stress in old age. In most cases these considerations will have been insufficiently explored, so standards for the general population must involve adequate safety factors.

2. ICNIRP Guidelines for Limiting Exposure to Time Varying Electric, Magnetic, and Electromagnetic Fields (up to 300 GHz)

International recommendations of health based guidance on limiting exposure needs an assessment of possible adverse health effects on the basis of established scientific and medical knowledge. This assessment should be free from vested interest.

ICNIRP as an independent scientific body, comprising all essential scientific disciplines, like dosimetry, biophysics, in vitro and in vivo biology, medicine and epidemiology, is qualified to carry out this task. ICNIRP works with the World Health Organization (WHO) to assess health effects of exposure to NIR and uses the results of this assessment to draft health-based exposure guidelines.

ICNIRP is the formally recognized non-governmental organization in NIR protection for the WHO, the International Labor Organization (ILO) and the European Union (EU), and maintains a close liaison and working relationship with other international bodies engaged in the field of non-ionizing radiation protection. A review of general activities of the ICNIRP has been published recently [2].

Recently, the ICNIRP adopted guidelines for limiting EMF exposure up to 300 GHz [7]. The guidelines are a revision and replacement of the IRPA Guidelines issued in 1988 and 1990. The new guidelines were developed since 1994 by the Commission, its Standing Committees and external experts. In early 1997 the draft was sent to IRPA and international experts for comments and an advanced draft was prepared taking into consideration of about 50 pages of comments. The final draft, after having included editorial and additional comments, was sent to the Health Physics Journal in September 1997. Until publication in April 1998 the Commission received additional comments and questions. In response of this, the Commission issued 17 questions and answers, which were published in Health Physics in summer 1998.

Biological effects reported as resulting from exposure to static and ELF fields and electromagnetic fields have been reviewed by UNEP/WHO/IRPA [18, 19]. These publications and a number of others [1, 10, 17] provide the scientific rationale for the

ICNIRP guidelines. Static magnetic fields are covered in the ICNIRP Guidelines issued in 1994 [6].

Two classes of guidance are presented: basic restrictions and reference levels.

Basic Restrictions. Restrictions on exposure to time varying electric, magnetic, and electromagnetic fields that are based directly on established health effects are termed "basic restrictions". Depending upon the frequency of the field, the physical quantities used to specify these restrictions are current density, specific energy absorption rate (SAR), and power density. Only power density in air, outside the body, can be readily measured in exposed individuals.

Different scientific bases were used in the development of basic exposure restrictions for various frequency ranges:

- Between 1 Hz and 10 MHz, basic restrictions are provided on current density to prevent effects on nervous system functions;
- Between 100 kHz and 10 GHz, basic restrictions on SAR are provided to prevent whole-body heat stress and excessive localized tissue heating; in the 100 kHz-10 MHz range, restrictions are provided on both current density and SAR;
- Between 10 and 300 GHz, basic restrictions are provided on power density to prevent excessive heating in tissue at or near the body surface.

Reference Levels. These levels are provided for practical exposure assessment purposes to determine whether the basic restrictions are likely to be exceeded. Some reference levels are derived from relevant basic restrictions using measurement and/or computational techniques, and some address perception and adverse indirect effects of exposure to EMF. The derived quantities are electric field strength, magnetic field strength, magnetic flux density, power density and currents flowing through the limbs. Quantities that address perception and other indirect effects are contact current and, for pulsed fields, specific energy absorption. In any particular exposure situation, measured or calculated values of any of these quantities can be compared with the appropriate reference level. Compliance with the reference level will ensure compliance with the relevant basic restriction. If the measured or calculated value exceeds the reference level, it does not necessarily follow that the basic restriction will be exceeded. However, whenever a reference level is exceeded it is necessary to test compliance with the relevant basic restriction and to determine whether additional protective measures are necessary.

The restrictions are different for the occupationally exposed population and the general public. The frequency dependence of the reference levels is consistent with data on both biological effects and coupling of the fields.

The safety factors used in the ICNIRP guidelines have a wide range of values from ~2 to >10 (depending upon the extent to which thresholds for health effects are known for direct and indirect field effects at various frequencies). Public guidelines include additional safety factors of 2 to 5 relative to occupational guidelines (depending upon the frequency and the relevant dosimetric parameters). As a consequence of having insufficient information on the biological and human health effects of EMF exposure, a rigorous basis does not exist for establishing safety factors. In the new ICNIRP guidelines, the safety factors vary considerably for fields in different frequency ranges and for different biological end points such as effects of direct versus indirect field interactions. In general, threshold field levels for indirect effects (e.g., responses to contact currents) are better defined than for direct effects, and hence less conservative safety factors are required.

The guidelines do not address product performance standards, or guidance precluding interference with medical devices, nor does the document deal with the measurement techniques for determining the reference values, or protective measures.

3. Dosimetric Considerations

3.1. APPLIED DOSIMETRIC MODELS IN THE ICNIRP GUIDELINES

Where appropriate, the reference levels are obtained from the basic restrictions by mathematical modeling and by extrapolation from the results of laboratory investigations at specific frequencies. They are given for the condition of maximum coupling of the field to the exposed individual, thereby providing maximum protection. The reference levels are intended to be spatially averaged values over the entire body of the exposed individual, but with the important proviso that the basic restrictions on localized exposure are not exceeded.

Reference levels for exposure of the general public have been obtained from those for occupational exposure by using various factors over the entire frequency range. These factors have been chosen on the basis of effects that are recognized as specific and relevant for the various frequency ranges. Generally speaking, the factors follow the basic restrictions over the entire frequency range, and their values correspond to the mathematical relation between the quantities of the basic restrictions and the derived levels.

3.1.1. *Low Frequency Fields*
Several computational and measurement methods have been developed for deriving field-strength reference levels from the basic restrictions. The simplifications that have been used to date did not account for phenomena such as the inhomogeneous distribution and anisotropy of the electrical conductivity and other tissue factors of importance for these calculations. The frequency dependence of the reference field levels is consistent with data on both biological effects and coupling of the field.

Magnetic field models assume that the body has a homogeneous and isotropic conductivity and apply simple circular conductive loop models to estimate induced currents in different organs and body regions, e.g., the head. More complex models use an ellipsoidal model to represent the trunk or the whole body for estimating induced current densities at the surface of the body [13].

If for simplicity, a homogeneous conductivity of 0.2 S m^{-1} is assumed, a 50-Hz magnetic flux density of 100 µT generates current densities between 0.2 and 2 mA m^{-2} in the peripheral area of the body [3]. According to another analysis [11], 60-Hz exposure levels of 100 µT correspond to average current densities of 0.28 mA m^{-2} and to maximum current densities of approximately 2 mA m^{-2}. More realistic calculations based on anatomically and electrically refined models resulted in maximum current densities exceeding 2 mA m^{-2} for a 100-µT field at 60 Hz. However, the presence of biological cells affects the spatial pattern of induced currents and fields, resulting in significant differences in both magnitude (a factor of 2 greater) and patterns of flow of the induced current compared with those predicted by simplified analyses [20].

Electric field models must take into account the fact that, depending on the exposure conditions and the size, shape, and position of the exposed body in the field, the surface charge density can vary greatly, resulting in a variable and non-uniform distribution of currents inside the body. For sinusoidal electric fields at frequencies below about 10 MHz, the magnitude of the induced current density inside the body increases with frequency.

The induced current density distribution varies inversely with the body cross-section and may be relatively high in the neck and ankles. The exposure level of 5 kV m^{-1} for exposure of the general public corresponds, under worst-case conditions, to an induced current density of about 2 mA m^{-2} in the neck and trunk of the body if the E-field vector is parallel to the body axis [3, 8]. However, the current density induced by 5 kV m^{-1} will comply with the basic restrictions under realistic worst-case exposure conditions.

For purpose of demonstrating compliance with the basic restrictions, the reference levels for the electric and magnetic fields should be considered separately and not additively. This is because, for protection purposes, the currents induced by electric and magnetic fields are not additive.

For the specific case of occupational exposure at frequencies up to 100 kHz, the derived electric fields can be increased by a factor of 2 under conditions in which adverse indirect effects from contact with electrically charged conductors can be excluded.

3.1.2. High Frequency Electromagnetic Field

At frequencies above 10 MHz, the derived electric and magnetic field strengths were obtained from the whole-body SAR basic restriction using computational and experimental data. In the worst case, the energy coupling reaches a maximum between 20 MHz and several hundred MHz. In this frequency range, the derived reference levels have minimum values. The derived magnetic field strengths were calculated from the electric field strengths by using the far-field relationships between E and H. In the near-field, the SAR frequency dependence curves are no longer valid; moreover, the contributions of the electric and magnetic field components have to be considered separately. For a conservative approximation, field exposure levels can be used for near-field assessment since the coupling of energy from the electric or magnetic field contribution cannot exceed the SAR restrictions. For a less conservative assessment, basic restrictions on the whole body average and local SAR should be used.

3.2. DOSIMETRIC NEEDS FOR STRONGLY COUPLED NEAR-FIELD EXPOSURE

As stated above, the derivation of the reference levels from the basic restrictions are given for the condition of maximum coupling of the field to the exposed individual, thereby providing maximum protection. For a less conservative assessment, basic restrictions on the induced current density, whole-body average and localized SAR should be used.

The determination of the induced current densities, the internal electric field strengths and of the SAR is a complicated function of various exposure parameters and the biophysical properties of the biological object. It should be recognized that determination of whether a particular electromagnetic field source will meet the exposure criteria poses technical difficulties, and can be done only by a qualified person, a laboratory, or a scientific body for a general class of equipment. The translation of biologically based restrictions into a practical system of standards includes knowledge of numerical modeling of the human body as well as of the field coupling to the body, including knowledge of the design of equipment and the principles and the practice of field measurement. Mainly due to these reasons the ICNIRP guidelines do not address product performance standards or guidance concerning computational methods or measurements techniques. It is the opinion of the ICNIRP that technical standard bodies (i.e., IEC, ISO, CIE, CENELEC) should perform the provision of technical advice concerning the practical implementation of standards.

In order to perform less conservative exposure assessments, near field exposure situations, localized and non-uniform field exposure are of special interest. Examples of typical EM sources with a strongly coupled near-field exposure are

- hand held mobile telephones,
- inductive heating equipment,
- dielectric heaters,
- antitheft devices and personnel identification systems,
- other high power devices used in the near field.

Significant recent dosimetric developments include the introduction of anatomically derived voxel-based electromagnetic models of the human body of various resolutions as well as varieties of effective numerical schemes, such as the impedance method, modified versions of the finite-difference time-domain (FDTD) method, the finite element method (FEM) and the scalar potential finite-difference (SPFD) method. The application of these methods has resulted in a significant progress in the accurate numerical modeling of low-frequency induction in humans [see, i.e., 4]. Problems may result from the near field coupling of the fields to the phantoms or body, especially in the RF-range. These methods should be further developed.

Experimental dosimetric methods may suffer from the influence of measuring probes on the field distribution inside the phantoms. As a consequence, calculations and measurements should be applied complementary. Combinations of the most recently developed experimental and computational dosimetric tools enables assessment of the internal field distribution with a precision better than 1 dB [9]. There is, however, a clear need to analyze systematically the general classes of equipment, examples of which are given above, in order to decide whether the application of the equipment meet the basic restrictions and to determine whether additional protective measures are necessary. This should be an important future task of technical standard bodies.

4. Acknowledgements

The present composition of the International Commission on Non-ionizing Radiation Protection (ICNIRP) is as follows: A. Ahlbom (Sweden); U. Bergqvist; (Sweden); J.H. Bernhardt, Chairman (Germany); J.P. Cesarini (France); F.R. De Gruijl (Netherlands); M. Grandolfo (Italy); M. Hietanen (Finland); A.F. McKinly, Vice-Chairman (UK); M.H. Repacholi , Chairman emeritus (Switzerland); D.H. Sliney (USA); J.A.J. Stolwijk (USA); L.D. Szabo (Hungary); M. Taki (Japan); T.S. Tenforde (USA); R. Matthes , Scientific Secretary (Germany). Standing Committee on "Epidemiology": A. Ahlbom, Chairman (Sweden); E. Cardis (France); M. Linet (USA); D. Savitz (USA); J.A.J. Stolwijk, Vice-Chairman (USA); T. Swerdlow (UK); Standing Committee on "Biology": F.R. De Gruijl, Vice-Chairman (Netherland); R. Owen (USA); R. Saunders (UK); T.S. Tenforde, Chairman (USA); L. Verschaeve (Belgium); B. Veyret (France). Standing Committee on "Physics": H. Bassen (USA); M. Grandolfo, Chairman (Italy); K. Jokela (Finland); C. Roy (Australia); M. Taki, Vice-Chairman.

5. References

1. Allen, S.G., Bernhardt, J.H., Driscoll, C.M.H., Grandolfo, M., Mariutti, G.F., Matthes, R., McKinlay, A.F., Steinmetz, M., Vecchia, P. and Whillock, M. (1991) Proposals for basic restrictions for protection against occupational exposure to electromagnetic non-ionizing radiations. Recommendations of an International Working Group set up under the auspices of the Commission of the European Communities, *Phys. Med.* **VII**, 77-89.
2. Bernhardt, J.H. and Matthes, R. (1997) Recent and future activities of the ICNIRP, *Radiation Protection Dosimetry* **72**, 167-176.
3. Commission on Radiological Protection (1997) Protection against low-frequency electric and magnetic fields in energy supply and use. Recommendations approved on 16th/17th February 1995. In: Berichte der Strahlenschutzkommission des Bundesministeriums für Umwelt, Naturschutz und Reaktorsicherheit. Heft 7. Stuttgart: Fischer.
4. Dawson, T. W. and Stuchly, M.A. (1998) Effects of skeletal muscle anisotropy on human organ dosimetry under 60 Hz uniform magnetic field exposure, *Phys. Med. Biol.* **43**, 1059-1974.
5. FDA (1993) Good laboratory practice for non-clinical laboratory studies. Food and Drug Administration, US Department of Health and Human Services. *Fed. Reg.* 21 CFR Ch. 1 (4-1-93 Edition), Part 58, 245-258.

6. International Commission on Non-Ionizing Radiation Protection (1994) Guidelines on limits of exposure to static magnetic fields, *Health Phys.* **66,** 100-106.
7. International Commission on Non-Ionizing Radiation Protection (1998) Guidelines for limiting exposure to time-varying electric, magnetic, and electromagnetic fields (up to 300 GHz), *Health Phys.* **74,** 494-522.
8. International Labor Organization (1994) *Protection of Workers from Power Frequency Electric and Magnetic Fields,* Geneva: International Labor Office; Occupational Safety and Health Series. No. 69.
9. Kuster, N. and Balzano, Q. (1996) Experimental and numerical dosimetry, in Kuster, N., Balzano, Q., and Lin, J.G. (eds); *Mobile Communications Safety.* London: Chapman Hall, pp. 13-64.
10. Michaelson, S.M. and Elson, E.C. (1996) Modulated fields and "window" effects, in Polk, C. and Postow, E., (eds.), *Biological effects of electromagnetic fields.* Boca Raton, FL: CRC Press, pp. 435-533.
11. National Academy of Science/National Research Council (1996) Possible health effects of exposure to residential electric and magnetic fields. Washington. DC: National Academy Press.
12. NTP (1992) Specification for the conduct of studies to evaluate the toxic and carcinogenic potential of chemical, biological and physical agents in laboratory animals for the National Toxicology Program (NTP), Attachment 2, August 1992 (Including modifications through 9/95), Available from the National Institute of Environmental Health Sciences, Environmental Toxicology Program, PO Box 12233, Research Triangle Park, NC 27709, USA.
13. Reilly, J.P. (1992) *Electrical Stimulation and Electropathology.* Cambridge, MA: Cambridge University Press.
14. Repacholi, M.H. and Stolwijk, J.A. (1991) Criteria for evaluating scientific literature and developing exposure limits, *Radiat. Prot. In Australia* **9,** 97-84.
15. Repacholi, M.H. and Cardis, E. (1997) Criteria for EMF health risk assessment, *Radiat. Prot. Dosim.* **72,** 305-312.
16. Repacholi, M.H. (1998) Low-level exposure to radiofrequency electromagnetic fields: health effects and research needs, *Bioelectromagnetics* **19,** 1-19.
17. Tenforde, T.S. (1997) Interaction of ELF magnetic fields with living systems, in Polk, C. and Postow, E. (eds.), *Biological Effects of Electromagnetic Fields.* Boca Raton, FL: CRC Press, pp. 185-230.
18. United Nations Environment Programme/International Radiation Protection Association/World Health Organization (1987) *Magnetic Fields,* EHC 69 (Geneva: WHO).
19. United Nations Environment Programme/International Radiation Protection Association/World Health Organization (1993) *Electromagnetic Fields in the Frequency Range 300 Hz to 300 GHz,* EHC 137, (Geneva: WHO).
20. Xi, W. and Stuchly, M.A. (1994) High spatial resolution analysis of electric currents induced in men by ELF magnetic fields, *Appl. Comput. Electromagn. Soc. J.* **9,** 127-134.

REQUIREMENTS FOR THE PROTECTION OF PERSONS AND THE ENVIRONMENT FROM ELECTROMAGNETIC FIELDS OF HIGH POWER TRANSMITTERS IN THE FEDERAL REPUBLIC OF GERMANY

K. W. HOFMANN
Forschungsinstitut für Hochfrequenzphysik
Forschungsgesellschaft für Angewandte Naturwissenschaften
Neuenahrer Str. 20
D - 53343 Wachtberg - Werthhoven

1. Introduction

Scarcely a week passes where an article does not appear in the daily newspaper or in a journal in which a threat for persons and the environment is reported. This threat seems so insidious since no one can see it, feel, smell or taste it - the so-called „electrosmog" which is spreading unobserved over the countryside. „Electrosmog" is the vogue word used to describe electric, magnetic or electromagnetic pollution. As the complex theory of propagation is commonly unknown or little understood to the greater public, these radiation fields are widely seen as extremely dangerous.

This feeling is increased, because many transmitting stations – broadcast- and TV-transmitters, and increasingly base stations for mobile communication - are situated very often within highly populated regions. Military operated stations may also be located in the vicinity of civilian settlements. Under the assumption that these installations are very often equipped with high power transmitters, conflicts between the operating units or agencies and the public concerned are likely to arise. In these cases not only possible harmful effects on persons are discussed but also questions of electromagnetic compatibility of electric devices and electronic circuits as they are found in private households, doctor's practices, small industries or shopping centres.

Taking this into consideration, all users of these radiation fields have a high degree of responsibility when emitting the radiation energy generated by their transmitters.

The following sections shall give an insight into which regulations or guidelines have to be observed and what has to be done to convert the rules given therein into practice in order to safely operate a high power radar installation.

As an example, the Forschungsgesellschaft für Angewandte Naturwissenschaften (FGAN) (Research Establishment for Applied Science) may serve, which is tasked by the German Ministry of Defence to operate various kinds of Radar systems for research purposes.

B.J. Klauenberg and D. Miklavcic (eds.), Radio Frequency Radiation Dosimetry, 523-530.
© 2000 *Kluwer Academic Publishers. Printed in the Netherlands.*

2. Regulations and Standards

In order to be permitted to operate high frequency transmitter stations besides the technical requirements a lot of conditions and rules have to be complied with. Essentially it is required to know the physical characteristics of the emitted radiation, namely

- frequency
- modulation
- polarization
- and the radiated power

and, furthermore, the geometrical dimensions of the radiating antenna, the geographic co-ordinates of the site and the direction of the emission as well as the radiation pattern of the antenna in azimuth and elevation.

All this data have to be known to request permission from the national authority in order to put a transmitter into operation. Exceptions are made for only few installations. These apply mostly to military transmitters. Here should be mentioned that the military itself is responsible for its installations, but close co-operation with the civilian authorities is recommended.

In a „Certificate of Safety" it is certified that radiation emitted by the transmitting device under consideration will not cause adverse health effects for „the general public and the neighbourhood". The certificate is issued with respect to the 26th Ordinance Implementing the Federal Emission Control Act (26BImSchV) of December 1996 of the Federal Government. This ordinance represents the basic requirements to safeguard the protection of persons being exposed to electric, magnetic and electromagnetic fields. But the certificate is only valid for the *General Public,* it is not related to workers in their workplaces. For these purposes other regulations have to be considered. Particularly with respect to military installations special directives are provided by the German Ministry of Defence. However, military and civil regulations do not differ very much. In the basic restrictions and reference levels they are similar or even identical.

Some of the most important documents are listed below:

2.1. GENERAL PUBLIC

- *26th Ordinance Implementing the Federal Emission Control Act (EMF Ordinance - 26th BImSchV) of 16 December 1996*
  (Federal Law Gazette (BGBl.) I p. 1966)
  This is the basis for all German regulations relevant to the protection of persons and follows the ICNIRP Guidelines. The ordinance was brought into force on 1 January 1997, but may be revised in the near future.
- *DIN VDE 0848 Safety in Electrical, Magnetic and Electromagnetic Fields*
  Part 1 : Definitions, Methods for Measurement and Calculation

In July 1998 a draft has been brought to the attention of the public for discussion. This process has not yet finished.

Part 2 : Protection of Persons in the Frequency Range 0 Hz to 300 GHz

Here still exists a draft of October 1991, which has meanwhile become obsolete. Part 2 of this standard is currently revised, and a draft will be published not before 1999.

Part 3-1 : Protection of Persons with Active Body Implants in the Frequency Range 0 Hz to 300 GHz (here: cardiac pacemakers

A draft has been worked out this year, which is being brought to the attention of the public at about the end of this year.

-*Unfallverhütungsvorschriften der Berufsgenossenschaften.*(UVV)
(Accident-Protection order of the Professional Associations)
Regulations for Restriction of Human Exposure to Electric,
Magnetic and Electromagnetic Fields at the Workplace
There exists already a first draft of December 1997. [1]

2.2. MILITARY REGULATIONS

- *Regulations of the Bundeswehr for Protection of Individuals against Damaging Effects of Radiofrequency Electromagnetic Fields (RF Radiation)*
- *(Bestimmungen der Bundeswehr zum Schutz von Personen vor schädigenden Wirkungen hochfrequenter elektromagnetischer Felder (HF Strahlen))*

These special regulations for occupational safety and health have been issued on 10 February 1992 by the Federal Ministry of Defence and are based on the VDE-Standards DIN VDE 0848/ Parts 2 (October 1991) und 4 (October 1989), which are not longer valid. Therefore, this regulation will have to undergo a revision.

Because of the steadily increasing international co-operation in all fields it is of importance to have good knowledge of regulations beyond the national frame. Some of them are - for example the European standards - partly obligatory for national application. These are:

2.3. EUROPEAN STANDARDS

- *ENV 50166 Part 1: Human Exposure to Electromagnetic Fields. Low Frequency (0 Hz to 10 kHz)*
This is a pre-standard, which is still under revision
- *ENV 50166 Part 2: Human Exposure to Electromagnetic Fields. High Frequency (10 kHz to 300 GHz)*
This is a pre-standard, which is still under revision

- European Council Recommendation

A proposal for a Council recommendation „On the Limiting of Exposure of the General Public to Electromagnetic Fields 0 Hz - 300 GHz" was published in June 1998 by the Commission of the European Communities. It has already become a Bundestags-Vorlage (Document of the German Parliament) and could perhaps serve as a supplement to the 26BImSchV.

2.4. INTERNATIONAL MILITARY GUIDELINES/REGULATIONS

For the military there are also drafts of regulations for the protection of personnel from exposures to RF-fields, which have been established for fitting best military needs under consideration of best care for the health of personnel. These are for instance:

- STANAG 2345 med: Evaluation and Control of Personnel to Radiofrequency Fields : 3 kHz to 300 GHz

- NATO Naval Radio and Radar Radiation Hazard Manual

Both are under discussion at the moment and shall get new input from this present-day NATO Advanced Research Workshop.

Besides the regulations for the protection of persons it must be observed that the transmitting station under consideration shall not influence or even disturb the operation of other electric devices or electronic circuits. This is to be seen from the standpoint of Electromagnetic Compatibility. Here also exist a lot of standards and regulations, for instance:

2.5. ELECTROMAGNETIC COMPATIBILITY

- EN 50082 - 1 Electromagnetic Compatibility
Generic Emission Standard
Part 1:Residential, Commercial and Light Industry
Here are tabulated permissible levels for residential and business areas and small industries
- DIN VDE 0848 Part 5:Safety in Electrical, Magnetic and
Electromagnetic Fields: Protection against Explosion
This draft of April 1998 contains rules which shall help to avoid ignition of explosive atmospheres by unintended irradiation by rf fields in the frequency range 10 kHz - 30 GHz

Simply the multitude of all these documents with sometimes differing contents makes the application of them not very easy. Of great help, and a necessary condition, will be the above mentioned Certificate of Safety, for which fixed radio transmitters are tested for compliance with the limits to persons and cardiac pacemakers. In the case of implementation of all necessary requirements, this official certificate is issued by the *Regulatory Authority for Telecommunications and* Posts and states therewith that the emitted field strengths are in compliance with the safety limits where the non- transient presence of persons may be assumed [2]. The Regulatory Authority checks not only the radiation fields emitted by the respective facility but also investigates all contributions of other sources effecting the area of concern. However, it must be recognised again that these statements are restricted only to areas with uncontrolled access by the public.

In regions which are situated within the area that can be controlled by the organization which operates the radiation sources, the Accident-Protection-Order (UVV) of the Berufsgenossenschaften (see 2.1) will have to be applied in the future. In the past, the German standard DIN VDE 0848 Part 2 defining permissible levels for controlled and uncontrolled exposure areas had then entered a factor 5 higher for power densities, respectively, $\sqrt{5}$ for field strengths in comparison to the limits for the general public [3]

3. Calculation and Measurements

The knowledge of the relevant regulations is a basic condition for taking a transmitter facility into operation. Its application to the respective transmitter station implies that appropriate calculations and measurements have to be implemented. As reference the German Standard DIN VDE Part 1 (Definitions, Methods for Measurement and Calculation) is adhered to.

When there is only one transmitter to be considered, calculation of its radiation field might be very simple. But it will be more difficult at a multiple emitter location with equipment of different frequencies, modulations or polarisation. Also, for this case, the DIN VDE 0848 Part 1 gives rules how to proceed. But there can occur considerable differences when the results of theoretical calculations are compared with measured values. These can be caused by reflections from buildings, cars or other obstacles. Therefore, as far as the evaluation of radiation protection is concerned, the results of measurements should be preferred. And measurements will have the advantage that they very often have a higher degree of acceptance by the people who might have been, or will be, exposed to this radiation.

The Research Establishment for Applied Science (Forschungsgesellschaft für Angewandte Naturwissenschaften FGAN) may serve as an example how one can or is obliged to proceed in the case of radiation safety. The Research Establishment is tasked by the German Ministry of Defence to operate various transmitting systems for research purposes on its site in Wachtberg-Werthhoven near Bonn/Germany. The research work is focussed on radar systems and methods.

A total of 16 transmitters are installed on the site, operating in a frequency range from about 4 MHz up to 220 GHz. The spectrum of relevant peak powers starts at 10 W and ends at the maximum peak power of 5 MW, laboratory equipment and

528

communication installations not included. This indicates the complexity of evaluating exposure data.

Therefore, it is to be examined which systems need to be evaluated by calculation and measurement and on the other hand which ones can be checked by simple estimation of power densities or field strengths.

Amongst all this equipment the Radar-Big-Dish-Antenna of the Research Institute for High Frequency Physics (Forschungsinstitut für Hochfrequenzphysik FHP), operated for wide-range-radar research (Space Object Identification) attracts most attention. It consists of a parabola antenna under radome with a diameter of 34 m and is operated in the Cassegrainian mode. The antenna is fully steerable over 360 degree in azimuth and from 0 to 90 degree in elevation. The maximum peak power of the transmitter is 5 MW, but at presently the peak power is limited to 1,5 MW. The electromagnetic fields radiated by this antenna have been carefully theoretically analyzed and experimentally verified.

The usual methods of calculation in this case cannot be applied to the requirements of radiation protection, because in the near-field of such an antenna the electromagnetic energy is radiated into the free space in a first approximation as a tube of rays, whose diameter is geometrically congruent with that of the antenna's aperture and which does not impinge on the ground. But in addition, the far-field pattern of the feedhorn within the Cassegrainian system has to be considered. In the spillover range between the main reflector and the Cassegrainian subreflector there exists a complex superposition of both, which can be measured in the antenna-forefield. This had been confirmed by complex and extended measurements.

The measurements have been performed on a so-called free field test range, where the influence of reflections in the terrain had been minimum. The results of calculation and measurement are shown in Figure 1.

Deviations from the theoretical results in the near vicinity of the antenna building are caused by the building itself and scattering effects from edges and the metal structure of the radome. Beginning at a distance of about 150m from the vertical antenna axis measured and calculated power densities are in good agreement. They diminish as expected with the square of the distance from the antenna and reach values before getting to the fence of the FGAN territory which are about a factor 8 below the permissible exposure level as given in the 26th Ordinance (26BImSchV). (This value has been taken for comparison, even that it does not apply, because here the limits for workers in their workplaces would be valid. But it is the policy of the Research Establishment to take the protection rules for the general public as a guideline).

Therefore, this radar station with its principally high output power can be classified not to cause harmful effects to the public and the neighborhood.

In a similar way other powerful radar systems have been investigated, whilst other ones have only be treated by calculated estimates.

The measurements with the 34-m-antenna as a source have been repeatedly conducted during long-time intervals. The first measuring series had been performed with an especially for these purposes assembled test equipment. In a second measuring period results of the first one could be confirmed by measurements taken by two mobile measuring stations of the German Bundeswehr as external experts [4].

Further measurements, repeatedly taken, have been and are performed with commercial test equipment. This simplifies the analysis of the data received, and - what should be seen as very important - the measuring procedure is comprehensible also by persons who are not experts in this field. This may be of advantage in reviewing the results and the risk analysis.

During the measurements it was always strongly observed, that the measuring probe had been kept far enough away from receiver/monitor and personnel to avoid coupling effects or interference.

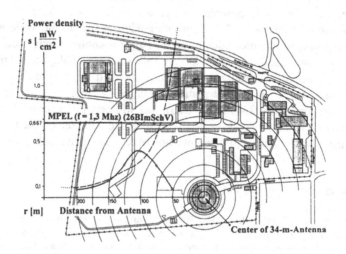

Figure 1. Site-plan of the FGAN (Buildings) Power Densities in the Fore-field of the 34-m-Antenna (- - - - calculated; ---o--- measured).

These measurements have been taken within and outside the FGAN-territory mainly for the reason of protection of persons. But the results have also been applied to the evaluation of radiation effects on electric devices and electronic circuits. This became important because there had been plans by the surrounding community to build a site for small industries and a shopping centre later on. On the basis of the stations with the most powerful transmitters independent external experts performed calculations with respect to electromagnetic compatibility. From their results it could be seen that there would be mutual incompatibilities, which initiated new plans, screening and protection measures for concerned installations outside the FGAN area.

530

4. Final Remark

Known or only assumed effects of non-ionizing radiation on biological systems have led to worldwide controversial and heated debate. Again and again the public has been „warned by inflammatory headlines" from harmful effects on persons and the environment. Very soon after the increased use of high frequency radiation fields, especially in military applications, permissible exposure levels had been established and consequently improved by new findings of all the research activities that followed. This process is still going on.

In the technically developed countries, the results of this research have been transformed into standards and guidelines for the protection of persons on a continuing basis. A leading role herein nowadays plays the International Committee on Non-Ionizing Radiation Protection (ICNIRP), whose guidelines meanwhile world-wide have become fundaments for the protection of persons against known adverse health effects. But these are only guidelines, and there is still a lack of obligatory standards or regulations that are internationally recognized. A standard that may have a chance to form the basis for such an international obligatory regulation is the NATO Standardization Agreement 2345 (STANAG), that will certainly get well-based inputs from the present-day Advanced Research Workshop and will convert the new findings presented here into regulations for the protection of exposed individuals.

5. References

1. Eggert, S., Krause, N., and Goltz, S. (1997) Regulations for the restriction of human exposure to electric, magnetic and electromagnetic fields at workplaces - A project of an accident protection order in Germany, *2nd World Congress for Electricity and Magnetism in Biology and Medicine*, Bologna, Italy.
2. Regulatory Authority for Telecommunications and Posts (1998) *Certificate of Safety - Electromagnetic Radiation and the Environment* (as of January 1998).
3. Information available from Regulierungsbehörde für Post und Telekommunikation, Postfach 8001, D 55003 Mainz
4. Hofmann, K.W. (1993) Radiofrequency radiation safety guidelines in the Federal Republic of Germany, in B.J. Klauenberg et al (eds.), *Radiofrequency Standards*, Plenum Press, New York, pp. 35-40
5. Hofmann, K.W. (1981) The determination of power densities around large microwave antennas, in J.C. Mitchell (ed.), *Aeromedical Review 3-81, Proceedings of a Workshop on the Protection of Personnel Against Radiofrequency Electromagnetic Radiation*, USAF School of Aerospace Medicine Brooks Air Force Base, San Antonio, 1981, pp. 54-67.

RADIOFREQUENCY (RF) EXPOSURE OF MOBILE COMMUNICATIONS IN HUNGARY AND EVALUATION RELEVANT TO EU AND NATIONAL STANDARD: BASE STATIONS AND HANDY DEVICES

G. THUROCZY, J. JANOSSY and N. NAGY
National Research Institute for Radiobiology and Radiohygiene
Department of Non-Ionizing Radiation,
H-1221 Budapest, Anna str.5, Hungary

1. Introduction

There are many questions related to the assessment of risk to the users and the general public from exposure to radiofrequency (RF) radiation from mobile transceivers including hand-held units. Because of the public awareness of rapid growth of cellular base stations in the human environment it is important to evaluate the real exposure to general population from the radiation of these stations. Many research studies, industrial projects and international concerted actions reflect the present state of knowledge concerning biological effects and potentially adverse health effects from (RF) electromagnetic (EM) radiation in the frequency range of today's wireless (mobile) communication technology. There are many questions related to the assessment of risk to the users and the general public from exposure to RF radiation from mobile transceivers including hand-held units [1, 2]. Scientists, the management of industry and suppliers and general users encounter the following questions which are in progress nowadays: (i) How can we estimate the exposure of humans from the wireless telecommunication system including the base stations, where the exposure may occur to the public, and hand held phones? (ii) What is the level of the electromagnetic absorption in the human body and especially in the head due to the mobile phone's exposure? (iii) What are the possible health consequences and risks of RF radiation at doses analogous to those produced by cellular telephones? (iv) Is there electromagnetic interference (EMI) with medical equipment, especially with implanted devices in the human body? [3]

1.1. WIRELESS COMMUNICATION IN HUNGARY

In many countries individuals or citizen's groups feel they have no authoritative source or agency from which to obtain the information they need to determine whether they or their children's health will be affected by the increasing exposure to EMF fields, which is especially true for countries like Hungary. One of the biggest radio broadcast stations

531

B.J. Klauenberg and D. Miklavcic (eds.), Radio Frequency Radiation Dosimetry, 531-539.

in Europe was installed in Hungary near to Solt with 2 MW power at RF frequency (534 kHz). The number of mobile communication base stations increased rapidly in the last few years. Altogether more than 2000 base station have been built including the digital (GSM), analog (NMT) mobile and radio local loop (RLL) systems. The number of GSM users in Hungary are more than 750.000 and the RLL users will be 250.000 approximately. Additional environmental problems are associated with Budapest, which is a big capital (2.5 millions), and the density of broadcast and mobile base stations is relatively higher than other areas.

2. Measurement and Exposure Assessment in the Environment

2.1. BASE STATIONS

During the last period, RF radiation was measured by our laboratory in the environment of more than hundred base (GSM and NMT) stations in Hungary. In the measurement commercially broadband RF monitors were used as Narda 8616 series, Wandel & Goltermann EMR 300 and Chauvin Arnoux C.A 43 field meter. The frequency spectrum of RF sources was evaluated by Advantest U4941 spectrum Analyzer with calibrated Rhode & Schwarz HL-040 Log-Periodic Dipole Antenna. The RF exposure levels at the living area of general public were collected and evaluated according to the European Prestandard (ENV 50166-2) and the Hungarian Exposure Standard (MSZ 16260-86, see Table 1 and Table 2).

TABLE 1: Hungarian MSZ 16260-86 Standard: "Safety levels of high frequency electromagnetic fields" in the frequency range 300 MHz-300 GHz.

Area	Power density (mW/cm^2) 300 MHz-300 GHz	
	Standing source	Rotating source
Uncontrolled (general public)	0.01	0.1
Controlled (occupational)-	0.1	1.0
Restricted in time (restricted in time for occupational)	$\sqrt{\dfrac{0.08}{hour}}$	$\sqrt{\dfrac{8}{hour}}$
Harmful (not allowed)	10	100

It was found that in most cases the exposure levels to humans at location accessible to public were many times below the strict Hungarian permissible RF exposure level for the general public (0.01 mW/cm²) and did not exceed a few microwatt/cm² except in special cases where the exposure was in the order of the permissible level but below the relevant ENV standard (0.40-0.45 mW/cm^2).

TABLE 2: Hungarian MSZ 16260-86 Standard: "Safety levels of high frequency electromagnetic fields" in the frequency range 30 kHz-300 MHz

Area	Electric field strength (V/m)		
	30 kHz-3 MHz	3-30 MHz	30-300 MHz
Harmless	3	3	3
uncontrolled (general public)	50	30	20
Controlled (occupational)	120	60	40
Restricted in time (for occupational)	960/hours	480/hours	320/hours
Harmful (not allowed)	1000	600	400

In Table 3 and Table 4 the figures of all measurements are shown nearby of base station antennas. The power density versus distance from the base station antennas were estimated and measured as shown on Figure1, 2.

TABLE 3. Power densities in μW/cm^2 vs. distance nearby to rooftop mounted GSM 900 MHz base station antennas with 25 W mean (15-40 W) EIRP. The measurements were made on the roof accessible to public at 1.5 m and 2.5 m height respectively, in the flat of top floor and at the roof of the nearest building.

Height of measurement	1.5 m			2.5 m		
(μW/cm^2) → Distance ↓	mean	CI (95%)	n	mean	CI (95%)	n
1-5 m	3.57	2.26-4.88	70	5.85	3.54-8.15	70
5-10 m	2.40	1.28-3.52	47	3.35	1.98-4.72	47
10-20 m	0.98	0.27-1.69	14	1.94	0.49-3.39	14
20-30 m	0.51	0.19-0.82	21	0.83	0.37-1.28	21
top floor	0.18	0.03-0.33	94	0.52	0.23-0.78	62
near building	0.44	0.20-0.68	39	-	-	-

TABLE 4. Power densities nearby to tower mounted GSM 900 MHz base station antennas with 250 W mean (200-450 W) EIRP. The measurements were made on the ground at 1.5 m height, 100 m and 200 m distance respectively.

Distance from the tower	100 m			200 m		
(μW/cm^2) $\rightarrow$ Tower height $\downarrow$	mean	CI (95%)	n	mean	CI (95%)	n
15 m	0.155	0.01-0.33	5	0.180	0.01-0.40	4
30 m	0.08	0.01-0.15	13	0.177	0.02-0.33	12
40 m	0.02	0.01-0.05	7	0.062	0.01-0.14	7
50 m	0.04	0.02-0.07	38	0.101	0.03-0.16	38
60 m	0.04	0.01-0.07	9	0.066	0.01-0.11	8

2.2 MOBILE HANDY DEVICES AND RLL SUBSCRIBER UNITS

The power density in the close environment of handheld devices was measured on 900 MHz GSM, 450 MHz NMT devices and 900 MHz RLL (Radio Local Loop) subscriber units. The measurements were performed with Wandel & Goltermann EMR 300 field meter system and Schmid & Partner ER3DV4R near-field electric field probe. The power density beyond the reactive near field at 4-5 cm for the radiated antenna reached 1 mW/cm^2 in case of GSM and 8-9 mW/cm^2 for NMT devices respectively (Figure 3). Experimental RF dosimetry for SAR (Specific Absorption Rate, W/kg) of laboratory animals and human head model were also performed using immersible miniature isotropic electric field probe developed by our research team.

3. Evaluation and Discussion

3.1. EXPOSURE OF BASE STATIONS AND PUBLIC RELATION

The general public has increasingly becoming aware of the potential dangers of exposure to base stations of mobile communication systems. Satisfactory methods and policy are needed to measure or predict deposited dose both in laboratory and in the field to provide advice on exposure limits which will protect the health of individuals, and will not affect the immediate or subsequent performance of services by the operators. The most important factors which are associated with the assessment of exposure emitted by the base stations are: transmitter power, frequency, antenna gain and pattern, location, channel number, nearby reflecting surfaces and conductive objects and the density of buildings in the exposed area. The typical transmitter power is between 10-100 W, the operation frequency range is 400-950 MHz (including the analogue systems), and the preferred sector antenna gain is between 10-20 dB [4].

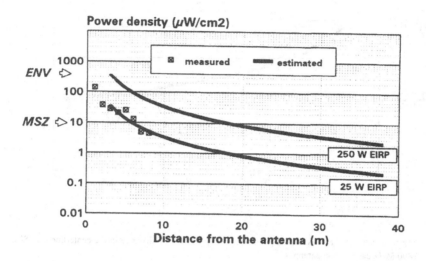

Figure 1. Measured and estimated power density (μW/cm²) vs. distance in the centerline of GSM 900 MHz base station antennas.

Commonly used height being 40-50 m in the case of tower-mounted antennas. The height of rooftop-mounted antennas is variable depending on the building where the base station was installed. According to our policy, before the use of base station by the operators we request information including the following data: 1) the location of antennas (tower/roof mounted); 2) the antenna characteristics (gain, pattern); 3) operation frequency; 4) transmitted power and channels (including ERP); 5) height of antennas from the ground and roof respectively; 6) buildings within 200 meters of antennas; 7) schools or kids institution within 200 meters of the antennas; 8) configuration of the terrain around the base station. The RF field measurements showed that the population is normally exposed to insignificant levels of RF radiation in the ambient environment of base stations. The exposure levels to humans at location accessible to public are many times below all recommended limits proposed in the different national and international guidelines [2, 5]. The typical RF power density in the exposed area does not exceed the few microwatts/cm². The limits recommended by

536

the guidelines are in the range of 0.4 mW/cm². The exposure level approaches these limits in the front of the sector antennas below the near-field zone only, which is usually not accessible to the general population. Inside the buildings can be expected to be reduced at least 10-20 dB due to attenuation caused by building materials in the walls and roof.

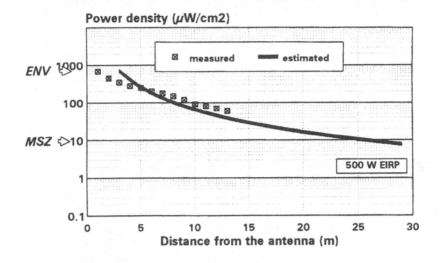

Figure 2. Measured and estimated power density (µW/cm²) vs. distance in the centerline of GSM 1800 MHz base station antennas.

3.2. EXPOSURE OF HANDHELD AND SUBSCRIBER UNITS

The evaluation of spatial distribution of absorbed energy from the exposure of cellular phones presents theoretical and experimental difficulties. The absorbed power generally expressed in terms of SAR (Specific Absorption Rate, W/kg) [6]. The portable devices are to be evaluated with respect to SAR. The evaluation of SAR distribution in the human head and hand of the users represents a very complex problem. The main difficulties came from the near-field radiation of handheld phone's antenna, the geometrical complexity and electrical inhomogeneity of the exposed human head [7, 8, 9]. Otherwise the RF radiation from the portable cellular phones makes many differences from the RF exposure relevance to the similar EM exposure as: unilateral localized energy absorption in the head, not continuous but occasional exposure duration, individual affects, not steady state but sharply increasing exposure, the range of absorbed energy many times higher than generally from the other EM sources, but the exposure is intentional by the individuals (Table 5.). There are various theoretical, computational and experimental investigations in order to estimate the near-field electromagnetic field of the radiated antenna and the peak SAR in the head with its spatial distribution in the brain tissue and the eye.

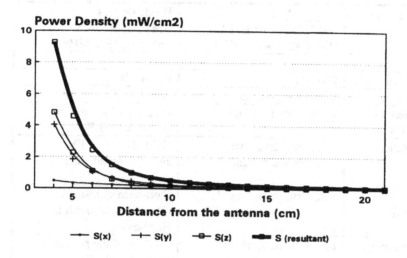

Figure 3. Power density near to the NMT (450 MHz) mobile hand-held device using helix antenna emitted 1 W (continuous) radiation.

Because of the near field exposure and inhomogeneity of exposed object the absorbed energy is very inhomogeneous and unilateral in the human head due to the reflection and local enhancement of SAR at the border of different tissue layers [10]. The computational models and experimental evaluation reported that about 40-70% of transceiver output power is absorbed in the head [11]. The measurements of hand-held devices indicate that the equipment with helix antennas produce higher peak power density close to the antennas and the field strength depends more strongly on the proximity to the antennas. The power density near to the antennas of handy devices some cases strongly exceed the permissible levels of relevant Hungarian standard for uncontrolled area (for general public) and even the EU permissible levels (Fig.3). Therefore the initiation of the EU CENELEC Specification for mobile telecommunication Equipment (MTE) [12] seems to be unavoidable. In contrast to the public awareness from the equipment of mobile communication systems the radiation of base stations are many times below of the hand-held devices.

TABLE 5. Comparison of public exposure to RF radiation emitted by base station and hand-held devices.

Description of source	Base stations NMT, GSM, RLL, DCS, DECT 450, 900, 1800 MHz	Handy devices NMT, GSM, DCS 450, 900, 1800 MHz
RF power	2 to 100 W	0.2 to 8 W
Exposition	far-field exposition	near-field exposition
Exposure to human body	whole body exposure	local exposition to head (more than 95 %)
Modulation	Depending on active channels and system	GSM, DCS , or FM analogue modulation
Continuity of exposure	long-term exposition	short-term exposition (250-400 min/month in Hungary)
Absorption in the body	Less than 10 % of the external RF exposure	40-70 % of the radiated power into the head and hand
SAR	SAR is in the range of nW/g in the whole body	SAR between 0.2-10 mW/g in the head and hand
Power density	generally below 1 $\mu W/cm^2$ in the living area	between 50-3000 $\mu W/cm^2$ close to the antenna
Prediction of exposure	spatial measurement and exposure assessment are difficult but available	exposure assessment and dosimetry are difficult (APC, DTX, differences in devices etc., user mode)
Exposure to public	not intentional by the individuals without benefit	intentional by the users with benefits
Psychological factors-	the antenna systems are visible to the eye and they are unfriendly in the area	small, user-friendly devices
Number of sources	approximately 3.000 in Hungary	approximately 750.000 users in Hungary (1998)
Standardization	EU CENELEC Prestandard ICNIRP Guidelines Hungarian Standard	EU Specification of MTE ICNIRP Guidelines FCC Guidelines

There have been many projects initiated to develop a new design of cellular handy phones. These efforts focused on the reduction of absorbed energy in the head by using specific antenna location on the device and new types of antenna design. The human body also significantly alters the radiation patterns and decrease the quality of communication. Therefore the system engineering aspects interfaced with the concepts of health safety and EMI requirements.

4. References

1. WHO. (1993) Electromagnetic Fields (300 Hz - 300 GHz), Geneva, World Health Organization, *Environmental Health Criteria* **137**.
2. ICNIRP (1996) Statement: Health issues related to the use of hand-held radiotelephones and base transmitters, *Health Physics* **70**, 587-593
3. Kuster, N., Balzano, Q., Lin, J. C. (1997) *Mobile Communications Safety*, Champan & Hall, London.
4. Petersen, R.C and Testagrossa, P.A. (1992) Radio-frequency electromagnetic fields associated with cellular-radio cell site antennas, *Bioelectromagnetics* **13**, 527-542.
5. Duchene, A.S., Lakey, J.R.A., and Repacholi, M.H. (1991) *IRPA Guidelines on Protection Against Non-Ionizing Radiation* Pergamon Press, New York.
6. Durney, C., Massoudi, H., and Isakner, M. (1985) *Radiofrequency Radiation Dosimetry Handbook (4th ed.)*, USAF School of Aerospace Medicine, Brooks AFB, Texas.
7. Cleveland, R.F. and Athey, T.W. (1989) Specific absorption rate (SAR) in models of the human head exposed to hand-held UHF portable radios, *Bioelectromagnetics* **10**, 173-187.
8. Anderson, V. and Joyner, K.H. (1995) Specific absorption rate levels measured in a phantom head exposed to radiofrequency transmissions from analog hand-held mobile phones, *Bioelectromagnetics* **16**, 60-69.
9. Jensen, M.A., and Rahmat-Samii, Y. (1995) EM interaction of handset antennas and a human in personal communications, *Proc.of IEEE* **83**, 7-16.
10. Stuchly, M.A. (1995) Mobile communication systems and biological effects on their users, *Radio Science Bulletin* **275**, 7-13.
11. Martens, L. (1994) Electromagnetic field calculations for wireless telephones, *Radio Science Bulletin* **272**, 9-11.
12. CENELEC (1997) *Considerations for Human Exposure to EMFs from Mobile Telecommunication Equipment (MTE) in the Frequency Range 30 MHz-6 GHz*, CENELEC, Secreteriat **SC 211/B**.

THE CZECH LIMITS AND THE EUROPEAN PRESTANDARDS

J. MUSIL
National Institute of Public Health
Srobarova 48
100 42 Praha 10
Czech Republic

1. Introduction

The last order in the Czech Republic concerning protection of health from the adverse effects of electromagnetic radiation [1] has been in force since November 1, 1990. The order determines the values of the maximum permissible irradiation and field ceiling levels for both occupational exposure and general population (Table 1). Instructions for determination of both the field and irradiation levels as well as categorization of the work and methodological guideline [2] are referenced. Standards for ELF exist at present in draft formerly.

2. Basic Rationale

Since 1965 the time factor together with the field level has been included in the Czech standard. The reason is that the field level alone is not a sufficiently good measure of the threshold of EMFs effectiveness - the organism has certain means of defense against hostile external influences. It has been demonstrated that for chemical and thermal injury and for ionizing radiation the function of the ability of organism to resist an effect is given by a product of two variables - quantitative value of that factor and duration of the exposure. The product is then called the irradiation W in our case. In practice it is necessary to insure that the irradiation does not exceed the maximum permissible value.

Values of the maximal permissible irradiation have been established not only on the basis of available proven knowledge of biological effects. Possible partial biological effects have been also considered in our country as well as theoretical and model studies, which could explain experimentally, found specific sensitivity of biosystems at the so called nonthermal levels. From the viewpoint of probable nonlinearity of effects, energy cannot be the sole factor at very short exposures and a field ceiling level shall be introduced instead. The frequency dependence of established maximum permissible irradiation is derived from the frequency dependence of the average SAR. We have

541

B.J. Klauenberg and D. Miklavcic (eds.), Radio Frequency Radiation Dosimetry, 541-544.
© 2000 *Kluwer Academic Publishers. Printed in the Netherlands.*

also introduced a certain safety factor because our knowledge can cover a certain average sensitivity of the organism at active field parameters only. A sufficient safety coefficient for the general population allows for more sensitive population groups (pregnant women, children´s population, aged and sick people, etc.).

For survey measuring we prefer broadband isotropic devices. The general principles of measuring however, are stated also for more simple meters. For mixed fields at a number of frequencies for which there are different values of the maximum permissible irradiation levels, the fraction of the irradiation incurred within each frequency interval should be determined, and the sum of all such fractions should not exceed unity.

3. Comparison with the European Prestandard

In contrast to the European prestandard [3] the quantities used as predictors of exposure or as basic restrictions, which are not directly measurable, have no place in our practical hygienic limits and checking their compliance under field conditions.

It is not acceptable also to neglect the possible (although at the present time not unanimously verified) long term action of low ("athermal") field levels. That is the reason for including the time factor into the Czech standard. The duration of exposure "t" is derived from the character and purpose of eventually existing or planned buildings, eventually from the character of utilization of the surrounding premises and from the operation time schedule of transmitters impinging the given area. In residential built-in areas usually t = 24 h, in lived-in recreational areas also 24 h, in uninhabited recreational areas 12 h, in other premises the time may be shorter or limited administratively. Adjustments are possible in individual cases depending on the operation time schedule of individual transmitters.

The smooth crossing of limit values between particularly established frequency bands complicate the evaluation process under the real field conditions of measuring when the appropriate measuring probes do not exist as yet.

Selective meters are allowed as secondary means for purposes of checking survey measurements, but broadband measuring apparatus with isotropic antenna is preferred. It is not possible also to presume at final evaluation average values which are ascertained within the dimensional area of body. For hygienic purposes the worst conditions are assumed in the course of checking. From the same reason one cannot correctly presume during evaluation rms values but rss values.

There must be as a permanent component part of the documentation, from the level of introductory design, data which characterize the distribution of the calculated field levels (field intensity E or average power density S) in the surroundings as satisfactory (or next lower) for the period of exposure of the population of concern in the given area (eventually characterizing also the distribution of calculated limit field levels for the population in the given environs if such a situation be actually possible). The data mentioned must agree with what is expected from known sources in the area. At the same time also the eventual influence of transmitters from other localities (background) must be taken into account.

4. Conclusions

For short-term exposures up to few minutes duration as the first time interval our standard and CENELEC standard are in order of magnitude agreement at least.

For the middle time interval (defined by a few minutes on one and by tens minutes on the other hand) the agreement is even better, and in some frequency bands the slopes are even overlapped. This should be regarded as a very important circumstance and, in industrial applications, even as a basic one. Experience has shown that average exposure time during one shift does not exceed tens of minutes.

The comparison of standards in both the short and middle time intervals has shown that the Czech standards also attribute here the decisive role to thermal effects.

The third time interval (long-term exposures) is the most controversial: while in the CENELEC the limiting factor is the field level, in the Czech Republic it is regarded as absolutely necessary not to exceed a certain constant irradiation level. That is, for a longer exposure, the field level must be lower.

The emphasis on prevention is significant here or the so called ALARA (as low as reasonably achievable) principle. The late impact of the field influence may manifest itself after very long periods only, so, especially at low levels, it is difficult to relate it solely to the field action. Any application in hygiene practice therefore requires great caution as well as carefully performed clinical and epidemiological studies.

5. References

1. *Order Concerning the Protection of Health from the Adverse Effects of Electromagnetic Radiation.* No. 408, Ministry of Health of the Czech Republic (1990).
2. *Checking Compliance With Maximum Admissible Values of General Population's Irradiation in the Vicinity of AM, FM, TV and Radar Stations and Radio Relay Systems* (1993) Methodological Guideline of the Chief Hygienic Officer of the Czech Republic to Supplement of Order No. 408. AHEM, edit. NIPH Praha, Suppl. No. 3, June 12, 1993.
3. *Human Exposure to Electromagnetic Fields. High Frequency (10 kHz to 300 GHz).* European Prestandard ENV 50166-2, CENELEC (1995).

544

Table 1
THE MAXIMUM PERMISSIBLE IRRADIATIONS W, CEILING FIELD LEVELS E_c, H_c, S_c, WORKDAY-SHIFT E_{sh}, H_{sh}, S_{sh} AND/OR CALENDAR DAY LEVELS E_{cal}, H_{cal}, S_{cal} AND SHORTER PERIOD LEVELS E_t, H_t, S_t (t in hours)

a) OCCUPATIONAL

f (MHz)	0.06 - 3	>3 - 30	>30 - 300	>300
W_{Ep} [(V/m)^2.h]	50000	7000	800	-
E_{Cp} (V/m)	500	300	100	-
E_{sh} (V/m)	80	30	10	-
E_{tp} (V/m)	$(5/t)^{0.5} \ast 10^2$	$(0.7/t)^{0.5} \ast 10^2$	$(8/t)^{0.5} \ast 10$	-
W_{Hp} [(A/m)^2.h]	200	-	-	-
H_{Cp} (A/m)	50	-	-	-
H_{sh} (A/m)	5	-	-	-
H_{tp} (A/m)	$(2/t)^{0.5} \ast 10$	-	-	-
W_{Sp} (Wh/m^2)	-	-	-	$8 \ast K_1$ [1)]
S_{Cp} (W/m^2)	-	-	-	26.5
S_{sh} (W/m^2)	-	-	-	1
S_{tp} (W/m^2)	-	-	-	$(8/t) \ast K_1$ [1)]

b) GENERAL POPULATION

f (MHz)	0.06 - 3	>3 - 30	>30 - 300	>300
W_{Eo} [(V/m)^2.h]	5000	700	100	-
E_{Co} (V/m)	180	80	30	-
E_{cal} (V/m)	14	5	2	-
E_{to} (V/m)	$(0.5/t)^{0.5} \ast 10^2$	$(7/t)^{0.5} \ast 10$	$(1/t)^{0.5} \ast 10$	-
W_{Ho} [(A/m)^2.h]	20	-	-	-
H_{Co} (A/m)	15	-	-	-
H_{cal} (A/m)	0.9	-	-	-
H_{to} (A/m)	$(0.2/t)^{0.5} \ast 10$	-	-	-
W_{So} (Wh/m^2)	-	-	-	$1.2 \ast K_2$ [2)]
S_{Co} (W/m^2)	-	-	-	2.5
S_{cal} (W/m^2)	-	-	-	0.05
S_{to} (W/m^2)	-	-	-	$(1.2/t) \ast K_2$ [2)]

1) $K_1 =$	2) $K_2 =$	
1	1	for stationary anntenna or emitters
2.5	5	for mechanically scanning antenna
120	360	for rotating antenna
A/3	A	for electronically scanning antenna (A - angle of scanning in degrees)

SUMMARY OF SESSION J: THE DOSIMETRY HANDBOOK

P. A. MASON
*Air Force Research Laboratory, Human Effectiveness Directorate,
Directed Energy Bioeffects Division, Brooks AFB, TX, 78235 and
Veridian Engineering, Inc., San Antonio, TX, 78216.*

The two reports in this session provide fascinating insight into the past, present, and future of the *Radiofrequency Radiation Dosimetry Handbook*. Mr. John C. Mitchell presents a historical perspective on the origins of the handbook. As stated by Mr. Mitchell, the handbook was essential to "assist biological scientist to: (1) design better experiments, (2) more correctly interpret existing biological effects studies, and (3) extrapolate from animal studies performed under various laboratory conditions to likely human response under realistic (real world) RFR exposures." Development of the four editions of the dosimetry handbook was clearly a passion of several individuals working as a coherent group. Two important qualities of the handbook were that it was written at a level that was easy to comprehend by the novice researcher and the data withstood the test-of-time. Specific absorption rate (SAR) values predicted by more current computer hardware and software are similar to those presented in the dosimetry handbook. Testimonial to the importance of the Radiofrequency Radiation Dosimetry Handbook is that more than 2000 hardcopies have been distributed and it is one of the most referenced publications in bioelectromagnetic research.

Over the past several years, there has been discussion about producing a newer version of the dosimetry handbook. Advances in technologies utilized in dosimetry measurements permit better mapping of the electromagnetic field (EMF) and assessment of the biological responses to EMF exposure. Increased computer power permits the prediction of localized SAR values at higher resolutions. Dr. John M. Ziriax presents a peek into the possible future of the dosimetry handbook. It is critical that future versions of the dosimetry handbook remain user friendly and maintain the high quality of scientific data reported in the four previous versions. The content will be expanded to cover the frequency range from 0 to 300 GHz and include the tremendous advances in EMF dosimetry since 1986. The new handbook will include data for researchers conducting *in vivo* or *in vitro* experiments, epidemiologists, electrical engineering professionals, medical professionals, and those involved in compliance testing and establishing exposure standards. Dr. Ziriax outlines the creation of an internationally accepted EMF dosimetry handbook that explains how EMF dosimetry measurements and calculations should be performed. Development of the handbook will employ the Internet as an open international forum. Technologies such

B.J. Klauenberg and D. Miklavcic (eds.), Radio Frequency Radiation Dosimetry, 545-546.
© 2000 *Kluwer Academic Publishers. Printed in the Netherlands.*

as hypertext markup language (HTML), World Wide Web (WWW), virtual reality markup language (VRML), and JAVA^{TM} will be incorporated to permit sharing of ideas and results on an international basis. This international EMF dosimetry project could serve as a common group for harmonizing the EMF exposure standards that are currently unique to each country.

The need for a new version of the dosimetry handbook is evident. Mr. Mitchell predicts that questions will continue to be raised concerning the way in which the emitted energy from new EMF devices interacts with biological systems. In his closing remarks, Mr. Mitchell writes that evaluation of a system's effectiveness and operational safety will continue to be a challenge to those developing and operating this new technology.

HISTORICAL PERSPECTIVE ON THE RADIO FREQUENCY RADIATION DOSIMETRY HANDBOOK

Yesterday, Today, and Tomorrow

J. C. MITCHELL
USAF Senior Executive (Retired)
Department of the Air Force
U S Air Force Research Laboratory (AFMC)2

1. Introduction

The major difficulty in assessing the biological effects of Radio Frequency Electromagnetic Radiation (RFR) is to relate correctly the biological response being measured to the actual radiation dose causing the response. This must be accomplished while maintaining control of all other variables. It was sometime in 1972 that I had a meeting with Dr. Curtis C. Johnson and Dr. Arthur W. Guy, Department of Electrical Engineering, University of Washington, Seattle, Washington to discuss some of the more obvious dosimetry problems that were showing up in many biological effects studies being reported in the literature. In subsequent discussions with Curtis Johnson, we came up with the idea of developing a "dosimetry handbook". The original idea was to develop sufficient data in some type of "handbook" format to assist biological scientists to: (1) design better experiments, (2) more correctly interpret existing biological effects studies, and (3) extrapolate from animal studies performed under various laboratory conditions to likely human response under realistic (real world) RFR exposures. We felt that if such data could be developed and published for all to use, that the whole field of bioelectromagnetic research might be significantly improved. In early 1973, Curtis Johnson and Bill Guy met with me and other members of our staff at the School of Aerospace Medicine (USAFSAM) to further explore what might be done. Over the next twelve to eighteen months this idea for a handbook was discussed with many other people at numerous research centers and I pressed our organization for the necessary approval and funding to proceed. Subsequently in 1975 a decision was made at the USAF School of Aerospace Medicine to award a contract to the University of Utah to begin the development of the *"Radiofrequency Radiation Dosimetry Handbook"*. Dr. Curtis C. Johnson, who had moved from the University of Washington to the University of Utah, was selected to head up the University team and Mr. Stewart J. Allen was selected as the Project Scientist for the Air Force. In the following ten years, four editions of the handbook were produced; the first was published in 1976 and the last in 1986. More than 2000 copies have been distributed throughout the world and

B.J. Klauenberg and D. Miklavcic (eds.), Radio Frequency Radiation Dosimetry, 547-553.
© 2000 *Kluwer Academic Publishers. Printed in the Netherlands.*

this handbook remains as one of the most used and referenced publications in bioelectromagnetic research.

2. Background Information

It has been almost sixty years since the development of RADAR (radio detecting and ranging). Very soon after the first systems became operational, questions were raised regarding radiation exposures and possible harm to humans. This led to theoretical and empirical studies of all kinds to assess the biological effects of radio frequency and microwave electromagnetic radiation. In most of the early experimental studies, the biological response was recorded in terms of the radiation levels emitted by the exposure device or in terms of the free-field exposure levels incident on the test subjects.

The U. S. Air Force became acutely aware of the potential hazards of these new radiation emitting systems and during the decade 1950 to 1960 they played a major role in a tri-service program to assess the biological effects of microwave radiation. The three United States services (Army, Navy, and Air Force) expended more than fifteen million dollars over a period of about five years to establish safety guidelines regarding human exposures to radio frequency and microwave radiation. The research results were published in annual proceedings in 1957 [1], 1958 [2], and 1959 [3].

On the basis of these tri-service studies, a "ten milliwatt" (10 mW/cm^2) exposure level was established and it became widely accepted as a reasonable "microwave radiation" safety guideline for most operational situations in the military. The tri-service data base was so well accepted that the Air Force felt no need for further research and by 1961 most all USAF sponsored research was terminated. Shortly thereafter, the U. S. Air Force began to develop a series of high-powered radar systems that operated in the 3 to 30 MHz frequency range. The program offices for these systems were confident that all public exposures from such systems were well within acceptable safety levels, but public persistence for empirical data led the developers to ask the U. S. Air Force School of Aerospace Medicine (USAFSAM) for biological effects evaluations. As such questions were explored, it became evident that little, if any, useful biological effects data were available at the frequencies of interest. This prompted the school (USAFSAM) in 1968 to initiate a series of biological studies to assess the effects of HF band (3-30 MHz) electromagnetic radiation.

From the onset, it was realized that the relatively long wavelengths of 3-30 MHz radiation would require some new and unique tools to conduct meaningful research using laboratory animals and make it relevant to human exposures. Several different ideas were explored before we settled on a closed stripline design. The idea for this design and the original specifications were derived from some work that had been done by S. B. Cahn [4], published in March 1955. Subsequently, a one-tenth scale model was developed and tested for us by Glenn Skaggs [5] at the Naval Research Laboratory in Washington, D. C. After some preliminary studies, the first-of-its-kind large-scale TEM mode test chamber (HF band stripline) was constructed by personnel at

USAFSAM during the period May 1969 to February 1970 [6]. This rectangular coaxial device had dimensions of approximately 1.5 m by 2.75 m by 10 m with a 2 m wide center conductor. The instrumentation system, consisting of portable E- and H-field probes and fixed E-field probes, was developed by personnel at the National Bureau of Standards in Boulder Colorado [6]. Up to 12 animals as large as rhesus monkeys could be exposed simultaneously to uniform fields ranging in power density up to 700 mW/cm^2. The fact that it was a closed and balanced system with matched impedance at each end allowed accurate absorbed power measurements in the test subjects and this became very important for experimental validation of theoretical data in the dosimetry handbook.

It is to be noted that most of the biological effects studies conducted by USAFSAM from 1969 through 1975 were never published in the open literature. They were performed for and with the financial support of the project offices developing these new "over-the-horizon-backscatter" radar systems. The work was performed to assure ourselves (the Air Force) that these systems could be operated without harm to our Air Force personnel or to those people residing in the communities surrounding the Air Force radar sites. As a side benefit of these studies, important new data were developed concerning the absorption of 3-30 MHz radiation. These data also led to the publication of the first "frequency-dependent" RFR exposure standard in the United States in November 1975 (Air Force Regulation 161-42) [7]. By the time the fourth edition of the dosimetry handbook was published in 1986, most of the world's RFR exposure standards embodied this frequency dependent aspect.

3. RFR Dosimetry Handbook (September 1976) [8]

This first "Radiofrequency Radiation Dosimetry Handbook" provided a series of theoretical curves to enable the user to estimate the specific absorption rate (SAR) in humans at a given frequency as a function of the incident power density and to make corresponding calculations for animals exposed under relevant laboratory conditions. It also provided a "Rules of Thumb" section for making quick estimates of power absorption, and a table summarizing some typical values. This section served as a good educational tool for biological scientists who were just beginning to deal with the dosimetry problems of bioelectromagnetic research. The data in this first handbook were based on calculations assuming both ellipsoidal and spheroidal homogeneous models of man and animals. It was clearly recognized that spheroids and ellipsoids were very crude models of humans but it turned out to be a very good starting place. Using the HF band stripline at USAFSAM reasonably good experimental correlations had already been obtained by Stewart Allen [9] for a sitting rhesus monkey compared to the theoretical values using an ellipsoidal model at low frequencies. These studies gave us confidence that we were on the right track. The data in this handbook were calculated using three different methods: (1) a perturbation technique, valid for a homogeneous body that is small compared to the wavelength of the incident electromagnetic radiation; (2) the extended boundary condition method, a numerical calculation that is

550

useful for homogeneous bodies up to frequencies near resonance; and (3) a geometrical optics approximation, valid for an object that is large compared to the wavelength of the electromagnetic radiation. A brief description of each method is given in the handbook.

This first handbook, developed in only nine months, was a great start but, it was lacking in several areas. Of most importance was the fact the models used did not allow calculations at the higher frequencies of interest, especially 2.45 GHz where many of the animal experiments were being performed. Secondly, no experimental data were included to substantiate the theoretical data. As it turned out, this lack of experimental data gave researchers a challenge to test the theoretical data and, in time the data in the handbook stood the test. For example, this first handbook gives SAR's of 1.1 W/kg for a medium-size mouse exposed at 1.5 GHz to an incident power density of 1 mW/cm^2, 0.035 W/kg for an average size man exposed at a frequency of 2 GHz, and 0.23 W/kg at a resonant frequency of 75 MHz. These values show a close comparison to the results in the fourth edition of 1.1 W/kg for a medium size mouse exposed at a frequency of 1.5 GHz (0% difference), 0.028 W/kg for an average size man exposed at a frequency of 2 GHz (-20%), and a resonant frequency of 75 MHz (0%) and 0.25 W/kg (+9%) for the average size man.

4. Second Edition (May 1978) [10]

The second edition of the handbook extended the frequency coverage from 10 MHz to 100 GHz and included many other improvements. It included: (1) SAR calculations above resonance for prolate spheroid models, (2) predictable scatter patterns for prolate spheroid models, (3) localized SAR predictions, (4) metabolic rates and heat-stress predictions, (5) dielectric constants as a function of frequency for various tissues, (6) ground-plane effects, and (7) calculated and measured SAR comparisons based on experimental data published by Bill Guy, Om Gandhi, Stewart Allen, and Richard Olsen.

This edition of the handbook was very well received and became an instant hit among the scientific community. With its distribution at the International Microwave Power Institute symposium in June 1978, there became a real excitement that this document provided the information necessary to interpret RFR bioeffects data from animal experiments and apply the results to predict likely effects in humans across a broad range of frequencies. The experimental data showed excellent agreement with the calculated values. The data were bracketed by the SAR in rodents measured by Om Gandhi at the University of Utah and shown to be slightly below that predicted and the SAR in rodents measured by Stewart Allen at the School of Aerospace Medicine shown to be slightly above the theoretical values.

In the final analysis, I believe this edition of the handbook had a major impact on the development and acceptance of ANSI C95.1-1982 [11], the safety guideline for RFR exposures. Of the 32 bioeffects papers cited in the standard, 28 were performed in the frequency range of 0.9 to 3 GHz and 15 were performed at 2.45 GHz. In 26 of the 32 experiments rats, mice, or hamsters were used. Without the data provided by the

"*Radiofrequency Radiation Dosimetry Handbook*", the results of these experiments could not have been generalized to human exposures over the 300 KHz to 100 GHz frequency range. It is to be noted that the ANSI standard incorporated frequency dependence and used SAR as a common denominator for assessing the biological effects. It limited the average whole-body absorption in humans to 0.4 W/kg or less and the spatial peak SAR to 8 W/kg as averaged over any one gram of tissue.

5. Third Edition (August 1980) [12]

The main thrust of the third edition was the presentation of new data on the radiation absorption in humans and animals in the near-field zone of RFR sources. It included: (1) a complete section on dosimetric measurement techniques, (2) newly published data on complex dielectric constants, (3) calculations of SAR for circularly and elliptically polarized electromagnetic waves, (4) near-field SAR calculations, and (5) explanations of qualitative near-field dosimetry. In my opinion, this edition, in a significant way, introduced the complexities of the dosimetry problem in the real world. It provided in one place the nature of these electromagnetic fields of interest, the importance of understanding how temperature relates to bioeffects assessments, the importance of an accurate description of the dosimetric measurement technique and a tutorial on dosimetry requirements including a suggested "research data sheet" to make replication of important biological effects experiments possible. Also included is a section on RFR exposure devices commonly used for bioeffects studies including the pitfalls to watch for and a description of common techniques for measuring the SAR in biologic systems and phantom models.

6. Fourth Edition (October 1986) [13]

The fourth edition provides a convenient compilation of information contained in the previous editions, including updated tables of published data, and adds significant new information. After a brief introduction, there is a section on how to use this handbook. This is followed by an extensive chapter on the basics of electromagnetic radiation. Chapter 4 gives information on dielectric properties including updated data from previous editions. Theoretical dosimetry is discussed in Chapter 5, including a brief discussion of methods used for dosimetric calculations, data for models of biological systems, and a tabulated summary of published work in theoretical dosimetry. Chapter 6 contains new calculated average-SAR data for spheroidal models at frequencies well beyond resonance and for near-field conditions. Chapter 7 contains a complete history of experimental dosimetry and a discussion of experimental techniques. Chapter 8 contains updated experimental dosimetric data. Chapter 9 contains information on very low frequency radiation dosimetry, which is not found in any previous edition. The information on heat-response calculations in Chapter 10 is compiled and updated from

552

previous editions. This edition ends with a discussion of RFR safety standards that were current at the time of publication.

Dr. Curtis Johnson died just before the second edition was published. As I recall the first discussions I had with him in 1972 and our subsequent decisions to develop the "*Dosimetry Handbook*", I believe he would be very pleased with the way it turned out. The original idea was to provide substantial information in a "handbook" form that would allow biological scientists to: (1) design better experiments, (2) interpret existing bioeffects data, and (3) make correct extrapolations from animal studies conducted under laboratory conditions to likely human response under realistic RFR exposures. I feel these goals were achieved! The acceptance and use of the "*Radiofrequency Radiation Dosimetry Handbook*" is well established. The fourth edition references 182 primary authors; among them were the "who's who" in RFR research up to the time of publication. It has become one of the most, if not the most, cited references in bioelectromagnetic research.

7. Future Needs And How Do We Get There From Here?

It has been thirteen years since the publication of the last edition of the *Radiofrequency Radiation Dosimetry Handbook*. This NATO Advanced Research Workshop may well point the way to develop a new state-of-the-art dosimetry handbook which, in its final form, undoubtedly will be provided in a book, on a compact disc, and available on the world-wide web. Bioelectromagnetic research continues to provide many new challenges and radiation dosimetry is still at the heart of the problem.

The future needs in military operations undoubtedly will include some specialized applications of electromagnetic energy. As these new technologies are developed, questions will continue to be raised concerning the way in which the emitted energy interacts with biological systems. The dosimetric requirements will be identified by those doing the systems research. I can foresee new requirements arising from many different types of systems including those with unique pulse profiles and including the magnetic component of systems using high powered electromagnets. Evaluation of the system's effectiveness and operational safety will continue to be a challenge to those developing and operating this new technology. It will also remain a challenge to provide an accurate assessment of any and all reported biological effects to understand cause and effect.

8. Acknowledgements

As I look back on the history of the development of the four editions of the dosimetry handbook beginning in 1972 and ending in 1986, I am reminded of many talented and dedicated people who made the whole thing possible. I cannot begin to name them all for I am sure I would forget some even if I tried. On the other hand, I would be remiss if I did not name at least those persons whom I consider made the most significant contributions to this project. In alphabetical order they are: Stewart Allen, Peter Barber, Carl Durney, Om Gandhi, Bill Guy, William Hurt, Magdy Iskander, Curtis Johnson (deceased), and Habib Massoudi.

Also, I want to acknowledge these other collaborators, contributors, and reviewers who served in special ways. They are in alphabetical order: Eleanor Adair, Kenneth Foster, James Lords, Luis Lozano (deceased), Donald McRee, Richard Olsen, David Ryser, Herman Schwan, and Tom Tenforde.

In the preparation of this paper I have also used information from the notes of Stewart Allen who reminisced about the development of the handbooks in a verbal presentation at a conference honoring Dr. Carl H. Durney upon his retirement from the University of Utah last year. My personal thanks to"Stu" for the use of the notes and for his critique of this paper.

9. References

1. First Tri-service conference on biological hazards of microwave radiation (1957) Rome Air Development Center, NY.
2. Proceedings of the Second Tri-service conference on biological effects of microwave energy (1958) University of Virginia, Charlottesville, VA.
3. Susskind, C. (1959) Proceedings of the third annual tri-service conference on biological effects of microwave radiating equipment.
4. Cahn, S.B. (1955) Problems in strip transmission lines, IRE Group on Microwave Theory and Techniques, **MIT-3**, 246-251.
5. Skaggs, G.A. (1970) High frequency exposure chamber for radiobiological research, Radar Techniques Branch, Radar Division, Naval Research Laboratory, Washington, D. C. (Unpublished).
6. Mitchell J.C. (1970) *A Radiofrequency Radiation Exposure Apparatus, USAFSAM-TR-70-43*, Brooks Air Force Base, TX.
7. U. S. Air Force Regulation 161-42 (1975), Radiofrequency Radiation Health Hazard Control .
8. Johnson, C.C., Durney, C.H., Barber P.W., and Massoudi, H. (1976) University of Utah; Allen, S.J. and Mitchell, J.C. *Radiofrequency Radiation Dosimetry Handbook, USAFSAM-TR-76-35*, Brooks Air Force Base, TX.
9. Allen S.J. (1976) *Measurement of Radiofrequency Power Absorption in Monkeys, Monkey Phantoms, and Human Phantoms Exposed to 10-50 MHz Fields, USAFSAM-TR-76-5*, Brooks Air Force Base, TX.
10. Durney C.H., Johnson C.C., Barber P.W., Massoudi H., Iskander M.F., Lords, J.L., Ryser, D.K., University of Utah; Allen S.J. and Mitchell, J.C. (1978) *Radiofrequency Radiation Dosimetry Handbook (Second Edition), USAFSAM-TR-78-22*, Brooks Air Force Base, TX.
11. ANSI C95.1-1982 (1982) Safety levels with respect to human exposure to radiofrequency electromagnetic fields, 300 KHz to 100 GHz. American National Standards Institute.
12. Durney, C.H., Iskander, M.F., Massoudi, H., University of Utah; Allen S.J. and Mitchell J.C. (1980) *Radiofrequency Radiation Dosimetry Handbook (Third Edition), USAFSAM-TR-80-32*, Brooks Air Force Base, TX.
13. Durney C.H., Hassoudi, H., and Iskander M.F. (1986) University of Utah, *Radiofrequency Radiation Dosimetry Handbook (Fourth Edition), USAFSAM-TR-85-73*, Brooks Air Force Base, TX.

DOSIMETRY MEASUREMENTS AND MODELING: INTERACTIVE PRESENTATIONS IN THE NEW DOSIMETRY HANDBOOK

J. M. ZIRIAX[1], P. A. MASON[2], W. D. HURT[3], J. A. D'ANDREA[1], M. A. ARCE[4] and J. F. PETRI III[5].
[1]Naval Health Research Center Detachment at Brooks AFB, TX 78235, [2]Veridian, Inc., San Antonio, TX 78216, [3]Directed Energy Bioeffects Division, Brooks AFB, TX 78235, [4]GTE (GSC), Brooks AFB, TX 78235, [5]CACI (ARG), Brooks AFB, TX 78235.

1. Introduction

The fourth edition of the Radiofrequency Radiation (RFR) Dosimetry Handbook was published in 1986 [1]. Just as the three that proceeded it, it became an essential reference for laboratories studying the bioeffects of RFR. As described by John Mitchell in this volume [2], it was the product of a United States Air Force effort to advance the area of RFR bioeffects dosimetry.

In 1986, a paper book was the only practical means for distributing the dosimetry handbook. Since 1986 not only have there been important developments in RFR dosimetry as described in other chapters in this volume, but a new powerful media for presenting information has exploded onto the world scene. This new electronic media raises many possibilities, not only for the next version of the Handbook, but also for scientific literature in general. These changes will effect not only the content, but also how a document will be used and even how it might be created.

In the future, scientific literature will increasingly be an electronic literature. When testifying before the United States Congress, the Director of the National Library of Medicine, Donald A.B. Lindberg, M.D., reported that as of September 1998, 273 medical journals were available online [3]. Compared to only 40 fourteen months before. Dr. Lindberg referred to getting the complete text of all medical journals available online as the "Holy Grail." The time is right for an electronic version of the RFR dosimetry handbook.

The next handbook should maintain the high quality of scientific results reported in the previous four editions. In addition to the hardcopy version, the next handbook could be a dynamic tool tailored to the needs of each user. The purpose of this paper is to present a snapshot of some of the possibilities. Without a doubt, options will continue to expand. The rapid evolution of the personal computer and the Internet are at the core of the continuing development of electronic media. Therefore, if a new handbook is to keep apace of changes in media and the very active area of RFR bioeffects dosimetry, it

B.J. Klauenberg and D. Miklavcic (eds.), Radio Frequency Radiation Dosimetry, 555-564.

cannot be allowed to sit static for several years between revisions. First, let us consider some technologies that may be applied to a new handbook and to scientific literature in general and then how they might be applied to the creation of both an electronic and hardcopy handbook.

2. Hypertext Markup Language

The first technology is hypertext markup language or HTML. HTML was the first language of the World Wide Web (WWW) browser. It was a system for encoding formatted text and graphics, plus a way for the reader to jump from one predefined location in the text, a link, to a new location which may exist in another file, on another computer anywhere on the Internet. This frees the reader to follow his or her own information needs rather than reading the text in the order chosen by the author. Furthermore, it allows the author to direct the reader to other material immediately rather than just referencing it.

Unlike paper texts, HTML documents may be searched automatically. In fact, a whole Internet industry has arisen to provide searches of Internet documents for the cost of viewing advertisements. HTML can be used to cause the remote WWW sever to do more then just return search results. The user can fill out forms including credit card information, send and receive e-mail and many other activities. While HTML documents are typically accessed on the Internet, they may also be stored and accessed on a local storage device.

HTML is still evolving. Recent revisions include mathematical notation and mapping links to areas on images among other improvements. The evolution of HTML has been matched by powerful developments in its counterpart, the WWW server.
The next version of the dosimetry handbook could be radically different from the fourth edition even as it currently exists on the WWW. The new handbook could include more links to other parts of itself, to other relevant Internet documents. It could be easily copied and customized, and updated.

3. WWW Servers

Users are largely unaware of the WWW servers either where the servers are located or when users switch from one to another. This is by design, since the user sees only the page currently displayed by the WWW browser on their local computer. However, developments on the WWW servers have increased their potential offerings to users of WWW. The server not only delivers HTML pages as requested, but can also run programs according to instructions given through the HTML pages sent to the browser.

The ability to search the text of the electronic handbook on demand makes traditional predefined index unnecessary. Elements of the WWW server have also become available as options to the browser. With PC based search tools, such as Alta

Vista Discovery, a stand-alone PC can search the handbook. This gives the handbook something it never had before, an index.

An electronic handbook need not be limited to the type of content that can be printed. Large, otherwise unpublishable, data sets, an archive of published data, multicolor images, and computer programs will be part of an electronic handbook. Unlike a paper-only handbook, the new handbook can be delivered inexpensively over the Internet from many possible WWW sites.

4. Virtual Reality Markup Language

Virtual reality markup language or VRML works as an add-on to any WWW browser. It provides three-dimensional animation and sound in which the user can control the point of view. It connects with HTML and Java, described below. VRML could be used to allow a user to construct three dimensional models, "fly-through" a volume of data, or to construct a three dimensional interface to the dosimetry handbook.

5. Java

Java is a programming language add-on for a browser. A Java program can be stored on a WWW server or locally. One of the best features of Java is that the same compiled Java program can run on any computer with a Java runtime environment. Java can be mixed with HTML and VRML. As with any computer language it is possible to write entire applications in Java.

The handbook contains many formulas and graphs that show the behavior of RFR under a variety of conditions. Java applications could create all of these and more on the fly according to the immediate requirements of the user. An example is shown in Figure 1. This Java application or applet, as they are called, performs the prolate spheroid calculations for a number of pre-constructed and user-definable models. Certainly, over the years many laboratories have created other software tools that would be useful to the dosimetry community at large. These could be translated into Java and become part of the electronic handbook.

With Java applets, it is also possible for an electronic handbook to provide a gateway to "super"-computer-based dosimetry modeling and analyze the results as well. An applet for exploring five-dimensional data is shown in Figure 2. The silhouette of the monkey head at the bottom is used to select the horizontal slice of the data to be displayed in the two images at the top. The image on the left shows the tissue type according to the legend on the far left. The image on the right shows the specific absorption rate (SAR) as color coded according to the scale on the far right. When a location is selected in either image, the point in both images starts flashing and the tissue type and SAR are displayed at the top of the applet.

558

6. JINI

JINI is only 400kb of Java code. When connected to a network, JINI devices introduce themselves and their services to the other JINI devices on the network. The goal of JINI is to allow this process to proceed without the configuration issues that characterize current network systems. Java applets can then be used to control JINI devices through the network. Just as programs could be exchanged, Java applets for controlling JINI-based devices could also be shared via an electronic handbook. This kind of sharing makes the documentation and replication of experimental procedures much more precise than can be achieved by text-only description.

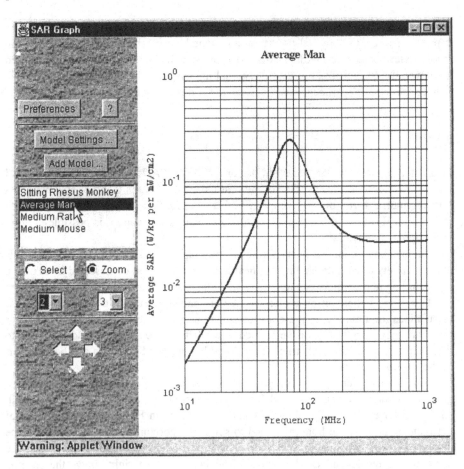

Figure 1. An example window created by a Java applet. This applet produces the calculated planewave average SAR in a number of standard and user-defined prolate spheroids. Edwin Grubbs of Cymitar Technology Group, Inc., San Antonio, Texas, created this applet from a program written by Don Hatcher at NHRC-Det.

7. Powerful Personal Computers

Enormous increases in the speed and storage capacity of personal computers have fueled a revolution, tasks that used to require specialized and expensive hardware and software are easily performed on inexpensive desktop computers. Recent advances include multi-gigabyte memory capacities, multiple processors, a variety of large-capacity storage media, and even clustering. The most powerful computers are actually many smaller computers connected together by dedicated high-speed networks and specially written software. Recently, a number of PC-based clustering technologies have been developed. The Beowulf system is built from a cluster of PC's running GNU/Linux and developed by NASA. Another system under development is HPVM developed by the National Computational Science Alliance at University of Illinois at Urbana-Champaign. Moreover, there are others as well. While these systems benefit from high-speed dedicated networking hardware, they will also work with standard network hardware. This allows the daytime desktop PC workstations to become a supercomputer at night.

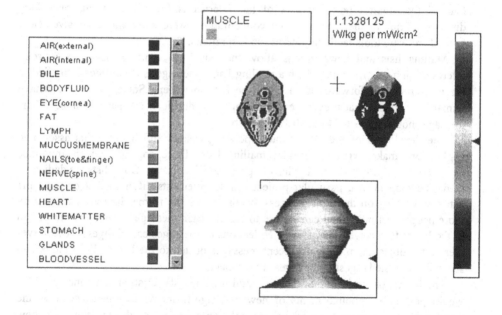

Figure 2. Output of a Java applet that allows the user to explore the SAR output of a finite difference-time domain (FD-TD) program. On the left is a list of tissue types and their associated color codes. The slider allows the user to view the entire list. A profile of the model, a rhesus monkey head, is shown in the bottom center of the figure. Above it and to the left is an image of the coded tissue types for the slice selected in the profile. To the right are the color-coded SAR values. Slices are selected by clicking on the profile. Clicking on either the tissue slice or the SAR slice will display the tissue type and SAR in the boxes at the top and the pointer at the right to move to the SAR color at that point. Edwin Grubbs (Cymitar Technology Group, Inc., San Antonio, Texas) wrote this applet.

Recently, several research groups, including the Navy-Air Force team at Brooks AFB, have constructed detailed biological models to be used in calculating RFR dose. Although, models of the human and rhesus monkey are still works in progress, it is apparent that for these models, computer memory requirements (See Table 1) are within the reach of small PC clusters and in some cases of single PCs.

TABLE 1. FD-TD Memory Required for various models developed at Brooks AFB

Model	RAM Memory Required
Rat model	17 MB
Monkey Head	140 MB
Rhesus Monkey	2.1 GB
Human (5 mm thick slices)	~4.0 GB
Human (1 mm thick slices)	~18 GB

8. Internet Collaborations

One of the greatest potential uses of the Internet is for collaborations over long distances. Prior to the Internet, such collaborations were slow and expensive. Now several software tools facilitate Internet collaborations.

Mailing lists and news groups allow individuals to send messages to a group interested in a specific topic. With a mailing list, a message is electronically mailed to the list server that forwards it on to all the list's members. Some mailing lists have human moderators that preview each message and decide if all, part, or none of the message should be posted to the list's subscribers.

The developers of the GNU/Linux operating system and many of its associated applications make extensive uses of mailing lists. There are lists, which announce developments and lists discussing how to promote GNU/Linux. Other lists are dedicated to the developers of a particular project. These latter lists differ in that only people actively working on the project access the list, while the former lists are accessed by more people. An individual can choose to receive each message as soon as it is posted to the list or to receive only periodic collections of messages called digests. The server creates the digest from the most recent messages posted to the list, adds an index and sends the collected digest to the digest subscribers.

The handbook project could have several mailing lists. First, an announcement list would post only announcements of new and significant developments such as the starting of a new project, an appeal for help on a project or the location of a new section. Next would be a list for the Editorial Board and the members of the program committee. This list would allow members of these committees to communicate with each other on the management of the handbook project overall. The Developers' list would serve as a discussion area for advocating and planning new projects or sections of the handbook. Once a project had sufficient interest the interested parties would, through the project leader, announce the new project on the announcement list and a new mailing list would be created for those working on that project. The project leader would control that list.

Eric S. Raymond [4] has described this process in his article "The Cathedral and the Bazaar." Although he is concerned with writing software, his observations concerning the process apply to writing a handbook as well. As needed, I have taken the liberty of translating some of his major guidelines to writing a handbook:

(1) Every good work starts by scratching a writer's personal itch. Write something useful to you and your organization.

(2) Good writers know what to write. Great ones know what to rewrite (and reuse).

(3) ``Plan to throw one away; you will, anyhow."

(4) Treating your audience as co-authors is your least-hassle route to rapid document improvement and effective rewriting.

(5) Release early. Release often. Listen to your customers.

(6) Given a large enough audience and co-author base, almost every problem will be characterized quickly and the fix obvious to someone

(7) It is clear that one cannot write from the ground up in bazaar style. One can edit and improve in bazaar style, but it would be very hard to originate a project in bazaar mode. Your nascent author community needs to have something to start with.

The fourth edition will make an excellent starting point for the International Electromagnetic Field (EMF) Dosimetry Project.

9. The Electronic Handbook

Deciding how to apply these technologies to the electronic handbook will be the job of the developers working on each project. Here are some speculations. First, hypertext in the form of HTML will allow the handbook to be searched electronically. As an electronic document, the handbook can reside in many places on the Internet and even on individual computers. Second, graphs and formulas will be incorporated into Java applets with increased functionality including user customizations. Third, calculations and figures will be exportable to other applications. These suggest only a few of the software tools that could be incorporated into the handbook. Fourth, HTML links to related WWW sites would mean that the user is not limited to the contents of the handbook. Rather, the handbook would be a jumping off point to other relevant electronic documents. Fifth, the ability to use a WWW browser's point-and-click interface to execute complex tasks on the remote WWW server can be exploited in a number of ways, these include searching databases and accessing supercomputer resources all from a desktop PC. Finally, a WWW server could deliver copies of the handbook as HTLM pages or in a printable format. The cost of printing would become the only cost for distributing the electronic handbook and that cost would be controlled and born completely by the end user. The low cost of distribution would make frequent incremental or evolutionary revisions of the handbook a practical option.

10. Creating The New Handbook

Past handbooks were created by those who receive the contract from the U.S. Air Force and with the contract monitor's approval. The next edition of the handbook could be created in a very different way. The process could be opened up to all interested parties as part of Internet collaboration. The potential advantages of this alternative are several:

(1) With more people and organizations involved the topic areas could be greatly expanded.

(2) With little or no distribution costs, new revisions could be released more frequently. The new handbook would evolve rather than leap forward all at once.

(3) As an inclusive process, interested parties work on sections important to them and their organizations.

(4) Each user will be able to pick and chose the sections relevant to their needs to create a custom handbook.

(5) Finally, each section of the handbook would be written and reviewed by the people most interested in that section. The quality of the handbook would be the joint responsibility of all that contribute to it and use it.

11. The Open Handbook

The first four editions of the handbook were created when the Air Force contracted with a group to do the work. The content was described in the contract and the authors were paid by the contract. The finished product was then presented as a completed technical report.

In order for the next handbook to be created through an open evolutionary process, the authors will not be working according to the specifications of a single fund source. Rather, the creators will work on the new handbook because of the advantages to themselves and their organizations in playing a role in creating an internationally accepted EMF dosimetry handbook. Participants will most likely be volunteers. This is similar to the system used by scientific journals in which the editors and reviewers are also volunteers.

The most important requirement for creating an open handbook is an active community of participants. Areas of sufficient concern to the EMF dosimetry community will become projects. A project would start with a discussion and the selection of a volunteer leader. The leader must be acceptable to the community. The leader's first job would be to define the starting point for the project. At this point, a mailing list would be created as the primary means for contributors to communicate with each other and the project leader. The starting point could be either a section of the current handbook or text from leader. The leader controls the project and is responsible for maintaining and encouraging those contributing to the project.

Additions and revisions submitted by the group are added, modified or rejected by the project leader. The leader posts a stable version on the International EMF Dosimetry Project's web site for those wishing to use the current version either for downloading or browsing. The leader also posts a developer's version that is frequently updated with input from the leader and the working group. A project exists as long as parties contribute.

A likely causality of the evolutionary approach to the new handbook is the notion of "editions." Instead, the handbook is likely to have versions that have fractional increments such as Version 4.1.2. This would allow a gradual evolution of the new handbook from the existing electronic version, 4.0.0. As sections are re-written or updated, the version could be incremented. Users could download updates over the Internet or just use the web site or its mirrors.

Starting from version 4.0.0 has advantages. First, it exists immediately. Second, it will provide a common starting point from which to first update existing sections. The relatively simpler process of updating the existing document will encourage the development of an electronic culture. This culture will be needed for the more ambitious projects to come.

12. The Handbook: What's New?

The Air Force Radiofrequency Radiation Dosimetry Handbook was created to change the area of RFR bioeffects research. It succeeded. The document envisioned at the NATO "Advanced Research Workshop on Radio Frequency Radiation Dosimetry and Its Relationship to the Biological Effects of Electromagnetic Fields" in Gozd Martuljek, Slovenia will start with the fourth edition. However, it is likely to evolve into a quite different document. The fourth edition of the handbook represented the contribution of a few individuals and the resources of the United States Air Force. The next handbook will be created by an international group of scientists. Its ultimate form will be determined by the needs of the scientists and institutions concerned with EMF dosimetry and who take an active role in the evolution of the new handbook.

13. Notes

HTML is a product of World Wide Web Consortium. JAVA and Jini are products of Sun Microsystems, Inc. Mountain View California. AltaVista is a Trademark of Digital Equipment Corporation.

14. Disclaimer

Views presented are those of the authors and do not reflect the official policy or position of the Department of the Navy, Department of the Air Force, Department of

564

Defense, or U.S. Government. Trade names of materials and/or products of commercial or non-government organizations are cited as needed for precision. These citations do not constitute official endorsement or approval of the use of such commercial materials and/or products.

15. Acknowledgements

Research was funded by U.S. Navy Naval Medical Research and Development command under Work Unit 61153N MR04101.001-1603, and U.S. Air Force Materiel Command, Brooks AFB, TX, 78235, Air Force Office of Scientific Research awards (1995,1996) to Mr. William Hurt and U.S. Air Force contract F41624-96-C-9009. Edwin Grubbs of Cymitar Technology Group, Inc., San Antonio, TX, provided Java programs in Figures 1 and 2.

16. References

1. Durney, C.H., Massoudi, H., and Iskander, M.F. (1986) *Radiofrequency Radiation Dosimetry Handbook. USAF School of Aerospace Medicine Report, USAFSAM-TR-85-73*, Brooks Air Force Base, TX.
2. Mitchell, J.C. (1998) Historical Perspective on the Radio Frequency Radiation Dosimetry Handbook, in B.J. Klauenberg and D. Miklavčič (eds.) *Radio Frequency Radiation Dosimetry and Its Relationship to the Biological Effects of Electromagnetic Fields*. Kluwer Academic Publishers B.V. Dordrecht, The Netherlands, pp. 551-558.
3. Lindberg, D.A.B. (1998) Testimony before Congress as reported in *Gratefully Yours*. M. F. Brdlik (ed.) National Library of Medicine, Bethesda, Maryland. http://www.nlm.nih.gov/pubs/nlmnews/nlmnews.html.
4. Raymond, E. S. (1998) *The Cathedral and the Bazaar*. http://www.redhat.com/redhat/cathedral-bazaar/.

APPENDIX

The International EMF Dosimetry Project was created during this NATO/ARW meeting in Gozd Martuljek, Slovenia. Attendees devoted several days designing this project and below are the products of their tremendous efforts.

INTERNATIONAL EMF DOSIMETRY PROJECT

Mission Statement:
Establish an **International Resource** that provides state-of-the-science knowledge on EMF dosimetry.

Potential Users:
- Industry (compliance testing)
- Standard setting bodies (rationale for standards)
- Communications (compliance testing)
- Medical government (compliance testing – EMF interference)
- Independent Research Laboratories
- Universities
- Military

Overall Project Goals:
Promote the field of EMF dosimetry by the creation of an internationally accepted EMF dosimetry handbook and software that describe how EMF dosimetry measurements and calculations should be performed. Employing an open international forum, the Internet, the project should proceed more rapidly, at less cost, and the results internationally accepted. The project could serve as the common ground for harmonizing the EMF exposure standards that are currently unique to each country. The project has a Program Management Committee and an Editorial Board.

Program Management Committee: Members will develop yearly and long-term project goals and oversee their timely completion.

Editorial Board: Members will review: 1) yearly and long-term project goals developed by Program Management Committee, 2) content to be included in new Dosimetry Handbook, and 3) funding proposals. Review of funding proposals would be to facilitate inter-laboratory research, avoid duplication of effort, and maximize research dollars.

PROPOSED FORMAT OF NEW EMF DOSIMETRY HANDBOOK

Keep in the spirit of the 4th edition of the RFR Dosimetry Handbook and explain how to use dosimetry techniques in the laboratory setting.

565

B.J. Klauenberg and D. Miklavcic (eds.), Radio Frequency Radiation Dosimetry, 565-570.
© 2000 *Kluwer Academic Publishers. Printed in the Netherlands.*

New handbook should be user friendly from introductory level up to advanced research level.

Three operational levels within the handbook:
- Read about existing dosimetry research data
- Interact with software providing global access to dosimetry data and programs
- Forum for establishing contacts (i.e., referral list for services)

PROPOSED CONTENT OF NEW EMF DOSIMETRY HANDBOOK

New handbook should include data for "in vivo" and "in vitro" settings and epidemiologists, as well as for the electrical engineering professionals and those involved in establishing exposure standards and compliance testing.

Sections should be divided by frequency range as required by applicable dosimetry techniques.

Topics in 4th edition of the RFR Dosimetry Handbook (to be updated as necessary):
Basics of Electromagnetics
Permittivity Properties
Theoretical Dosimetry
Calculated Dosimetric Data
Experimental Dosimetry
Thermal Responses of Man and Animals
Tissue Stimulation Responses of Man and Animals
EMF Safety Standards

General topics to be added:
Derivation of exposure standards
Harmonization of exposure standards
Harmonize exposure standards across frequency range (e.g., millimeter waves to lasers)
Parameters for EMF dosimetry
Parameters for exposure assessment
Standard protocols for exposure assessment
Ultra-wide-band exposures
Near-field exposures (cellular telephones and others)
Partial body exposures
Grounded exposures
Static fields
Pulsed fields
Cavity exposure systems/multi-mode environment
Spark discharge/contact currents/induced currents
Experimental thermometry

Reference values
Prognosis on the development of cellular communications over the next decade
Rules of thumb/red flags
Precision vs. accuracy discussion

Modeling topics to be updated or added:
Include benchmark models of spheres, cubes and human models for validation of techniques and for inter-laboratory comparison

Describe which computer models and codes are available and how to request access

Develop research plan to develop additional models:
- Female adult human
- Child
- Pregnant female (with fetuses in different phases)
- Phantoms with pacemakers and other metal implants

Basic description of numerical methods (FD-TD, MoM, FEM, MMP, and others)

SAR values
- Influence of clothing, including protective clothing, on SAR values, especially at millimeter wave frequencies
- Whole body versus peak localized SAR values for different sources
- Compare calculated SAR values with exposure standards
- Develop thermoregulatory models to link SAR values to temperature rise
- Permittivity values (which database would be used?)
- New permittivity values needed
- Effects of permittivity values on SAR calculations

Current Density values
- Whole body vs. local current density
- Dielectric values
- Inductive vs. capacitive current
- Theoretical vs. experimental values
- Compare current density values with exposure standards

Computer Algorithms topics to be updated or added:
Preferred algorithm for each type of exposure (e.g., near- and far-field)

Advantages and disadvantages of each method for the specified kind of exposure

Pulsed fields & mechanical effects
Applicability of numerical methods in certain frequency ranges:
0-30 MHz (including the 300Hz – 10MHz intermediate range)

Quasi-static methods
- impedance-admittance-methods
- surface integration equation method
- finite integration technique (not the boundary value)
- MoM (for the sources)

High frequency range
- Mie Series
- MoM (FFT MoM, CG-FFT MoM)
- FEM (FETD), FVTD
- FDTD, $(FD)^2TD$, FIT
- EBCM, IEBCM, GMT (e.g., MMP)
- Genetic algorithms

Higher frequency range (> 5GHz)
- Quasi-optical techniques (ray tracing, GTD, UTD)

Material properties to be updated or added:

Electrical properties to be included:
- Include data for linear permittivity or electrical properties
- Develop research plan to obtain information on pathological tissue, anisotropy, high frequencies and non-linear responses

Thermal properties to be included:
- Specific heat capacity
- Thermal conductivity
- Density

Sources:
- Previous editions of the RFR Dosimetry Handbook
- Technical Reports
- Reference Man
- New sources to be identified

Physiological responses to consider:
- Blood perfusion/regional blood flow
- Sweat rate
- Evaporative heat loss
- Thermoregulatory responses and models
  Sources:
  - Existing hyperthermia data

Experimental procedures to be updated or added:

Review experimental design

Microdosimetry: Interfaced with theoretical studies

Review *in vitro* dosimetry and exposure systems
- Well-characterised systems
- Internal fields and samples

Describe obtaining internal measurements: E, H, current, and temperature
- Calibration procedures
- Estimation of errors
- Interpolation and extrapolation algorithms

Describe obtaining measurements of external fields
- Calibration procedures
- Full spectral characterisation, in space and time, of E- and H-fields
- Power density
- Estimation of errors

Use of experimental procedures to validate medical/computer models
- Phantoms

PROPOSED PUBLICATION FORMAT

NATO ARW Proceedings: With respect to the copyright agreement with Kluwer, determine if an electronic version of the NATO ARW proceedings could be placed on the WWW.

New International EMF Dosimetry Handbook: Inform potential publisher that exclusive copyrights will not be granted. The handbook will be in the public domain and available in electronic format to interested parties.

Handbook should be available in paper and electronic format. Electronic format will provide features such as links, index search, hyperlinks, calculation possibilities, etc.

Electronic version of the handbook will be located and managed on one server, and mirror sites should be established.

The WWW domain name will be purchased so that the handbook address will not change if the server is relocated.

Following WWW statistics will be followed:
- number of hits
- frequency of specific topics hits
- track domains from where the hits originate

Legal Issues to be Addressed
Limited liability
Copyright issues

SPEAKERS AT THE ADVANCED RESEARCH WORKSHOP
RADIO FREQUENCY RADIATION DOSIMETRY
Gozd Martuljek, Slovenia
October 12-16, 1998

Eleanor Adair
United States Air Force Research
Laboratory
AFRL/HEDR
8315 Hawks Rd. (Bldg. 1162)
Brooks AFB, TX 78235
USA
Phone: 210-536-4698
FAX: 210-536-3977
eleanor.adair@afrlars.brooks.af.mil

Robert Adair
Yale University
Department of Physics
P.O. Box 208121
New Haven, CT 06520
USA
Phone: 203-432-3370
FAX: 203-432-6125
adair@hepmail.physics.yale.edu

Jurgen Bernhardt
Bundesamt fur Strahlenschutz
Institut for Strahlenhygiene
Ingolstadter Landstrasse 1
D-85764 Oberschleissheim (BfS)
GERMANY
Phone: +49-89-316-03220
FAX: +49-89-315-2987
0006175887@mcimail.com

Philip Chadwick
National Radiological Protection Board
Chilton, Didcot
Oxon OX11 ORQ
UNITED KINGDOM
Phone: +44-1235 822731
FAX: +44-1235 833891
phil.chadwick@nrpb.org.uk

Alessandro Chiabrera
University of Genoa
ICEMB at DIBE
Via Opera Pia 11A
16145 Genova
ITALY
Phone: +39-10-353-2757
FAX: +39-10-353-2777
chiabrera@dibe.unige.it

Chung-Kwang Chou
Motorola, Inc.
Corp. Research Lab
8000 W. Sunrise Blvd. Rm. 2107
Plantation, FL 33322
USA
Phone: 954-723-5387
FAX: 954-723-5611
ECC017@email.mot.com

John D'Andrea
United States Navy
NMRDC – DET
8301 Navy Road
Brooks AFB, TX 78235
USA
Phone: 210-536-6527
FAX: 210-536-6537
john.dandrea@navy.brooks.af.mil

Guglielmo D'Inzeo
Univ. of Rome "La Sapienza"
Dept. of Electronic Engineering
Via Eudossiama 18
00184 Roma
ITALY
Phone: +39-6-44585853
FAX: +39-6-4742647
dinzeo@tce.ing.uniroma.it

571

John de Lorge
486 Citation Dr.
Cantonment, FL 32533
USA
Phone: 850-478-2581
Johndelorg@aol.com

Kenneth R. Foster
Univ. of Pennsylvania
Dept. of Bioengineering
220 S. 33rd St.
Philadelphia, PA 19104
USA
Phone: 215-898-8534
FAX: 610-896-0620
kfoster@seas.upenn.edu

Peter Gajšek
Slovenian Institute of Quality and
Metrology (SIQ)
Non-ionizing Radiation Laboratory
Tržaška cesta 2
1000 Ljubljana
SLOVENIA
Phone: +386 61 1778480
FAX: +386 61 1778444
peter.gajsek@siq.si

Youri Grigoriev
State Research Center of Russia
Center of Electromagnetic Safety
46 Zhivopisnaya St.
123182 Moscow
RUSSIA
Phone: +7 (095) 193-01-87
FAX: +7 (095) 190-3590
SEMS.1@g23.zelcom.zu

Peter Dimbylow
National Radiological Protection Board
Chilton, Didcot
Oxon OX11 ORQ
UNITED KINGDOM
Phone: +44-1235-822769
FAX: +44-1235-833891
peter.dimbylow@nrpb.org.uk

Camelia Gabriel
Microwave Consultants Ltd.
17B Woodford Rd.
London E18 2EL
UNITED KINGDOM
Phone: +44 181 989 5055
FAX: +44 181 989 6658
c.gabriel@ukonline.co.uk

Om Gandhi
University of Utah
Electrical Engineering
Merril Eng. Bldg., Rm. 3280
50 S. Central Campus Dr.
Salt Lake City, Utah 84112
USA
Phone: 801-581-7743
FAX: 801-581-5281
gandhi@ee.utah.edu

Klaus Hofmann
Forschungsinstitut fur
Hochfrequenzphysik (FGAN)
FHP-AuS
Neuenahrer Strabe 20
D-53343 Wachtberg-Werthhoven
GERMANY
Phone: +49-(0)228-9435250
FAX: +49-(0)228-340951
kwhofmann@fgan.de

William Hurt
United States Air Force Research
Laboratory
AFRL/HEDR
8308 Hawks Rd. (Bldg. 1184)
Brooks AFB, TX 78235
USA
Phone: 210-536-3167
FAX: 210-536-3977
william.hurt@afrlars.brooks.af.mil

Eugene Khizhnyak
Temple University Medical School
3420 N. Broad St.
Philadelphia, PA 19140
USA
Phone: 215-707-4012
FAX: 215-707-4324
eugen@nimbus.temple.edu

B. Jon Klauenberg
United States Air Force Research
Laboratory
AFRL/HEDR
8315 Hawks Rd. (Bldg. 1162)
Brooks AFB, TX 78235
USA
Phone: 210-536-4837
FAX: 210-536-3977
b.jon.klauenberg@afrlars.brooks.af.mil

Tapani Lahtinen
Kuopio University Hospital
Dept. of Radiotherapy & Oncology
FIN-70210 Kuopio
FINLAND
Phone: FIN+17+172910
FAX: FIN+17+172907
tapani.lahtinen@kuh.fi

Michel Israel
Physical Factors Department
National Centre of Hygiene
Med. Ecology and Nutr.
15 Dimitr Nestorov St.
1431 Sofia
BULGARIA
Phone: (003592) 596154
FAX: (003592) 9581277

Johnathan Kiel
United States Air Force Research
Laboratory
AFRL/HEDR
8315 Hawks Rd. (Bldg. 1162)
Brooks AFB, TX 78235
USA
Phone: 210-536-3583
FAX: 210-536-4716
johnathan.kiel@afrlars.brooks.af.mil

Niels Kuster
ETH Zurich
Lab of EMF & Microwave Elect.
IFH-ETZ, ETH Gloriastr. 35
Zurich CH-8092
SWITZERLAND
Phone: +41 1 632 2737
FAX: +41 1 632 1057
Kuster@ifh.ee.ethz.ch

David Land
University of Glasgow
Dept. of Physics & Astronomy
University Avenue
Glasgow G12 8QQ
UNITED KINGDOM
Phone: 044 0141 330 4703
FAX: 044 0141 330 5881
d.land@physics.gla.ac.uk

574

John Leonowich
Battelle Pacific NW National
Laboratory
502 Stagecoach Court
Richland, Washington 99352
USA
Phone: 509-375-6849
FAX: 509-375-6936
john.leonowich@afrlars.brooks.af.mil

Marko Markov
EMF Therapeutic
4 Squares Bus Center
1200 Mountain Creek Rd., Ste. 160
Chattanooga, TN 37405
USA
Phone: 423-876-1883
FAX: 423-876-1851
msmarkov@emftherapy.com

James Merritt
United States Air Force Research
Laboratory
AFRL/HEDR
8315 Hawks Rd. (Bldg. 1162)
Brooks AFB, TX 78235
USA
Phone: 210-536-4703
FAX: 210-536-3977
james.merritt@afrlars.brooks.af.mil

Kjell Hansson Mild
National Institute for Working Life
Non-Ionizing Radiation
P.O. Box 7654
Umea S-907 13
SWEDEN
Phone: 46 90 7865098
FAX: 46 90 7866508
mild@niwl.se

Alexander Lerchl
University of Muenster
Institute of Reproductive Medicine
Domagkstr. 11, D-48129
Muenster
GERMANY
Phone: +49 251 8356447
FAX: +49 251 8356093
lerchl@uni-muenster.de

Patrick Mason
Veridian
AFRL/HEDR/VEDA
8315 Hawks Rd. (Bldg. 1162)
Brooks AFB, TX 78235
USA
Phone: 210-536-2362
FAX: 210-534-2919
patrick.mason@afrlars.brooks.af.mil

Damijan Miklavčič
University of Ljubljana
Faculty of Electrical Engineering
Laboratory of Biocybernetics
Tržaška 25
SI-1000 Ljubljana
SLOVENIA
Phone: +386 61 1768 456
FAX: +386 61 1264 658
damijan@svarun.fe.uni-lj.si

John Mitchell
16750 Worthington
San Antonio, TX 78248
USA
Phone: 210-493-1642
JMITC 42332@aol.com

Michael Murphy
United States Air Force Research
Laboratory
AFRL/HEDR
8315 Hawks Rd. (Bldg. 1162)
Brooks AFB, TX 78235
USA
Phone: 210-536-4833
FAX: 210-536-3977
michael.murphy@afrlars.brooks.af.mil

Konstantina Nikita
National Technical Univ. of Athens
Dept. of Electrical & Computer
Engineering
Iroon Polytechniou 9
Zografos 15773, Athens
GREECE
Phone: +301 772 2285
FAX: +301 772 3557
knikita@cc.ece.ntua.gr

John Osepchuk
Full Spectrum Consulting
248 Deacon Haynes Rd.
Concord, MA 01742
USA
Phone: 978-287-5849
FAX: 978-318-9303
JMOsepchuk@compuserve.com

Vladislav Petin
Russian Academy of Medical Scientists
Medical Radiological Research Center
Korolev Str. 4, 249020 Obninsk
Kaluga Region
RUSSIA
Phone: 7(095)956-1439
FAX: 7(095)956-1440
mrrc@obninsk.ru

Jan Musil
National Institute of Public Health
National Reference Lab for the
Nonionizing EM Field
100 42 Praha 10
Srobarova 48
CZECH REPUBLIC
Phone: (4202)67082918
FAX: (4202)67082918
jmusil@bbs.szu.cz

Richard Olsen
NHRC – DET
8301 Navy Rd.
Brooks AFB, TX 78235
USA
Phone: 210-536-6535
FAX: 210-536-6439
richard.olsen@afrlars.brooks.af.mil

Andrei Pakhomov
McKesson Bio Services
8308 Hawks Rd. (Bldg. 1168)
Brooks AFB, TX 78235
USA
Phone: 210-536-5599
FAX: 210-536-5382
andrei.pakhomov@afrlars.brooks.af.mil

Patrick Reilly
The Johns Hopkins University
Applied Physics Consultant
12516 Davan Dr.
Silver Springs, MD 20904
USA
Phone: 301-680-9151
FAX: 301-680-1902
patrick.reilly@jhuapl.edu

Michael Repacholi
World Health Organization
Office of Global & Integrated
Environmental Health
CH-1211 Geneve 27
SWITZERLAND
Phone: +41 22 791 3427
FAX: +41 22 791 4123
repacholim@who.ch

Dina Šimunić
University of Zagreb
Dept. Radiocommun & Microwaves
Unska 3
Zagreb HR-10000
CROATIA
Phone: +385 1 6129 606
FAX: +385 1 6129 606
dina.simunic@fer.hr

Stanislaw Szmigielski
Military Institute of Hygiene &
Epidemiology
Center for Radiobiology & Radiation
Safety
4 Kozielska ST
01-163 Warsaw
POLAND
Phone: +(4822) 8380129
FAX: +(4822) 8381069
szmigielski@wihe.waw.pl

Gyorgy Thuroczy
Natl. Res. Inst. for Radiobiology &
Radiohygiene
Dept. of Non-Ionizing Radiation
H-1775 Budapest POB 101
HUNGARY
Phone: +36 1 226 5331
FAX: +36 1 226 5331
thuroczy@hp.osski.hu

Marat Rudakov
St. Petersburg State Inst. of
Electric Tech.
Dept. A9
1 – St. Krasnoarmeyskaya
198005, St. Petersburg
RUSSIA
Phone: +7 812 430-71-77,
+7 812 110-12-73
FAX: +7 812 316-15-59
slaa@noise.spb.su

Laszlo Szabó
Nat. Res. Inst. of Radiobiology &
Radiohygiene
Dept. for Non-Ionizing Radiation
H-1775 POB 101
Budapest H-1775
HUNGARY
Phone: +361 226 5331
FAX: +361 226 5331
h7391lsza@ella.hu

Richard Tell
Richard Tell Assoc., Inc.
8309 Garnet Canyon Lane
Las Vegas, NV 89129
USA
Phone: 702-645-3338
FAX: 702-645-8842
rtell@radhaz.com
www.radhaz.com

Santi Tofani
Servizio di Fisica Sanitaria
Laboratorio di Fisica Sanitaria
Via Aldisio, 2 Azienda ASL 9
10015 Ivrea
ITALY
Phone: +39-125-616325
FAX: +39-125-251012
fisica.sanitaria@eponet.it

Paolo Vecchia
National Institute of Health
Physics Laboratory
00161 Roma
Viale Regina Elena 299
ITALY
Phone: +39 06 4990 2857
FAX: +39 06 493 87075
vecchia@iss.infn.it

John Ziriax
United States Navy
Microwave Dept., NHRC – DET
8301 Navy Rd.
Brooks AFB, TX 78235
USA
Phone: 210-536-6530
FAX: 210-536-6439
john.ziriax@afrlars.brooks.af.mil

Thomas Walters
Veridian
AFRL/HEDR/VEDA
8305 Hawks Rd.
Brooks AFB, TX 78235
USA
Phone: 210-536-4768
FAX: 210-536-2603
thomas.walters@afrlars.brooks.af.mil

PARTICIPANTS AT THE ADVANCED RESEARCH WORKSHOP
RADIO FREQUENCY RADIATION DOSIMETRY

Dirk Adang
Royal Military Academy Optronics &
Microwaves
OPTO-Electronics & Microwaves
Avenue de la Renaissance 30
1000 Brussels
BELGIUM
Phone: 32-2-737-6563
FAX: 32-2-737-6212
Dirk.Adang@omra.rma.ac.be

Katarina Beravs
Institute "Jozef Stefan"
Jamova 39
1000 Ljubljana
SLOVENIA
Phone: +386 61 177 3696
FAX: +386 61 219 385
katarina.beravs@ijs.si

Michael Burkhardt
Lab for EMF & Microwave Electronics
Gloriastr 35
ETH Zurich, CH-8092 Zurich
SWITZERLAND
Phone: +41 1 632 5121
FAX: +41 1 632 1057
burkhard@ifh.ee.ethz.ch
www.ifh.ee.ethz.ch/biem/index.html

Jean Claude DeBouzy
Ministere de la Defense
Biophysics/NMR/Non ionizing
Radiations
24 Avenue des maquis du Gresivaudan
F-38702 La Tronche Cedex
FRANCE
Phone: 33-476-63.69.39
FAX: 33-476.63.69.40
101513.2106@compuserve.com

Anna Maria Alloe
Telecom Italia Mobile
Via Vitorchiano 4
00189 Rome
ITALY
Phone: +39-06-3900 4 330
FAX: +39-06-3900 4 617
aaloe@tim.it

Dave Baron
Holaday Industries, Inc.
14825 Martin Dr.
Eden Prairie, MN 55344
USA
Phone: 612-934-4920
FAX: 612-934-3604
baron006@tc.umn.edu

David Cukjati
Laboratory of Biocybernetics
Faculty of Electrical Engineering
University of Ljubljana
Tržaška 25
1000 Ljubljana
SLOVENIA
Phone: +386 61 1768 456
FAX: +386 61 126 46 58
david@svarun.fe.uni-lj.si

Etienne Degrave
Staff of the Medical Service
Military Hospital Queen Asstrid
Bruynstreet 2
B-1120 Brussels
BELGIUM
Phone: (32)2/264.40.41
FAX: (32)2/264.40.48
Etienne.degrave@smd.be

Robert Gardner
Directorate of Safety, Environment &
Fire Policy
Ministry of Defence
Rm. 750 St Giles Court
1 St Giles High St.
London WC2H 8LD
UNITED KINGDOM
Phone: +44 171 305 1007
FAX: +44 171 305 1144

Natasa Kitak
Non-ionizing Radiation Laboratory
SIQ, Trzaska c.2
1000 Ljubljana
SLOVENIA
Phone: +386 61 1778180
FAX: +386 61 1778444
natasa.kitak@siq.si

Micaela Liberti
University of Rome "La Sapienza"
Electronic Engineering Dept.
Via Eudossiana 18
00184 Rome
ITALY
Phone: +39 06 44 585 420
FAX: +39 06 4742647
liberti@die.ing.uniroma1.it

Richard Miller
United States Air Force Research
Laboratory
AFRL/HEDR
8308 Hawks Rd. (Bldg. 1184)
Brooks AFB, TX 78235
USA
Phone: 210-536-2091
FAX: 210-536-1899
richard.miller@afrlars.brooks.af.mil

Jan Helge Halleraker
Royal Norwegian Navy
Naval District Vestlandet
Medical Dept.
N-5078 Haakonsvern
NORWAY
Phone: +47 55 502150
FAX: +47 55 502153

Roman Kubacki
Military Institute of Hygiene and
Epidemiology
01-163 Warszana
ul. Kozielska 4
POLAND
Phone: +48 22 6817148
FAX: +48 22 8104391

David Mayes
DBPS
Institute of Naval Medicine
Crescent Rd.
Alverstoke, Gosport, PO12 2DL
UNITED KINGDOM
Phone: +44 0 1705 768140
FAX: +44 0 1705 768150

Lluis M. Mir
UMR 1772 CNRS
Institute Gustave-Roussy
39, Rue Camille Desmoulins
F-94805 Villejuif Cedex
FRANCE
Phone: +33 (0) 1 42 11 47 92
FAX: +33 (0) 1 42 11 52 76
luismir@igr.fr

Steinar Nestas
Royal Norwegian Navy
SFK/T-TKK-EMC
Postbox 3, K.4-2
N-5078 Haakonsvern
NORWAY
Phone: +47 55 504364
FAX: +47 55 504300

Emanuele Piuzzi
University of Rome "La Sapienza"
Electronic Engineering Dept.
Via Eudossiana 18
00184, Rome
ITALY
Phone: +39 06 44 585 420
FAX: +39 06 4742647
piuzzi@die.ing.uniroma1.it

Berthold Roemer
German Armed Forces Sci. Inst. for
Protection Tech, NBC Protection
High Power Microwave & Field
Coupling
Humboldtstrabe, D-29633 Munster
GERMANY
Phone: +049 5192136462
FAX: +049 5192136355

Simona Bruna
ICEMB at DIBE-UNI. of GENOA
Via Opera Pia 11 A
16145 Genoa
ITALY
Phone: +39 010 3532031
FAX: +39 010 3532290
simo@dibe.unige.it

Mojca Pavin
University of Ljubljana
Faculty of Electrical Engineering
Tržaška 25
1000 Ljubljana
SLOVENIA
Phone: +386 61 1768 456
FAX: +386 61 126 46 58
mojca@svarun.fe.uni-lj.si

Marko Puc
University of Ljubljana
Faculty of Electrical Engineering
Laboratory of biocybernetics
Tržaška 25
1000 Ljubljana
SLOVENIA
Phone: +386 61 1768 456
FAX: +386 61 126 46 58
mac@svarun.fe.uni-lj.si

Paola Russo
University of Ancona
Dept. of Electronics
Via Brecce Bianche
60131 Ancona
ITALY
Phone: +39 071 2204459
FAX: +39 071 2804334
p.russo@ee.unian.it

Vladimir Stepanov
The State Research Center of Russia
Institute of Biophysics
Zhivopisnaya Str., 46
Moscow, 123182
RUSSIA
Phone: 7-095-190-54-97
FAX: 7-095-190-35-90
ibphgen@rcibph.dol.ru

Bart J.A.M. van Leersum
TNO Physics & Electronics Laboratory
Electromagnetic Effects Group
2509 JC, The Hague
P.O. Box 96864
THE NETHERLANDS
Phone: +31 70 374 0358
FAX: +31 70 374 0653
Leersum@fel.tno.nl

Jonna Wilen
National Institute of Working Life
Dept. of Technical Hygiene
Box 7654, S-907 13
Umea
SWEDEN
Phone: +46 (0) 90-786 96 20
FAX: +46 (0) 90-786 65 08
jonna@niwl.se

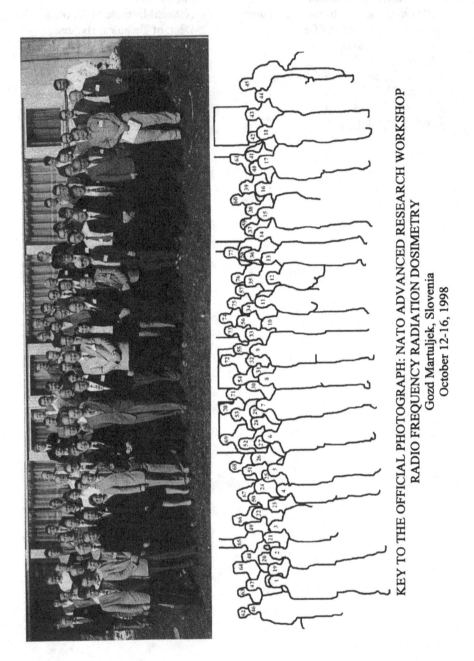

KEY TO THE OFFICIAL PHOTOGRAPH: NATO ADVANCED RESEARCH WORKSHOP RADIO FREQUENCY RADIATION DOSIMETRY

Gozd Martuljek, Slovenia

October 12-16, 1998

1. Igor Likar
2. Peter Gajšek
3. Niels Kuster
4. Camelia Gabriel
5. Marc Pirou
6. Michael Murphy
7. Franc Bratkovič
8. Jon Klauenberg
9. Damijan Miklavčič
10. Katarina Beravs
11. Patrick Mason
12. Youri Grigoriev
13. Eleanor Adair
14. John D'Andrea
15. M. Šinkovec
16. Michel Israel
17. Om Gandhi
18. John Osepchuk
19. Peter Dimbylow
20. Robert Adair
21. Laszlo Szabo
22. Paolo Vecchia
23. Guglielmo D'Inzeo
24. Dina Šimunić
25. John Leonowich
26. J. Patrick Reilly
27. Kjell Hamson Mild
28. Richard Olsen
29. Alexander Lerchl
30. John de Lorge
31. Marko Markov
32. Eugene Khizhnyak
33. Santi Tofani
34. David Land
35. C.K. Chow
36. William Hurt
37. John Ziriax
38. Thomas Walters
39. James Merritt
40. Klaus Hofmann
41. Philip Chadwick
42. Thuroczy Gyorgy
43. Jurgen Bernhardt
44. Stanislaw Szmigielski
45. Andrei Pakhomov
46. Vladimir Stepanov
47. Jan Musil
48. Richard Tell
49. Tapani Lahtinen
50. Alessandro Chiabrera
51. David Baron
52. Marat Rudakov
53. Mike Repacholi
54. Kenneth Foster
55. Etienne Degrave
56. Jean Claude DeBouzy
57. Lluis M. Mir
58. Micaela Liberti
59. Emannuele Piuzzi
60. Paola Russo
61. Marko Puc
62. Anna Maria Aloe
63. Nataša Kitak
64. Bob Gardner
65. Michael Burkhardt
66. Dirk Adang
67. Simona Bruna
68. Berthold Roemer
69. Jan Helge Halleraker
70. Steinar Nestas
71. Dave Mayes
72. Bart Van Leersum
73. Jonna Wilen
74. Vladislav Petin
75. Roman Kubacki
76. Richard Miller
77. Andrej Pajnič

INDEX

585

588

471, 492, 504, 508-509, 511, 513-516, 520-521, 524-526, 530-531, 534, 539, 541, 543, 553, 555

Health risk assessment, 22-25, 27-28, 259, 514, 521

Heart, 90, 138

heart rate, 18

heat loss, 87, 105, 208, 346-347, 351-353, 363, 370, 473, 568

heat production, 346-347, 349, 376

High peak power, 397, 402, 406-407

Hot spot, 49, 141, 154, 297, 354, 370, 481

Human body, 29, 39-40, 42-43, 52, 60, 62, 93-94, 97, 103, 109, 111-115, 120-121, 123, 128, 131, 142-143, 155, 222, 224, 239, 248, 254, 257-258, 285, 291, 294-295, 297-299, 323, 357, 494, 506, 519, 531, 538-539

Human Factors and Medicine Panel, 9

Human model, 49, 51, 73, 82, 112, 120, 138, 154-155, 196, 299, 449, 461, 471, 473, 475, 478, 567

Hyperthermia, 51, 85, 120-121, 131, 134, 139, 152, 155, 349, 351-352, 354, 357, 375, 382, 388, 391, 409, 418, 493, 498, 568

Hypothalamus, 209, 210, 348, 351, 362, 495, 497-498

ICNIRP, 9, 10, 17, 22, 30, 56, 62, 84, 120, 129, 131, 218, 221, 225-226, 257, 261, 268, 299, 303-306, 324, 326, 334, 501, 506, 513, 515-517, 519-520, 524, 530, 538-539

IEEE, 6, 10, 16-17, 19, 30, 51-52, 55, 61-62, 83-84, 95, 101, 106, 120-121, 131, 140, 153-155, 197, 205, 218, 226, 237, 249, 254-255, 287-289, 291-292, 298-299, 301-307, 319, 335-340, 342, 364-368, 372-373, 381-382, 390, 401-404, 406, 418, 441-442, 445, 447, 471, 481, 492, 498, 501-503, 505-511, 539

IEEE Standards Coordinating Committee, 6, 16, 17, 301, 307, 373

Immune system, 493- 498

Induced current density, 82, 114-117, 128-129, 133, 239, 321-324, 326-328, 330-333, 513, 518-519

Induced current limits, 298, 304, 368

Induced currents, 82, 109, 114-117, 120, 128-131, 133, 137, 140, 143, 222-224, 239-240, 245, 257-258, 262, 265, 293-295, 298-299, 302, 304-306, 321-324, 326-333, 368, 451, 513, 517-519, 566

Infrared, 14, 87-88, 105, 142, 148-152, 155, 197-198, 355, 364, 391, 398, 402, 508, 511

INIRC, 224

Institute of Electrical and Electronic Engineers (IEEE), 6, 10, 16-17, 19, 30, 51-52, 55, 61-62, 83-84, 95, 101, 106, 120-121, 131, 140, 153-155, 197, 205, 218, 226, 237, 249, 254-255, 287-289, 291-292, 298-299, 301-307, 319, 335-340, 342, 364-368, 372-373, 381-382, 390, 401-404, 406, 418, 441-442, 445, 447, 471, 481, 492, 498, 501-503, 505-511, 539

Integral equation, 134-135, 137, 139, 409, 411-415, 418

Internal currents, 291

Internal e-field, 49-50, 283

International Commission on Non-Ionizing Radiation Protection (ICNIRP), 9-10, 17, 22, 30, 56, 62, 84, 120, 129, 131, 218, 221, 225-226, 257, 261, 268, 299, 303-307, 324, 326, 334, 501, 506, 513, 515-517, 519-521, 524, 530, 538-539

International Electrotechnical Commission (IEC), 267, 269, 292, 506, 519

International EMF Project, 10, 21-23, 27, 217, 307, 451